SEED PROTEINS

SEED PROTEINS

Edited by

PETER R. SHEWRY

and

ROD CASEY

SPRINGER SCIENCE+BUSINESS MEDIA, B.V.

Library of Congress Cataloging-in-Publication Data

Seed proteins / edited by Peter R. Shewry and Rod Casey.
p. cm.
Includes index.
ISBN 978-94-010-5904-6 ISBN 978-94-011-4431-5 (eBook)
DOI 10.1007/978-94-011-4431-5
1. Seed proteins. 2. Seed storage compounds. I. Shewry, P. R. (Peter R.) II. Casey, R. (Roderick)
QK898.P8S442 1999
575.6'846--dc21 98-54491

ISBN 978-94-010-5904-6

Printed on acid-free paper

TABLE OF CONTENTS

FRONTISPIECE

Thomas Burr Osborne (1859-1929) was one of the founding fathers of protein chemistry. He worked at the Connecticut Agricultural Experiment Station from 1886 till 1928, publishing some 250 papers including studies of seed proteins from 32 plant species. Oil portrait provided by the Connecticut Agricultural Experiment Station.

FOREWORD

Seeds provide more than half of the world's intake of dietary protein and energy and thus are of immense economic, cultural and nutritional importance. Proteins can account for up to 40% of the dry weight of various types of seeds, thereby making a large contribution to the nutritional quality and processing properties of seeds. It is, therefore, not surprising that seed proteins were among the first plant components to be systematically studied, some 250 years ago, and have been a major focus of research over the past 100 years. The properties and behaviour of seed proteins pervade modern life in numerous ways. For example, legume and cereal proteins are used in the production of a wide range of meat-free foods; the process of bread-making is dependent on the physical chemical properties of wheat seed proteins; and in developed, as well as developing, countries, nutritional deficiencies among vegetarian diets are avoided through balancing legume and cereal seeds as sources of dietary proteins. Understanding seed proteins, in order to improve their composition and properties and to increase their concentrations, will thus continue to be an important research objective for the future.

The present volume represents the culmination of a long-discussed plan of the editors, to bring together the best international authorities in order to compile a definitive monograph on biological, biochemical, molecular and genetic aspects of seed proteins. The aim has been to include not only the major crop species, but also a wider spectrum of plants of biological and evolutionary interest. Similarly, although there is an emphasis on the storage proteins which have the greatest impact on seed utilization, a range of other proteins are covered including many which may contribute to the protection of the seed against pests and pathogens.

An enormous amount of work has gone into the individual chapters and we are grateful to the authors of these for all their efforts. In addition, we are indebted to our colleagues at IACR-Long Ashton and the John Innes Centre for their assistance, particularly Mrs. Valerie Topps, Mrs. Pat Baldwin (IACR-Long Ashton) and Miss Tarn Dalzell (John Innes Centre) for skilled secretarial help. Finally, we wish to thank Mr. Ken Williams (IACR-Long Ashton) for photographs including the front cover of the volume.

Peter R. Shewry and Rod Casey

F. Takaiwa, Department of Cell Biology and Molecular Biology, National Institute of Agrobiological Resources, Tsukuba, Ibaraki 305, Japan.

A.S. Tatham, IACR-Long Ashton Research Station, Department of Agricultural Sciences, University of Bristol, Long Ashton, Bristol BS41 9AF, UK. Fax: 01275 394281 E-mail: arthur.tatham@bbsrc.ac.uk

E.J.M. Van Damme, Katholieke Universiteit Leuven, Laboratory for Phytopathology and Plant Protection, Willem de Croylaan 42, 3001 Leuven, Belgium. Fax: 32 16 322976 E-mail: Els.VanDamme@agr.kuleuven.ac.be

A.L. Vettore, Centro de Biologia Molecular e Engenharia Genética, Universidade Estadual de Campinas, 13083-970 Campinas, SP, Brazil.

J.A. Yunes, Centro de Biologia Molecular e Engenharia Genética, Universidade Estadual de Campinas, 13083-970 Campinas, SP, Brazil.

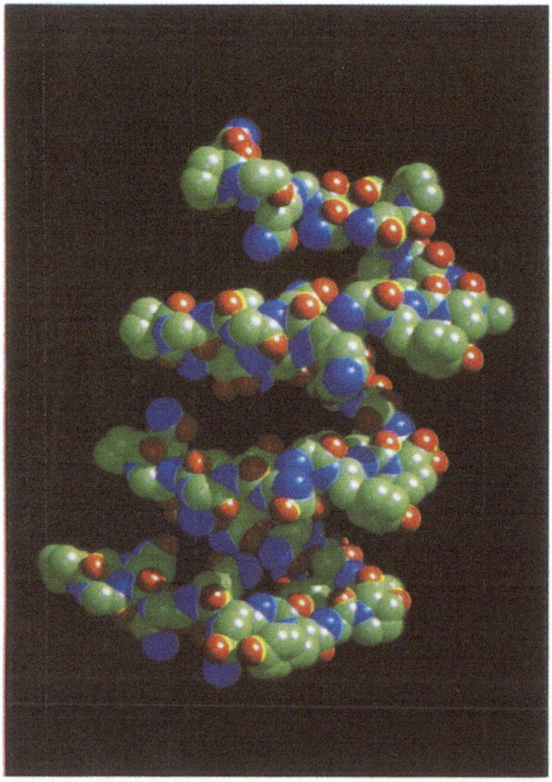

Plate 1 Molecular modelling of the β-spiral structure formed by the repetitive sequences present in the HMW subunits of glutenin The figure was kindly provided by Dr. O. Parchment and Dr. D. Osguthorpe of the University of Bath, UK.

A.

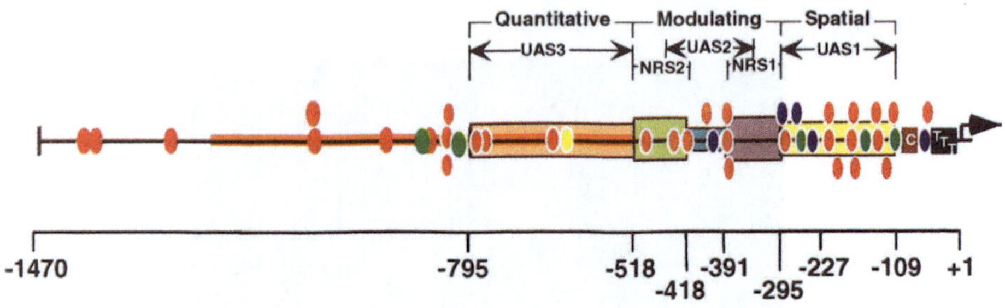

Quantitative
Modulating
Spatial
UAS3
UAS2
UAS1
NRS2
NRS1
-1470
-795
-518
-418
-391
-295
-227
-109
+1

B.

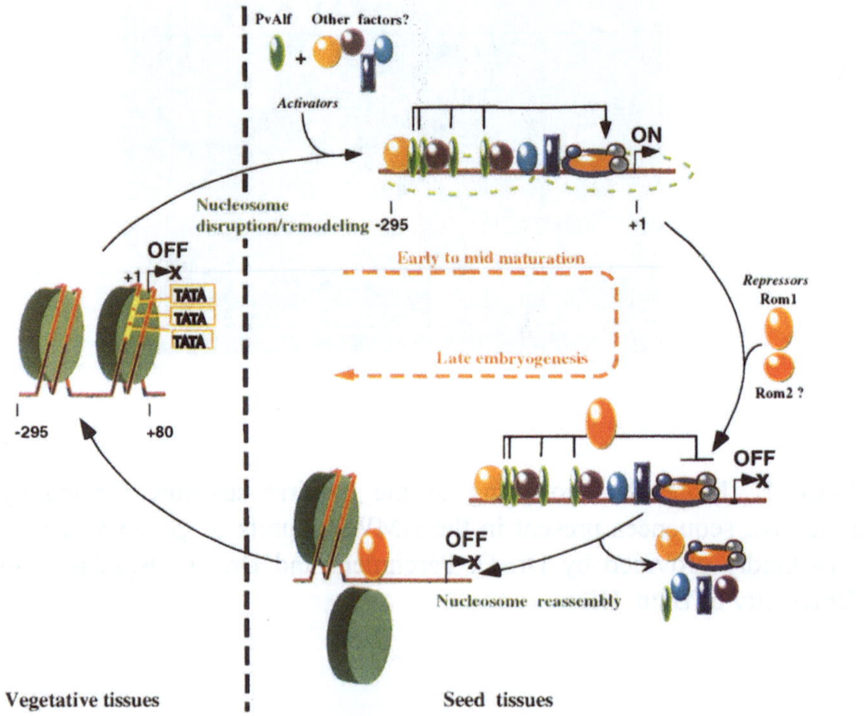

PvAlf
Other factors?
Activators
ON
Nucleosome disruption/remodeling
-295
+1
OFF
TATA
TATA
TATA
-295
+80
Early to mid maturation
Late embryogenesis
Repressors
Rom1
Rom2 ?
OFF
OFF
Nucleosome reassembly
Vegetative tissues
Seed tissues

Plate 2 Regulatory elements in the phaseolin (*phas*) promoter and their assembly into chromatin. A. Diagram of the various regions of the *phas* promoter discussed in the text. UAS1, UAS2 and UAS3 are upstream activating sequences; NRS1 and NRS2 are negative regulatory sequences. UAS1 includes three domains: a seed specific enhancer (SSE), a minimal seed specificity element (MSS) and a minimal promoter (MP). The orange bar denotes the 5' matrix attachment region. There are three TATA-regions (T) and one CAAT-like element (C). Each of the colored ovals represents a consensus *cis*-element: blue, RY elements (CATGCAC/A); green, E sites (CANNTG); yellow, pollen box (GAATTTGTGA); red, various other *trans*-acting factor binding elements predicted on the basis of sequence comparison. B. Model for the involvement of chromatin in regulation of expression from the *phas* promoter. In vegetative tissues, the TATA regions are rendered inaccessible to TATA-binding protein by a rotationally and translationally-positioned nucleosome. Nucleosome disruption (by remodeling or displacement) during embryogenesis permits access by transcriptional activators and assembly of the basal transcription apparatus. Reassembly of the nucleosome, perhaps assisted by regulation of maturation (Rom) factors, occurs late in embryogenesis, terminating transcription. For details, see text.

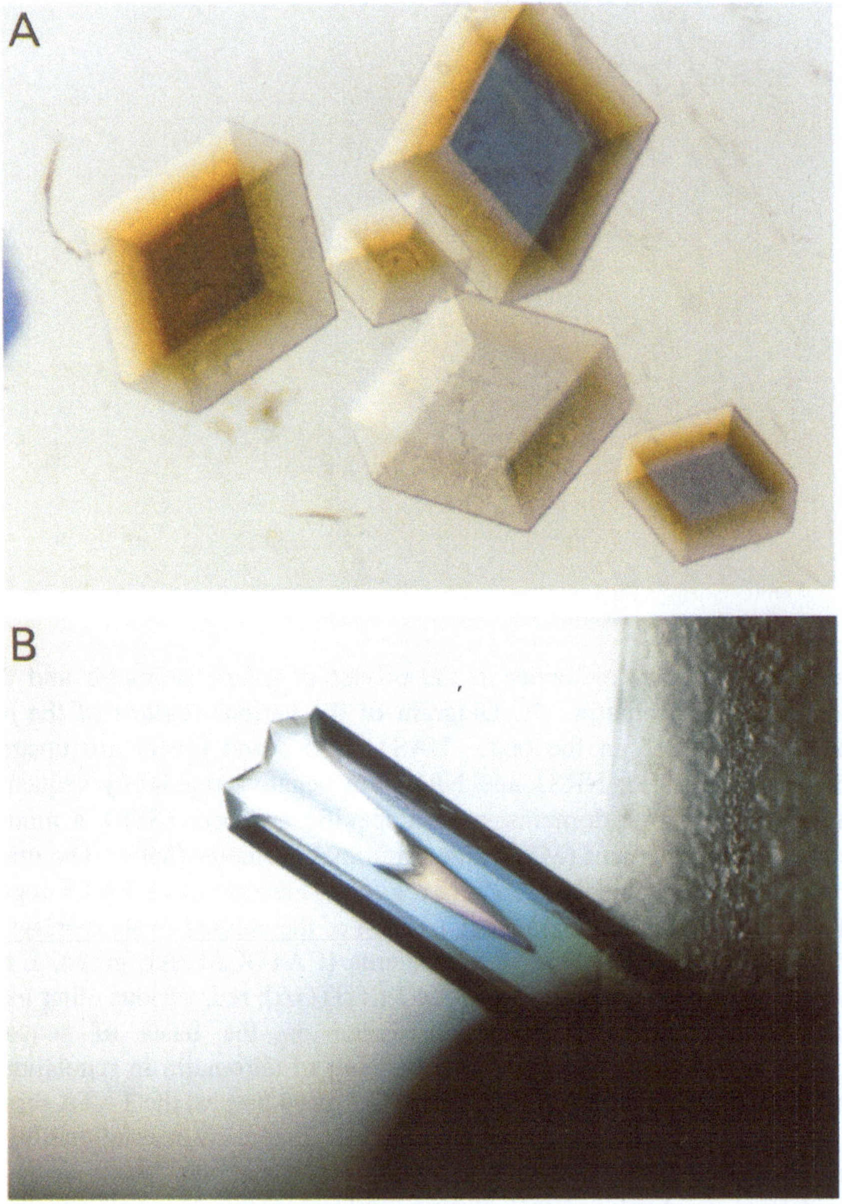

Plate 3 A are rhombohedral crystals of canavalin like those grown by Sumner and Howell (1936). It is the most common crystal form. B is a single crystal of the hexagonal modification of the canavalin crystals, and the form from which the structure of canavalin was first elucidated by X-ray diffraction analysis. Unlike the crystals first grown by Sumner and Howell, these particular crystals were not grown on earth but on the U.S. Space Shuttle in the course of macromolecular crystallization experiments in microgravity, in space.

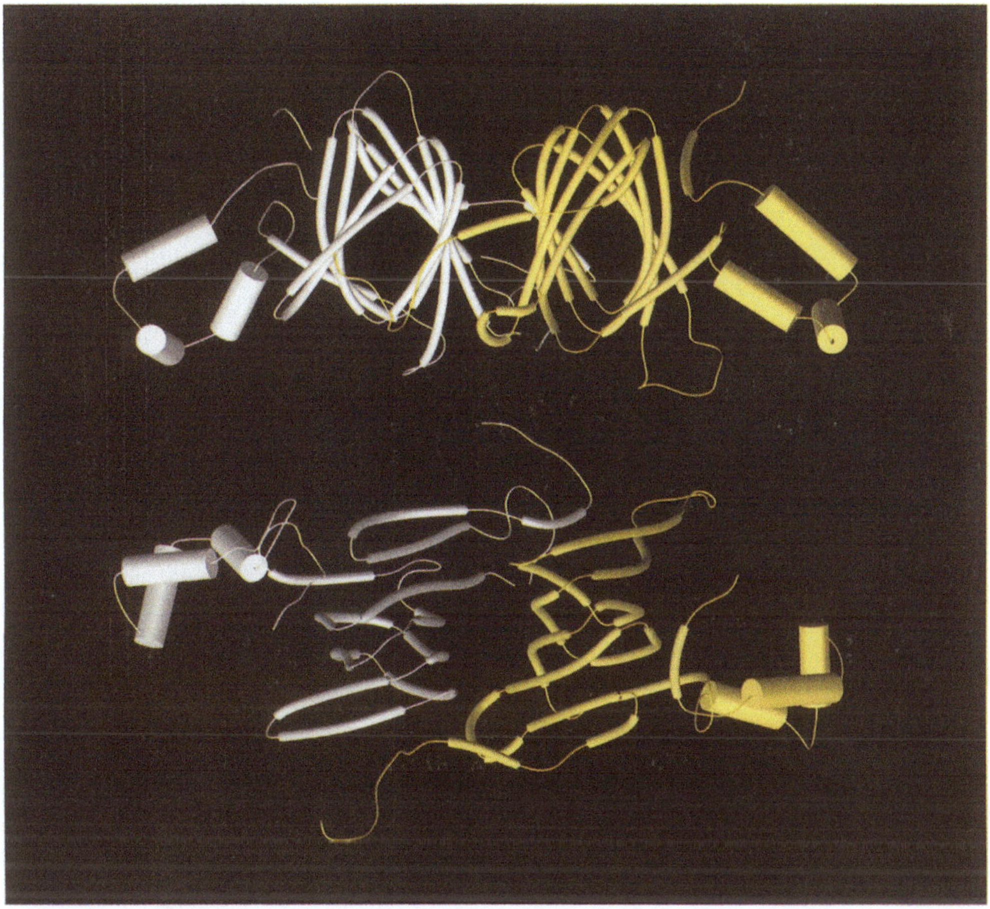

Plate 4 Computer-generated schematic diagrams of the canavalin subunit showing a view along the 3-fold axis (top) and a view along the pseudo-dyad axis (bottom) of the molecule with the amino-terminal domain in yellow and the carboxyl-terminal domain in white.

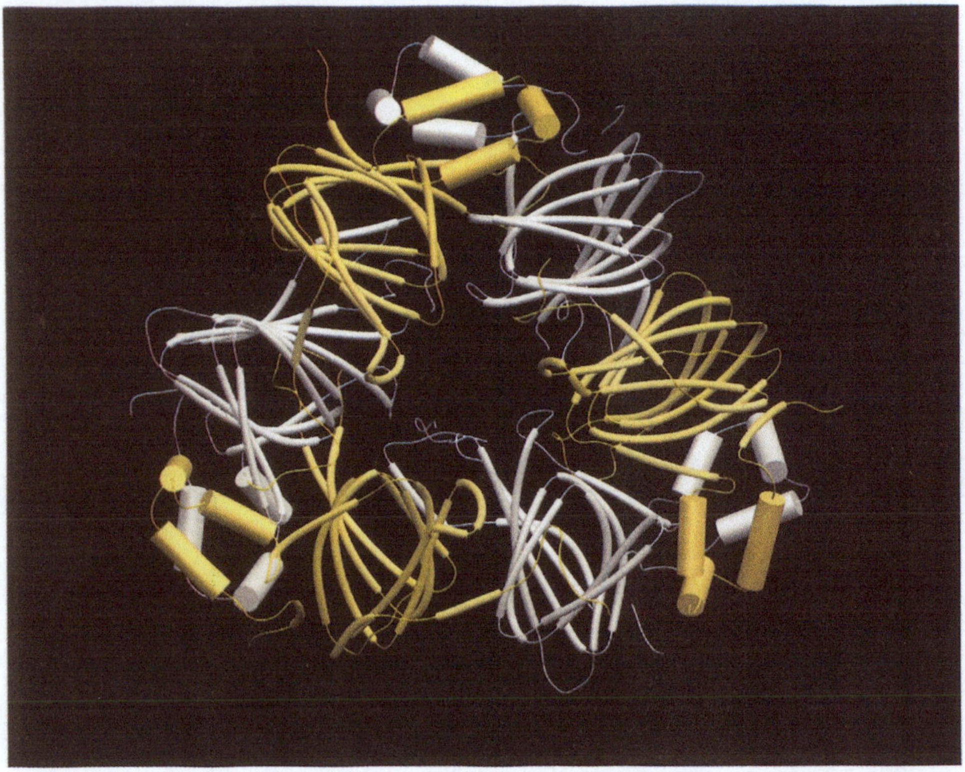

Plate 5 Computer-generated schematic diagram of the entire canavalin trimer with amino-terminal domains in yellow and carboxyl-terminal domains in white. The strands of the Swiss roll β-barrels that make up the cores of the two domains and the α-helices and extended chain elements forming the subunit-subunit interfaces are clearly evident. The central hole along the triad is about 18 Å in diameter, and the entire molecule has a diameter of about 86 to 88 Å.

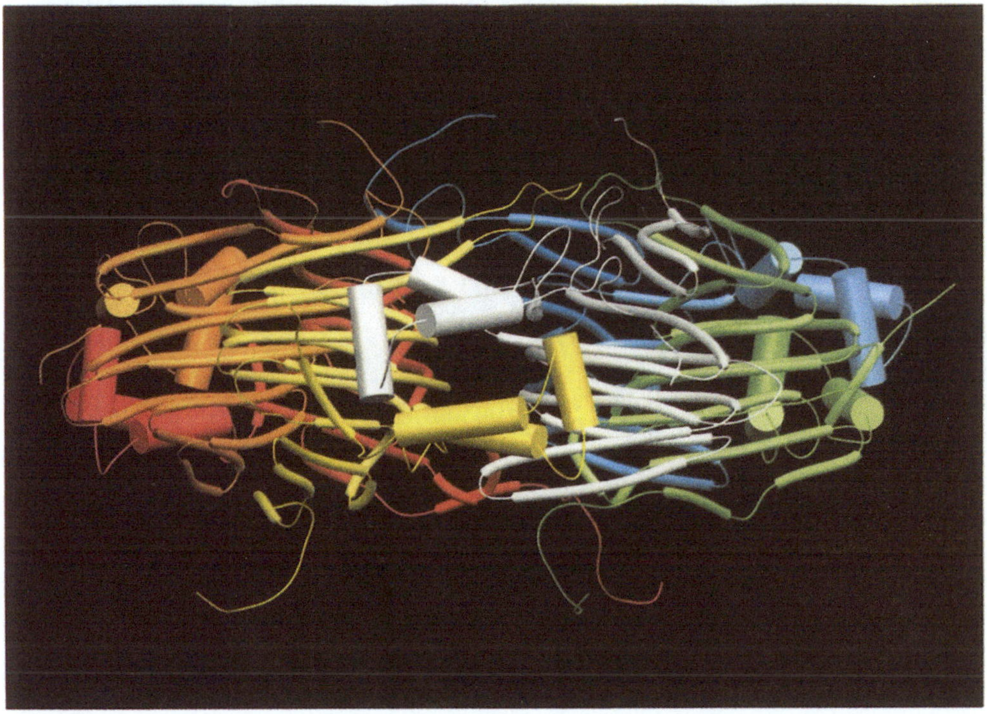

Plate 6 The entire canavalin trimer is shown here viewed perpendicular to the triad axis and along one of the pseudo dyads of the trimer. The domains within subunits are here colored yellow-white, blue-green, and red-orange. The weaving of the β strands is evident over the surface of the molecule as is the interdigitation of the helical subunit interfaces. The disk shaped molecule has a thickness of about 35 Å to 40 Å.

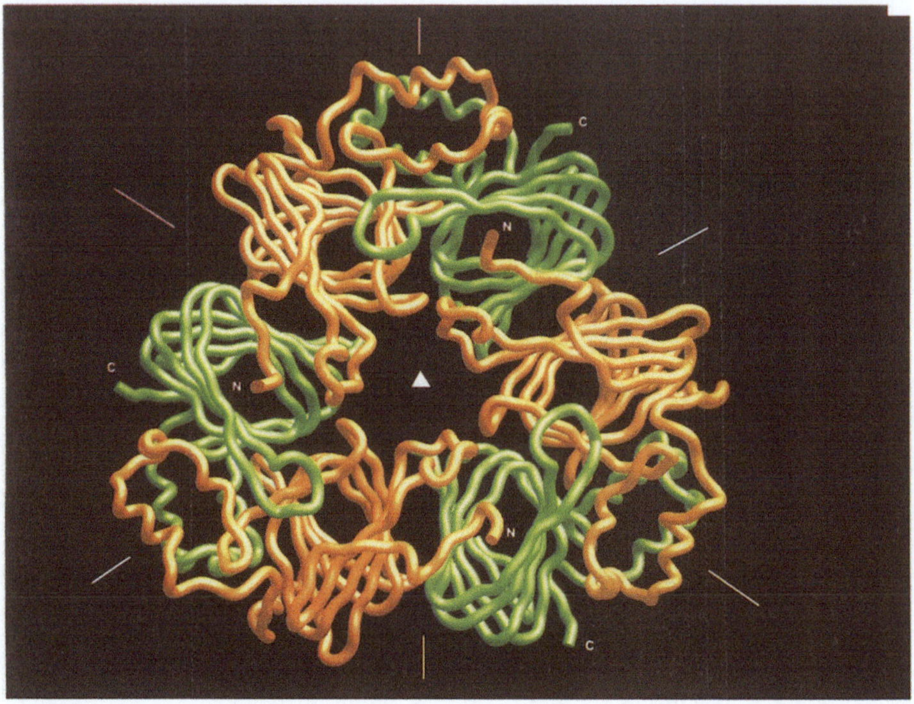

Plate 7 Schematic ("backbone-worm") representation of the phaseolin trimer, based on the X-ray structure of Lawrence *et al.* (1994). The *N*-terminal portions of the subunit polypeptide are coloured orange, whilst the *C*-terminal portions are coloured green. The *N*- and *C*-terminii of each polypeptide are labelled. The location of the three-fold axis perpendicular to the plane of the Plate is indicated by the white triangle, whilst the locations of the pseudo-twofold axes are indicated by white lines. The latter axes lie in the plane of the paper and occur both as intra-subunit axes (relating the *N*- and *C*-terminal modules of the same polypeptide) and as inter-subunit axes (relating *N*- and *C*-terminal modules of neighbouring polypeptides). The polypeptide link between helix 4 and *C*-terminal strand A' is absent in the structure. The Figure was produced using GRASP (Nicholls *et al.*, 1991).

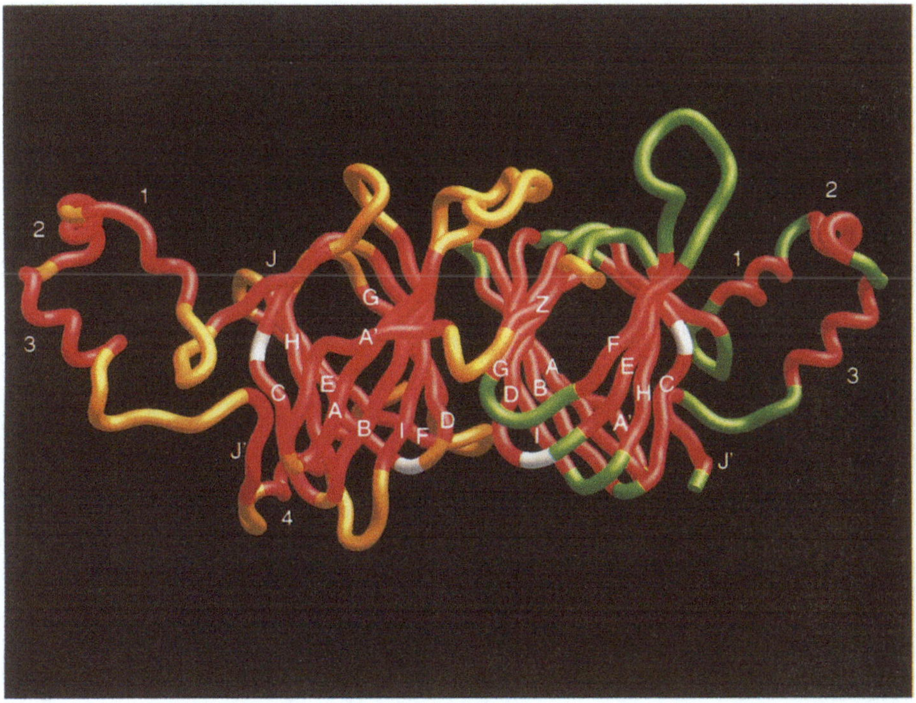

Plate 8 Schematic ("backbone-worm") representation of a single phaseolin subunit, based on the X-ray structure of Lawrence *et al.* (1994). The colour scheme follows that of Plate 7, except regions of the polypeptide adopting a defined secondary structure are now highlighted in red and are labelled according to the convention described in the text. The Pro/Gly motif is highlighted in white in both the *N*- and *C*-terminal domains: the Pro residue forms a bulge in the centre of strand C and the Gly residue forms part of the turn between strands D and E. The Figure was produced using GRASP (Nicholls *et al.*, 1991).

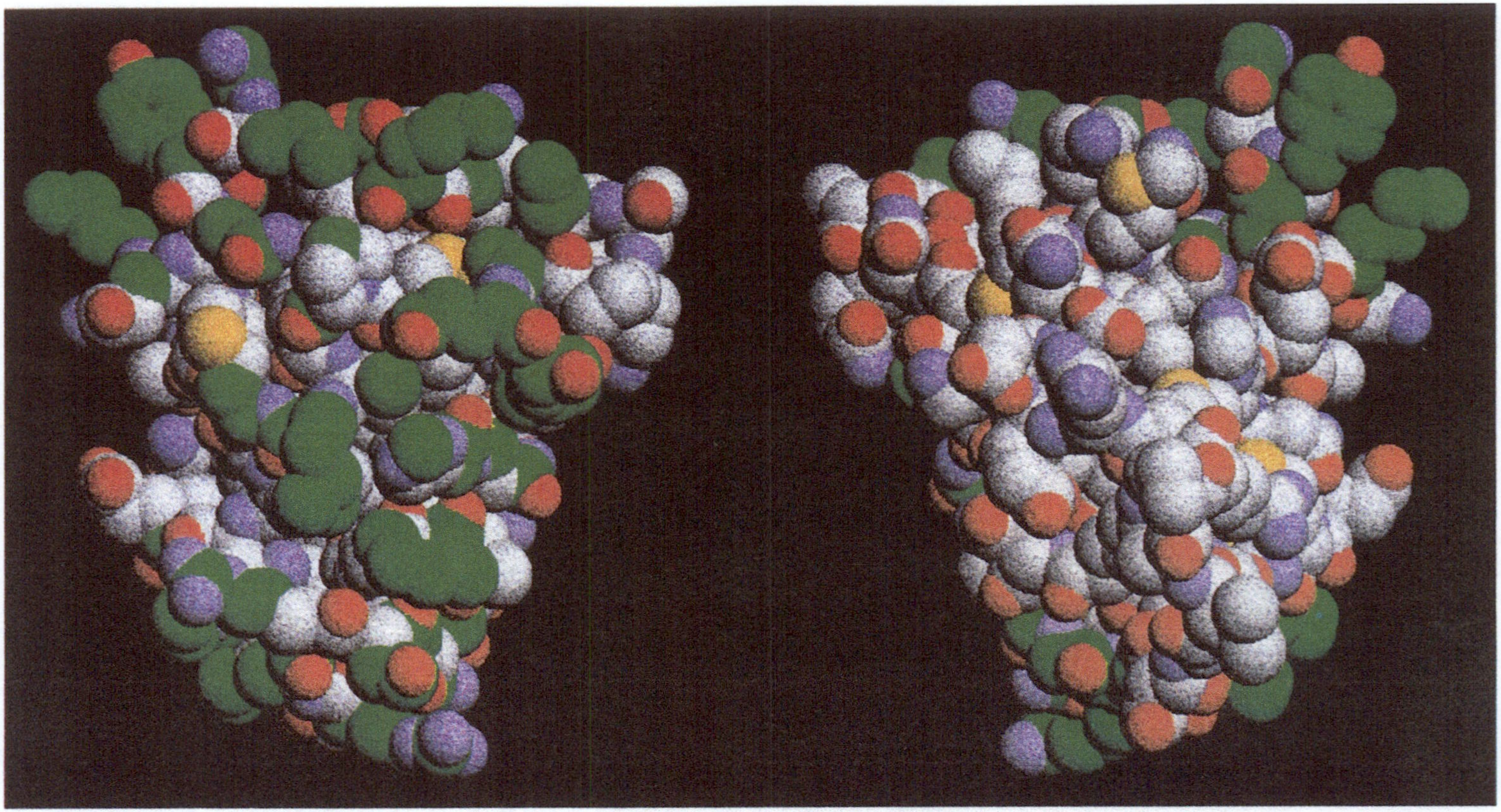

Plate 9 Space-filling model of SFA-8 depicting surface hydrophobic patches. Rotation of the structure by 180° shows that SFA8 has both polar and apolar faces. Atoms are coloured as follows: carbon and sulphur atoms involved in hydrophobic patches (green), other carbon (white), nitrogen (blue), oxygen (red) and other sulphur (yellow). The model is taken from Pandya *et al.* (1998).

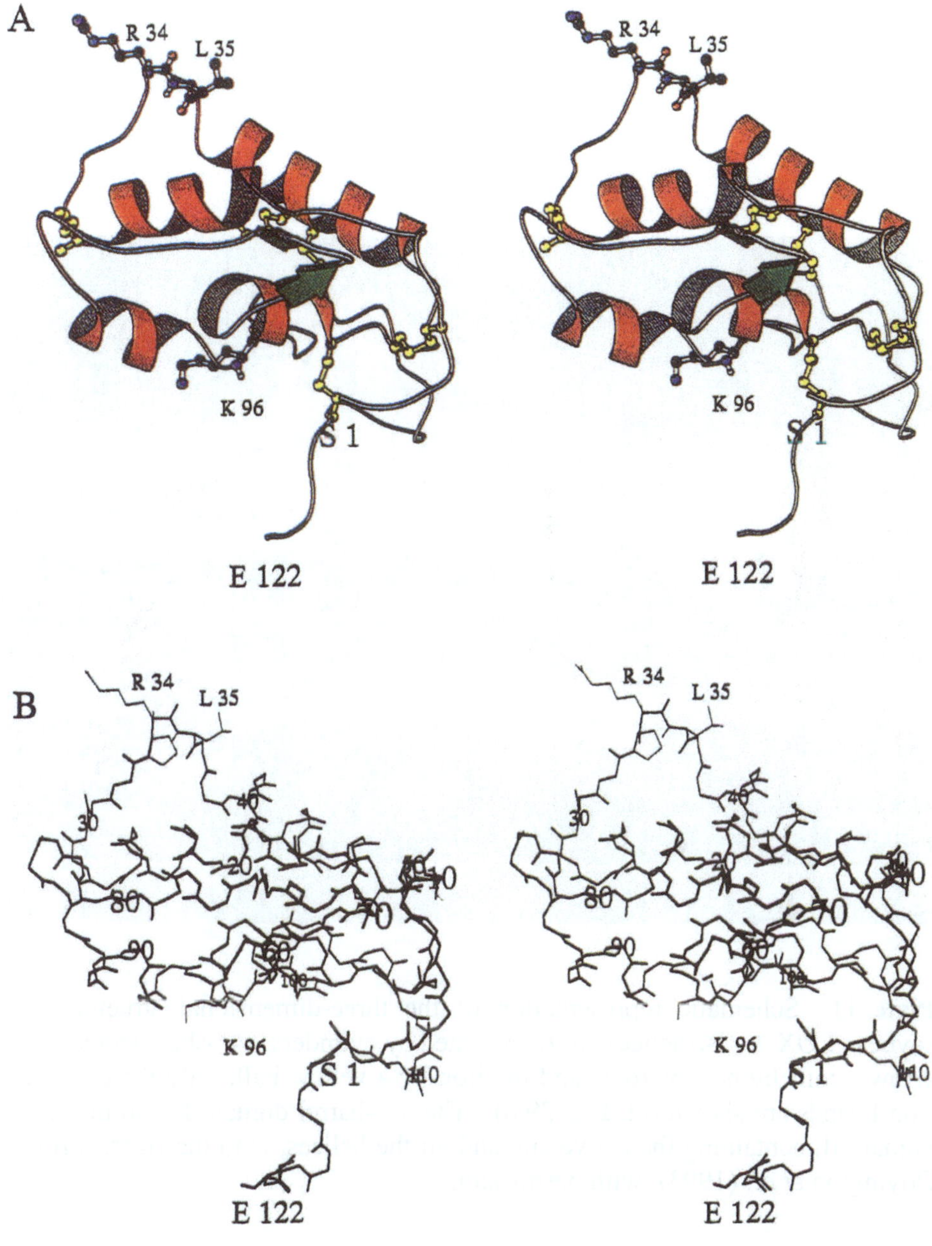

Plate 10 Ribbon drawing of RBI showing the secondary structure, disulphide bridges and side chains of residues serine (S1), arginine (R34), leucine (L35), lysine (K96) and glutamate (E122). Reproduced with permission from Biochemistry (Strobl *et al.*, 1995).

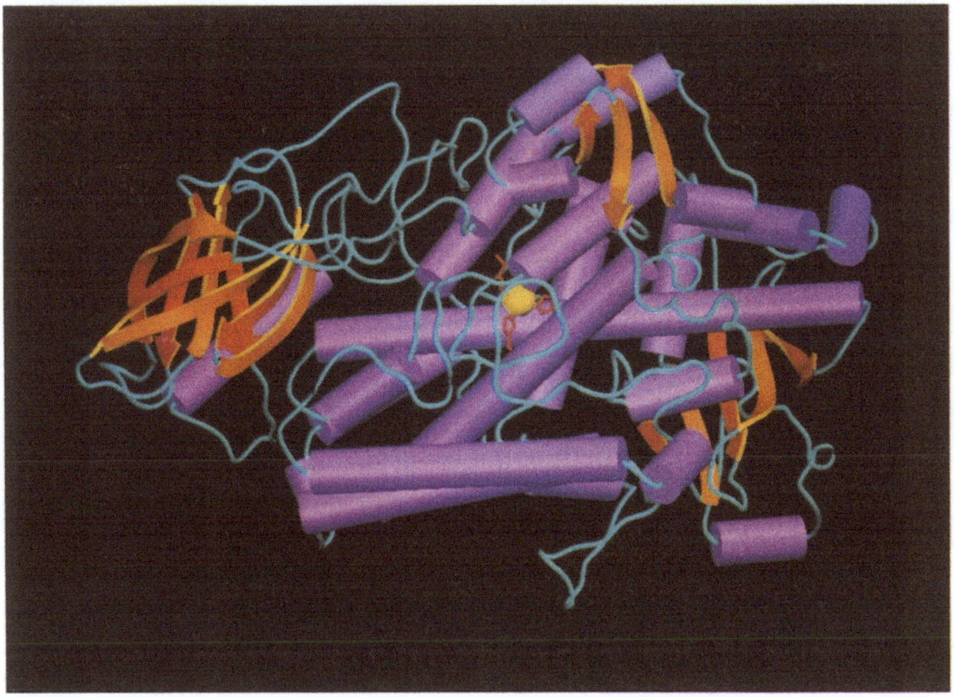

Plate 11 Schematic representation of the three-dimensional structure of soybean LOX-1. ✂-helices are represented by cylinders, ♋-sheet strands by arrows, coils by narrow rods, and the iron by a yellow ball; only three of the iron ligands are shown (cf. Fig. 29.6). The ♋-barrel domain I is to the left; domain II, containing the active site and all the helices, is to the right. From Boyington *et al*. (1993), with permission.

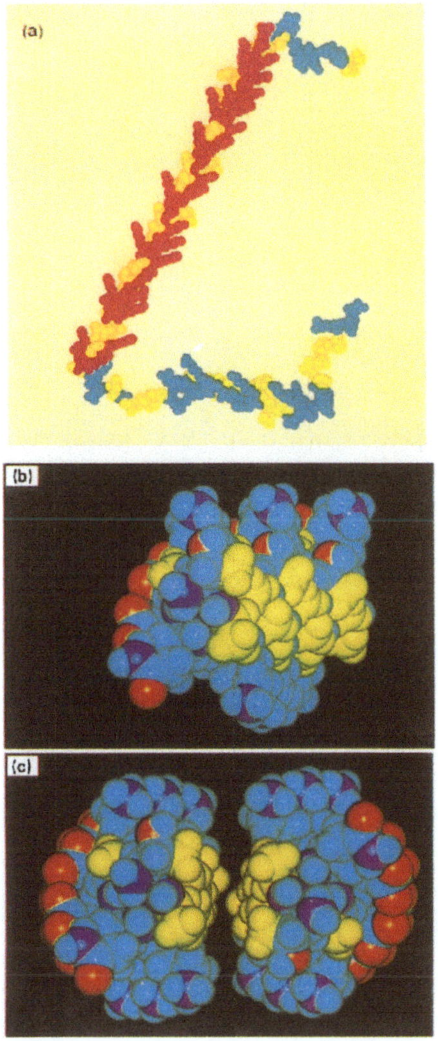

Plate 12 Predicted structure of the Group 3 LEA repeat. Panel "a" shows the predicted secondary structure of the Group 3 LEA D-7 protein of cotton. The region containing the tandemly repeated 11-mer sequence motif is shown as a long, magenta α-helix. The hydrophobic residues, including the *N*-terminal methionine (top right) are shown in yellow. Other residues are blue. Panel "b" presents an oblique view from the *N*-terminus down the long axis of an α-helix comprising three 11-mer consensus Group 3 LEA repeat units. The methyl residues of the threonine and alanine side-chains are shown in yellow, and appear as a distinct hydrophobic stripe along one side of the helix. Alll other carbon and hydrogen atoms are light blue, nitrogen atoms are dark blue and oxygen atoms red. Panel "c" shows two of the helices illustrated in "b" juxtaposed to form a dimer by association of their hydrophobic faces. Reproduced from Dure (1993) by kind permission of the author and publishers.

19. Globulins of the Pseudocereals: Amaranth, Quinoa, and Buckwheat.

MAGDALENA SEGURA-NIETO[1], PETER R. SHEWRY[2], and OCTAVIO PAREDES-LÓPEZ[1].
[1]Departmento de Ingtenierie Genetica del Centro de Investigacion y de Estudios Avanzados del IPN, Km 9.6, Libramiento Norte Carretera, Irapuato-Leon, Apdo. Postal 629, Irapuato, Gto CP 36500, Mexico. [2]IACR-Long Ashton Research Station, Department of Agricultural Sciences, University of Bristol, Long Ashton, Bristol BS41 9AF, UK.

1. INTRODUCTION

Amaranth, quinoa and buckwheat are dicotyledonous crops which, although not closely related to each other or to grasses, are often termed pseudocereals because they produce small grain-like seeds. Amaranth (*Amaranthus* spp., Amaranthacea), and quinoa (*Chenopodium quinoa*, Chenopodiacea) are both natives of Meso America and South America, respectively, and were important crops in pre-Colombian times. In contrast, buckwheat (*Fagopyrum esculentum*, Polygonaceae) is thought to have originated in China (Campbell, 1995).

Three species of grain amaranth were domesticated and grown in certain zones in Northwest, Central and South America: *A. cruentus* (*A. paniculatus*) in Central and Southern regions in Mexico and Central America, *A. hypochondriacus* (*A. leuocarpus* / *A. frumentaceus*) in the North West and Central regions of Mexico, and *A. caudatus* (*A. edulis* /*A. mantegazzianus*) in the Andes (NAS, 1984, Sauer, 1995). Quinoa was also domesticated in the Andes with remains dating from about 5000 BC found in Peru, and was an important crop in the Inca empire, while a related species, huauhzontli (*C. berlandieri*) achieved importance to the north in Aztec Mexico (Galwey, 1995). Both declined after the Spanish conquest, their cultivation being discouraged due to their association with the traditional cultures and religions. As with quinoa, the association of amaranth with Aztec society and ritual led to decline after the Spanish conquest. However, in recent years both amaranth and quinoa have been the subject of renewed interest, in their native countries and in North America, Europe and Asia. This is partly due to their high

P.R. Shewry and R. Casey (eds.), Seed Proteins, 453–475.

protein quality, a character which is shared by buckwheat and makes the pseudocereals attractive targets for the expanding health food market in developed countries and potentially important sources of dietary protein for developing countries.

2. AMARANTH

2.1 Biochemical and physicochemical characteristics

Amaranth grain is considered among the 36 non-conventional crops with most potential as food and feed resources, due to its high protein content (13-18% on dry weight basis) (Singhal and Kulkarni, 1988), with more lysine and a better essential amino acid composition than most cereals (NRC, 1984). In terms of digestibility, amaranth grain has been assigned a high chemical score of 73 in the recomended FAO/WHO scoring patterns (Becker *et al.*, 1981).

2.2 Solubility and abundance

Amaranth seed proteins have been obtained from both defatted and whole flour, by differential extraction procedures based on solubility criteria, originally described by Osborne (1924) and subsequently modified by others. Two types of procedure have been used to extract salt-soluble globulins. In some cases, albumins and globulins were extracted together, followed by dialysis against solutions of low ionic strength or water, allowing the globulins to be recovered as precipitates (Konishi *et al* ., 1985; Paredes-López *et al.*, 1988; Mora-Escobedo *et al.*, 1990; Gorinstein *et al.*, 1991; Barba de la Rosa *et al.*, 1992a, Soriano-Santos *et al.*, 1992). In other procedures globulins were extracted as a separate fraction with salt solution after extraction of albumins (Bressani and García-Vela, 1990; Segura-Nieto *et al.*, 1992).

There are several important differences in the methods used to prepare amaranth globulins, which probably affect both their quality and quantity. These include the type of salt used in the extraction buffer, and its concentration. The final ionic strength used in most studies was about 0.5M, and exceptionally 1.0M, with 0.4-1.0M NaCl being most widely used. However, some workers used other salts: 0.1M Na_2HPO_4, 0.04M Na_2SO_4 or 0.04M K_2SO_4. The pH in most cases was around 7.0-7.5, except for one study where the authors measured the effect of pH (7 to 10) (Soriano Santos *et al.*, 1992). The extraction time (1-18 hr.) also affected globulin recovery.

The protein distribution in Table 1 confirms that different extraction procedures lead to clear differences between the yields of each protein group.

In nearly all cases, the albumin and globulin fractions together represent the most abundant group of seed proteins, varying from 39.55% up to 82.0% of the total with the conditions used. This is true for the three cultivated amaranth species. However, the globulin fraction alone represents less than 20% of the total in14 out of the 18 reports shown in Table 1 (Abdi and Sahib, 1976; Konishi *et al.*, 1985, Duarte-Correa *et al.*, 1986; Bressani and García-Vela, 1990b; Barba de la Rosa *et al.*, 1992a; Segura-Nieto *et al.* 1992; Soriano-Santos *et al.*, 1992). In contrast, all the studies, with the exception of Bressani and García-Vela (1990), show that albumins are the most abundant protein group in the seeds accounting for 20-65% of the total protein.

Bressani and García-Vela (1990) extracted the seeds initally with water, called method A in Table 1, and showed low concentrations of albumins in all three cultivated species (8.5% in *A hypochondriacus*, 8.9% in *A. cruentus* and 7.7% in *A. caudatus*), with globulins forming the major fraction.

2.3 Amino acid composition

The use of different methods for globulin extraction is also reflected in the amino acid compositions of the fractions obtained. Table 2 shows that, in terms of essential amino acids, amaranth globulins contain higher concentrations of lysine (3.7 to 7.6%) (Konishi *et al.*, 1985; Mora-Escobedo *et al.*, 1990; Bressani and García-Vela, 1990; Barba de la Rosa *et al.*, 1992a; Segura-Nieto *et al.*, 1992) than cereal grains (Peterson, 1978). This also appears to be the case with the levels of sulphur-containing amino acids (3.1 to 7.1%) which are higher than in legume seeds (1.3% and 2.8%) (Hurich *et al.*, 1977; Smith and Circle, 1981, respectively) although Marcone and Yada (1991) reported lower values.

In general, the amino acid compositions reported for amaranth globulins show high concentrations of aspartic acid+asparagine, glutamic acid+glutamine, and arginine, amino acids which are often present in high proportions in storage proteins.

2.4 Electrophoretic analysis

Figure 1B shows a total globulin fraction of *A. hypochondriacus,* separated by SDS-PAGE in one dimension (1D) and by isoelectric focusing followed by SDS-PAGE to give a 2D separation. These analyses reveal heterogeneity in both the size and isoelectric point (pI) of the component polypeptides, with similar patterns being detected with globulin fractions from defatted flour as shown in Figure 1A (Segura-Nieto *et al.*, 1992). It has also been reported that the electrophoretic profile of the globulins does not change when they are extracted either from whole or defatted flour, although the total

Table 1 Changes in the proportions of amaranth seed proteins extracted with various solvents and procedures*

Albumins	Globulins	Albumins + Globulins	Prolamins	Glutelins	Residue	References
						A. hypochondriacus
48.9	13.7	62.6	7.2	8.5	21.7	Abdi and Sahib, 1976
		56.9	1.6•	22.4°	8.5	Konishi *et al.*, 1985 (w)
		59.4	1.0•	29.4°	12.6	Konishi *et al.*, 1985 (nw)
65.0	17.0	82.0	11.0	7.0	nd	Duarte-Correa *et al.*, 1986
		39.5	3.2	39.2	18.1	Paredes-López *et al.*, 1988 (W)
		29.2	1.3	40.3	29.2	Paredes-López *et al.*, 1988 (D)
8.5	38.5	47.0	9.1	29.1	14.8	Bressani and García-Vela (A), 1990
19.2	18.1	37.3	2.0	42.3	18.4	Bressani and García-Vela (B), 1990
		57.3	1.4	25.8	12.1	Gorinstein *et al.*, 1991
		51.6	1.2	42.2	2.8	Barba de la Rosa *et al.*, 1992
50.0	15.6	65.6	2.0	30.4	2.0	Segura-Nieto *et al.*, 1992
						A. cruentus
61.9	18.3	80.2	15.5	4.3	nd	Soriano-Santos *et al.*, 1992 (K)
33.0	17.5	50.5	6.9	1 7	28.8	Soriano-Santos *et al.*, 1992 (Br)
8.9	44.2	53.1	10.4•	29.9°	nd	Bressani and García-Vela (A),1990
20.0	19.1	39.1	2.8•	46.3o	nd	Bressani and García-Vela (B) 1990
						A. caudatus
		69.2	3.1•	21.1°	3.6	Konishi *et al.*, 1985 (nw)
7.7	40.6	48.3	9.2•	32.6o	nd	Bressani and García-Vela (A), 1990
22.9	20.5	43.4	1.7•	44.3	nd	Bressani and García-Vela (B), 1990

*, Percent of extracted protein; w, waxy; nw, non waxy; W, whole flour; D, defatted flour; A, initial extraction with salt; B, Initial extraction with water, Br, Protein quantified by Bradford's method (1976) ; K , Protein quantified by Kjeldhal's method: n.d, non described;•, Prolamin+Prolamin-like; °, Glutelin+Glutelin-like.

Table 2 Amino acid composition(g/100g protein) of *A. hypochondriacus* globulins extracted under different conditions, compared with globulins from other species

Amino Acid	A	B	C	D	E	F	Pea[+]	Soybean*	Oat**	FAO/WHO/UNU‡ (Ch)	FAO/WHO/UNU‡ (Ad)
Cys	5.5	2.4	4.1	0.7	3.8	1.3	0.6	1.4	1.0		
His	3.0	3.1	2.0	2.9	1.1	2.6	3.4	2.2	2.1	1.9	1.6
Ile	4.6	4.0	2.6	4.6	4.1	3.8	4.0	4.5	4.6	2.8	1.3
Leu	6.3	6.8	3.8	6.3	5.5	5.7	8.1	6.8	7.1	6.6	1.9
Lys	3.7	4.6	5.7	4.4	6.5	7.6	8.0	5.9	2.8	5.8	1.6
Met	1.2	3.1	2.9	1.9	3.3	1.8	0.7	1.4	0.9		
Phe	6.1	4.3	4.5	8.1	4.8	2.4	6.7	4.4	5.0		
Thr	3.2	3.6	2.2	3.6	4.0	3.4	2.9	3.8	4.7	3.4	0.9
Trp	n.d	n.d	0.6	n.d	n.d	n.d	n.d	1.1	n.d	1.1	0.5
Tyr	3.4	3.6	3.4	2.8	4.2	3.9	4.8	3.4	3.4		
Val	4.7	4.3	2.8	5.1	4.5	4.1	3.7	4.7	6.1	3.5	1.3
Ala	3.4	3.7	2.7	3.3	3.9	3.2	2.3	4.0	5.8		
Arg	9.7	10.4	11.7	13.8	9.2	12.3	12.3	7.4	6.3		
Asx[a]	11.6	8.1	8.4	8.2	8.4	9.0	11.7	10.6	8.8		
Glx[b]	16.5	17.2	29.6	15.7	16.8	18.8	21.3	18.5	18 3		
Gly	4.1	5.6	4.5	3.7	6.4	6.61	1.9	4.0	7.2		
Pro	4.0	5.2	n.d.	5.0	3.8	3.3	3.6	5.5	4.7		
Ser	5.8	5.3	3.2	4.9	4.8	5.6	4.0	5.0	6.7		
Met + Cys	6.7	5.5	7.0	2.6	7.1	3.1	1.3	2.8	1.9	2.5	1.7
Phe + Tyr	9.5	7.9	7.9	10.9	9.0	6.3	11.5	7.8	8.4	6.3	1.9

[a] Asx include Asp and Asn; [b] include Glu and Gln, nd, not determined. Ch, children; Ad, adult; A, Konishi; *et al*., 1985;
B, Mora-Escobedo *et al.* 1990; C, Bressani and García-Vela, 1990, D, Marcone and Yada, 1991; E, Barba de la Rosa *et al*, 1992; F, Segura-Nieto *et al*., 1992; [+],Hurich *et al*., 1977, from Gatehouse *et al*., 1984; *, Smith and Circle, 1981;**, Peterson, 1978; ‡ FAO/WHO/UNU, 1985

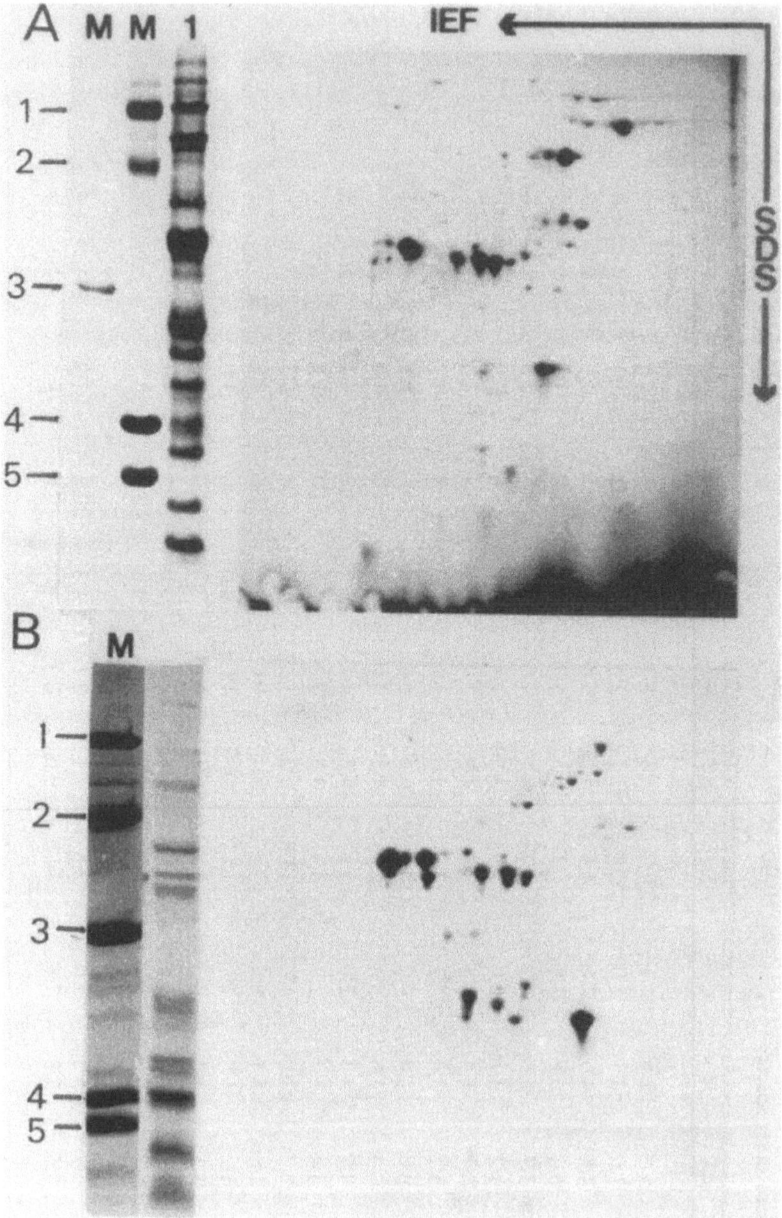

Fig. 1. One and two dimensional electrophoretic patterns of proteins from defatted seed meal (A) and a total globulin fraction (B) from *A. hypochondriacus*. M, M_r markers; 1, 1-D analysis by SDS-PAGE.

protein content is slightly lower after treatment with organic solvents (Paredes-López *et al.*, 1988; Barba de la Rosa *et al.*, 1992a).

The heterogeneity in pI of the amaranth globulins clearly displayed in the 2D gels (Figure 1 B) suggests the presence of several isoforms in each group of proteins (Segura-Nieto *et al.*, 1992; 1994), as described previously for 7S and 11S globulins from legumes (Casey *et al.*, 1986).

2.5 Sedimentation coefficients

Sedimentation coefficient is one of the major criteria used to identify and characterize seed globulins. Sedimentation analyses have therefore been performed with total amaranth globulins as shown in Figure 2A. These revealed the presence of three discrete groups (shown by the continuous line), with sedimentation coefficients of 1.2-2.0S, 8-10S and 12.7-13S, respectively (Konishi *et al.*, 1985; Yamauchi *et al.*, 1991; Barba de la Rosa *et al.*, 1992b; Segura-Nieto *et al.*, 1992).

The main component with low sedimentation coefficient (2-3S), may correspond to a globulin that has been described previously in wheat, oat and pea by Robert *et al.* (1985) and reported to share similarities in glycosylation and antigenic determinants with vicilin. It remains to be demonstrated whether these low molecular weight "globulins" of amaranth are real and correspond to processed globulins or to degradation products of them.

The 8S and 13S globulin components of amaranth are similar to the 7S and 11S globulins, respectively, of pea (Figure 2A, Table 3) and are therefore referred to as 7S-type and 11S-type (Olguín- Martínez, 1992).

SDS-PAGE of the amaranth globulin fractions separated on the basis of their sedimentation coefficients (Figure 2B) (Segura-Nieto *et al.*, 1992), reveals five groups of polypeptides. The first group, of M_r 50-75,000, corresponds to the α, α' and β subunits of 7S-like globulin as indicated in Table 3, while the second (M_r 30-37,500) and third (M_r 18-27,500) groups correspond to the acidic and basic subunits, respectively, of the 11S type globulins (Figure 1B). The fourth and fifth groups comprise low molecular weight polypeptides of M_r 14-17,000 and M_r 18-30,000. The latter correspond to components of about 1.4-1.9S and 4.6S, respectively, and could be related to the 2-3S globulins that have been described in some cereals and legumes such as pea (Robert *et al.*, 1985) and soybean (Yamauchi *et al.*, 1991) (see also Chapter 24). More studies are needed at the gene level in order to identify which of the different polypeptides correspond to distinct gene products, and which originate by post-translational modification (proteolytic processing, glycosylation) or degradation during preparation.

A total globulin preparation has also been fractionated by gel filtration and ion exchange chromatography to give major globulin forms with apparent

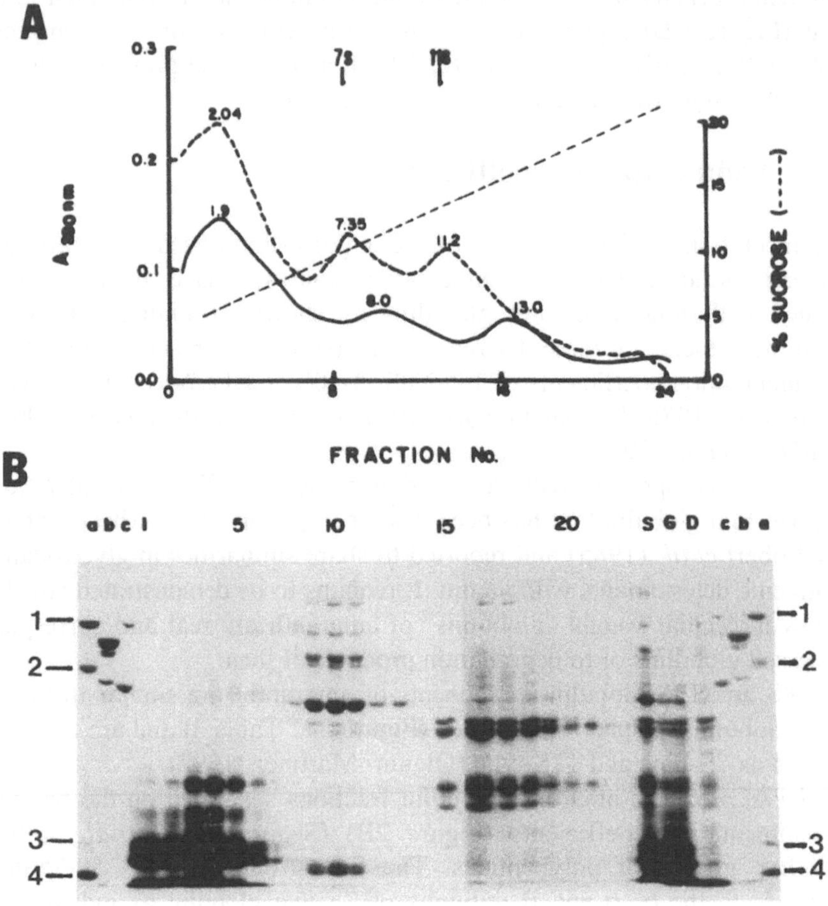

Fig. 2. Sedimentation pattern of amaranth globulins and the polypeptide compositions of the fractions on SDS-PAGE. A shows the sedimentation profile of amaranth globulins (continuous line). The sedimentation pattern of protein markers of known sedimentation coefficient are marked with dashed lines. The half arrows indicate the positions of pea 7S and 11S globulins. B shows SDS-PAGE patterns. Lanes a-c correspond to molecular weight markers. Lanes 1-23 contain all the fractions from the sucrose gradient. Lanes S, G and D correspond to proteins from defatted meal, and total globulins before and after dialysis, respectively. Reprinted with permission from publishers and authors, from Segura-Nieto, *et al.* (1992).

Table 3 Molecular weights of 7S-type and 11S-type native amaranth globulins and their subunits

7S-Type			11S-Type				Reference
Native		Subunits	Native		Subunits		
M_r (x1000)	(S)	M_r (x1000)	M_r (x1000)	(S)	A M_r(x1000)	B M_r(x1000)	
nd	nd	nd	440	(12.7)	36.0-32.0	20.0-18.0	a
nd	nd	nd	398-337	nd	nd	nd	b
nd	(10.0)	52, 37, 34, 28, 23	nd	(12.7)	37.5-31.5	26.5-20.5	c
nd	(8.0)	75, 61, 51, 39	nd	(13.0)	36.0-30.0	22.0-18.0	d
nd	nd	nd	400-330	(11.9)	34.0-32.0	24.0-22.0	e

a, Konishi *et al.*, 1985; b, Marcone and Yada, 1991; c, Barba de la Rosa *et al.*, 1992b; d, Olguín-Martínez, 1992; e, Chen and Paredes-López, 1997; nd, non described.

molecular weights of 440,000 (Konishi *et al*., 1985) and 398,000 (Marcone and Yada 1991) and a minor species of M_r 337,000 (Marcone and Yada, 1991). Konishi *et al.* (1985) have also described a component with a sedimentation coefficient of 12.7S comprising at least four kinds of subunits, of M_r 36,000, 32,000, 20,000 and 18,000.

Recently, Chen and Paredes-López (1997) used different procedures to isolate amarantin, the 11S type globulin, comprising two heterogeneous forms of M_r 330,000 and 400,000. These showed the same subunit composition after reduction, consisting of polypeptides of M_r 32-34,000 and 20-24,000, respectively (see Table 3).

Although several authors have tried to purify amaranth globulins by gel filtration, various problems have been encountered including the presence of high molecular weight protein aggregates, discrepancies in the molecular weights and cross contamination of the 8S by the 13S groups of subunits, probably resulting from dissociation of hexameric 13S into trimers.

2.6 Molecular association/dissociation

A common characteristic of most globulins is association/dissociation under non-denaturing conditions, and amaranth globulins are no exception. Freeze drying of the 7S-type globulins results in aggregation from 8S to 10S (Barba de la Rosa *et al*., 1992b). The effect of cryoprecipitation on aggregation increases as the temperature decreases, as reported also for soybean11S and peanut globulins (Konishi *et al*., 1985). Amaranth globulin also undergoes a substantial change in surface charge density as a function of pH. The turbidity test in the pH range 3 to 8 gave a pI of 5.0 while at alkaline pH the oligomeric globulin dissociated to monomers (Konishi *et al*., 1985). On the other hand, Marcone and Yada (1992) reported that aliphatic hydrophobicity was greatest in the pH range 4.5 to 6.8 due to changes in tertiary structure, resulting in exposure of internal aliphatic amino acids. The pI of the globulin was 4.8 which corresponds to the highest value of aliphatic surface hydrophobicity. Hydrophobicity and charge frequency constitute the two main physical features influencing functional properties such as solubility (Konishi and Yoshimoto, 1989; Marcone and Yada 1992). 11S globulins are dimers composed of specific acidic (M_r ~35,000) and basic (M_r ~20,000) polypeptides linked by a disulphide bond. The position of the disulphide bond between soybean glycinin subunits has been determined by Staswick *et al.* (1984a). He found that only one disulphide bond was involved and this was located in analogous positions between the cysteine residue at position 7 of the basic polypeptide and the cysteine residue at position 86 of the acidic subunit. All 11S-type globulins so far analyzed have this disulphide bond in similar positions. However, 11S globulins contain more than 2 cysteine residues (6-8 in total), which could also participate in the association/dissociation phenomena, in addition to non-covalent interactions.

Marcone and Yada (1991, 1992) showed that amaranth globulins may be isolated and purified to homogeneity by gel filtration on Sephacryl S-300 and anion exchange chromatography on Separom hema-Bio 1000 DEAE. They identified two heterogeneous forms, which they called major and minor species, with apparent M_r of 398,000 and 337,000, respectively. Electrophoretic analysis using both native PAGE and SDS-PAGE of each globulin species showed only slight differences in their polypeptide compositions, but appreciable variation in mobility and dissociation behaviour were detected on native-PAGE. The principal protein fraction contained tetrameric and hexameric oligomers held together by weak non-covalent forces with distinct numbers of charged groups, resulting in the difference in protein mobility. The amino acid compositions of the two oligomers showed more cysteine in the minor component (hexamer) with higher molecular weight, than in the major component (tetramer), suggesting that exchange of sulphydryl-disulphide groups occurred.

2.7 Functional Properties

In addition to the great interest in plant proteins as a source of essential amino acids for nutrition, their potential utilization in foodstuffs depends to a large extent on their functional properties. These are derived from physicochemical characteristics such as molecular weight, amino acid composition and sequence, conformation, net surface charge and effective hydrophobicity (Caer et al., 1990). The ability of proteins to bind and adsorb different food components (carbohydrates, water, lipids, and flavors) and to form fibres, gels, foams and emulsions are of fundamental importance for products such as cheeses, sausages, dressings, and puddings (Kinsella and Phillips, 1989).

2.8 Hydrophobicity, surface charge, and emulsifying activity

Konishi and Yoshimoto (1989) studied the emulsifying activity of crude amaranth globulins as a function of protein concentration and hydrophobicity, in comparison with more conventional food proteins. They showed that the emulsifying activity was about twice of that of soybean protein isolates, but lower than that of casein and bovine serum albumin when determined at 0.4% protein concentration. They also found that both emulsion stability after 24 h and emulsifying activity varied with pH, the maximum emulsion stability being at pH 7.5 and above pH 10, as shown in panel A of Figure 3. In

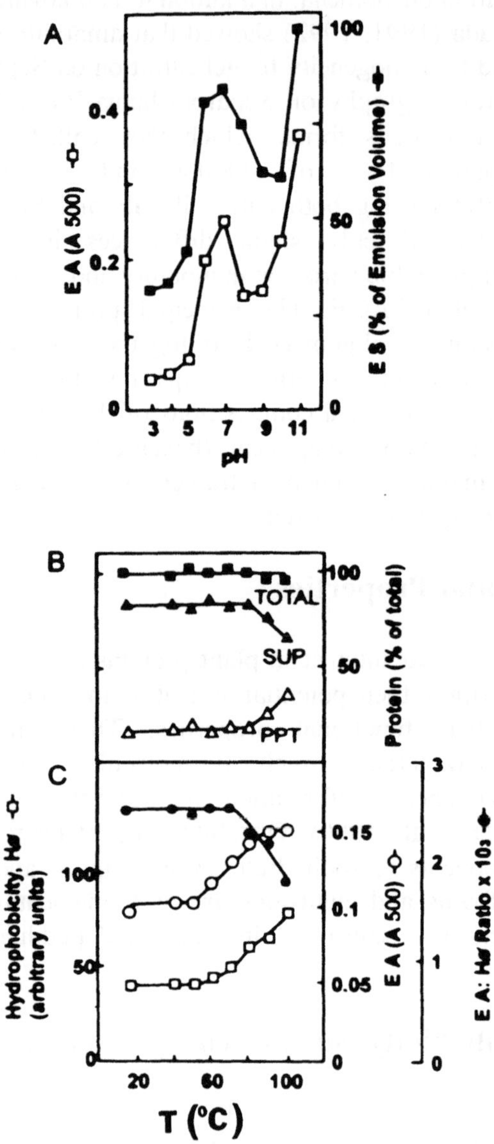

Fig. 3 Amaranth globulin as a heat stable-emulsifying agent. A shows the effect of pH on the emulsifying activity (EA), and emulsion stability (ES) of crude amaranth globulins. B shows the effect of temperature (T °C) on the solubility. C shows the effect of heat treatment on the relationship between hydrophobicity (H) and EA. Reproduced with permission from publishers and authors, from Konishi and Yoshimoto (1989)

general, the hydrophobicity of proteins increases upon denaturation or unfolding due to heat treatment. Panel B of Figure 3 shows that the solubility of an amaranth globulin solution (0.12%) remained constant after heating for 15 min at temperatures up to 70°C. The effect of heat treatment on the relationship between emulsifying activity and hydrophobicity showed that both increased significantly, by 2 and 1.5 fold at 100°C, respectively, compared with the values at 17°C. The ratio of emulsifying activity to hydrophobicity remained constant up to 70°C, and then decreased gradually as shown in panel C of Figure 3. These results are partly explained by a decrease in protein solubility. These preliminary data show that amaranth globulins are relatively heat-stable and are therefore expected to be useful as emulsifiers with high nutritional value. Although in this study total globulins were used, it is possible that 7S-like globulins could have a greater ability to decrease surface tension and better emulsifying properties compared with 11S-type globulin, as described for pea globulins (Dagorn-Scaviner, 1987).

2.9 Quaternary structure

There is scarcity of information about the primary, secondary and tertiary structure of amaranth globulins. With regard to quaternary structure, Marcone *et al.* (1994) have demonstrated that amaranth globulin exists as a dodecameric structure made of subunits arranged into two annular hexagonal rings placed parallel to one another, resulting in a hollow oblate cylinder. This structure was observed using electron microscopy and computer image analysis techniques. Figure 4 shows an electron micrograph of a single globulin complex exhibiting six stain-excluding regions and a central stain-filled depression (hole). Panel B of Figure 4 shows a shaded surface representation of a three-dimensional computational model comprising 12 spheres of equal size and density, arranged in two parallel layers with a central six-fold axis of rotational symmetry.

2.10 Molecular biology and evolutionary relationships

The first molecular cloning of a cDNA encoding an 11S-type globulin from amaranth was recently reported by Barba de la Rosa *et al.* (1996). The deduced amino acid sequence (Figure 5) confirmed that the protein is synthesized as a precursor of M_r 59,000 which is processed into the acidic subunit of M_r 36,000 and the basic subunit of M_r 24,000.

This globulin appears to have similar features to other 11S-type globulins such as those from *Arabidopsis*, sunflower and pumpkin, and glycinin A5A4B3 from soybean (Momma *et al.*, 1985; Pang *et al.*, 1988; Vonder Harr *et al.*, 1988; Hayashi *et al.*, 1988). The amaranth globulin encoded by the

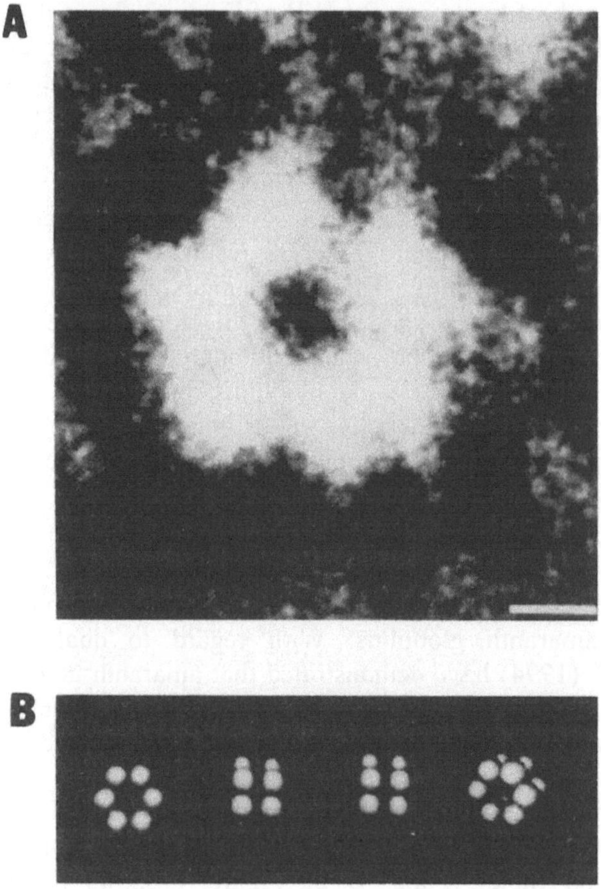

Fig. 4 Quaternary structure and model for the amaranth oligomeric 11S globulin. A. Electron micrograph of a single globulin complex showing six stain-free regions and a central stain-filled cavity. (Scale bar corresponds to 2.5 nm). B. Shaded surface representation of a three dimensional computational model, comprising 12 spheres of equal size and density, arranged in two parallel layers with a central six-fold axis of rotational symmetry. Reprinted with permission from publishers and authors, from Marcone *et al.* (1994).

cDNA has a high content of essential amino acids (Table 4), making it a good candidate to be overexpressed in transgenic plants and used as protein ingredient of high nutritional value. Table 5 compares the amino acid sequences around the post-translational processing sites of the 11S-type amaranth globulin and other 11S-type globulins (Hara-Nishimura *et al.*, 1995) showing some homology. The sequence of the amaranth 11S-type

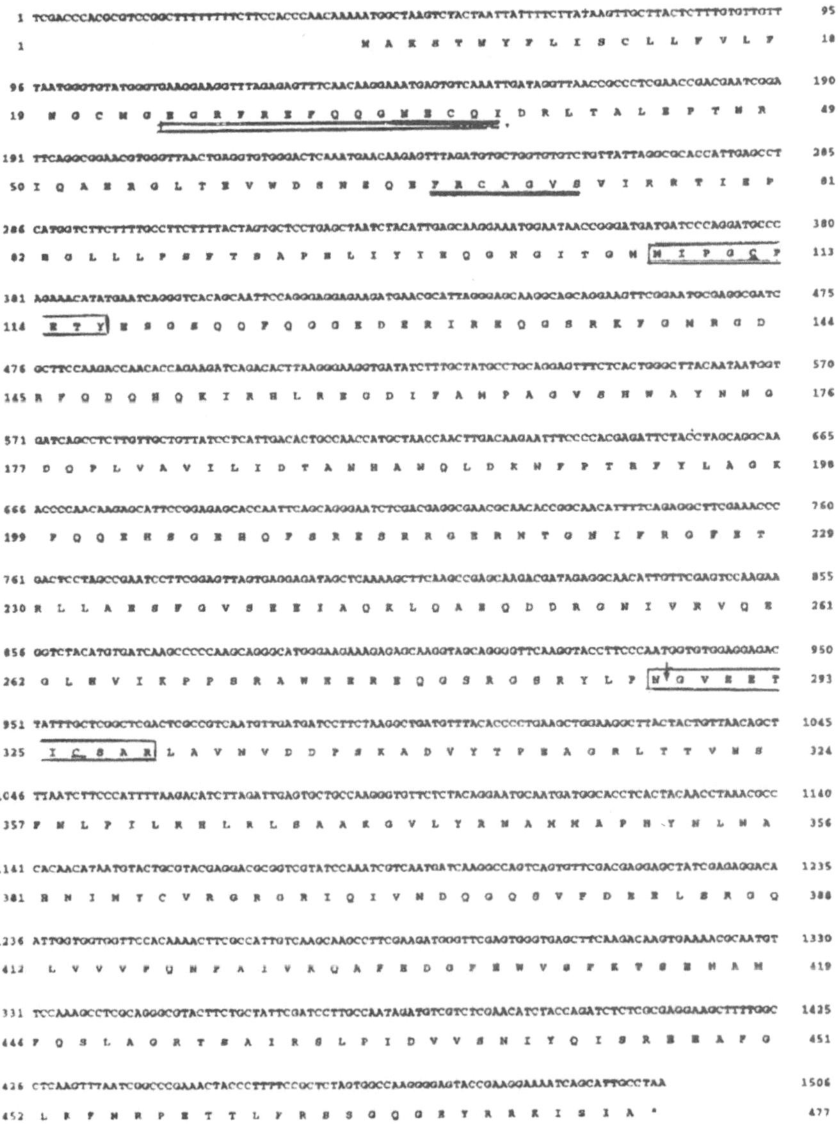

Fig. 5. The nucleotide and deduced amino acid sequences of the amaranthin cDNA (aglo1). The *N*-terminal sequence predicted for the M_r 59,000 proprotein is underlined twice. The two cysteine residues thought to form and intra-chain disulphide bonds are underlined. The putative proteolytic cleavage site between the acidic and basic polypeptides between Asn 288 and Gly 289 is indicated by an arrow. Taken from Barba de la Rosa *et al.* (1996).

globulin shows 44% identity with the 11S *Arabidopsis* globulin, 42% with cruciferin of oilseed rape, 41% with rice glutelin and oat globulin, 36% with sunflower helianthinin and 34% with both soybean and faba bean legumins (Barba de la Rosa *et al.*, 1996).

Table 4. Comparison of the amino acid composition (mol %) deduced from the amaranthin cDNA and FAO/WHO/UNU (1985) recommended levels of essential amino acids.

Amino Acid	Amaranthin	FAO/WHO/UNU Child	FAO/WHO/UNU Adult
*Cys	1.6		
His	2.4	1.3	1.2
Ile	5.6	2.1	1.0
Leu	6.7	5.0	1.4
Lys	4.0	4.0	1.1
Met	2.4		
Phe	6.5		
Thr	4.0	2.9	0.8
Trp	0.7	0.5	0.2
Tyr	2.6		
Val	5.4	3.0	1.1
Ala	6.0		
Arg	8.4		
Asp	3.7		
Asn	5.6		
Glu	8.9		
Gln	6.3		
Gly	8.4		
Pro	3.8		
Ser	7.0		
Met + Cys	4.0	1.7	1.1
Phe + Tyr	9.1	4.3	1.3

Taken from Barba de la Rosa *et al.* (1996).

3. QUINOA

Quinoa seeds contain, on average, about 12.6% protein on a dry weight basis (Cardoza and Tapia, 1979) with two major groups of storage proteins, the 2S albumins (see Chapter 24) and the 11S chenopodin (Brinegar, 1997). A detailed study of chenopodin has been reported by Brinegar and Goudan (1993), on which the following account is based.

Chenopodin has a typical 11S globulin structure, with a native M_r of about 320,000 and consisting of six pairs of acidic, or A, subunits (M_r 32-39,000) and basic, or B, subunits (M_r 22-23,000). Each subunit pair is

Table 5 Comparison of the amino acid sequences around post-translational processing sites of 11S-type globulins

Seed proteins	Sequence	Reference
	7 ↓	
Amaranth 13S (putative site)	--GSRGSRYLPN*GVEETICSAR--	Barba de la Rosa *et al*, 1996
Pumpkin 11S globulin	---YIESESESEN*GLEETIICTLR--	Hayashi *et al*., 1988
Pea Legumin A	--KGKSRRQGDN*GLEETVCTAK--	Lycett *et al*., 1984
Pea Legumin B	---EKQRSEERKN*GLEETICSAK--	Gatehouse *et al*., 1988
Sunflower 11S globulin HaG3	--QQGRRGGWSN*GVEETICSMK--	Vonder Harr *et al*., 1988
Soybean glycinin A1aBx	---RGSQSKSRRN*GIDETICTMR--	Negoro *et al*., 1985
Soybean glycinin A2B1a	---CQRQSKRSRN*GIDETICTMR--	Staswick *et al*., 1984
Soybean glycinin A3B4	---PRGRGCQTRN*GVEENICTMK--	Fukazawa *et al*., 1985
Soybean glycinin A5A4B3	---PRERGCETRN*GVEENICTLK--	Momma *et al*., 1985
Oat globulin	---SGQSWDQSEN*GLEENFCSLE--	Shotwell *et al*., 1988

Proprotein processing occurs on the C-terminal side of the asparagine residue indicated by the asterisk. The processing site was established by aligning the deduced precursor sequences with the N-terminal amino acid sequences of the respective mature proteins except for the amaranth protein (Barba de la Rosa *et al*., 1996). C, cysteine residue 7 of the basic subunit (indicated by an arrow) forms a disulphide bond with a corresponding cysteine residue in the acidic subunit (not shown).
Adapted from Hara-Nishimura *et al* (1995).

stabilized by inter-chain disulphide bonds and they form a single band of M_r about 55,000 when separated by SDS-PAGE without reduction. A minor band of M_r about 50,000 is also observed when separated under reducing conditions which could correspond to the unprocessed acidic/basic subunit precursor proteins. The amino acid composition determined for the whole protein and the purified acidic and basic subunits are typical of 11S storage globulins (Table 6), with high glutamate/glutamine, aspartate/asparagine, arginine, serine, leucine and glycine. In addition, the *N*-terminal sequence of the basic subunit shows clear homology with those of other 11S globulins, with 16 out of 21 residues being identical to the cruciferin sequence reported by Simon *et al.* (1985).

Table 6. Amino acid compositions (mol%) of chenopodin and its A and B subunits.

Amino acid	Chenopodin	A subunit	B subunit
Asp[a]	10.5	10.1	11.0
Thr[b]	3.9	3.8	4.2
Ser[b]	8.9	9.5	8.8
Glu[a]	14.8	16.0	13.2
Pro	4.9	5.0	4.0
Gly	8.7	10.2	6.8
Ala	5.3	5.1	8.0
Val	6.8	5.0	5.9
Cys[c]	0.8	0.4	0.7
Met	1.7	1.9	1.1
Ile	4.9	5.0	5.7
Leu	7.4	6.7	8.6
Tyr	2.9	2.5	3.6
Phe	4.0	3.6	4.3
Lys	3.2	3.2	2.9
His	3.0	3.7	1.4
Arg	9.7	8.7	9.8

[a], Asp and Glu include Asn and Gln, respectively
[b], Thr and Ser are corrected for partial degradation during acid hydrolysis
[c], Cys was determined as carboxymethyl cysteine
Trp was not determined
Taken from Brinegar and Goudan (1993).

4. BUCKWHEAT

Buckwheat seeds contain about 18% protein (Pomeranz *et al.*, 1975), over 40% of which is salt-soluble globulins (Belozersky, 1975). The major storage globulin is a 13S legumin-type protein but recent work suggests that vicilin-like proteins may also be present.

The 13S globulin has an M_r of about 280,000 (Belozersky, 1975) and has a typical "legumin" structure, consisting of six subunit pairs each comprising a large acid subunit (M_r 32-43,000) and a small basic subunit (M_r 20-23,000) (Radovic' *et al.*, 1996). The pairs of subunits are associated by inter-chain disulphide bonds and separation by SDS-PAGE under non-reducing conditions gives about eight major bands of M_r ranging from about 43,000 to 68,000, corresponding to unreduced subunits pairs (Radovic' *et al.*, 1996). The homology of the buckwheat 13S with 11S "legumin" type globulins from other species was recently demonstrated by *N*-terminal sequence analysis of a basic subunit of M_r about 26,000 (Rout *et* al., 1997). Eleven of the first seventeen residues were identical to the *N*-terminal sequences of soybean glycinins G1, G2 and G3 (Nielsen *et al.*, 1989) and ten residues identical to the sequences of pea legumins A1 and A2 (Lycett *et al.*, 1984).

Although there is no doubt that the buckwheat 13S globulin belongs to the legumin group of seed globullins, pulse chase labeling of developing cotyledons failed to show that the acidic and basic subunits originated from a single precursor protein (Maksimovic' *et al.*, 1996) as described for other legumins (see chapter 8). *In vitro* synthesis using mRNA from developing seeds also resulted in the synthesis of individual acidic and basic subunits and the authors concluded that they were encoded by separate mRNAs. These studies remain to be confirmed but provide the first evidence for the separate synthesis of individual acidic and basic subunits of a legumin-type protein. This mode of synthesis would, of course, have consequences for the combination of individual acidic and basic subunits into pairs, as this is usually determined by their synthesis as a single precursor protein.

Buckwheat globulins also contain a group of subunits of M_r about 57-58,000 (Radovic' *et al.*, 1996) which were initally thought to be components of the 13S globulin (Dunaevsky and Belozersky, 1989; Rout and Chrungoo, 1996). However, subsequent studies have shown that these subunits can be separated from the 13S globulin subunits by sucrose gradient ultracentrifugation, forming a separate peak of about 8S (Radovic' *et al.*, 1996). They also differ from the 13S globulin in tissue location, being mainly, but not exclusively, located in the endosperm while the 13S protein is cotyledon-specific (Radovic' *et al.*, 1996; Rogl and Javornik, 1996), and in accumulating earlier than the 13S globulin subunits in the developing seed (Maksimovic' *et al.*, 1996). Radovic' *et al.* (1996) suggested that the 8S globulin belongs to the vicilin group of 7-8S globulins but final proof of this will depend on the determination of amino acid sequences. The same authors also showed, based on the sedimentation equilibrium profiles of total salt-soluble proteins, that the 13S and the 8S groups of globulins accounted for about 75% and 15% of the total fraction, respectively. The remaining 10%

consisted of low M_r (14-16,000) components which are probably water soluble albumins (see Takumi *et al.*, 1995). It is probable that at least some of these components belong to the 2S albumin family as discussed in Chapter 24.

ACKNOWLEDGMENTS

MSN and OPL gratefully acknowledge financial support from Consejo Nacional de Ciencia y Tecnología de México. The authors also acknowledge the authors and publishers permission to reproduce copyrighted material.

IACR recives grant-aided support from the Biotechnology and Biological Sciences Research Council of the United Kingdom.

REFERENCES

Abdi, H., and Sahib, M.K. 1976 Distribution of lysine in different legumes and some species of Amaranthus seeds. *Journal of Food Science and Technology*, **13**, 237-239.

Barba de la Rosa, A.P., Gueguen, J., Paredes-López, O., and Viroben, G. 1992a Fractionation procedures, electrophoretic characterization, and amino acid composition of amaranth seed protein. *Journal of Agricultural and Food Chem*istry, **40**, 931-936.

Barba de la Rosa, A.P., Paredes-López O, and Gueguen J. 1992b Characterization of amaranth globulins by ultracentrifugation and chromatographic techniques. *Journal of Agricultural and Food Chem*istry, **40**, 937-940.

Barba de la Rosa, A.P., Herrera-Estrella, A., Utsumi, S., and Paredes-López, O. 1996 Molecular characterization, cloning and structural analysis of a cDNA encoding and amaranth globulin. *Journal of Plant Physiol*.ogy, **149**, 527-532.

Becker R., Wheeler E.L., Lorenz K., Stafford A.E., Grosjean O.K., Betschard A. and Saunders R.M. 1981 A compositional study of amaranth grain. *Journal of Food Science*, **46**,1175-1180.

Belozersky, M.A. 1975 Isolation and characterization of buckwheat seed 13S globulin. In: *Biosynthesis of Storage Proteins*. Nauka: Mosco, pp 152-156.

Bressani R. and García-Vela L.A. 1990 Protein fractions in amaranth grain and their chemical characterization. *Journal of Agricultural and Food Chem*istry, **38**, 1205-1209.

Brinegar, C. 1997 The seeds storage proteins of quinoa. In: *Food Proteins and Lipids* (ed. Damodaran) Plenum Press, New York pp. 109-115.

Brinegar, C. and Goudan, S. 1993 Isolation and characterization of chenopodin, the 11S seed storage protein of quinoa (Chenopodium quinoa) *Journal of Agricultural and Food Chem*istry, 41, 182.

Caer D., Baniel A., Subirade M., Gueguen J., and Colas B. 1990 Preparation and physicochemical properties of glycosilated derivatives of pea legumin. *Journal of Agricultural and Food Chemistry*, **38**, 1700-1706.

Campbell, C.G. 1995 Buckwheat *Fagopyrum esculetum* (Polygonacea). In *Evolution of Crop Plants*. Longman Scientific Technical, Harlow Essex . (eds. J. Smartt and N.W. Simmonds) pp. 409-412.

Cardoza, A.and Tapia M. 1979 Valor nutricio. In: *Quinoa y Kañiwa*. M. Tapia ed. Serie libros y Materiales educativos No. 49CIID-IICA, Bogotá.

Casey, R., Domoney, C., and Ellis, N. 1986 Legume storage proteins and their genes, in *Oxford Survey of Plant Molecular and Cell Biology* Vol **3**,1-95.

Chen, S.F. and Paredes-López, O. 1997 Isolation and characterization of the 11S globulin from amaranth seeds. *Journal Food Biochem.*istry. **21**, 53-65.

Dagorn-Scaviner, C., Gueguen, J., and Lefebvre, J. 1987 Emulsifying properties of pea globulins as related to their adsorption behaviors. *Journal of Food* Sci.ence, **52**, 335-341.

Duarte-Correa, A., Jokl, L., and Carlsson, R. 1986 Amino acid composition of some Amaranthus sp. grain protein and of its fractions . *Archivos Latinoamericanos de la Nutrición.* **26**, 466-476.

Dunaevsky, Y.E. and Belozersky, M.A. 1989 Proteolysis of the main buckwheat seeds at the early stage of germination. *Physiologia Plantarum* **75**, 424-428.

FAO/WHO/UNU. 1985 Energy and Protein Requirements. *Technical report Series* No. 724. Geneve: WHO.

Fukazawa, C., Momma, T., Hirano, H., Harada, K., and Udaka K. 1985 Glycinin A3B4 mRNA. *Journal of Biol.ogical Chem*istry. **260**, 6234-6239.

Galwey, N.W. 1995 Quinoa and relatives Chenopodium spp. (Chenopodiacea). In: *Evolution of Crop Plants* Longman Scientific Technical, Harlow Essex . J. Smartt and N.W. Simmonds eds. pp. 41-46.

Gatehouse, J.A., Brown D., Gilroy, J., Levasseur M., CastletonJ., and Ellis, T.H.N. 1988 Two genes encoding "minor" legumin polypeptides in pea (*Pisum sativum* L.). *Biochemical Journal.* **250**,15-24

Gorinstein, S., Moshe, R., Greene, L.J., Arruda, P. 1991 Evaluation of four *Amaranthus* species through protein lectrophoretic patterns and their amino acid composition. *Journal of Agricultural and Food Chemistry,* **39**, 851-854.

Hara-Nishimura, I., Shimada, T., Hiraiwa, N., and Nishimura, M. 1995 Vacuolar processing enzyme responsible for maturation of seed proteins. *Journal of Plant Physiology*, **145**, 632-640.

Hayashi, M., Mori, H., Nishimura, M., Akazawa, T., Hara-Nishimura . I. 1988 Nucleotide sequence of cloned cDNA coding for pumpkin 11S-globulin β-subunit. *European Journal of Biochemistry,* **172**, 627-632.

Hurich, J., Parzysz, H., Przbylska, J. 1977 Amino acid patterns of albumin and globulin fractions in seeds of wild and primitive forms *Pisu, Pisum Newsletter.* **9**, 10.

Kinsella , J.E. and Phillips L.G. 1989 Structure: function relationship in food proteins, film and foaming behavior, in *Food Proteins* , (eds. Kinsella, J.E. and W.G. Soucie). American Oil Chemists Society. Ch 4.

Konishi, Y., Fumita, Y., Ikeda, K., Okuno, K., and Fuwa, H. 1985 Isolation and characterization of globulin from seeds of *Amaranthus hypochondriacus*, L. *Agricultural and Biological Chemistry*, **49**, 1453-1459

Konishi, Y. and Yoshimoto, N. 1989 Amaranth globulin as a heat-stable emulsifying agent. *Agricultural and Biological Chemistry*, **53**, 3327-3328.

Lycett, G.W., Croy, R.R.D., Shirsat, A.H. and Boulter, D. 1984 The complete nucleotide sequence of a legumin gene from pea (*Pisum sativum* L.). *Nucleic Acid Research,* **12**, 4493-4506.

Makisimovic', V.R., Varkonji-Gasic', E.I., Radovic', S.R. and Savic', A.P. 1996 The biosynthesis of 13S buckwheat seed storage protein. *Journal of Plant Physiology*, **147**, 759-761.

Marcone, M. and Yada, R.Y. 1991 Isolation, purification and characterization of the oligomeric seed globulin from *Amaranthus hypochondriacus*. *Agricultural and Biological Chemistry*, **55**, 2281-2289.
Marcone, M.F. and Yada, R.Y. 1992 Study of the charge profile and covalent subunit association of the oligomeric seed globulin from *Amaranthus hypochondriacus*: *Journal of Agricultural and Food Chemistry,* **40**, 385-389.
Marcone, M., Beniac, D.R., Harauz,,G., and Yada, R.Y. 1994 Quaternary structure and model for the oligomeric seed globulin from *Amaranthus hypochondriacus*: K343. *Journal of Agricultural and Food Chemistry,* **42**, 2675-2678.
Momma, T., Negoro. T., Hirano, H.. Matsumoto, A., Udaka K., and Fukazawa, C. 1985 Glycinin $A_5A_4B_3$ mRNA: cDNA cloning and nucleotide sequencing of a splitting storage protein subunit of soybean. *European Journal of Biochem*istry, **149**, 491-496.
Mora-Escobedo. R., Paredes-López, O., and Ordorica-Falomir, C. 1990 Characterization of albumins and globulins from amaranth. *Lebensmittel Wissenschaft Technologie,* **23**, 484-487.
National Research Council 1984 *Amaranth, Modern prospects for an ancient crop*. National Academy Press Washington D.C.
Negoro, T., Momma, T., and Fukasawa, C. 1985 A cDNA clone encoding a glycinin A-1a subunit precursor of soybean. *Nucleic Acid Research,* **13**, 6719-6731.
Nielsen, N.C., Dickinson, C.D., Cho, T.-J., Thanh, V.H., Scallon, B.J., Fischer, R.L., Sims,T.L., Drews, G,N, and Goldberg, R.G. 1989. Characterization of the glycine gene family in soybean. *Plant Cell,* **1**, 313-328.
Olguín-Martínez. L.E. 1992 Characterization of Globulins from Amaranth Seeds (*Amaranthus hypochondriacus*, L.). M.Sc. Thesis. Instituto Politécnico Nacional. Escuela Nacional de Ciencias Biológicas, México, D.F., México.
Osborne, T.B. 1924 The vegetable proteins. In *Monographs on Biochemistry*, 2nd edition, eds. Loongmans, Green and CO., New York.
Pang, P.P., Pruitt, R.E., and Meyerowitz. 1988 Molecular cloning, genomic organization, expression and evolution of 12S seed storage protein genes of *Arabidopsis thaliana*. *Plant Molecular Biology*, **11**, 805-820.
Paredes-López ,O., Mora-Escobedo, R., and Ordorica-Falomir, C. 1988 Isolation of amaranth protein. *Lebensmittel Wissenschaft Technologie,* **27**, 59-61.
Peterson, D.M. 1978 Subunit structure and composition of oat seed globulin, *Plant Physiol*ogy, **62**, 506-509.
Pomeranz. Y., Marshall, H.G., Robbins, G.S. and Gilbertson, J.T. 1975 Protein content and amino acid composition of maturing buckwheat (*Fagopyrum esculentum* Moench) *Cereal Chemistry*, **52**, 479-485.
Radovic'. S.R., Maksimovic', V.R. and Varkonji-Gasic', E.I. 1996 Characterization of buckwheat seed storage proteins. *Journal of Agricultural and Food Chemistry,* **44**, 972-974.
Robert, L.S., Adeli K., and Altosaar, I. 1985 Homology among 3S and 7S globulins from cereals and pea. *Plant Physiology,* **78**, 812-816.
Rogl, S. and Javornik, B. 1996 Seed protein variation for ifentification of common buckwheat (*Fagopyrum sculentum* Moench) cultivars. *Euphytica*, **87**,111-117
Sauer, J.D. 1995 Grain amaranths *Amaranthus* spp. (Amaranthaceae). In *Evolution of Crop Plants* (eds J Smartt and N.W. Simmonds) pp. 9-10.
Segura-Nieto. M., Vázquez-Sánchez, N., Rubio-Velázquez, H., Olguín-Martínez, L., Rodríguez-Nester, C. and Herrera-Estrella L. 1992 Characterization of amaranth (*Amaranthus hypochondriacus* L) seed proteins. *Journal of Agricultural and Food Chemistry,* **40**, 1553-1558.

Segura-Nieto, M., Barba de la Rosa, A.P. and Paredes-López, O. 1994. Biochemistry of amaranth proteins. in: *Amaranth: Biology, Chemistry and Technology*. (ed. O. Paredes-López) CRC Press, Boca Ratón FL. Chap 5, pp 75-106.

Shotwell, M.A., Afonso, C., Davies, E., Chesnut, R.S., and Larkins B.A. 1988 Molecular characterization of oat seed globulins. *Plant Physiology*, **87**, 698-704.

Simon, A.E., Tenbarge, K.M., Scofield, S.R., Finkelstein, R.R. and Crouch M.L. 1985 Nucleotide sequence of a cDNA clone of *Brassica napus* 12S storage protein shows homology with legumin from *Pisum sativum*. *Plant Molecular Biology*. **5**, 191-201

Singhal, R.S. and Kulkarni, P.R. 1988 Review : Amaranths an underutilized resource. Int. *Journal of Food Science and Technology*, **23**, 125-140

Smith, A.K., and Circle, S.S. 1981 Chemical composition of the seed in *Soybeans: Chemistry and Technology*. AVI Publishing, Westport, CT. Vol. 1, Chap 3.

Soriano-Santos, J., Iwabuchi, S., and Fujimoto, K. 1992 Solubility of amaranth seed proteins in sodium sulfate and sodium chloride: the main factor in quantitative extraction for analysis. *International Journal of Food Science and Technology*. **27**, 337-346.

Staswick, P.E., Hermodson, M.A., and Nielsen, N.C. 1984a The amino acid sequence of the A2B1a subunit of glycinin. *Journal of Biological Chemistry*, **259**, 13424-13430

Staswick, P.E., Hermodson, M.A., and Nielsen, N.C. 1984b Identification of the cysteines which link the acidic and basic components of the glycinin subunits. *Journal of Biological Chemistry*, **259**, 13431-13435.

Takumi, K., Koga , T., Kanoh, M., Udaka, J., Tsuji, H. and Manabe, S. 1995 Immunochemical crossreactivity between globulins from buckwheat and indigo seeds. *Bioscience , Biotechnology and Biochemistry*, **59**, 1971-1972.

Vonder Harr, R.A., Allen R.D., Cohen, E.A., Nessler, C. L., and Thomas, T.L. 1988 Organization of the sunflower 11S globulin storage protein gene family. *Gene*, **74**, 433-443.

Yamauchi, F., Yamagishi, T. and Iwabuchi, S. 1991 Molecular understanding of heat-induced phenomena of soybean protein. *Food Review. International*, **7**, 283-322.

20. 7S Globulins of Cereals

ALAN L. KRIZ
DEKALB Genetics Corporation, 62 Maritime Dr., Mystic, CT 06355-1958, USA

1. INTRODUCTION

Storage proteins of the 7S globulin class have been identified in seeds of a number of cereals. In contrast to the situation in legumes, however, these proteins are present in relatively small amounts in cereals. In those cases where they have been examined, the cereal 7S globulins accumulate primarily in the embryo and aleurone layer (Yupsanis *et al.*, 1990). This is in contrast to the situation with the prolamins, the primary storage proteins of most cereals, which accumulate in the starchy endosperm. The 7S cereal globulins probably serve a role as seed storage proteins, and their hydrolysis during germination may provide an immediate source of nitrogen and carbon to the young seedling (Khavkin *et al.*, 1978; Kriz, 1989). Their role as storage proteins is consistent with the observation that in maize mutants such as *opaque2* and *floury2*, which condition a reduction in grain prolamin content, the levels of globulins are increased relative to those in normal seeds (Puckett and Kriz, 1991; Dierks-Ventling, 1982). It is presumed that this increase in globulins occurs to compensate for the decrease in levels of the primary storage proteins in the seed. The globulins are, however, not absolutely required for normal seed functions, as maize mutants completely lacking in 7S globulins exhibit normal patterns of seed development and germination (Kriz and Wallace, 1991).

Since 7S globulins represent a minor class of cereal storage proteins, they remain relatively uncharacterized as a group. In those situations where the genes encoding these proteins have been characterized, some interesting features regarding gene regulation have been identified which make them attractive candidates for studies of gene expression in cereal seeds. These features include highly regulated and coordinated expression patterns during seed development (Kriz, 1989; Wallace and Kriz, 1991; Heck *et al.*, 1993), positive regulation of expression by the phytohormone abscisic acid (ABA) (Williamson and Quatrano, 1988; Heck *et al.*, 1993; Paiva and Kriz, 1994),

P.R. Shewry and R. Casey (eds.), Seed Proteins, 477–498.

and the fact that, at least in maize and barley, each globulin component is encoded by only a single gene (Schwartz, 1979; Kriz, 1989; Heck *et al.*, 1993). This latter point makes the 7S cereal globulins unique among seed storage proteins in that most other classes of storage proteins are encoded by multigene families (Shotwell and Larkins, 1989).

Table 1 Amino acid compositions of cereal globulins as deduced from nucleotide sequences of cloned cDNAs or from hydrolysis of purified seed protein fractions

	Deduced from Nucleotide Sequence			Purified 7S Protein			
Amino Acid	Glb1-S[1]	Glb2[2]	Beg1[3]	PSVIC[4]	Barley Embryo[5]	Oat Embryo[6]	Pea Vicilin[7]
Glu	13.4	10.5	11.0	10.4	nd	nd	nd
Arg	12.1	9.1	12.9	6	5.9	9	4.6
Gly	9.5	9.8	8.6	5.1	10.8	12.5	4.8
Ser	7.6	11.0	6.9	8.8	6.8	8.9	7.7
Val	7.4	8.0	8.8	5.8	7.4	6.4	4.8
Ala	7.2	6.8	7.5	4.2	9.9	7	4.1
Leu	5.8	5.9	6.5	9.9	5.5	6.3	9.8
Pro	4.5	4.0	4.5	4.2	4.7	4.3	2
Phe	4.5	4.7	4.3	5.1	3.5	4.6	4.7
Lys	4.1	4.2	2.8	6.9	6.3	4.4	8.1
His	3.9	3.0	2.2	1.9	2.2	2.8	1.6
Gln	3.7	3.5	6.5	7.9	nd	nd	nd
Thr	3.5	3.3	3.2	2.8	4.8	3.4	2.6
Asp	3.5	4.2	4.1	5.5	nd	nd	nd
Ile	3.3	3.5	3.9	5.5	4.1	3.1	3.9
Tyr	2.5	2.1	1.5	2.1	2	2.3	1.9
Asn	2.3	2.3	3.0	8.1	nd	nd	nd
Met	0.6	1.9	0.9	0	1.5	0.3	0
Cys	0.6	1.2	0.4	0	1.2	1.3	0
Trp	0.2	0.9	0.4	0	nd	nd	nd
Asx[8]	5.8	6.6	7.1	13.6	9.7	7.6	18.9
Glx[8]	17.1	14.1	17.5	18.3	13.7	15.8	22.1

[1]Maize, GenBank M24845; [2]Maize, GenBank X53715; [3]Barley, GenBank M64372; [4]Pea, GenBank Y00722; [5]Burgess and Shewry, 1986; [6]Burgess *et al*, 1983; [7]Croy *et al*, 1980; [8]Summed for the cDNA clones, determined as combination of acid and amide forms in the hydrolysates; nd, not determined

2. AMINO ACID COMPOSITION OF 7S CEREAL GLOBULINS

The 7S cereal globulins are similar to other seed storage proteins in that they contain relatively high amounts of glutamate + glutamine and arginine (Table 1). These proteins therefore provide an excellent potential source of nitrogen to the germinating seed. Comparison with vicilin, the 7S globulin from pea, reveals that the cereal 7S globulins generally contain higher amounts of arginine, glycine, and alanine than the dicot proteins, and lower amounts of leucine, lysine, and aspartic + aspartate. The cereal globulins are low in methionine and cysteine, but differ from the dicot proteins in that they contain at least some sulphur amino acids.

3. MAIZE EMBRYO GLOBULINS

The 7S, or vicilin-like, globulins of cereals which have been characterized in greatest detail are those of maize. These are saline-soluble, water-insoluble components of M_r 60-70,000, 45,000 and 27-30,000, as determined by SDS-PAGE (Khavkin *et al.*, 1978; Dierks-Ventling, 1981; Cross and Adams, 1983; Kriz, 1989). The M_r 60-70,000 proteins associate to form complexes of M_r greater than 200,000, while the M_r 45,000 and M_r 27-30,000 components form complexes of M_r approximately 127,000 (Cross and Adams, 1983). The major globulin components are those of M_r 60-70,000, which are encoded by the *Glb1* gene, and the M_r 45,000 component, which is encoded by the *Glb2* gene (Kriz, 1989). These protein products, designated GLB1 and GLB2, are two of the most abundant proteins in maize embryos, collectively accounting for 10% to 20% of total embryo protein (Kriz, 1989). Both proteins possess characteristics of the 7S class of seed storage proteins with respect to accumulation profiles during seed development, amino acid composition, and polypeptide sequence. Both proteins, and their corresponding mRNAs, are localized exclusively in seed tissues. Expression of both *Glb1* and *Glb2* primarily occurs in embryo tissues (both scutellum and axis), and expression in the endosperm is attributed to that in the aleurone layer.

3.1 Solubility characteristics and purification of maize embryo globulins

The cereal 7S globulins possess solubility characteristics similar to those of the legume vicilins in that they are soluble in solutions of 0.5 to 1M NaCl, from which they may be precipitated by dialysis against H_2O. The protocol for extraction of maize embryo globulins is based on a modification of Danielsson's (1949) procedure developed by Khavkin *et al.* (1978). Mature kernels are soaked in water for 0.5 hours to overnight to soften, and embryos (germs) are manually excised with the aid of a sharpened spatula. The embryos are collected in liquid nitrogen, homogenized in 30 volumes of ice-cold acetone or petroleum ether, and stirred for 3 hour in the cold to defat the meal. The resultant meal is air-dried and added to 10 volumes of 0.05 M phosphate buffer, pH7, containing 1 M NaCl and 0.01M 2-mercaptoethanol. After stirring for 1 hour in the cold, the supernatant is subjected to sequential ammonium sulphate fractionation such that the precipitate is obtained from the 40-70% saturation cut. This pellet is resuspended in the original saline extraction buffer and dialyzed exhaustively against H_2O, during which time the 7S globulins precipitate in the dialysis tubing. The pellet is resuspended in buffered 1M NaCl. This procedure yields the whole globulin fraction as shown in Figure 1, lane 4. The GLB1 protein may be selectively fractionated from whole globulin by simply incubating the whole globulin fraction at 4°C overnight and collecting the precipitated GLB1 protein (Kriz, 1989).

3.2 Historical perspective on identification and characterization of maize embryo globulins

The presence of 7S globulins in maize seeds was first described by Danielsson (1949), and these proteins were initially characterized for reasons other than for their potential function as seed storage proteins. These early studies were concerned with identification of proteins associated with maize embryo development and differentiation. Because these studies, particularly those of Schwartz (1979), laid the basic foundation for further analysis of the maize embryo globulins, the work is worthy of mention here.

One of the first descriptions of changes in protein profile during the course of maize embryo development was provided by Khavkin *et al.* (1977). The primary goal of this work was to investigate the temporal changes in accumulation of specific proteins during maize embryo development and subsequent germination. The basic methods employed were double-immunodiffusion assay and cross-immunoelectrophoresis, in which an antiserum raised against a total protein extract from mature maize embryos was used to investigate qualitative and quantitative changes in protein profiles during seed development and germination. Although these methods may seem crude by modern standards, these authors identified several "embryonal antigens" which were specific to the developing embryo. Some of these proteins accumulated to high levels during the course of embryo development, beginning at 18 days after pollination (DAP), and were subsequently degraded upon the onset of germination. The authors suggested that the most abundant

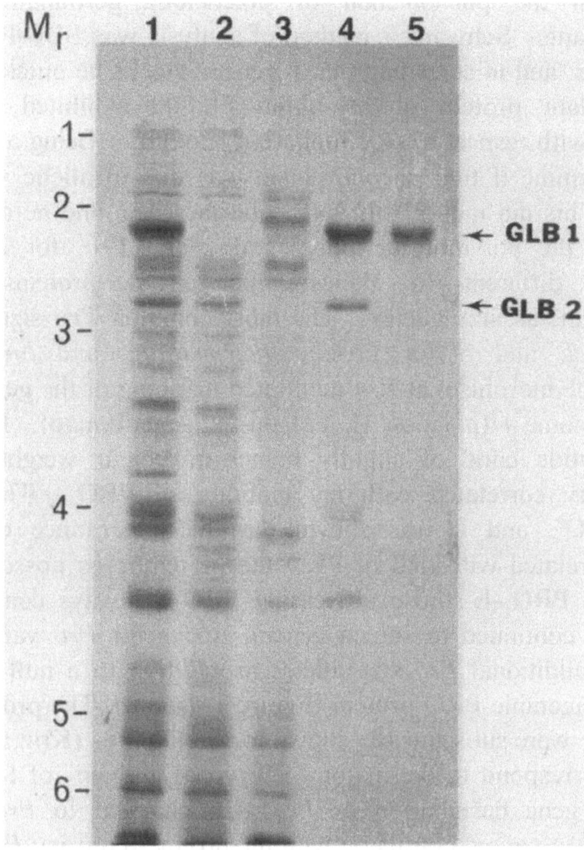

Fig. 1 SDS-PAGE analysis of maize embryo proteins. Total protein extracts from individual mature embryos of different *Glb* genotypes (lanes 1-3) or purified globulins (lanes 4-5) were subjected to electrophoresis in gels containing 12.5% acrylamide. Lane 1, *Glb1-S/S* (inbred line Va26); lane 2, *Glb1-0/0* (null for GLB1) ; lane 3, *Glb2-0/0* (null for GLB2); lane 4, whole globulin obtained from mature Va26 embryos; Lane 5, GLB1-S protein fractionated from whole globulin by cryoprecipitation at 4°C for 48 h. Positions of M_r standards are indicated on the left. 1, 97,400; 2, 68,000; 3, 43,000; 4, 25,700; 5, 18,400 and 6, 14,300. Reproduced from Kriz (1989) with permission of both the author and the publisher.

embryonal antigens corresponded to some form of protein reserve in the embryo, and in a subsequent study it was demonstrated that these abundant proteins did indeed behave as "classical" 7S globulins with respect to solubility characteristics (Khavkin *et al.*, 1978).

In a somewhat similar manner, Schwartz (1979) initiated a study of maize embryo proteins in an attempt to identify changes in polypeptide profile associated with the phenomenon of precocious germination in maize *viviparous* mutants. Schwartz's method of analysis was SDS-PAGE of total embryo proteins, and in screening maize genetic stocks he quickly recognized that an abundant protein of M_r about 60,000 exhibited a degree of polymorphism with respect to electrophoretic mobility. Being a geneticist, he set out to determine if this polymorphism was due to allelic variation at a single locus. This did indeed turn out to be the case, and he designated the gene encoding the polymorphic polypeptide band *Pro*, for *Protein*. He referred to the different *Pro* alleles, which encode proteins of different molecular weight, as size alleles. The most common *Pro* size alleles were designated *L*, *I*, and *S* for *Large*, *Intermediate*, and *Small* proteins, respectively. Polymorphism at *Pro* facilitated mapping of the gene to the long arm of chromosome 1 (position 1L-121 on the genetic map). He noted also that a polypeptide band of slightly higher molecular weight exhibited a relative mobility correlated with the mobility of PRO. This band was designated PRO', and it was shown that the inheritance of PRO' was completely correlated with that of PRO: that is, embryos possessing PRO-L always contain PRO'-L, those possessing PRO-S always contain PRO'-S, etc. Schwartz continued to screen genetic stocks for *Pro* variants, and he identified two additional *Pro* size alleles in addition to a null allele, which produced no detectable PRO protein (Figure 1, lane 2). The proteins encoded by the *Pro* gene were subsequently shown to be globulins (Kriz and Schwartz, 1986) which correspond to the major "embryonal antigen" of Khavkin *et al.* (1978). The gene designation of *Pro* was changed to *Prot* (Kriz and Schwartz, 1986) to avoid confusion with the proline-requiring *Pro* mutants of maize (Gavazzi *et al.*, 1973), and subsequently to the more descriptive designation of *Glb* (for *Globulin*; Kriz, 1989).

3.3 Synthesis, processing, and deposition of maize globulins

Globulin polypeptides are first detected in immature maize embryos between 10 and 15 days after pollination (DAP), and continue to accumulate throughout most of seed development (Figure 2) (Schwartz, 1979; Cross and Adams, 1983; Kriz, 1989; Thijssen *et al.*, 1996). Globulins are rapidly degraded during seed germination (Cross and Adams, 1983; Kriz, 1989) which is consistent with their role as seed storage proteins. As discussed in this section, the maize globulins share characteristics with globulin storage proteins of dicots with respect to post-translational modification and accumulation in membrane-bound protein storage vacuoles.

The maize *Glb1* proteins exhibit a rather complex pattern of synthesis. Two *Glb1*-encoded polypeptides, designated GLB1' and GLB1, are usually

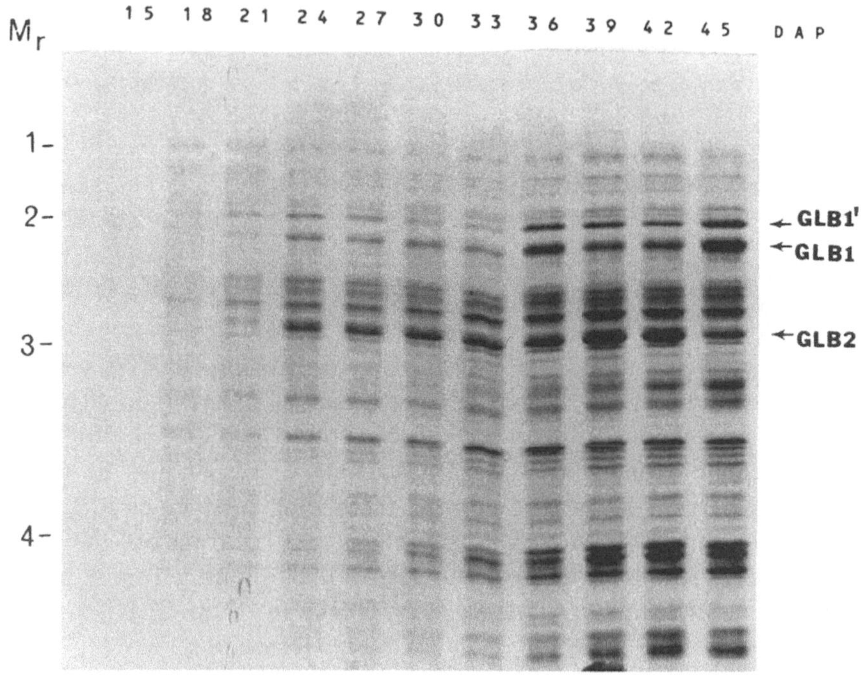

Fig. 2 Accumulation of major globulin components during embryo development. Individual embryos were excised from kernels of the maize inbred line W64A (*Glb1-L/L*) at 15 to 45 DAP. Positions of GLB1', GLB1, and GLB2 are as indicated. Positions of M_r standards are indicated on the left. 1, 97,400; 2, 68,000; 3, 43,000; 4, 25,700. Reproduced from Kriz (1989) with permission of both the author and the publisher.

present in maize embryos (Figure 2). The M_r of GLB1' is approximately 2000 greater than that of GLB1 (Kriz and Schwartz, 1986). Analysis of *Glb1* protein synthesis by *in vitro* translation of embryo RNA and *in vivo* radiolabelling of embryo proteins in pulse-chase experiments indicated that GLB1 is derived from GLB1' by a post-translational processing event (Kriz and Schwartz, 1986). These studies also demonstrated that GLB1' is derived from a short-lived precursor, proGLB1', which is in turn a processed form of the primary translation product (Figure3). The observation that the primary translation product is of a lower molecular weight than that of the first processing intermediate detected *in vivo* suggests that the primary translation product is modified by some post-translational processing event (Kriz and Schwartz, 1986). Although this is probably a type of glycosylation, the nature of this modification remains to be elucidated. Both GLB1' and GLB1 react positively with a general glycoprotein stain (Kriz and Schwartz, 1986), but neither protein binds concanavalin A (Kriz, 1989). The processing of GLB1' to GLB1 is controlled by the unlinked gene *Mep* (*Modifier of embryo*

Fig. 3 Diagrammatic representation of GLB1 protein processing events, as determined by pulse-chase radiolabelling experiments (Kriz and Schwartz, 1986), *N*-terminal amino acid sequence analysis of GLB1-S (Kriz, 1989), and nucleotide sequence analysis of a cDNA clone corresponding to *Glb1-S* (Belanger and Kriz, 1989). The primary translation product of 573 amino acids is processed as follows: A) co-translational cleavage of the signal peptide; B) cleavage of prosequence to form GLB1' during first 5 h post-translation; C) MEP-catalyzed cleavage of GLB1' to form GLB1, between amino acids 86 and 87, between 5 h and 72 h post-translation.

protein; chr 5, near *Pr*; Schwartz, 1979), and embryos homozygous for the recessive *mep* allele accumulate GLB1' and very little, if any, GLB1. Although many seed globulins undergo post-translational proteolytic processing (Shewry *et al.*, 1995), the *Glb1* proteins are unique in that a single gene (*Mep*), for which a non-functional allele is available, has been demonstrated to control one of these processing events.

Maize globulins accumulate in protein storage vacuoles (PSV) in both the scutellum and embryonic axis (Thijssen *et al.*, 1996). By using whole globulin antiserum and light microscopy immunolabelling techniques, the temporal deposition of globulins was analyzed in developing embryos. Globulins were first detected in the scutellar node at 10 DAP, and in the coleoptile and leaf primordia in addition to the scutellar node at 17 DAP. By 30 DAP, globulins were present in the root primordia, coleorhiza, leaf primordia, coleoptile, and all cells of the scutellum except for the epidermis and provascular tissues. Subcellular localization of globulins was performed by transmission electron microscopic analysis of immunolabelled sections of scutellum and root cortex cells. Globulins were found in PSV, where they were either evenly distributed or localized in electron-dense inclusions within the PSV. Because the antiserum used for these studies recognized all globulin components (see Figure 4, lane 3), no distinction could be made as to any differential distribution of globulin components within the PSV.

The maize globulin proteins are apparently transported to the PSV by way of the endoplasmic reticulum (ER) (Figure 4). The processing intermediate

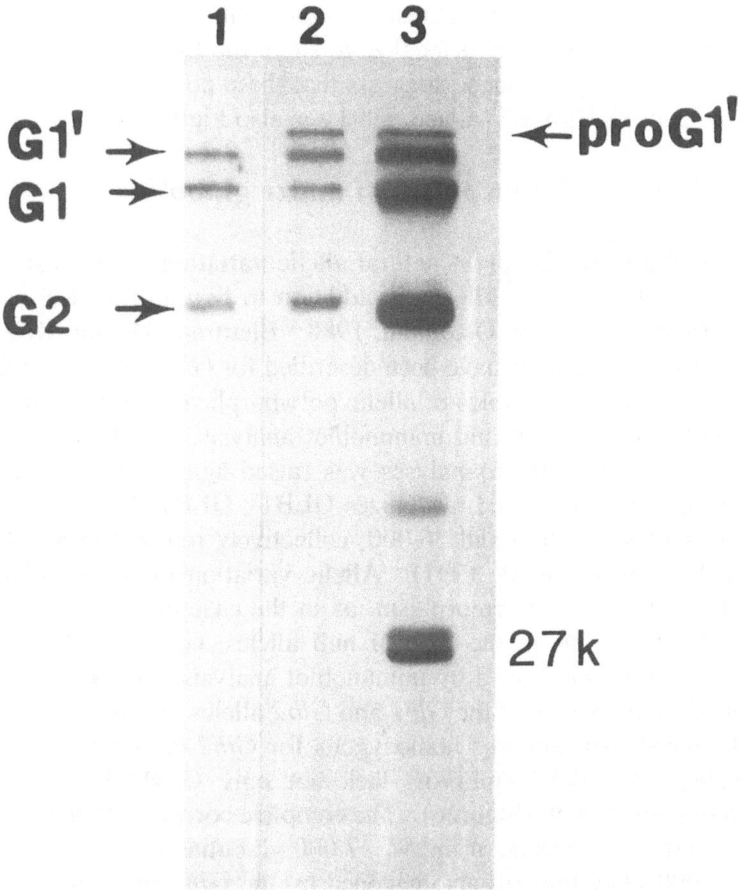

Fig. 4 Association of embryo globulins with different subcellular fractions. Developing embryos (27 DAP) were subjected to subcellular fractions in which ER and PSVs were isolated through discontinuous sucrose gradients as described by Chrispeels *et al.* (1982). Protein extracts were subjected to immunoblot analysis in which antiserum to whole globulin was used as probe. GLB1`, GLB1, and GLB2 are present in both PSVs and ER, and the processing intermediate proGLB1` is present only in the ER and is presumably converted to GLB1` in the Golgi. The lower molecular weight globulins of M_r 27K are apparent in the whole embryo extract (lane 3) but are not present in either the PSV or ER fractions.

proGLB1` is present in ER fractions of immature maize embryos (Figure 4, Lane 2), but not in the PSV fraction (Figure 4, lane 1). This suggests that processing of proGLB1` to GLB1` is complete by the time the proteins reach the vacuole. It is also apparent from the analysis in Figure4 that both GLB1` and GLB1 are present in the ER, suggesting that the *Mep* gene product is active in the ER. It is of interest to note that the M_r 27,000 globulin components do not appear to be localize to either the ER or the PSV (Figure4, lane 3), in that genetic evidence suggests that these polypeptides are products of the *Glb2* gene (Kriz and Wallace, 1991;see also Figure6).

3.4 Allelic polymorphism of maize globulins

Both *Glb1* and *Glb2* exhibit natural allelic variation: in the case of *Glb1*, several electrophoretic size alleles, in addition to two null alleles, have been described (Schwartz, 1979; Osterman, 1988; Bhattramakki and Kriz, 1996). No electrophoretic variants have been described for *Glb2*, but two null alleles have been identified. Examples of allelic polymorphism in *Glb1* and *Glb2*, as determined by SDS-PAGE and immunoblot analysis, are shown in Figure 5. The antiserum used in these analyses was raised against the whole globulin fraction (Figure 1, lane 4) and recognizes GLB1', GLB1, GLB2, and the two polypeptides of M_r 27,000 and 26,000, collectively referred to as M_r 27,000 globulins (Puckett and Kriz, 1991). Allelic variation in embryo globulins is exhibited as either size polymorphism, as in the case of *Glb1-S* and *Glb1-L* proteins (Figure 5), or, in the case of null alleles, complete absence of the protein product as determined by immunoblot analysis (Figure5b, Figure6b). Designations and sources of the *Glb1* and *Glb2* alleles are listed in Table 2. It should be noted that embryos homozygous for *Glb2-0*, *Glb2-N1*, as well as heterozygous *Glb2-0/N1* embryos, lack not only GLB2 but also the M_r 27,000 globulins as well (Figure6). The complete correspondence of the *Glb2* null phenotype and absence of the M_r 27,000 globulins strongly suggests that the M_r 27,000 globulins are also encoded by the *Glb2* gene, and it is likely that these proteins are derived from GLB2 by proteolytic processing (Kriz and Wallace, 1991).

The variation in globulin phenotype shown in Figures 5 and 6 represents the only observable phenotype for plants of different *Glb1* or *Glb2* genotypes. No differences in kernel or embryo size, development, or germination are apparent among the various genotypes, even in plants of the constitution *Glb1-0/0 Glb2-0/0*, which completely lack embryo globulins (Figure5). This indicates that maize embryo globulins serve a secondary role to the abundant endosperm zeins as kernel storage proteins. The fact that these null traits have been maintained in the homozygous state for numerous generations, with no noticeable effect on plant performance, indicates that maize globulins are not necessary for normal seed development or germination. Indeed, the fact that the *Glb2-N1* null allele occurs in the inbred line H99, a public line still in use for research purposes, is indicative of the non-deleterious effect of the GLB2 null phenotype. The dispensability of seed globulins to normal maize

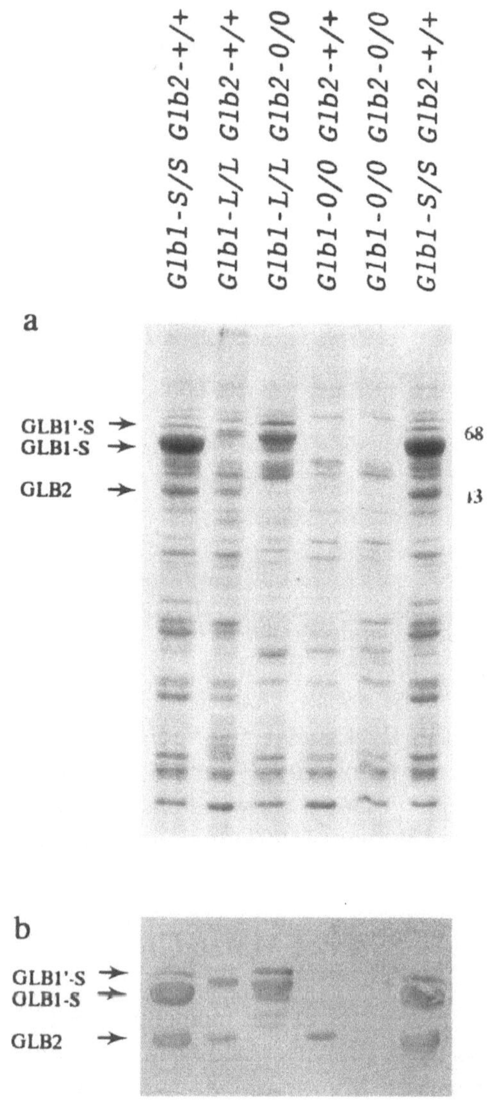

Fig. 5 SDS-PAGE (a) and immunoblot analysis (b) of protein extracts from mature maize embryos of different *Glb* genotypes, as indicated. Antiserum to whole globulin was used for this analysis. Note the absence of GLB1 in the *Glb1-0/0* lanes, the absence of GLB2 in the *Glb2-0/0* lanes, and the complete absence of globulins in the *Glb1-0/0 Glb2-0/0* double null sample. Reproduced from Kriz and Wallace (1991) with permission of both the author and the publisher.

functions is likely one of the reasons such a large amount of allelic polymorphism has occurred in the *Glb1* and *Glb2* genes.

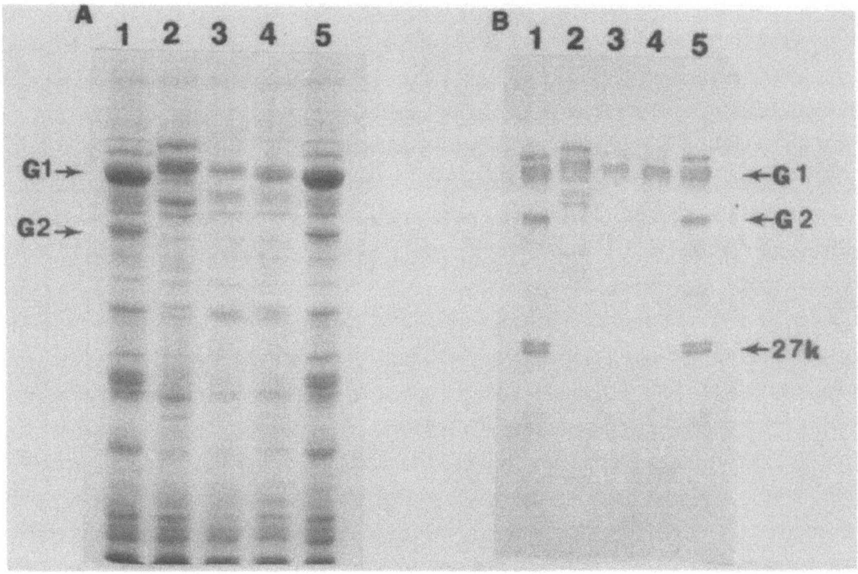

Fig. 6 Identification of an independent *Glb2* null allele. Total protein extracts from individual embryos of different *Glb2* genotypes were subjected to SDS-PAGE (A) and immunoblot analysis in which antiserum to whole globulin was used as probe (B). Lanes 1 and 5, *Glb2-+/+* (inbred line Va26); lane 2, *Glb2-0/0*; lane 3, *Glb2-N1/N1*; lane 4, *Glb2-0/N1* (heterozygous null). Note that the sample in lane 4 shows allelism of *Glb2-0* and *Glb2-N1*. Reproduced from Kriz and Wallace (1991) with permission of both the author and the publisher.

3.5 Molecular Analysis of Maize Globulin Sequences

To characterize globulin genes at the molecular level, full-length cDNA clones corresponding to *Glb1* (Belanger and Kriz, 1989) and *Glb2* (Wallace and Kriz, 1991) were isolated from a lambda expression library constructed from mRNA from mid-development maize embryos by screening with whole globulin antiserum. Confirmation that these clones corresponded to *Glb1* and *Glb2* was obtained by comparison of the deduced amino acid sequences with amino-terminal protein sequence determined directly from GLB1 (Kriz, 1989; Belanger and Kriz, 1989) and from GLB2 (Wallace and Kriz, 1991). Comparison of the deduced amino acid sequences with the direct amino-terminal protein sequence analysis indicated that both proteins possess amino-terminal transmembrane signal sequences, which is common for seed storage proteins (Shotwell and Larkins, 1989) and consistent with the presence of both proteins in membrane-bound protein storage vacuoles. From analysis of

Table 2 Summary description of maize *Globulin* alleles

Gene	Allele	Phenotype[1]	M_r[2]	Source	Clone	GenBank Accession	Reference
Glb1	S	Small	55,100	inbred line Va26	cDNA	M24845	Belanger and Kriz (1989)
					genomic	X59084	Belanger and Kriz (1991)
	L	Large	56,300	inbred line W64A	genomic	X59083	Belanger and Kriz (1991)
	0	null	-	Black Beauty Popcorn	genomic	X59085	Schwartz (1979) Belanger and Kriz (1991)
	N1Hb	null	-	Illinois High Protein	genomic	U28017	Bhattramakki and Kriz (1996)
	I	Intermediate	nd	inbred line Oh43	-		Schwartz (1979)
	V	Very small	58,000[3]	genetic stock	-		Osterman (1988)
	$S^{Pueblo\ 430}$	Small GLB1	nd	Pueblo 430	-		Schwartz (1979)
	$S^{Zapaluta\ Chica}$	lacking GLB1	nd	Zapaluta Chica	-		Schwartz (1979)
Glb2	standard		47,400	inbred line Va26	cDNA	X53715	Wallace and Kriz (1991)
	0	null	-	genetic stock	-		Kriz (1989)
	N1	null	-	inbred line H99	-		Kriz and Wallace (1991)

[1] Nomenclature of Schwartz (1979) for relative protein size: [2] Mature protein, deduced from nucleotide sequence
[3] estimated by SDS-PAGE; nd, not determined

the deduced *Glb1* protein sequence, it could be concluded that proteolytic processing of the primary translation product, including *Mep*-catalyzed cleavage of GLB1` to form GLB1, is limited to the amino-terminal region of the polypeptide, as depicted in Figure3.

To determine the differences between *Glb1* alleles, genomic clones corresponding to the size alleles *Glb1-S* and *Glb1-L* (Belanger and Kriz, 1991), as well as to the null alleles *Glb1-0* (Belanger and Kriz, 1991) and *Glb1-N1Hb* (Bhattramakki and Kriz, 1996), were isolated and characterized at the nucleotide sequence level. A diagrammatic representation of the structure of these alleles, along with a summary of the differences between them, is shown in Figure7. All five genes possess the same basic structure of 5 exons and 4 introns. Comparison of the different alleles indicates that *Glb1-L* and *Glb1-0* differ from *Glb1-S* by small oligonucleotide insertions within their respective protein coding regions (Belanger and Kriz, 1991). In the case of *Glb1-L*, a 36 base insertion in the last exon is in frame and results in the translation of a larger protein, whereas an 11 bp insertion in the first exon of *Glb1-0* results in a translational frameshift beginning at codon 85 which is followed shortly by a termination codon at position 123. In the case of *Glb1-N1Hb*, a single base insertion in Exon 3 results in a change in the reading frame beginning at codon 232, which is followed by a termination codon at position 408 (Bhattramakki and Kriz, 1996). The deduced amino acid sequences for each of the cloned *Glb1* alleles are compared in Figure8.

Although the defects in both *Glb1-0* and *Glb1-N1Hb* are due to the introduction of translational frameshift mutations which could result in the production of truncated *Glb1* proteins, in neither case is any *Glb1* protein detected by GLB1-specific antibodies (Belanger and Kriz, 1989; Bhattramakki *et al.*, 1996). *Glb1* transcripts are present at substantially reduced steady-state levels in embryos homozygous for either *Glb1-N1Hb* (Bhattramakki *et al.*, 1996) or *Glb1-0* (Belanger and Kriz, 1989). It has been shown, however, that *Glb1-0* is transcribed at a rate similar to that of *Glb1-L* (Belanger and Kriz, 1991). It is likely that the presence of the premature stop codon in either transcript results in premature uncoupling of the mRNA from the translational apparatus, which in turn results in degradation of the defective mRNA, as is apparently the case for the *Phaseolus* phytohemagglutinin *Pdlec* null allele (Voelker *et al.*, 1990). No *Glb2* transcripts are detected in embryos homozygous for the *Glb2-0* allele (Wallace and Kriz, 1991). The nature of the molecular defect has not been determined for either of the *Glb2* null alleles *Glb2-0* or *Glb2-N1*.

Both *Glb1* and *Glb2* encode polypeptides which possess hydrophobic *N*-terminal regions with characteristics of transmembrane signal sequences (Belanger and Kriz, 1989; Wallace and Kriz, 1991). The *Glb2* product does not contain an extensive prosequence, as does the *Glb1* gene product; comparison of the deduced amino acid sequence for *Glb2* with that

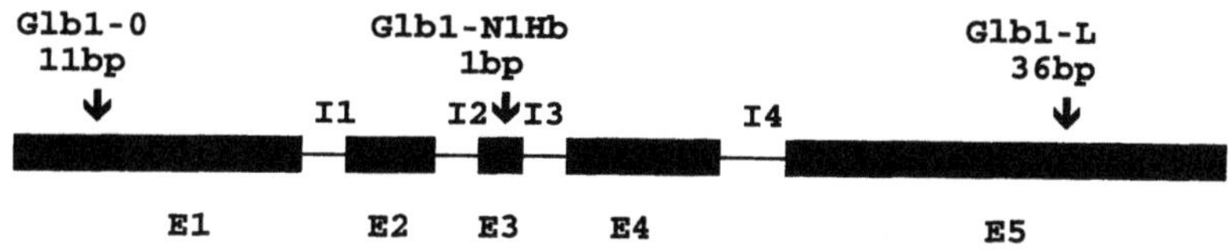

Exon:	1	2	3	4	5
S	565	176	81	291	913
L	545	176	81	291	952
0	556	176	81	291	952
N1Hb	555	176	82	294	952

Intron:	1	2	3	4
S	87	114	90	103
L	99	114	82	108
0	99	100	91	104
N1Hb	87	114	81	108

Fig. 7 Structural organization of *Glb1* alleles. Exons are shown as dark boxes, and introns are shown as narrow lines. The lengths of each exon and intron in base pairs is shown for each allele. Positions of allele-specific insertions in *Glb1-L*, *Glb1-0*, and *Glb1-N1Hb*, relative to the sequence of *Glb1-S*, are indicated. The lengths of Exon 1 and Exon 5 include the 5' untranslated region and the 3' untranslated region, respectively. Modified and redrawn from Belanger and Kriz (1991).

determined for the mature protein indicates that this protein contains a prosequence of only four amino acids (Wallace and Kriz, 1991). The mature proteins encoded by *Glb1* and *Glb2* exhibit approximately 28% amino acid identity over the length of the mature GLB2 protein (Figure 9). This is a significantly lower degree of homology than that observed between *Glb1* and the barley homologue *Beg1* (see below).

4. 7S GLOBULINS OF BARLEY, WHEAT, AND OATS

Danielsson (1949) identified globulins with sedimentation coefficients of approximately 8S in barley, wheat, and oats; the sedimentation coefficient of the oat fraction was subsequently demonstrated to be 7.1S (Burgess *et al.*, 1983). In each of these, as in maize, the 7S globulins are localized primarily to the embryo (Burgess and Shewry, 1986) and aleurone (Quatrano *et al.*,

```
                                ↓                                                                      ↓
GLB1S     MVSARIVVLLAVLLCAAAAVASSWEDDNHHHHGGHKSGRCVRRCEDRPWHQRPRCLEQCREEEREKRQERSRHEADDRSGEGSSEDEREREQEKEEKQKDR
GLB1L     MVSARIVVLLATLLCAAAAVASSWEDDNHHHHGGHKSGQCVRRCEDRPWHQRPRCLEQCREEEREKRQERSRHEADDRSGEGSSEDERE--QEKE-KQKDR
GLB1NIHb  MVSARIVLLLATLLCAAAAVASSWEDDNHHHHGGHKSGRCVRRCEDRPWHQRPRCLEQCREEEREKRQERSRHEADDRSGKGSSEDEREREQEKEEK--DR
GLB10     MVSARIVVLLATLLCAAAAVASSWEDDNHHHHGGHKSGQCVRRCEDRPWHQRPRCLEQCREEEREKRQERSRHEADDRSGEGSSrarrrmsesrrrrnrrt

GLB1S     RPYVFDRRSFRRVVRSEQGSLRVLRPFDEVSRLLRGIRDYRVAVLEANPRSFVVPSHTDAHCICYVAEGEGVVTTIENGERRSYTIKQGHVFVAPAGAVTY
GLB1L     RPYVFDRRSFRRVVRSEQGSLRVLRPFDEVSRLLRGIRDYRVAVLEANPRSFVVPSHTDAHCICYVAEGEGVVTTIENGERRSYTIKQGHVFVAPAGAVTY
GLB1NIHb  RPYVFDRRSFRRVVRSEQGSLRVLRPFDEVSRLLRGIRDYRVAVLEANPRSFVVPSHTDAHCICYVAEGEGVVTTIENGERRSYTIKQGHVFVAPAGAVTY
GLB10     ggrtcstgaafvawsgasrgp◆

GLB1S     LANTDGRKKLVITKILHTISVPGEFQFFFGPGGRNPESFLSSFSKSIQRAAYKTSSDRLERLFGRHGQDKGIIVRATEEQTRELRRHASEGGHGPHWPLPP
GLB1L     LANTDGRKKLVITKILHTISVPGEFQFFFGPGGRNPESFLSSFSKSIQRAAYKTSSDRLERLFGRHGQDKGIIVRATEEQTRELRRHASEGGHGPHWPLPP
GLB1NIHb  LANTDGRKKLVITKILHTISVPGEFQFFFGPGreeprivpvelqqehpescvqdlerpsgeavreawagqgdhrachggadpraaaprlggrprpalapaa

GLB1S     FGESRGPYSLLDQRPSIANQHGQLYEADARSFHDLAEHDVSVSFANITAGSMSAPLYNTRSFKIAYVPNGKGYAEIVCPHRQSQGGESERERGKGRRSEEE
GLB1L     FGESRGPYSLLDQRPSIANQHGQLYEADARSFHDLAEHDVSVSFANITAGSMSAPLFNTRSFKIAYVPNGKGYAEIVCPHRQSQGGESERERDKGRRSEEE
GLB1NIHb  vrrvarplqppgpaaqhrqparaalrgrraqlprprrarrqrllrqhhrgvheraivqhpfvqdrlraerqglrrdrvpappvagrrerararqgqeerrr

GLB1S     EE-SSEEQEEVGQQGYHTIRARLSPGTAFVVPAGHPFVAVASRDSNLQIVCFEVHADRNEKVFLAGADNVLQKLDRVAKALSFASKAEEVDEVLGSRREKGF
GLB1L     EEESSEEQEEAGQQGYHTIRARLSPGTAFVVPAGHPFVAVASRDSNLQIVCFEVHADRNEKVFLAGADNVLQKLDRVAKALSFASKAEEVDEVLGSRREKGF
GLB1NIHb  rrriv◆

GLB1S     LPGPKESGGHEEREQEEEEREERHGGRGERERHGREEREKEEEEREG------------RHGRGRREEVAETLIRMVTARM
GLB1L     LPGPEESGGHEEREQEEEEREERHGGRGERERHGREEREKEE-EREGRHGGREEREEEERHGRGRREEVAETLMRMVTARM
```

Fig. 8 Alignment of deduced amino acid sequences for the polypetides encoded by *the Glb1-S, Glb1-L, Glb1-N1Hb*, and *Glb1-0* alleles. Gaps have been inserted to facilitate alignment. Amino acid substitutions, relative to the *L* allele, are highlighted. The presumed signal sequence clevage site is indicated by an arrow (↓) and the position of Mep-catalyzed cleavage, which produces the mature from of the protein, is indicated by a bold arrow (🡇). The altered sequences of the *Glb1-N1Hb* and *Glb1-0* polypeptides, as conditioned by frameshift mutations, are indicated in lower case. The position of the premature stop codon for each of these null sequences is indicated by a diamond (◆).

1986; Yupsanis *et al.*, 1990). This is consistent with the observation of Danielsson (1949) that the 8S globulin of wheat is enriched in the embryo. In oats, the 7S seed globulin fraction primarily consists of polypeptides of approximately M_r 50-55,000, with minor components of approximately M_r 40,000 and 70,000 (Burgess *et al.*, 1983). Embryo globulin components in barley, wheat and rye exhibit immunological cross-reactivity with antibodies raised against the oat 7S embryo globulin fraction (Burgess and Shewry, 1986). The major 7S globulin components in each of these cereals are polypeptides of approximately M_r 50-55,000; the M_r 40,000 component is lacking in rye, and the M_r 70,000 component is present at very low levels in wheat (Burgess and Shewry, 1986).

The 7S globulins from each of these cereals are similar to those of dicots in their *N*-terminal amino acid sequences (Quatrano *et al.*, 1986; Yupsanis *et al.*, 1990; Heck *et al.*, 1993), and in that the subunits do not appear to be associated by disulphide bonds (Burgess *et al.*, 1983; Quatrano *et al.*, 1986;Yupsanis *et al.*, 1990). In addition, the M_r 50,000 component of wheat embryos also exhibits cross-reactivity with antibodies raised against pea vicilin (Quatrano *et al.*, 1986) and the maize GLB1 protein (Belanger and Kriz, 1989).

```
                signal sequence                                                        mature protein
                            ↓                                                                 ↓
G1     1 MVSARIVVLLAVLLCAAAAVASSWEDDNHHHHGGHKSGRCVRRCEDRPWHQRPRCLEQCREEEREKRQERSRHEADDRSGEGSSEDEREREQEKEEKQKD
                           ↓                                                                           ↓E
G2     1 MKVPVLLLLVSLCFSLALA-----------------------------------------------------------------WQTDTESGSG

G1   101 RRPYVFDRRSFRRVVRSEQGSLRVLRPFDEVSRLLRGIRDYRVAVLEANPRSFVVPSHTDAHCIGYVAEGEGVVTTIENGERRSYTIKQGHVFVAPAGAV
          RPY     S R   RS QG  RVL  F            YRVA LEA PR F  PSH DA     V EG         G
G2    30 -RPYHYGEESFRHWTRSRQGRFVLERFTHELLEDAVGN-YRVAELEAAPRAFLQPSHYDADEVMFVKEGGASSCCSAAGRGSRSASGRATSWSSPRAPS

G1   201 TYLANTDGRKKLVITKILHTISVPGEFQFFFGPGGRNPESFLSSFSKSIQRAAYKTSSD-RLERLFGRHGQDKGIIVRATEEQTRELRRHASEGGHGPHW
                 R            S          P      S  SS S              R  R   R         R       EL R  S GG
G2   128 CTRPTRTSRSGSASSCSSAPSSPRLDASRSSSPSEARAPSP-SSASSATTLSRRRSTLAGRSGRKCSRSRARGRSRRRPRNRSG-ELSRSCSRGGRSSRS

G1   300 PLPPFGESRGPYSLLDQRPSIANQHGQLYEADARSFHDLAEHDVSVSFANITAGSMSAPLYNTRSFKIAYVPNGKGYAEIV-----CPHRQSQGGESE--
              G S                                                        S                     CPH  S GG  S
G2   226 EGGDSGSSSSKWEIKPSSLTDKKPTHSNSHGRHYEITGDECPHLRLLDMDVGLANIARGSMMAPSYNTRANKIAIVLKGQGYFEMACPHV-S-GGRSSPR

G1   393 RERGKGRRSEEEEEESSEEQEEV-GQGYHTIRARLSPGTAFVV---PAGHPFVAVASRDSNLQIVCFEVHADRNEKVFLAGADNVLQKLDRVAKALSFASK
         RERG GR  EEEEE  EEQ    GQ     R   S      V   PAGHP   VA  D NL   CFEV A    KVFLAG    LQK DR AK L F
G2   323 RERGHGR--EEEEEREEEQGGGGGQKSRSYRQVKSRIREGSVIVIPAGHPTALVAGEDKNLAVLCFEVNASFDDKVFLAGTNSALQKMDRPAKLLAFGAD

G1   489 AEE--VDEVLGSRREKGFLPGPKESGGHEEREQEEEEREERHGGRGERERHGREEREKEEEEREGRHGRGRREEVAETLLRMVTARM♦
          EE  VD V G      FL GP
G2   422 -EEQQVDRVIGAQGDAVFLRGPQSHRVSSV♦
```

Fig. 9 Alignment of GLB1 and GLB2 amino acid sequences deduced from nucleotide sequences corresponding to *Glb1-S* (G1; GenBank M24845) and *Glb2* (G2; Genbank X53715). Sequences are aligned from their amino termini which contain hydrophobic transmembrane signal sequences. The beginning of actual sequence homology is near the amino terminus of each mature protein (arrows). Residues that are identical between the two sequences are indicated in the middle line. Small gaps (-) have been introduced to maximize homology.

The 7S globulins of oat are localized in membrane-bound protein bodies (Burgess and Miflin, 1985). These are probably distinct from those that contain the 12S globulin and prolamin storage proteins, as these proteins are localized in the starchy endosperm tissue (Lending *et al.*, 1989). Indeed, sucrose density gradient centrifugation demonstrated the presence of two populations of protein bodies in oat seeds, with the prolamin-containing protein bodies exhibiting a density slightly greater than that of the globulin-containing protein body population (Burgess and Miflin, 1985).

5. AMINO ACID SEQUENCE COMPARISONS

The only cereals other than maize in which 7S globulins have been characterized in any detail at the molecular level are barley and wheat. Interestingly, the identification of recombinant DNA clones corresponding to 7S globulins in both wheat and barley was the result of screening for genes which are inducible by the phytohormone abscisic acid (ABA). In the case of wheat, a short cDNA sequence obtained from an embryo cDNA library was shown to have sequence homology to 7S globulins of dicots (Quatrano *et al.*,

1986) and subsequently to the maize *Glb1* sequence (Belanger and Kriz, 1989). It was also noted that antiserum specific to maize GLB1 recognized a protein of M_r 50,000 in wheat seed extracts (Belanger and Kriz, 1989). In barley, a cDNA clone from an ABA-treated aleurone library was characterized and found to exhibit substantial similarity to maize *Glb1* (Heck *et al.*, 1993). The gene corresponding to this clone was subsequently designated *Beg1*, for *Barley Embryo Globulin*. The sequence of the short wheat cDNA and the barley *Beg1* cDNA sequences exhibit a high degree of similarity (68/91 amino acids) over a portion of the *Beg1* sequence. The protein coding sequence of a genomic clone reported to correspond to the wheat 7S globulin, however, is identical to that of the *Beg1* cDNA sequence at the nucleotide level. This is very difficult to explain, particularly since the cDNA sequences do not exhibit absolute identity (Heck *et al.*, 1993). This suggests, at least to this author, that the nucleotide sequence for the wheat gene (GenBank M81719) may actually be that of the barley *Beg1* gene, resulting from the wheat genomic library being contaminated with a barley library. The discussion below comparing the 7S globulins at the amino acid level therefore relates only to the maize and barley sequences.

The entire *Beg1* protein coding sequence, with alignment to the mature maize GLB1S protein sequence, is shown in Figure 10. The deduced amino acid sequence for the entire *Beg1* protein corresponds to a M_r 72,000 protein, and that of the mature form of the protein to a M_r 52,000 protein. The *N*-terminus of the mature protein is assigned on the basis of sequence identity with the sequences determined for the 7S barley aleurone globulin holoprotein and the M_r 40,000 subunit (Yupsanis *et al.*, 1990; Heck *et al.*, 1993). The primary translation product contains an *N*-terminal sequence of 23 amino acids with the hydrophobic characteristics of a transmembrane signal sequence. A striking feature ofBEG1 is a series of seven imperfect repeats of 16 amino acids, beginning at residue 71 relative to the initiator methionine (indicated by alternating single and double underlines in Figure 8). Such hydrophilic, highly charged *N*-terminal repeats are common among globulin storage proteins (Dure, 1990). The *N*-terminus of the mature BEG1 protein is located within the last (seventh) of these repeats. As suggested by Heck *et al.* (1993), the M_r 70,000 barley aleurone component described by Yupsanis *et al.* (1990) may correspond to the *Beg1* polypeptide from which the amino-terminal prosequence has not been removed, while the M_r 50,000 component may represent the mature form of the protein; it is also suggested that the M_r 40,000 component may be derived from the M_r 50,000 component by *C*-terminal proteolytic processing.

Direct comparison of the BEG1 and GLB1S sequences for this discussion begins with the *N*-terminus of BEG1 and eight residues in from the *N*-terminus of the mature GLB1S protein (Figure 10). Gaps of 6, 12, and 25 amino acids were introduced to facilitate alignment, as well as 5 minor gaps of a single amino acid each and a 3-amino acid gap. When these gaps are taken into account, the two mature proteins exhibit 63% amino acid identity along their entire sequences. Clearly, these two proteins are highly related, and they are more closely related to each other than is the maize *Glb1* gene to the maize *Glb2* gene (see Figure9). It should also be noted that the positions of introns within the reported wheat genomic sequence, which may actually be a barley genomic sequence as discussed above, correspond with those of the maize *Glb1* genes (analysis not shown). Therefore, the similarities between the maize and barley (wheat?) genes are conserved at the level of gene structure as well as at the protein coding level. Both *Glb1* and *Begl* polypeptides exhibit similarity to pea vicilin, with approximately 28% amino acid identity in the regions encoding each mature protein (Belanger and Kriz, 1989; Heck *et al.*, 1993). Like *Glb1*, *Begl* is encoded by a single-copy gene, which has been mapped to chromosome 4 by using wheat-barley substitution lines (Heck *et al.*, 1993).

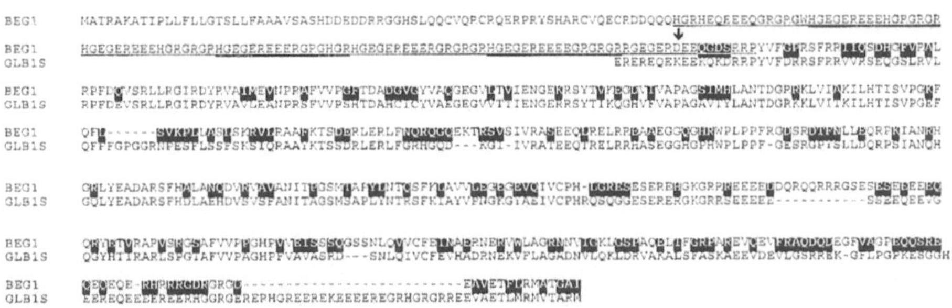

Fig. 10 Alignment of deduced amino acid sequences for the polypeptides encoded by the maize *Glb1-S* gene (GenBank M24845) and the barley *Begl* gene (GenBank M64372). The sequence of the *Begl* polypeptide is presented from the initiator methionine, and that of *Glb1-S* is presented from the amino terminus of the mature protein. The amino terminus of mature BEG1 is indicated by a bold arrow, and the putative site of signal sequence cleavage is indicated by a narrow arrow. Occurrences of a 16 amino acid imperfect repeat in the BEG1 prosequence are indicated by alternating single and double underlines. Amino acid substitutions in the mature BEG1 protein, relative to the GLB1-S sequence, are shaded. Gaps have been introduced to facilitate alignment.

There is at least one other 7S globulin gene in barley in addition to *Begl*. An mRNA corresponding to the maize *Glb2* gene has been identified in barley embryos by using the maize clone as a probe (Heck *et al.*, 1993), but clones corresponding to this sequence have not been isolated. Characterization of the barley *Glb2* homologue at the nucleotide sequence level would provide additional insight into the relationships between 7S globulins among cereals.

It would also be of interest to pursue cDNA clones corresponding to the 7S globulins in oat and rye, and to characterize a full-length cDNA clone from wheat embryos to gather a more complete story on the relatedness of these proteins, at the gene level, among cereals. An analysis of 7S globulin genes from rice would also be in order; although a class of 7S proteins has been reported in rice embryos (Horikoshi and Morita, 1975), no information is available as to the relatedness of these polypeptides to the "classical" 7S proteins described here. It will also be of interest to examine the 7S globulin genes from other cereals with respect to their regulation by ABA; the inducibility of7S gene expression by ABA has been well-documented in maize (Paiva and Kriz, 1994; Liu and Kriz, 1996), barley (Heck *et al.*, 1993), and wheat (Williamson and Quatrano, 1988), and this is likely to be the case in other cereals as well.

Given the unique characteristics of 7S cereal globulins with respect to gene copy number, ABA regulation, and dispensibility to normal seed function, these proteins provide an excellent model system for the study of gene expression in seed development.

REFERENCES

Belanger, F.C. and Kriz, A.L. (1989) Molecular characterization of the major maize embryo globulin encoded by the *Glb1* gene. *Plant Physiology*, **91**, 636-643.

Belanger, F.C. and Kriz, A.L. (1991) Molecular basis for allelic polymorphism of the maize *Globulin-1* gene. *Genetics*, **129**, 863-872

Bhattramakki, D. and Kriz, A.L. (1996) Nucleotide sequence analysis of a novel *globulin1* null allele from the Illinois High Protein strain of maize. *Plant Molecular Biology*, **32**, 1215-19.

Burgess, S.R. and Miflin, B.J. (1985) The localization of oat (*Avena sativa* L.) seed globulins in protein bodies. *Journal of Experimental Botany*, **36**, 945-54.

Burgess, S.R. and Shewry, P.R. (1986) Identification of homologous globulins from embryos of wheat, barley, rye and oats. *Journal of Experimental Botany*, **37**, 1863-71.

Burgess, S.R., Shewry, P.R., Matlashewski, G.J. et al. (1983) Charasteristics of of oat (*Avena sativa* L.) seed globulins. *Journal of Experimental Botany*, **34**, 1320-32.

Chrispeels, M.J., Higgins T.J.V., Craig, S. and Spencer, D. (1982) Role of the endoplasmic reticulum in the synthesis of reserve proteins and the kinetics of their transport to protein bodies in developing pea cotyledons. *Journal of Cell Biology*, **93**, 5-14.

Cross, J.W. and Adams, W.R. (1983) Embryo-specific globulins from *Zea mays* L. and their subunit composition. *Journal of Agricultural and Food Chemistry*, **31**, 534-8.

Croy, R.R.D., Gatehouse, J.A., Tyler, M.E., and Boulter, D. (1980) The purification and characterization of a third storage protein (convicilin) from the seeds of peas (*Pisum sativum* L.). *Biochemical Journal*, **191**, 509-16.

Danielsson, C.E. (1949) Seed globulins of the *Gramineae* and *Leguminoseae*. *Biochemical Journal*, **44**, 387-400.

Dierks-Ventling, C. (1981) Storage proteins in *Zea mays* (L.): interrelationship of albumins, globulins and zeins in the Opaque-2 mutation. *European Journal of Biochemistry*, **120**, 177-82.

Dure, L. III (1990) An unstable domain in the vicilin genes of higher plants. *New Biologist*, **2**, 487-93.

Gavazzi, G., Racchi, M.L. and Tonelli, C. (1973) A mutation causing proline requirement in *Zea mays*. *Theoretical and Applied Genetics*, **46**, 338-45.

Heck, G.R., Chamberlain, A.K. and Ho, D.T.-H. (1993) Barley embryo globulin 1 gene, *Beg1*: characterization of cDNA, chromosome mapping and regulation of expression. *Molecular and General Genetics*, **239**, 209-18.

Horikoshi, M. and Morita, Y. (1975) Localization of γ-globulin in rice seed and changes in γ-globulin content during seed development and germination. *Agricultural and Biological Chemistry*, **39**, 2309-14.

Khavkin, E.E., Misharin, S.I. and Ivanov, V.N. (1977) Embryonal antigens in maize caryopses: the temporal order of antigen accumulation during embryogenesis. *Planta*, **135**, 225-31.

Khavkin, E.E., Misharin, S.I., Markov Ye.Yu. and Peshkova. A.A. (1978) Identification of embryonal antigens of maize: globulins as primary reserve proteins of the embryo. *Planta*, **143**, 11-20.

Kriz, A.L. (1989) Characterization of embryo globulins encoded by the maize *Glb* genes. *Biochemical Genetics*, **27**, 239-51.

Kriz, A.L. and Schwartz, D. (1986) Synthesis of globulins in maize embryos. *Plant Physiology*, **82**, 1069-75.

Kriz, A.L. and Wallace, N.H. (1991) Characterization of the maize *Globulin-2* gene and analysis of two null alleles. *Biochemical Genetics*, **29**, 241-54.

Lending, C.R., Chesnut R.S., Shaw, K.L. and Larkins, B.A. (1989) Immunolocalization of avenin and globulin storage proteins in developing endosperm of *Avena sativa* L. *Planta*, **178**, 315-24.

Liu, S. and Kriz, A.L. (1996) Tissue-specific and ABA-regulated maize *Glb1* expression in transgenic tobacco. *Plant Cell Reports*, **16**, 158-62.

Osterman, J.C. (1988) An allele of the *Prot* locus in maize is a variant for the site of protein processing. *Biochemical Genetics*, **26**, 463-74.

Paiva, R. and Kriz, A.L. (1994) Effect of abscisic acid on embryo-specific gene expression during normal and precocious germination in normal and *viviparous* maize (*Zea mays*) embryos. *Planta*, **173**, 110-16.

Puckett, J.L. and Kriz, A.L. (1991) Globulin gene expression in *opaque-2* and *floury-2* mutant maize embryos. *Maydica*, **36**, 161-67.

Quatrano, R.S., Litts, J.C., Colwell, G. et al. (1986) Regulation of gene expression in wheat embryos by ABA: characterization of cDNA clones for the Em and putative globulin proteins and localization of the lectin wheat germ agglutinin, in *Molecular Biology of Seed Storage Proteins and Lectins* (eds. L.M. Shannon and M.J. Chrispeels), American Society of Plant Physiologists, Rockville, MD, pp. 127-36.

Schwartz, D. (1979) Analysis of the size alleles of the *Pro* gene in maize - evidence for a mutant protein processor. *Molecular and General Genetics*, **174**, 233-40.

Shewry, P.R., Napier, J.A., and Tatham, A.S. (1995) Seed storage proteins: structure and biosynthesis. *Plant Cell*, **7**, 945-56.

Shotwell, M.A. and Larkins, B.A. (1989) The biochemistry and molecular biology of seed storage proteins, in *The Biochemistry of Plants: A Comprehensive Treatise*, Vol. 15 (ed. A. Marcus), Academic Press, New York, pp. 297-345.

Thijssen, M.H., Spoelstra, P. and Emons, A.M.C. (1996) Immunodetection and immunolocalization of globulin storage proteins during zygotic and somatic embryo development in *Zea mays*. *Physiologia Plantarum*, **98**, 539-49.

Voelker, T.A., Moreno, J. and Chrispeels, M.J. (1990) Expression analysis of a pseudogene in transgenic tobacco: a frameshift mutation prevents mRNA accumulation. *Plant Cell*, **2**, 255-61.

Wallace, N.H. and Kriz, A.L. (1991) Nucleotide sequence of a cDNA clone corresponding to the maize *Globulin-2* gene. *Plant Physiology*, **95**, 973-5.

Williamson, J.D. and Quatrano, R.S. (1988) ABA-regulation of two classes of embryo-specific sequences in mature wheat embryos. *Plant Physiology*, **86**, 208-15.

Yupsanis, T., Burgess, S.R., Jackson, P.J. and Shewry, P.R. (1990) Characterization of the major protein component from aleurone cells of barley (*Hordeum vulgare* L.). *Journal of Experimental Botany*, **41**, 385-92.

21. Molecular Phylogenies and Structural Diversification of Gymnosperm and Angiosperm Storage Globulins

K.-P. HÄGER AND H. FISCHER
Department of Plant Systematics, University of Bayreuth, Universitätsstr. 30, D-95440 Bayreuth

1. INTRODUCTION

In recent years, a considerable amount of data has improved our knowledge of 11S (legumin) and 7S (vicilin) seed storage globulins. Most of this information comes from studies in crop plants representing only a small taxonomic range within the huge division of angiosperms; the main interest has focused on a handful of plant families (Chapters 9-20). Nevertheless, the wide distribution of legumins and vicilins in vascular plants has been confirmed now by sequence data from lower angiosperm clades, Gnetales, conifers, *Ginkgo*, cycads, and ferns. However, given a common ancestry of legumins and vicilins rooted somewhere in the plant kingdom, we may still be dealing with the upper branches of the legumin and vicilin evolutionary trees.

Structural similarities between vicilins and legumins have gained much attention, and two different models of their evolution from a common ancestor are discussed in the next two chapters. Within the two classes, surprisingly few comprehensive evolutionary studies have been performed. Gene duplications, sequence divergence of orthologous and paralogous genes, and differences in evolutionary rates between lineages of genes and organisms have played a role in seed globulin evolution; however, we are far from understanding it in its entirety. Moreover, establishing evolutionary relationships permits discrimination of derived features from ancestral ones and thus can also have impact on general models of storage protein evolution.

This and the next two chapters consider evolutionary relationships in 11S and 7S storage globulins. As the dataset of legumins is larger and covers a broad taxonomic range, this chapter lays emphasis on legumins, in particular on those of gymnosperms and lower angiosperms. A phylogenetic analysis is presented, and for the major groups recognized a synopsis of properties is

P.R. Shewry and R. Casey (eds.), Seed Proteins, 499-515.

given. Conspicuous structural features are described and their implications for legumin evolution are discussed, as are the most essential traits of vicilin relationships.

2. LEGUMINS (11S GLOBULINS)

2.1 Evolutionary relationships of legumins

We examined 55 legumin-encoding nucleotide sequences using a distance-based approach and constructed a "legumin tree" to determine evolutionary relationships. The analysis comprises the available gymnosperm, monocot and dicot data. The extent to which the species or families are sampled varies; some of them are thoroughly studied with respect to their legumins, while from others only few or single sequences are known. Nevertheless, the data set covers all but one of the five major taxa of extant seed plants (cycads, *Ginkgo*, conifers, gnetophytes, angiosperms).

Since the analysis produces an unrooted tree, an outgroup is necessary to recognize evolutionary trends. A most adequate outgroup would be a non-seed plant legumin sequence; such a sequence is not yet available, although 11S storage globulin fractions have been reported from ferns and lycopods (Templeman *et al.*, 1987, DeCamp *et al.*, 1994). Thus, there are two possibilities of rooting the legumin tree: a) setting a root between gymnosperms and angiosperms, considering the extant members of both groups to have evolved in parallel from a common ancestor, or b) defining the most ancient gymnosperm representative as an outgroup, implying that the gymnosperms gave rise to the ancestors of angiosperms. The latter alternative, with *Ginkgo* in the basal position, appears more suitable based on results, e.g., of morphological (Nixon *et al.*, 1994) and molecular phylogenetic (Chase *et al.*, 1993) studies.

Within the resulting tree (Fig. 1), legumins from *Ginkgo*, Coniferales, Gnetales and angiosperms form distinct and consecutive clusters, with the angiosperm legumins forming two subsets. Except for the Coniferales legumins, these clusters are statistically well-supported by maximal bootstrap values of 100%. Overall, plant relationships are also reflected in the tree topology.

The Conifer legumins are closest to those from *Ginkgo*, with the Pinaceae legumins being clearly distinct from those of Taxaceae, Taxodiaceae and Cupressaceae (i.e., *Taxus*; *Metasequoia* and *Cryptomeria*; *Calocedrus*). A gene duplication to be dated before, or at the very beginning of, Taxodiaceae/Cupressaceae diversification gave rise to two legumin subfamilies. Orthologs from both legumin subfamilies are ubiquitous in

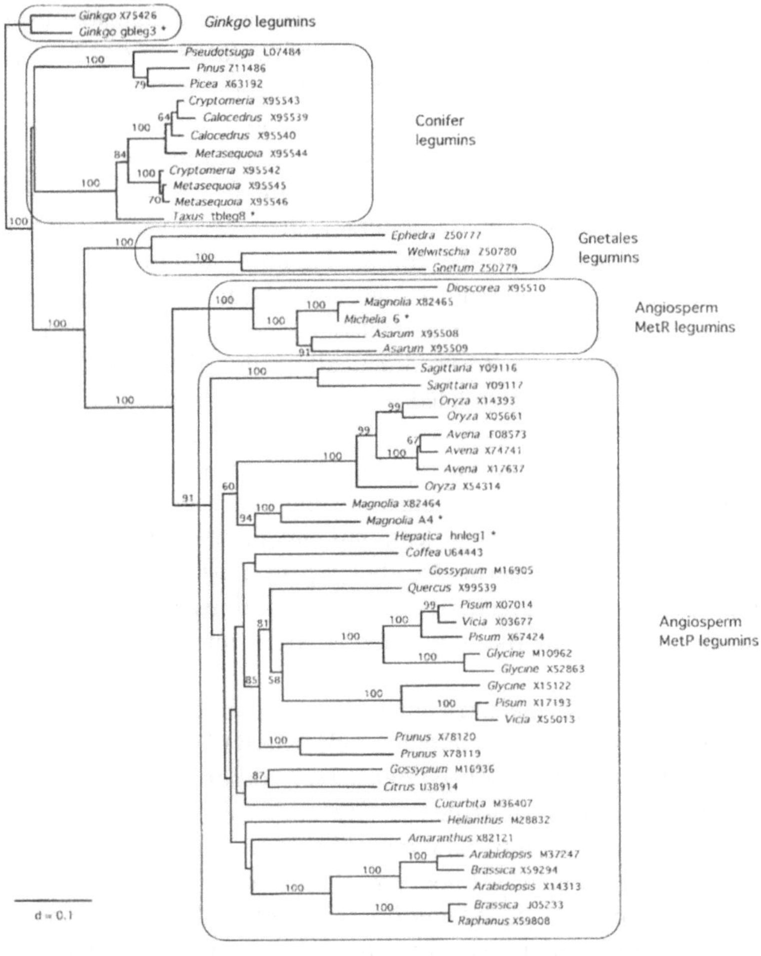

Fig. 1 Distance-based phylogenetic analysis of gymnosperm and angiosperm legumin-encoding sequences; genus names are followed by the GenBank accession numbers. Sequences with asterisks: own results, unpublished. d = dissimilarity. Only 1st and 2nd codon positions were included and sequence portions coding for variable regions were omitted; the alignment is available on request. Distances were calculated according to the Kimura algorithm; tree-construction was done with the Neighbor-Joining method. The PHYLIP (Felsenstein, 1993) program package was used throughout. Numbers at branches are booststrap values (100 replicates) indicating statistical significance of the respective nodes (100 = highest significance). Bootstrap values lower than 50 are not given.

Taxodiaceae (not shown), which confirms the relatively old age of the gene duplication involved. Thus, the principle of common legumin subfamilies, the respective members of which are traceable in related species, previously known only from angiosperms, is realized in gymnosperms as well. Assuming approximately homogeneous rates of evolution in conifer legumins, the distances observed suggest that the duplication event leading to both subfamilies may have occurred contemporarily with intergeneric divergence in the Pinaceae (Fig. 1), about 135 million years ago (Florin, 1963).

Compared to angiosperm legumin subfamilies, e.g., from Fabaceae (Leguminosae) or Brassicaceae (Cruciferae), there are conspicuously smaller distances between and within the conifer legumin subfamilies. Taking into account that genera of gymnosperms are about as old as orders of angiosperms, the rate of accepted nucleotide substitutions within legumin genes from conifers appears to be drastically reduced relative to that from angiosperms. This may be correlated with the generally conservative mode of evolution typical of gymnosperms, or by an accelerated rate of nucleotide substitutions concomitant with the rapid radiation of angiosperms, or both.

The legumins from Gnetales constitute the sister group to the legumins from angiosperms. The gnetalean legumins are in several respects the most aberrant within the gymnosperms (see below). In most concepts of seed plant systematics the three monogeneric gnetalean families are discussed as the closest living relatives to the angiosperms. (For references, see Friedman, 1994)

Within angiosperms, the most ancient gene duplication gave rise to the methionine-poor (MetP) and methionine-rich (MetR) legumins (Fischer *et al.*, 1996). While the former have been extensively described from various crop plants, most of which are regarded as evolutionarily advanced ("eudicots"), the MetR legumins have until now only been found in the relatively ancient lineages of Dioscoreales (monocot: *Dioscorea*), Magnoliales, and Aristolochiales (both dicot: *Magnolia, Michelia*; *Asarum*). None of the known gymnosperm legumin sequences is associated with the MetR legumins. The conclusion from this distribution is that the corresponding gene duplication must have taken place in the early evolution of angiosperms, before the separation of monocotyledonous and dicotyledonous plants.

All the remaining angiosperm legumin sequences belong to the MetP group. At its base, legumins of the aquatic monocotyledonous plant *Sagittaria* form a group of their own. This either means that *Sagittaria* is the most ancient of the plants represented in the MetP group, or that these legumins form an old separate subgroup of the MetP legumins, which would be confirmed if orthologous genes were to be found in other plants.

Within the remaining MetP legumins, one group comprises those from monocots (rice, oats) as well as from dicots with putatively primitive features (*Magnolia*, *Hepatica*) (Cronquist, 1981). The other group comprises legumins from the bulk of "advanced dicot" plants ranging from coffee tree to oilseed rape. This clustering widely corresponds to current notions of plant classification, in that the dicots may not be a homogeneous group but some of them may be more closely related to the monocots than to most other dicots.

Within the dicot cluster, there are few statistically significant subgroups; these relationships cannot be established with certainty. Additionally, there are indications of several unrecognized gene duplications involved in the MetP legumin evolution: the sequence relationships do not always reflect known plant relationships, the most conspicuous case being the separation of *Prunus* and the legume family (both rosid) by *Quercus* (hamamelidid). Only the assumption of a legumin gene duplication preceding the divergence of both plant groups can resolve this inconsistency of gene tree and species tree.

Remarkably, significant differences in branch lengths suggest rate heterogeneity within MetP legumins, with those from annual herbaceous plants evolving more rapidly than those from woody perennials. Interestingly, corresponding variation in evolutionary rates has also been described for the plastidal *rbc*L gene (Bousquet *et al.*, 1992).

2.2 Major groups of legumins - a compilation of characteristics

2.2.1 The legumins of *Ginkgo*.

Ginkgo biloba, the single extant species of the ancient Ginkgoales, was the first gymnosperm from which legumin-like proteins were demonstrated (Jensen and Berthold, 1989); they are the major fraction of the megagametophyte globulins. The subunits are made up mainly of M_r 28,000 α–polypeptides and M_r 20,100 β–polypeptides, being significantly lower in M_r than those from angiosperm legumins, consistent with a low oligomer sedimentation coefficient of 9.6 S (Arahira and Fukazawa, 1995). Charge polymorphism and sequence microheterogeneity of a β–polypeptide indicated the presence of a legumin multigene family in *Ginkgo* (Häger *et al.*, 1992).

Analysis of nucleotide sequences predicts an old gene duplication giving rise to two legumin gene subfamilies, represented in Fig. 1 by *GbLeg1* and *GbLeg3*. From subfamily I, which comprises about 45 gene copies per haploid genome (Häger and Müller, unpublished), two genes (Häger *et al.*, 1995), a cDNA (Arahira and Fukazawa, 1995) and a further dozen legumin encoding sequences have been characterized. They show 98-99% identity in the coding, the 3'- and 5'-flanking, and the intron sequences, and there is evidence for a tandem array of the genes. Obviously, they are subjected to intensive homogenization. Subfamily II comprises only two genes with about 83% identity to subfamily I. The lack of Asn-Gly processing sites and of extended variable regions in both subfamilies may be viewed as ancestral states in legumin evolution (see below).

2.2.2 Coniferales legumins.

Legumin-like proteins, generally of the cystalloid type (insoluble in dilute saline), are known from species of the Pinaceae:*Pseudotsuga*, *Pinus*, *Picea*, *Larix* and *Cedrus* (e.g., Green *et al.*, 1991; Gifford, 1988; Jensen and Lixue, 1991). Usually, legumins are the major seed proteins (50-80 % of total), with subunits consisting of M_r 32-35,000 α–polypeptides and M_r 20-23,000 β–polypeptides. Intraspecific diversity of Pinaceae legumins has been suggested by isoelectric focusing (IEF) of *Pinus* legumins (Allona *et al.*, 1992) and by Southern blot analysis of genomic DNA from *Pseudotsuga* (Leal and Misra, 1993). There are two major legumin subfamilies in Pinaceae (Häger and Müller, unpublished), neither of them related to those from *Ginkgo*. Subfamily I legumins seem to be restricted to *Pinus*, *Picea*, *Pseudotsuga* and *Larix*; cDNAs reported for the former three taxa (Baker *et al.*, 1996; GenBank accession X63192; Leal and Misra, 1993) clearly belong to subfamily I. Subfamily II legumins are considerably more conserved than subfamily I members. To date, we have found no legumin-encoding sequences in species from *Abies*, *Tsuga* and *Keteleeria*.

Crystalloid-type legumins are the predominant storage proteins in Cupressaceae seeds, as in *Calocedrus decurrens* and a further six species, representing the Cupressoideae and Callitroideae (Häger and Dank 1996). Despite denaturing conditions, the extracted legumins are preferentially found as stable subunit-oligomers; under reducing conditions, oligomer dissociation and subunit cleavage results in polypeptides of M_r 32-34,000 and M_r 23-25,000. Legumin cDNAs from *Calocedrus decurrens* predict the absence of the post-translational Asn-Gly processing site.

Solubility properties, oligomer stability, and constituent polypeptides show legumins from Taxodiaceae to be very similar to those from Cupressaceae. Legumin cDNAs from *Cryptomeria* and *Metasequoia* (Wind and Häger, 1996; Häger and Wind, 1997) represent two gene subfamilies, one of them additionally comprising legumins from *Calocedrus* (Fig. 1). Both subfamilies have specific post-translational processing sites of the respective predicted prolegumins. Cysteine residues located in variable region (VR) I, exposed to the subunit surface, may account for oligomer stability and solubility properties of the proteins.

Polyglutamate-stretches in VRI appear as a diagnostic feature of conifer legumins. This contrasts with angiosperm legumins: their VRI regions tend to be rich in glutamine, whereas glutamate repeats are preferentially located in the hypervariable region (HVR). Assuming a functional or structural significance inherent to VRI glutamate repeats in conifers, those in the HVR of angiosperms may be viewed as analogous in terms of convergent evolution.

2.2.3 The legumins of Gnetales.

Legumins, ranging from M_r 300,000 - 350,000 on native polyacrylamide gel electrophoresis (PAGE), are the main storage globulins of Gnetales, with *Ephedra* legumins showing the most complex subunit pattern; at least four different β–chains were detected (Conte, 1994a). Serological relationships were found to be weak or absent among the Gnetales (Conte, 1994b). Gnetales legumins share some peculiar insertions relative to most other gymnosperm and angiosperm legumins. The most prominent one is a one-residue insertion at the otherwise highly conserved β–chain *N*-terminus, in close vicinity to the α–β–cleavage site. Another insertion creates short Gln-stretches in the putative *N*-domain helix 3 - J′ link (Lawrence *et al.*, 1994; and see Chapter 22). Interestingly, a similar, certainly independent, insertion at the same position can be observed in A-legumins of the Fabaceae. All Gnetales legumins known have α–β–cleavage sites of the angiosperm consensus type (Fig. 2).

2.2.4 Angiosperms - the methionine-rich (MetR) legumins.

To date, legumins of this type have been found only in species belonging to Dioscoreaceae, Aristolochiaceae, and Magnoliaceae, but not in the well-studied crop plants of e.g., Poaceae (Gramineae) or Fabaceae. It is possible that the respective genes have been lost in the latter plant families; however, there is a very low degree of nucleotide identity between MetR and MetP genes (e.g., *Magnolia*: less than 55%), which can well result in a lack of detection when, for example, hybridizing a gene or cDNA library with a MetP legumin-specific probe.

In contrast to their methionine-poor counterparts with methionine proportions of 0-2 mol%, MetR legumins are characterized by 3-4 mol% methionine (Fischer *et al.*, 1996). Considering the efforts to increase the Met content of seed proteins, the MetR legumins may serve as a model; the additional Met residues are preferentially situated in the β–chain, often replacing the functionally similar isoleucine and valine residues. The variable regions - which otherwise tolerate sequence and length variation - are not involved in methionine storage, probably for reasons of maintenance of secondary and higher order structure. β–chain glycosylation has been demonstrated in *Magnolia*, but is not a consistent feature of the MetR legumins as a whole.

2.2.5 Angiosperms - the methionine-poor (MetP) legumins.

A deficiency in methionine is observable throughout this subfamily; the highest proportions are found in Brassicaceae legumins which have about 2 mol% Met. This can be viewed as a Brassicacean specialization in connection with their increased requirement of sulphur for the production of glucosinolates characteristic of this plant family.

Considering storage protein evolution, care has to be taken with possible parallel developments in the variable regions, which tolerate extensive sequence variation as well as insertions and deletions. An example of convergent evolution in variable regions is demonstrated by the *Sagittaria* legumin ssleg15 (Chen and Fischer, 1998) and cereal high molecular weight (HMW) glutenins (see Chapter 3); in ssleg15 direct degenerate repeats of the motif XGQXGXXP stretch over 358 residues, significantly resembling those which are characteristic for the HMW glutenins. Nevertheless, there is clearly no traceable evolutionary relationship between the proteins.

In general, the MetP legumins comprise the best-studied legumins, as they are often directly linked to the economic value of crop plants. They have extensively been described and reviewed (e.g., Shewry, 1995, and references therein), and are given due attention in the preceding chapters.

2.3 Post-translational processing sites: Diversity *vs.* invariance

Post-translational cleavage of angiosperm prolegumins is regularly directed to an Asn-Gly site adjacent to the *C*-terminal side of the HVR and being part of the cleavage region consensus sequence given in Fig. 2. This consensus region, which was viewed as an endoprotease recognition site (Borroto and Dure, 1987), and in particular the Asn-Gly junction to be cleaved, have been assumed to be highly conserved during legumin evolution (Dickinson *et al.*, 1989; and references therein; and see also Chapters 9,12 and 13).

In gymnosperm prolegumins regions homologous to the consensus and located at the same position can be recognized as well. Comparisons reveal some of the prolegumins to contain the conventional Asn-Gly site (Fig 2a), which was confirmed to be cleaved in *Metasequoia*. In contrast, prolegumins from *Ginkgo* and *Pseudotsuga* predict unusual Asn-Asn, Asn-Ser or Asn-Asp cleavage sites, respectively (Fig. 2b), and actual cleavage has been shown for a legumin from *Ginkgo* (Häger *et al.*, 1992). These sites are still located at the same position as that strictly conserved in angiosperm legumins. A third group of prolegumins, however, widespread in the Taxodiaceae, Cupressaceae and Taxaceae, deviates conspicously from the former patterns (Fig. 2c): cleavage is directed to an Asn-Thr or Asn-Ser site located in the HVR, making a HVR stretch an integral part of the resultant β–polypeptide (Häger and Wind, 1997). This cleavage mode is not found in any angiosperm legumin. Overall, diversity of cleavage sites proves to be a characteristic feature of gymnosperm prolegumins.

		———— *H V R* ———— $NG^{LE}_{ID}ET^{I}_{F}C^{S}_{T}$
a)	*Metasequoia*	..GQILSLAEEDTDDDSRPAE-------**N**GLAQLFCN..
	Taxus	..GQVVSFEEASNEVGSNA---------**N**GVEQLFCN..
	Pinus	..YAGRQRQWGREDSE------------**N**GVEELVCP..
b)	*Ginkgo*	..PSDDYEREREREG-------------**NN**VEEFYCS..
	Ginkgo	..RSERETEYLDSTAAFKS---------**N**SVEEFYCS..
	Pseudotsuga	..QRDNRMMQQLHETHNSFANENE----**N**DVEEVVCA..
c)	*Metasequoia*	..PGRGAPLFF**N**TFAEDSEEDEREGLNPGGLHQFYCN..
	Calocedrus	..GGEGAPLFL**N**TVAEASEDMKSWGINPGGLHQFYCN..
	Taxus	..RDSLVPSDS**N**SVQEDEEER-------EGFHQFYCN..

Fig. 2 Summary of post-translational cleavage site diversity in gymnosperm prolegumins. Post-translational processing has been determined (first sequence of each group) or is predicted to occur on the *C*-terminal side of Asn residues printed in bold. On the upper right, the cleavage region consensus sequence from angiosperms is aligned as a reference. Due to size variation in HVRs, gaps have been introduced to maximize alignment.

This agrees with the properties of the endoprotease(s) involved in angiosperm prolegumin cleavage: specifity is essentially determined by the Asn residue at the *N*-terminal side of the cleavage site, without requirement of a Gly at the *C*-terminal side (for review see Müntz, 1996 and also Chapters 12 and 13). Hence, ubiquity of the Asn-Gly site in angiosperm prolegumins does not result from the selective pressure of a highly specific cleavage enzyme. Considering the properties of the cleavage enzyme(s), the phylogenetic age of gymnosperms and the diversity of cleavage modes they use, fixed Asn-Gly cleavage (as observed in prolegumins from angiosperms and Gnetales) can be viewed as an advanced state in evolution.

Assuming an ancestor of recent legumins devoid of a distinct interdomain linker, acquisition of the HVR may have contributed to storage function as well as to the exposure of potential cleavage sites to the molecular surface, making them accessible to cleavage enzyme(s) and thus promoting the structural divergence of legumins from their progenitors.

2.4 Glycosylation signals and glycosylation

In constrast to vicilins, many of which have covalently bound glycans, legumins are generally considered as non-glycosylated. Almost all angiosperm

MetP legumin sequences available are devoid of Asn-X-(Thr/Ser) glycosylation signals. Glycosylated legumins have been shown only for *Lupinus* (Duranti *et al.*, 1995), where the Asn-Ser-Ser signal is located near the *N*-terminus of the α–chain, and the MetR legumin from *Magnolia* (Fischer *et al.*, 1995), which is glycosylated in the *C*-terminal region of its β–chain. In each case these signal sequences are not conserved in the legumins of related species; thus glycosylation of both legumins results from incidental mutations leading to signal sequences accessible to the cell's glycosylation machinery.

The situation is different in gymnosperms; except for *Ginkgo*, the gymnosperm legumin gene sequences regularly predict glycosylation signals (Fig. 3). All known Gnetales legumins share a conserved Asn-X-Thr site located in the α–chain, with an additional one being present only in *Gnetum*. The Pinaceae legumins contain a predicted glycosylation signal at the very *C*-terminal end of the β–chain, near but not homologous to that found in *Magnolia*. However, glycosylation of Pinaceae legumins could not be proved. Asn-Ile-Thr or Asn-Ile-Ser signals at strictly conserved positions are found in all legumin β–chains of Taxaceae, Cephalotaxaceae, Taxodiaceae, and Cupressaceae. Glycosylation has been experimentally verified in species from the last two families, strongly suggesting that the respective signals regularly act as carbohydrate attachment sites (Wind and Häger, 1996; Häger and Dank, 1996).

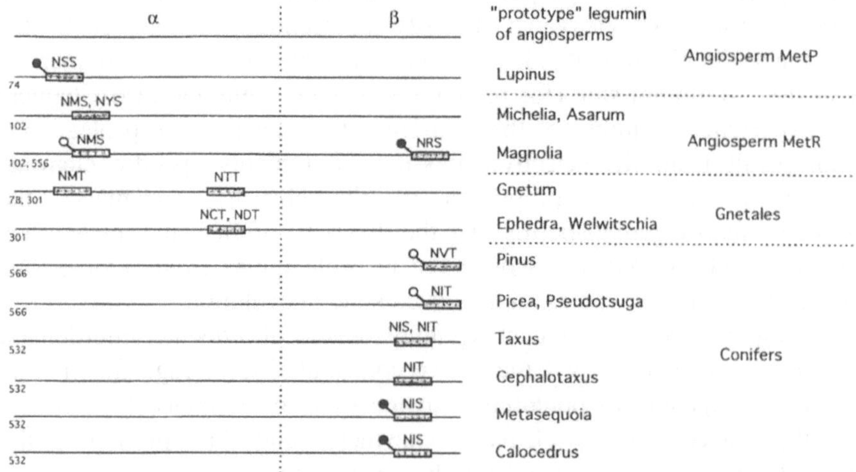

Fig. 3 Glycosylation signals and glycosylation in angiosperm and gymnosperm legumins. Relative position of glycosylation signals is indicated by boxes with the signal sequences given above. Numbers on the left specify signal positions, referring to numbering in an alignment given in Häger and Wind (1997). Open circles: not glycosylated; filled circles: glycosylated; others: not tested.

The ubiquitous occurrence of glycosylation signals and the occasional glycosylation of gymnosperm legumins resemble the situation in vicilins, which are also glycosylated, although more frequently. The preferential localization of glycosylation sites (actual or putative) in the legumin β–chain corresponds to the bias of vicilins to contain glycosylation signals, or to be glycosylated, in their *C*-terminal domain. It may be speculated that these features are relics of an ancestral state in seed globulin evolution.

2.5 Intron/exon structures of legumin genes

Ginkgo legumin genes (Häger *et al.*, 1995) are composed of five exons, with intron I and II interrupting the α–polypeptide, introns III and IV the β polypeptide encoding region. From this pattern, all other gymnosperm and angiosperm legumin gene structures can be derived by loss of one, two or even three introns. The positions and phases of the introns, if present, are strictly conserved (see also Chapter 23).

Analysis of legumin and vicilin intron/exon-structure, based on an alignment given in Lawrence *et al.* (1994) or Shutov *et al.* (1995) reveals homology of legumin gene introns I, II, III with vicilin gene introns I, II, IV, and close matching of legumin gene intron IV with vicilin gene intron V. These correspondences, as discussed in the next two chapters and in Häger *et al.* (1996), provide strong evidence for the common origin of vicilins and legumins and advocate the presumed duplication of an ancestral 2-intron-module, giving rise to a common two-domain ancestor of legumins and vicilins; other scenarios imply independent, parallel gains or shifts of introns, which in our opinion are less probable.

The 3′-portions of legumin genes coding for the legumin β–polypeptides from species representing six conifer families invariably reveal the same intron-exon-pattern as found in *Ginkgo*, with introns III and IV present, thus confirming intron IV as ancestral with respect to extant seed plants (Häger *et al.*, 1996). A more advanced state of legumin gene evolution is evident in the legumin gene of *Welwitschia*, which lacks introns I and III. The legumin genes of the most ancient angiosperm taxa studied so far, *Dioscorea*, *Magnolia*, and *Michelia*, lack intron IV. To date, this is also true for all other angiosperm legumin genes. Thus, loss of intron IV marks the transition between gymno- and angiosperms in legumin gene evolution.

Two-intron legumin genes have been shown to occur in several angiospermous families (see earlier chapters): intron I is absent from the B legumin genes of field bean and the J/K genes of pea, whereas intron III is lacking from a sunflower legumin gene and the cruciferin A genes from rape. Subsequent elimination of legumin gene introns is evident within the Ranunculaceae family (Fischer and Lang, unpublished): starting from a three-intron pattern in the most primitive taxa, there have been several intron loss events resulting in two-intron and one-intron legumin genes, respectively.

In conclusion, the presence of four introns seems to be ancestral in seed plant legumin genes. Intron number generally appears stable within the major legumin groups shown in Fig. 1, with no evidence in favour of intron additions, but obviously there is a long-term trend towards intron loss.

3. VICILINS (7S GLOBULINS)

The distribution of vicilins is often regarded as more restricted than that of legumins. This is probably due to difficulties in recognizing vicilins in contrast to legumins, which are readily identifiable by their specific interchain disulphide bonds (Boulter and Croy, 1997) in SDS-PAGE profiles. In angiosperms, vicilins probably are as widely distributed as legumins and often constitute the major seed protein fractions in monocotyledonous and lower dicot plants (Fischer, unpublished; and see earlier chapters). Isolation of vicilin-encoding transcripts or genes from a wide range of angiosperms therefore should be successful and is highly desirable, as vicilin data are available only from five families, with the vast majority of sequence data coming from the Fabaceae.

3.1 Evolutionary analysis of vicilins

Among vicilins, there are two major variants in size: the "normal" vicilins with subunits of about M_r 50,000, which have been extensively described from several legumes, and the "large" vicilins of the convicilin and β-conglycinin type, the subunits of which have additional *N*-terminal extensions of about 80-190 residues in length (see Chapters 9 and 13). These extensions generally show a low degree of sequence conservation but, in some cases, also the presence of peculiar conserved CXXXC motifs. Besides "normal" and "large" vicilins, the analysis shown in Fig. 4 also includes a sucrose-binding protein of soybean; it is associated with both seed and vegetative plant tissues involved in sucrose uptake (Grimes *et al.*, 1992) but is highly similar to cacao vicilin (Braun *et al.*, 1996; and see Chapter 23). For the study of spermatophyte vicilin evolution, a spore vicilin of the fern *Matteuccia struthiopteris* provides an unambiguous outgroup.

As the vicilin tree (Fig. 4) reflects adequately the relationships of the major plant lineages (cycads, conifers, monocots, dicots), the respective proteins can be regarded to represent orthologs. Within the Fabaceae several gene duplications are evident, but obviously have occurred after the tribal differentiation of the family; the vicilins from the Phaseoleae tribe (*Phaseolus*,

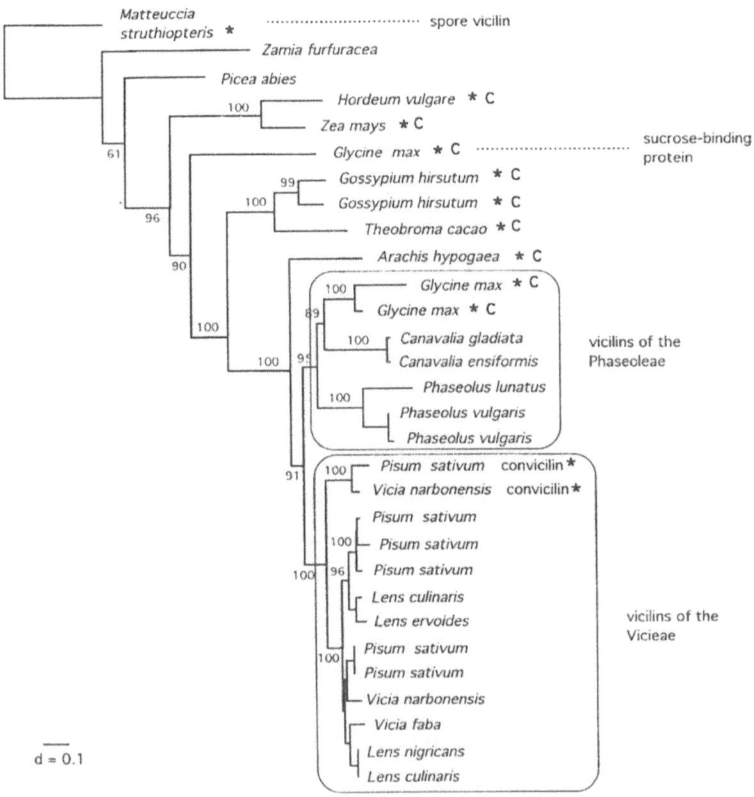

Fig. 4 Distance-based phylogenetic analysis of gymnosperm and angiosperm vicilin sequences. Asterisks following the species names mark "large" vicilins; "C" indicates the presence of CXXXC motifs in the *N*-terminal extension. d = dissimilarity. Only conserved regions were included; the alignment is available on request. Distances were calculated according to the Dayhoff PAM matrix; tree-construction was done with the Neighbour-Joining method. The PHYLIP program package was used throughout; numbers at branches are booststrap values (100 replications). Those lower than 50 are not given. GenBank accession numbers of the sequences from top to bottom are as follows: Z54364, Z50791, L47744, M64372, X59083, L06038, M19378, M16891, X62625, L38853, X17698, M13759, X06733, X59467, U01122, J01263, X03004, X06398, Z71986, X67429, X67428, J01260, Z48436, Z48438, X14076, Y00722, Z71987, Y00462, Z48437, Z48439.

Glycine, Canavalia) are distinct from those from the Vicieae tribe (*Vicia, Pisum, Lens*) and from peanut, which belongs to the tribe Aeschynomeneae (Polhill and Raven, 1981).

The evolutionary origin of the *Glycine* sucrose-binding protein (gmsbp) has been assumed to predate the origin of vicilins (Shutov *et al.*, 1996), and a gmsbp-like protein was suggested as an ancestor of recent vicilins and legumins (Braun *et al.*, 1996; and see Chapter 23). The phylogenetic analysis in Fig. 4, however, shows gmsbp to be rooted in relatively recent angiosperm vicilin history; its position basal to the dicot vicilin sequences suggests it to have originated before the differentiation of the dicot plant families included in the analysis. Our results indicate gmsbp to be a vicilin that has acquired a new function, with its expression not restricted to the storage tissues of seeds.

Regarding the relationships between "normal" and "large" vicilins, the latter are not only more widely distributed but also appear ancestral, as indicated by the fern vicilin structure and as also suggested by Dure (1990) and Shutov *et al.* (1995). Vicilins from *Zamia* and *Picea* have no extensions; this situation may indicate loss events in gymnosperms, rather than independent gains in ferns and angiosperms.

Except for the Fabaceae-Vicieae, monocot and dicot angiosperms uniformly possess "large" vicilins with CXXXC motifs. This ubiquity suggests this specific type of *N*-terminal extension to have originated before the separation of monocots and dicots; a more exact estimation is impossible with the limited dataset currently available.

Within the Fabaceae the distribution of "normal" and "large" vicilins is rather complex and suggests that "normal" vicilins developed independently in several lineages: in the Vicieae, the "normal" vicilins of *Vicia* and *Pisum* can be traced back to one gene duplication separating them from the convicilins. (From *Lens* vicilins, only partial sequences lacking the 5′-coding region are available; however, as they occupy positions within the Vicieae "normal" vicilin subfamily, it may be assumed that they also lack an *N*-terminal extension.) The situation is less clear in the Phaseoleae, where "large" vicilins seem to be confined to *Glycine* and show CXXXC motifs, while only "normal" ones are reported from *Canavalia* and *Phaseolus*. Either there have been additional, unrecognized vicilin gene duplications in the Phaseoleae, which may be detected if more data are collected, or the Phaseoleae vicilins have a complex history of loss and gain of *N*-terminal extensions. Possibly, "large" vicilins can gradually be transformed into "normal" ones by subsequent deletions - and *vice versa* by insertions. The great variability in size and sequence of *N*-terminal extensions favours an interpretation of them as analogous to other seed storage globulin variable regions.

Overall, the study of vicilin evolution in gymnosperms and angiosperms promises to yield exciting results, but demands an increased dataset covering a wider range of taxa in more detail.

ACKNOWLEDGEMENT

The authors appreciate support by the Deutsche Forschungsgemeinschaft.

REFERENCES

Allona, I., Casado, R. and Aragoncillo, C. (1992) Seed storage proteins from *Pinus pinaster* Ait.: homology of major components with 11S protein from angiosperms. *Plant Science*, **87**, 9-18.

Arahira, M. and Fukazawa, C. (1995) *Ginkgo* 11S seed storage protein family mRNA: unusual Asn-Asn linkage as post-translational cleavage site. *Plant Molecular Biology*, **25**, 597-605.

Baker, S.S., Rugh, C.L., Whitmore, F.W. and Kamalay, J.C. (1996) Genes encoding 11S-globulin-like proteins are expressed in the megagametophyte soon after fertilization in Eastern White Pine (*Pinus strobus* L.). *International Journal of Plant Science*, **157**, 453-61.

Borroto, K. and Dure, L. III (1987) The globulin seed storage proteins of flowering plants are derived from two ancestral genes. *Plant Molecular Biology*, **8**, 113-31.

Boulter, D. and Croy, R.R.D. (1997) The structure and biosynthesis of legume seed storage proteins: A biological solution to the storage of nitrogen in seeds. In *Advances in Botanical Research* (ed. J.A. Callow), Vol. **27**, Academic Press, San Diego, pp. 1-84.

Bousquet, J., Strauss, S.H., Doerksen, A.H. and Price, R.A. (1992) Extensive variation in evolutionary rate of *rbcL* gene sequences among seed plants. *Proceedings of the National Academy of Sciences of the USA*, **89**, 7844-48.

Braun, H., Czihal, A., Shutov, A.D. and Bäumlein, H. (1996) A vicilin-like seed protein of cycads: similarity to sucrose-binding proteins. *Plant Molecular Biology*, **31**, 35-44.

Chase, M.W., Soltis, D.E., Olmstead R.G., Morgan, D., Les, D.H., Mishler, B.D. Duvall, M.R., Price, R.A., Hills, H.G., Qiu, Y.-L., Kron, K.A., Rettig, J.H., Conti, E., Palmer, J.D. Manhart, J.R., Sytsma, K.J., Michaels, H.J., Kress, W.J., Karol, K.G., Clark, W.D., Hedren, M., Gaut, B.S., Jansen, R.K., Kim, K.-J., Wimpee, C.F., Smith, J.F., Furnier, G.R., Strauss, S.H., Xiang, Q.-Y., Plunkett, G.M., Soltis, P.S., Swensen, S., Williams, S.E., Gadek, P.A., Quinn, C.J., Eguiarte, L.E., Golenberg, E., Learn, G.H. Jr., Graham, S.W., Barrett, S.C.H., Dayanandan, S. and Albert, V.A. (1993) Phylogenetics of seed plants: An analysis of nucleotide sequences from the plastid gene *rbcL*. *Annals of the Missouri Botanical Garden*, **80**, 528-80.

Chen, L. and Fischer, H. (1998) Convergent evolution in seed storage proteins: The hypervariable region of a *Sagittaria* legumin shows structural similarity to wheat HMW glutenins. *Journal of Plant Physiology*, in press.

Conte, L. (1994a) Storage globulins in Gnetopsida. 1. Recognition of legumin-like proteins. *Giornale Botanico Italiano*, **128**, 839-43.

Conte, L. (1994b) Storage globulins in Gnetopsida. 2. A serotaxonomic study of legumin-like proteins. *Giornale Botanico Italiano*, **128**, 845-50.

Cronquist, A. (1981) An integrated system of classification of flowering plants. Columbia University Press, New York.

DeCamp, J.D., Stetler, D.A. and DeMaggio, A.E. (1994) Expression of sporophytic storage proteins in the corm of the quillwort (*Isoetes echinospora* Dur.) *Plant Physiology*, **106**, 1395-1402.

Dickinson, C.D., Hussein, E.H.A. and Nielsen, N.C. (1989) Role of posttranslational cleavage in glycinin assembly. *The Plant Cell*, **1**, 459-69.

Duranti. M., Horstmann, C., Gilroy, J. and Croy, R.R.D. (1995) The molecular basis for N-glycosylation in the 11S globulin (legumin) of lupin seed. *Journal of Protein Chemistry*, **14**, 107-10.

Dure, L. (1990) An unstable domain in the vicilin genes of higher plants. *New Biologist*, **2**, 487-93.

Felsenstein, J. (1993) PHYLIP (Phylogeny Inference Package) version 3.5c, Department of Genetics, Univeristy of Washington, Seattle.

Fischer, H., Haake, V., Horstmann, C. and Jensen, U. (1995) Characterization and evolutionary relationships of *Magnolia* legumin-encoding cDNAs representing two divergent gene subfamilies. *European Journal of Biochemistry*, **229**, 645-50.

Fischer, H., Chen, L. and Wallisch, S. (1996) The evolution of angiosperm seed proteins: A methionine-rich legumin subfamily present in lower angiosperm clades. *Journal of Molecular Evolution*, **43**, 399-404.

Florin, R. (1963) The distribution of conifer and taxad genera in time and space. *Acta Horti Bergiani*, **20**, 121-312.

Friedman, W.E. (1994) The evolution of embryogeny in seed plants and the developmental origin and early history of endosperm. *American Journal of Botany*, **81**, 1468-86.

Gifford, D.J. (1988) An electrophoretic analysis of the seed proteins from *Pinus monticola* and eight other species of pine. *Canadian Journal of Botany*, **66**, 1808-12.

Green, M.J., McLeod, J.K. and Misra, S. (1991) Characterization of Douglas fir protein body composition by SDS-PAGE and electron microscopy. *Plant Physiology and Biochemistry*, **29**, 49-55

Grimes, H.D., Overvoorde, P.J., Ripp, K., Franceschi, V.R. and Hitz, W.D. (1992) A 62-kD sucrose binding protein is expressed and localized in tissues actively engaged in sucrose transport. *The Plant Cell*, **4**, 1561-74.

Häger, K.-P., Jensen, U., Gilroy, J. and Richardson, M. (1992) The N-terminal amino acid sequence of the β-subunit of the legumin-like protein from seeds of *Ginkgo biloba*. *Phytochemistry*, **31**, 523-5.

Häger, K.-P., Braun, H., Czihal, A., Müller, B. and Bäumlein, H. (1995) Evolution of seed storage protein genes: legumin genes of *Ginkgo biloba*. *Journal of Molecular Evolution*, **41**, 457-66.

Häger, K.-P. and Dank, N. (1996) Seed storage proteins of Cupressaceae are homologous to legumins from angiosperms: molecular characterization of cDNAs from incense cedar (*Calocedrus decurrens* [Torr.] Florin). *Plant Science*, **116**, 85-96.

Häger, K.-P., Müller, B., Wind, C., Erbach, S. and Fischer, H. (1996) Evolution of legumin genes: loss of an ancestral intron at the beginning of angiosperm diversification. *FEBS Letters*, **387**, 94-8 (1996).

Häger, K.-P. and Wind, C. (1997) Two ways of legumin-precursor processing in conifers. *European Journal of Biochemistry*, **246**, 763-71.

Jensen, U. and Berthold, H. (1989) Legumin-like proteins in gymnosperms. *Phytochemistry*, **28**, 1389-94

Jensen, U. and Lixue, C. (1991) *Abies* seed protein profile divergent from other Pinaceae. *Taxon*, **40**, 435-40.

Lawrence, M.C., Izard, T., Beuchat, M., Blagrove, R.J. and Colman, P.M. (1994) Structure of phaseolin at 2.2 Å resolution. *Journal of Molecular Biology*, **238**, 748-76.

Leal, I. and Misra, S. (1993) Molecular cloning and characterization of a legumin-like storage protein cDNA of Douglas fir seeds. *Plant Molecular Biology*, **21**, 709-15.

Müntz, K. (1996) Proteases and proteolytic cleavage of storage proteins in developing and germinating dicotyledonous seeds. *Journal of Experimental Botany*, **47**, 605-22.
Nixon, K.C., Crepet, W.L., Stevenson, D. and Friis, E.M. (1994) A reevaluation of seed plant phylogeny. *Annals of the Missouri Botanical Garden*, **81**, 484-533.
Polhill, R.M. and Raven, P.H. (eds) (1981) *Advances in Legume Systematics*, Kew.
Shewry, P.R. (1995) Plant storage proteins. *Biological Reviews*, **70**, 375-426.
Shutov, A.D., Kakhovskaya, I.A., Braun, H., Bäumlein, H. and Müntz, K. (1995) Legumin and vicilin-like seed storage proteins: Evidence for a common single-domain ancestral gene. *Journal of Molecular Evolution*, **41**, 1057-69.
Shutov, A.D., Kakhovskaya, I.A., Bastrygina, A.S., Bulmaga, V.P., Horstmann, C. and Müntz, K. (1996) Limited proteolysis of β-conglycinin and glycinin, the 7S and 11S storage globulins from soybean [*Glycine max* (L.) Merr.]. *European Journal of Biochemistry*, **241**, 221-8.
Templeman, T.S., DeMaggio, A.E. and Stetler, D.A. (1987) Biochemistry of fern spore germination: globulin storage proteins in *Matteuccia struthiopteris* L. *Plant Physiology*, **85**, 343-9.
Wind, C. and Häger, K.-P. (1996) Legumin encoding sequences from the redwood family (Taxodiaceae) reveal precursors lacking the conserved Asn-Gly processing site. *FEBS Letters*, **383**, 46-50.

22. Structural Relationships of 7S and 11S Globulins

MICHAEL C. LAWRENCE
Biomolecular Research Institute, 343 Royal Parade, Parkville, Victoria 3052, AUSTRALIA

Although the proteins of plants early claimed the attention of many chemists, our present knowledge of them has not yet advanced beyond what is, in fact, a mere beginning of a serious study. Osborne (1909).

1. INTRODUCTION

It is now well recognized that an understanding of protein structure is fundamental to a proper understanding of protein function, and from this perspective the above quotation from Osborne's text "The Vegetable Proteins" remains a succinct description of our structural understanding of storage proteins even today. Indeed, it should be recalled that legume storage proteins were first isolated in the latter half of the last century, and that some of these proteins could be obtained in crystalline form[1]. Of course that was prior to the discovery of X-rays and high-resolution structural determination of these proteins could not be contemplated. Now, a full century later, detailed structural information is available for only two globulins, namely the closely-related 7S proteins from *P. vulgaris* (phaseolin, Lawrence *et al.*, 1990, 1994) and *C. ensiformis* (canavalin, Ko *et al.* 1993a,b). There remains no high-resolution detail for any member of the 11S family, despite preliminary crystallographic analysis of the respective 11S proteins of sesame (Hasegawa *et al.*, 1978), pumpkin (Hara *et al.*, 1976; Colman *et al.*, 1980), tobacco (Drenth and Wiebenga, 1955), Brazil nut (Drenth and Wiebenga, 1955; Kamiya *et al.*, 1983), hemp (Drenth and Wiebenga, 1955; Patel *et al.*, 1994) and coconut (Carr *et al.*, 1990). The crystals employed in these studies vary considerably in quality and none has yet led to any information at the atomic level. As a consequence, buoyed by the ever-increasing amount of primary sequence data, renewed effort has been made to investigate structural

[1] A review of the early work on seed globulins can be found in Osborne (1924).

P.R. Shewry and R. Casey (eds.), Seed Proteins, 517–541.

relationships between the 7S and 11S proteins (Argos *et al.*, 1985; Wright, 1988; Gibbs *et al.*, 1989; Lawrence *et al.*, 1994; Bäumlein *et al.*, 1995; Shutov *et al.*, 1995). It is the purpose of this chapter to review these latter investigations and to show how they lead to a measure of structural understanding of the 11S proteins.

2. THE OLIGOMERIC NATURE OF THE 7S AND 11S GLOBULINS.

Danielsson (1949), in an ultracentrifugation study of the seed globulins of more than thirty representatives of the Leguminosae, demonstrated that the legume storage proteins could be differentiated on the basis of sedimentation coefficient, with legumin-like proteins having sedimentation coefficients in the range 10.5 - 13.0 S and vicilin-like proteins having sedimentation coefficients in the range 7.0 - 9.0 S. The terms "7S" and "11S" have subsequently been used to describe all vicilin- and legumin-like proteins respectively, regardless of species. The difference in molecular weight is now understood to arise primarily from the differing oligomeric nature of the 7S and 11S proteins, with the 7S proteins being trimeric and the 11S being hexameric (for recent reviews, see Shewry, 1995; Boulter and Croy, 1997; Utsumi *et al.*, 1997; and see also preceding chapters). The trimeric nature of both phaseolin and canavalin has since been confirmed by X-ray crystallography (Lawrence *et al.*, 1990, 1994; Ko *et al.* 1993a,b). These molecules are disc-like in shape, approximately 90A in diameter and 35 Å in thickness, and have a large solvent-accessible channel running through the centre of the disc along the three-fold axis. As will be discussed later, it is most likely that all known vicilins share a similar oligomeric arrangement.

7S primary structural data became available in the early 1980's and it was readily established (Argos *et al.*, 1985; Doyle *et al.*, 1986) that the Leguminosae 7S globulins were related in sequence. This result was later extended to the wider spectrum of non-legume 7S proteins (Lawrence *et al.*, 1994). Less obvious was the existence of a weak internal sequence repeat within the 7S polypeptide (Gibbs *et al.*, 1989; Wright, 1988); the repeat reflecting no more than a 17-20% sequence identity between the *N*- and *C*-terminal portions of the sequence. This internal sequence duplication was subsequently shown (Lawrence *et al.*, 1990; Ko *et al.*, 1993a) to confer a pseudo-32 point group upon the 7S oligomeric structure, i.e. the *N*- and *C*-terminal portions of the 7S polypeptide are related to each other in the trimer by pseuo-twofold axes orthogonal to the three-fold axis (see next section).

The 11S polypeptide is synthesized as a single entity with at least one internal disulphide bond; subsequent proteolytic processing then yields a pair of disulphide-linked chains. These aspects of the 11S primary structure have been adequately reviewed (see for example Boulter and Croy (1997), and elsewhere in this volume) and will thus be described only briefly here. The site of the cysteine residues forming the disulphide bond appears conserved across all known 11S sequences and upon cleavage the *N*-terminal part of the precursor protein forms the "acidic" chain, whilst the *C*-terminal part forms the "basic" chain. *In vitro* studies have shown (Dickinson *et al.*, 1989) that in the case of soybean assembly of the 11S protein glycinin proceeds *via* a trimeric intermediate (proglycinin), with assembly into hexamers only being accomplished once the proteolytic processing of the polypeptide has occurred. In the absence of high-resolution structural information for the 11S proteins, the spatial arrangement of the subunits within the resulting hexamer is less clear. X-ray diffraction and electron microscopic experiments (Drenth and Wiebenga, 1955; Schepman *et al.*, 1972; Kamiya *et al.*, 1983) suggest that the Brazil nut 11S protein excelsin has six subunits arranged in two discs about a common three-fold axis. Perpendicular twofold axes then relate the subunits, conferring a 32 point group symmetry upon the oligomer. It is possible that the two-fold axes are only approximate, i.e. "pseudo" (Kamiya *et al.*, 1983). A 32 point group symmetry has also been established for crystals of edestin (Patel *et al.*, 1994) and helianthinin (Reichelt *et al.*, 1980). Cubic 23-point group packings have been observed at low resolution for cucurbitin (Colman *et al.*, 1980) with the corollary that if these molecules are hexamers then their respective monomers would be composed of two structurally similar modules. It should be noted that the 32 and 23 point groups are unrelated, and hence imply fundamentally different packings.

The dearth of high-resolution structural data for the 11S proteins can be attributed to two causes. First, the 11S proteins often have varying subunit composition, with the subunits arising from different genes. Secondly, the 11S proteins are characterized by hypervariable regions which may not adopt a defined secondary structure. Both these properties mitigate against high-quality crystals being obtained. Resolution of protein microheterogeneity can be problematic and it might be possible to overcome this difficulty *via* the use of recombinant material. There are two reports in the literature that bear some promise. Patel *et al.* (1994) have obtained crystals of natural edestin that diffract to 3.5 Å. Utsumi and co-workers (Utsumi *et al.*, 1993; Gidamis *et al.*, 1994; Utsumi *et al.*, 1996) have obtained crystals of recombinant and modified proglycinins that diffract to 2.9 Å. Their 6 Å electron density map of proglycinin (Utsumi *et al.*, 1996) suggest a "7S" nature for this protein, i.e. a disc-shaped molecule with three subunits arranged around a three-fold axis with pseudo 32-symmetry. It is possible to overlay the structure of phaseolin onto their map, but the 6Å resolution does not permit any detailed evaluation of the similarity of proglycinin and phaseolin. It is also not clear whether the final structure of proglycinin will provide information about how the hexamer is formed.

3. OVERVIEW OF 7S GLOBULIN X-RAY STRUCTURES

The X-ray structures of *Phaseolus vulgaris* β-phaseolin (Lawrence *et al.*, 1990) and *Canavalis ensiformis* canavalin (Ko *et al.*, 1993a; see chapter 11) are closely similar and can be summarized as follows. The native molecule is composed of three single-polypeptide subunits arranged around a 3-fold axis (Plate 7). The subunits themselves each consist of two closely-related structural modules, which reflect the internal sequence repeat. The modules formed by the *N*-terminal and *C*-terminal portions respectively of the polypeptide are related by pseudo 2-fold axes which confer a pseudo 32-point group symmetry upon the trimer. The modules themselves each consist of an anti-parallel β-barrel folding domain followed by an α-helical domain (Plate 8). The β-barrel domain has as its core a classical "jelly-roll" folding motif, the strands of which are labelled B through I in accordance with the convention established for this motif in viral capsid proteins (Rossmann *et al.*, 1983). Four additional anti-parallel strands then elaborate this barrel. Strands A' and A are immediately *N*-terminal to strand B in the polypeptide and elaborate the BIDG sheet. Strand J forms part of the CHEF sheet; it is split into two portions J and J' in order to accommodate the α-helical domain. The interface between *N*-and *C*-terminal modules is formed by a hydrophobic association of their respective BIDG β-sheets, and this interaction appears stabilized in part by the fourth additional strand, which is labelled Z. This strand Z adjoins strand G in the β-barrel of the neighbouring module, effectively cross-linking the barrels. In canavalin, the Z strand is found in both the *N*- and *C*-terminal β-barrels while in phaseolin it is found only in the *C*-terminal β-barrel.

The α-helical domain of each module consists of three helices, labelled 1, 2 and 3. The split in β-strand J appears to be facilitated by the presence of a wide β-bulge (Richardson, 1981) in the underlying strand C, the extra (bulge) residue in strand C being a proline (see Plate 8). The conformational nature of the polypeptide linking the modules appears different in phaseolin and canavalin. In the case of phaseolin the link between the N-terminal α-helical domain and the *C*-terminal β-barrel domain is formed in part by a helix 4 which extends from the *C*-terminus of strand J' across the rim of the disc towards the β-barrel interface on the opposite surface of the trimer. At this point the electron density disappears in the X-ray structure, re-establishing itself in the *C*-terminal A' strand. This disorder in the polypeptide encompasses a region of about 8 residues and is known to include protease sensitive sites (Nielsen *et al.*, 1988). In the case of the canavalin, the link between the *N*- and *C*-terminal modules appears intact in the sense that a Z-strand is present and its link onwards towards the *C*-terminal A' strand is observed; however there is no helix 4 and electron density is absent for 17 residues in this vicinity. It should be noted that canavalin was treated with trypsin prior to crystallization and that this region is most likely proteolytically degraded (Ng *et al.*, 1993).

In the case of both phaseolin and canavalin the assembly of the mature trimeric molecule is mediated by the α-helical domains: these domains protrude as hooks from the respective *N*- and *C*-terminal β-barrels and interact in a two-fold pseudo-symmetric fashion with the respective *C*- and *N*-terminal α-helical domains of the adjacent monomers (Plate 7).

4. A CANONICAL STRUCTURE FOR THE 7S GLOBULINS

What is particularly fascinating about the 7S X-ray structures is that the *N*- and *C*-terminal modules show remarkably similar tertiary structure despite a very limited sequence identity of *ca* 15-20 %. The rms deviation between the "core" α-carbon atoms of the *N*- and *C*-terminal modules of canavalin is only 1.04 Å (Ko *et al.*, 1993a). Based on this observation, re-analysis of the primary sequence data for the broader family of vicilins has led to some interesting conclusions. Table 1 presents an alignment of 7S sequences based on that presented by Lawrence *et al.* (1994), updated by the inclusion of further 7S sequences that have since become available. The alignment is overlaid with the positions of the secondary structural elements discerned in the phaseolin X-ray structure, as well as the position of the *C*-terminal Z-strand discerned in the structure of canavalin. The following are apparent:

(1) Whilst the 7S sequences vary in length, regions of significant insertion or deletion of one sequence with respect to the others are restricted to sites that map to loop regions of the phaseolin/canavalin secondary structure. These sites are as follows:

(a) The N-terminal FG loop. β-phaseolin shows a 7-residue insert at this site and α-phaseolin a 6-residue insert with respect to the remaining sequences.

(b) The N-terminal HI loop. α'β-conglycinin shows a 19-residue insert at this site with respect to the remaining sequences. (This insertion appears to be a duplication of the immediately preceding residues in this sequence, and hence the insertion could equally well be mapped structurally to the *N*-terminal GH loop).

Table 1 An alignment of 7S vicilin sequences.
The sequences were extracted from the GENPEPT protein sequence database, translated from GENBANK Release 101.0, and also from the Protein Research Foundation protein sequence database. The sequences aligned include, in order, *P. glauca* vicilin (X63191; Newton *et al.*, 1992), *Z. furfuracea* vicilin (Z50791; Braun *et al.*, 1996), *Zea mays* Glb1-L (X59083; Belanger and Kriz, 1991), *H. vulgare* globulin, BEG1 (M64372; Heck *et al.*, 1993), *G. hirsutum* α-globulin A (M19378; Chlan *et al.*, 1986), *G. hirsutum* α-globulin B (M16891; Chlan *et al.*, 1986), *T. cacao* vicilin (1818408A; McHenry & Fritz,1992), *P. vulgaris* α-type phaseolin (X02980; Slightom *et al.*, 1985), *P. vulgaris* β-type phaseolin (X03004; Slightom *et al.*, 1985), *P. sativum* vicilin (X14076; Higgins *et al.*, 1988), *P. sativum* convicilin (1713472A; Newbigin *et al.*, 1990), *V. narbonensis* vicilin (Z71987; unpublished), *V. faba* vicilin (Y00506; Bassüner *et al.*, 1987), *G. max* α-subunit, β-conglycinin (X17698; Sebastiani *et al.*, 1990), *G. max* α'-type β conglycinin (M13759; Doyle *et al.*, 1986), *G. max* CG4 β-conglycinin (S44893; Harada *et al.*, 1989), *C. ensiformis* canavalin (X59467; Ng *et al.* 1993), *M. struthiopteris* vicilin (Z54364; unpublished). For the sake of clarity and presentation, the highly variable *N*- and *C*-terminii of the sequences have been omitted from the alignment. Some residues (indicated by parentheses) have likewise been omitted in the long inserts of *Zea mays* Glb1-L globulin and *H. vulgare* globulin. The sequence numbering used for all proteins follows that of the relevant database, except for the two phaseolin sequences which follow that used in the reported X-ray structures (Lawrence *et al.*, 1990).
The location of the secondary structural elements of phaseolin are shown, labelled according to the established convention detailed in Section 20.3. Also shown is the position of the *C*-terminal Z-strand found in the structure of canavalin, but absent in phaseolin (see text). Residues that are totally conserved across all sequences presented are highlighted with a black background, whereas residues considered well-conserved in type across all sequences are highlighted with a grey background. The asterisks indicate the location of the Pro/Gly motif conserved both across species and across *N*- and *C*-terminal modules. The following software was used in the generation of the table: PILEUP (GCG 8.0; Devereux *et al.*, 1984), DCSE 3.4 (De Rijk & DeWachter, 1993) and ALSCRIPT 2.03 (Barton, 1993).

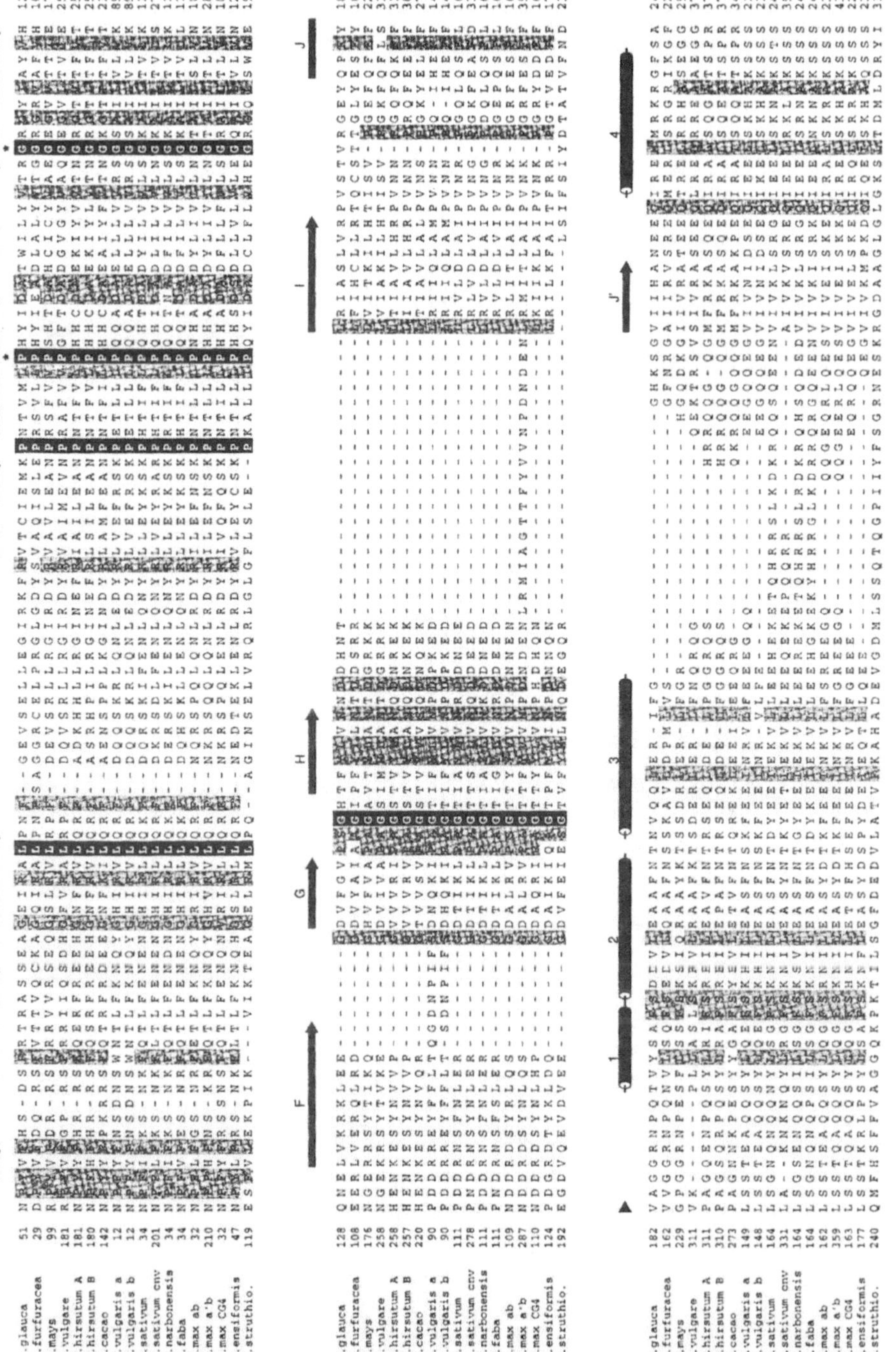

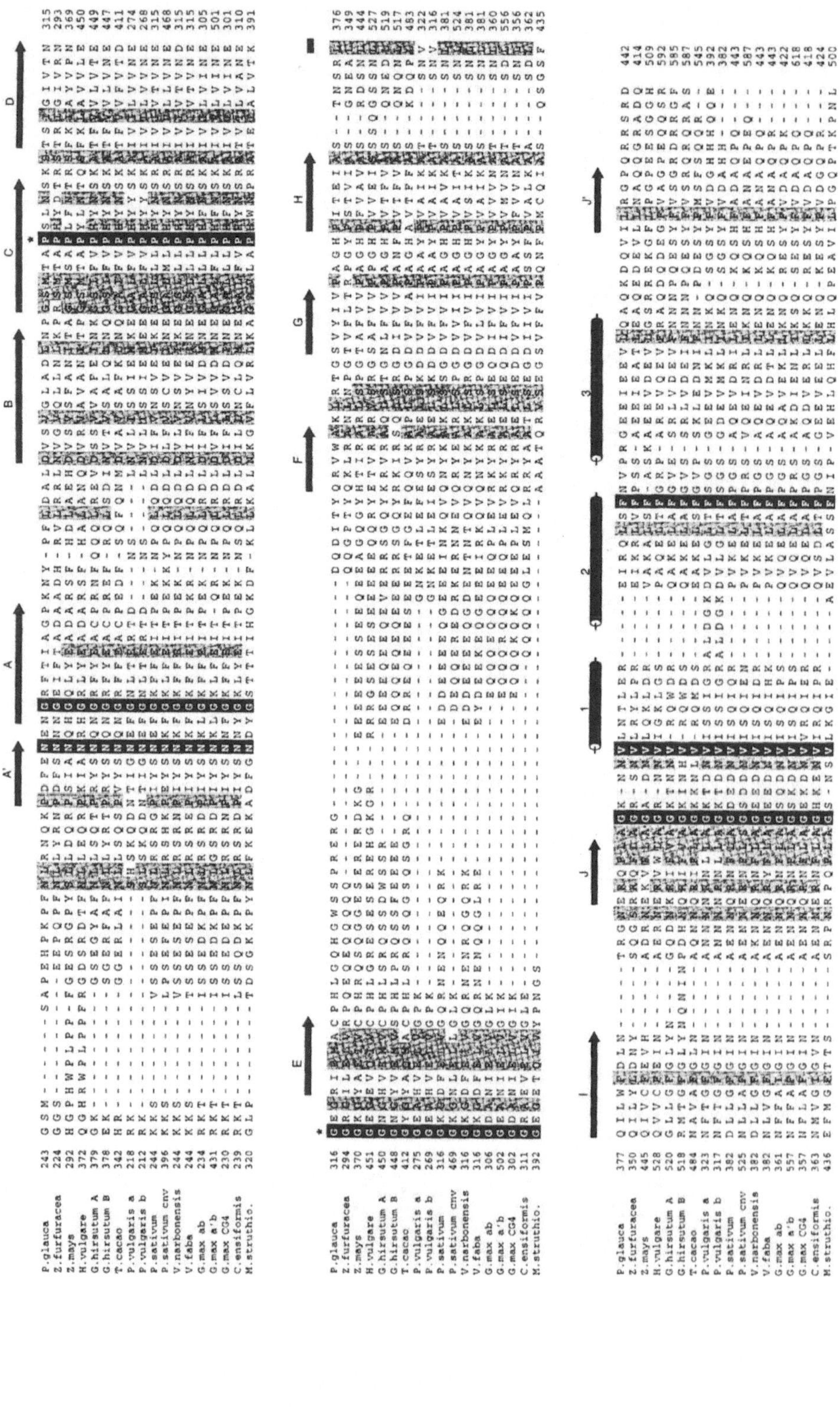
A'
A
B
C
D
E
F
G
H
I
J
J'
1
2
3
P.glauca
Z.furfuracea
Z.mays
H.vulgare
G.hirsutum A
G.hirsutum B
T.cacao
P.vulgaris a
P.vulgaris b
P.sativum
P.sativum cnv
V.narbonensis
V.faba
G.max ab
G.max a'b
G.max CG4
C.ensiformis
M.struthio.

(c) The region between *N*-terminal strand J and *N*-terminal helix 1. *H. vulgare* embryo globulin shows 6-residue deletion at this site.
(d) The region between *N*-terminal helix 3 and *N*-terminal strand J'. Considerable length and sequence variation is found in this region amongst all sequences considered. The shortest length of this polypeptide segment is 5 residues (*P. glauca* vicilin) whilst the longest is 23 residues (*P. sativum* vicilin).
(e) The region between helix 4 and the *C*-terminal Z-strand. The number of residues found in this region varies from 4 to 19 residues; any inference about the conformational nature of this region is most likely impossible given the variation observed in this region between the structures of phaseolin and canavalin.
(f) The *C*-terminal AB loop. This loop is 6-7 residues shorter in phaseolin compared to the remaining sequences.
(g) The *C*-terminal EF loop. Considerable length and sequence variation is found in this region amongst all sequences considered. The insert comprises predominantly charged residues and the length of this region varies from approximately 4 residues (*M. struthiopteris* vicilin) to approximately 47 residues (*H. vulgare* embryo globulin).
(h) The *C*-terminal IJ loop: *G. hirsutum* B shows an insert of 5 residues at this position.
(i) The *C*-terminal 1-2 loop. Phaseolin shows a 5-residue insert at this position.
(2) Two residues appear strictly conserved across both the *N*- and *C*-terminal portions of the alignment. These residues are the β-bulge-forming proline in strand C referred to in the preceding section, and a glycine residue forming the DE loop (see Plate 8). The strict conservation of the proline is important, given its occurrence at the nexus of the α- and β-domains. In the *N*-terminal region the separation between the proline and glycine is 13 residues in all sequences considered, whilst in the *C*-terminal region it is 14 residues. The strict conservation of a glycine at these sites appears to be associated exclusively with nature of the turn between the D- and E-strands.
(3) In addition to the Pro/Gly motif, a few residues are completely conserved across the entire alignment and a number show only conservative replacement. The locations of these residues have been examined (Ko *et al.*, 1993a; Lawrence *et al.*, 1994). On the whole they have been shown to map to sites in the polypeptide that have particular structural requirements, or alternatively to sites that form part of the subunit interface.

Taken together these observations favour the view that all 7S proteins considered share the same tertiary and quaternary structure as observed for phaseolin and canavalin, i.e. each polypeptide forms a pair of pseudo-symmetric α+β modules arranged around a three-fold axis. This highly-conserved modular structure can be considered "canonical" (Ko *et al.*, 1993a; Lawrence *et al.*, 1994), in the sense that it most likely represents the core folding topology of all 7S proteins. It should be noted that phaseolin and canavalin are amongst the shorter 7S sequences; others are elaborated *via* the inserts outlined above and also *via* considerable *N*-terminal extensions (e.g. convicilin). These elaborations are most likely tolerated as additional secondary/tertiary structural elements superimposed upon the canonical modular structure, rather than inducing a fundamentally different topology.

5. INTERPRETATION OF 11S SEQUENCES IN TERMS OF THE CANONICAL 7S STRUCTURE

Amino acid sequence relationships between the *C*-terminal portion of the 7S polypeptide and the basic chain of the 11S proteins were first identified by Argos *et al.* (1985), but at that stage any alignment between the *N*-terminal portions of the available 7S sequences and those of the 11S acidic chains was unconvincing. It was subsequently recognized that there were internal sequence homologies within the 7S proteins (Gibbs *et al.*, 1989; Wright, 1988), but there appeared little corresponding evidence for sequence homology between the acidic and basic chains for the 11S proteins. The 7S polypeptide had thus presumably arisen via duplication of an ancestral gene which encoded the equivalent of a single $\alpha+\beta$ module. The 11S protein may have arisen either via a similar duplication of the same ancestral gene (see Figure 1a) followed by considerable divergence of the acidic chain, or alternatively it may have arisen *via* the combination of the ancestral gene with a second unrelated gene. These possibilities are discussed in more detail in Chapter 23.

The advent of the X-ray structures for the 7S proteins phaseolin and canavalin led to a re-investigation of the 7S/11S sequence relationships. It was shown (Lawrence *et al.*, 1994) that an alignment of both the acidic and basic chains of the 11S sequences with respect to the 7S *N*- and *C*-terminal polypeptides could be constructed which would allow a direct mapping of the 7S structural elements onto the 11S sequences. The basis of this alignment is illustrated in Table 2, and includes the following observations:
(1) the Pro/Gly motif, shown above to be critical to the modular structure of the 7S proteins, appeared retained in all 11S sequences then available,
(2) insertions within the 11S sequences could on the whole be mapped to turn regions within the 7S structure, and
(3) a number of other individual residues judged as highly conserved within the 7S sequences appeared to be "globally" conserved within the 11S sequences.

(a)

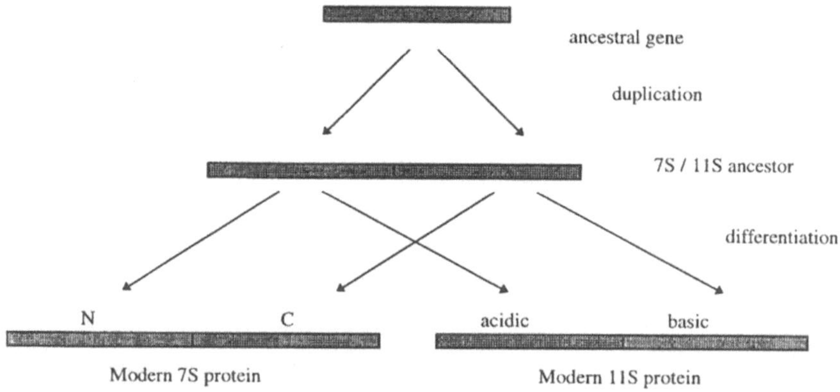

(b)

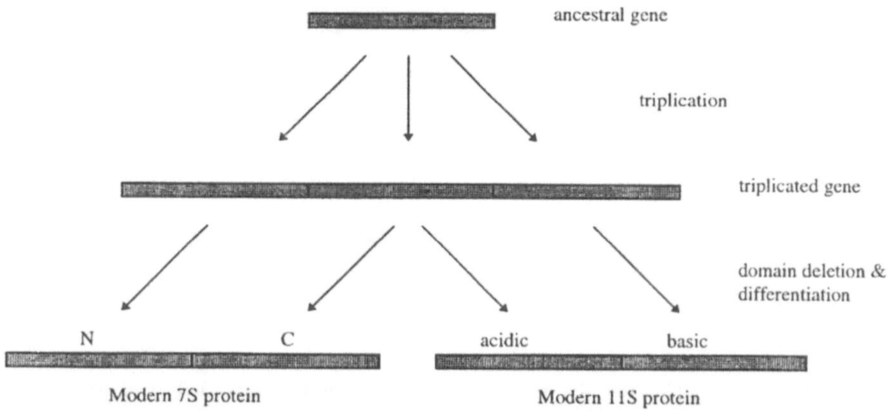

Fig. 1 Possible evolutionary pathways for the 7S and 11S globulins: (a) ancestral gene duplication (Gibbs *et al.*, 1989; Lawrence *et al.*, 1994), and (b) ancestral gene triplication (Shutov *et al.*, 1995). See also Chapter 23.

Table 2 A joint alignment of 11S sequences with that of the 7S protein phaseolin. 11S sequences are contained in the central block of 48 sequences. The uppermost sequence in the alignment is that of β-phaseolin and its alignment with respect to the 11S sequences below it follows the alignment given in Lawrence *et al.* (1994). The lowermost sequence in the alignment is again that of β-phaseolin, but with the strands corresponding to the *N*- and *C*-terminal modules interchanged; in this instance the alignment with respect to the 11S sequences is that given in Shutov *et al.* (1995). The positions of the secondary structural elements of phaseolin are shown together with both phaseolin alignments, as is the implied position of the *C*-terminal Z-strand of canavalin. The table does not imply an alignment of the phaseolin sequence at the top with its interchanged counterpart below. The extreme *N*- and *C*-terminii are deleted from all sequences, as are hypervariable regions within the 11S sequences - the sites of the latter deletions are indicated by short horizontal bars above and below the block of 11S sequences.

All sequences were extracted from GENPEPT protein sequence database, translated from GENBANK Release 101. The sequences aligned include, in order, *P. vulgaris* β-type phaseolin (X03004; Slightom *et al.*, 1985), *G. gnemon* 11S globulin (Z50779; unpublished), *W. mirabilis* 11S globulin (Z50780; unpublished), *E. gerardiana* 11S globulin (Z50777; unpublished), *P. glauca* 11S protein (X63192; Newton, unpublished), *P. menziesii* 11S protein (L07484; Leal and Misra, 1993), *P. strobus* globulin-1 (Z11486; Baker *et al.*, 1996), *C. japonica* legumin, CjLeg2 (X95542; Wind & Häger, 1996), *M. glyptostroboides* legumin, MgLeg18 (X95545; Häger and Wind, 1997), *M. glyptostroboides* legumin, MgLeg26 (X95546; Häger and Wind, 1997), *C. decurrens* legumin, CdLeg1 (X95539; unpublished), *C. decurrens* legumin, CdLeg2 (X95540; unpublished), *C. japonica* legumin, CjLeg31 (X95543; Wind & Häger, 1996), *M. glyptostroboides* legumin, MgLeg4 (X95545; Häger and Wind, 1997), *C. decurrens* legumin, CdLeg3 (X95541; unpublished), *G. biloba* (Z50778; Häger et al., 1995), *A. europaeum* leg10 (X95508; Fischer *et al.*, 1996), *A. europaeum* leg18 (X95509; Fischer *et al.*, 1996), *M. salicifolia* B14 legumin precursor (X82465; Fischer *et al.*, 1995), *D. caucasica* legumin (X95510; Fischer *et al.*, 1996), *V. faba* legumin A1 (X55013; Schlesier *et al.*, 1990), *V. faba* legumin A2 (X55014; Schlesier *et al.*, 1990), *V. narbonensis* legumin A (Z46803; unpublished), *V. sativa* legumin A (Z32835; unpublished), *P. sativum* legumin A2 (X17193; Rerie *et al.*, 1990), *G. max* G1 subunit (X15121; Sims & Goldberg, 1989), *G. max* G2 subunit (X15122; Thanh *et al.*, 1989), *G. max* G3 subunit (X15123; Cho & Nielsen, 1989), *G. max* G4 subunit (X52863; Xue *et al.*, 1992), *G. max* A3-B4 subunit (M10962; Fukazawa *et al.*, 1985), *V. faba* legumin B, LeB4 (X03677; Bäumlein *et al.*, 1986), *V. sativa* legumin B (Z32796; unpublished), *P.sativum* legumin J (X07014; Gatehouse *et al.*, 1988), *V. faba* legumin B3 (Z26489; Heim *et al.*, 1994), *A. sativa* 11S globulin, GLAV1 (X74740; Tanchak *et al.*, 1995), *A. sativa* 11S globulin, GLAV3 (X74741; Tanchak *et al.*, 1995), *M. salicifolia* A11 legumin precursor (X82463; Fischer *et al.*, 1995), *M. salicifolia* A2 legumin precursor (X82464; Fischer *et al.*, 1995), *G. hirsutum* legumin B, C134 (U43727; unpublished), *A. hypochondriacus* 11S globulin (X82121; Barba de la Rosa *et al.*, 1996), *P. amygdalus* prunin, (X78119; Garcia-Mas *et al.*, 1995), *Q.robur* legumin (X99539; unpublished), *G. hirsutum* legumin A (M69188; Galau *et al.*, 1991), *C. arabica* 11S globulin (U64443; unpublished), *S. sagittifolia* legumin (Y09117; unpublished), *H. annuus* 11S protein, G3-D1 (M22832; Von der Haar *et al.*, 1988), *B. napus* cruciferin (M16860; Simon *et al.*, 1985), *A. thaliana* cruciferin, ATCRU3 (U66916; unpublished), *C. pepo* 11S globulin (M36407, Hayashi *et al.*, 1988), *P. vulgaris* β-type phaseolin with N and C modules interchanged (X03004; Slightom *et al.*, 1985). The sequence numbering used for all proteins follows that of the database, except for the phaseolin sequences where the numbering follows that used in the reported X-ray structure (Lawrence *et al.*, 1990). The highly variable *N*- and *C*-terminii of the sequences have been omitted from the alignment.

Residues that are totally conserved across all 11S sequences presented are highlighted with a black background, whereas residues considered well-conserved in type across all 11S sequences are highlighted with a grey background. Shading of residues within the phaseolin sequences are in concordance with Table 1. The asterisks indicate the location of the Pro/Gly motifs. The following software was used in the generation of the table: PILEUP (GCG 8.0, Devereux *et al.*, 1984), DCSE 3.4 (De Rijk and DeWachter, 1993) and ALSCRIPT 2.03 (Barton, 1993).

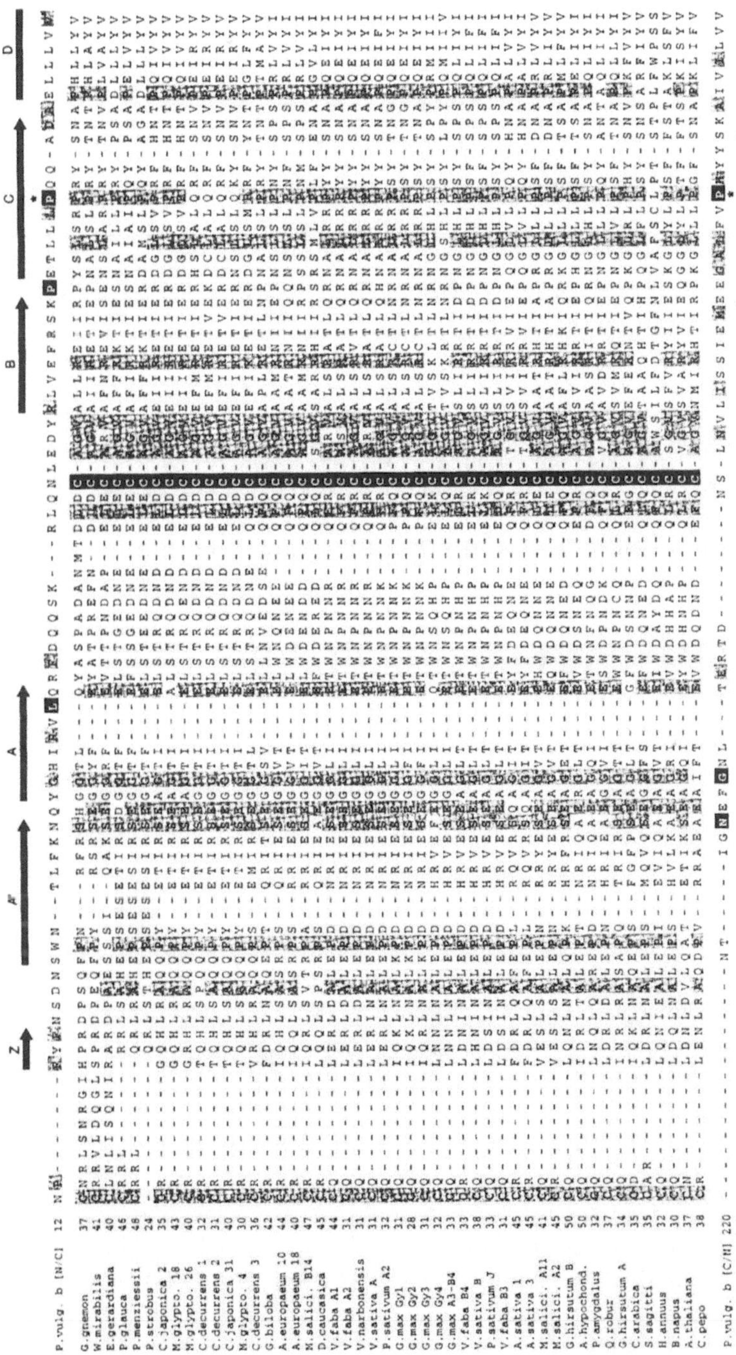

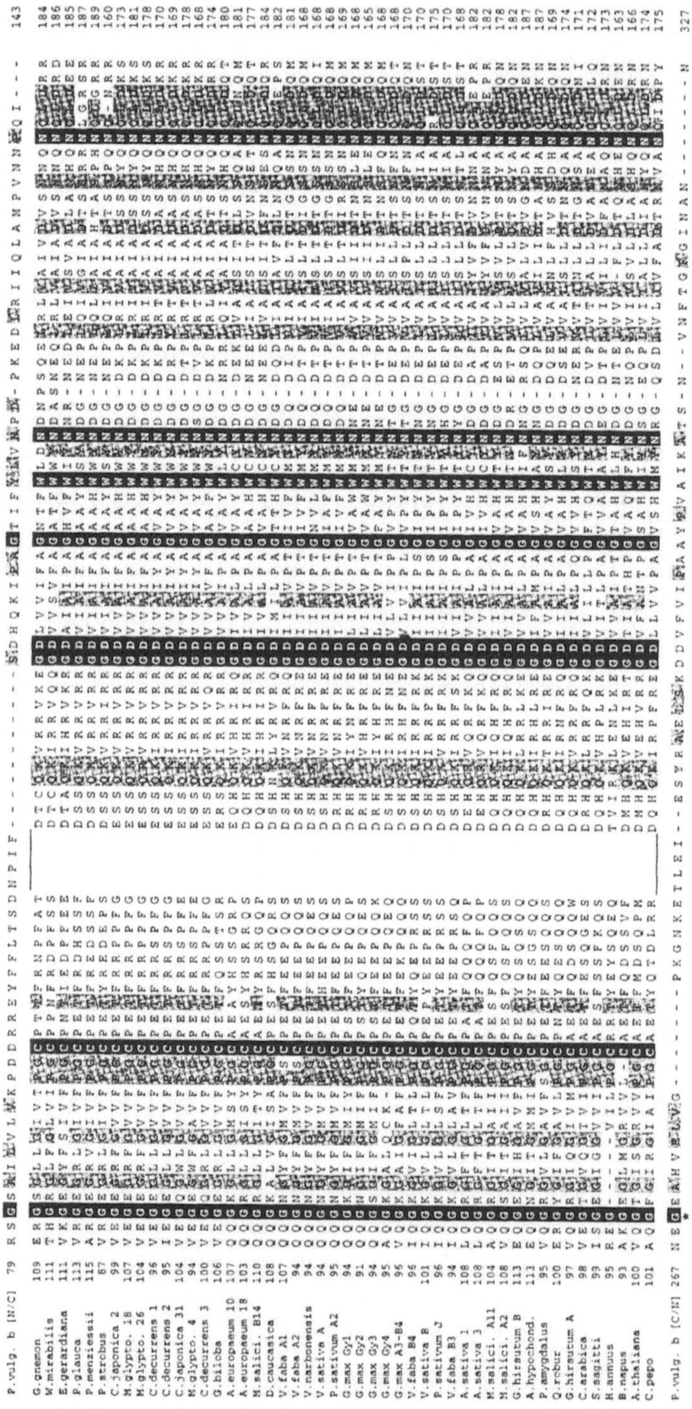

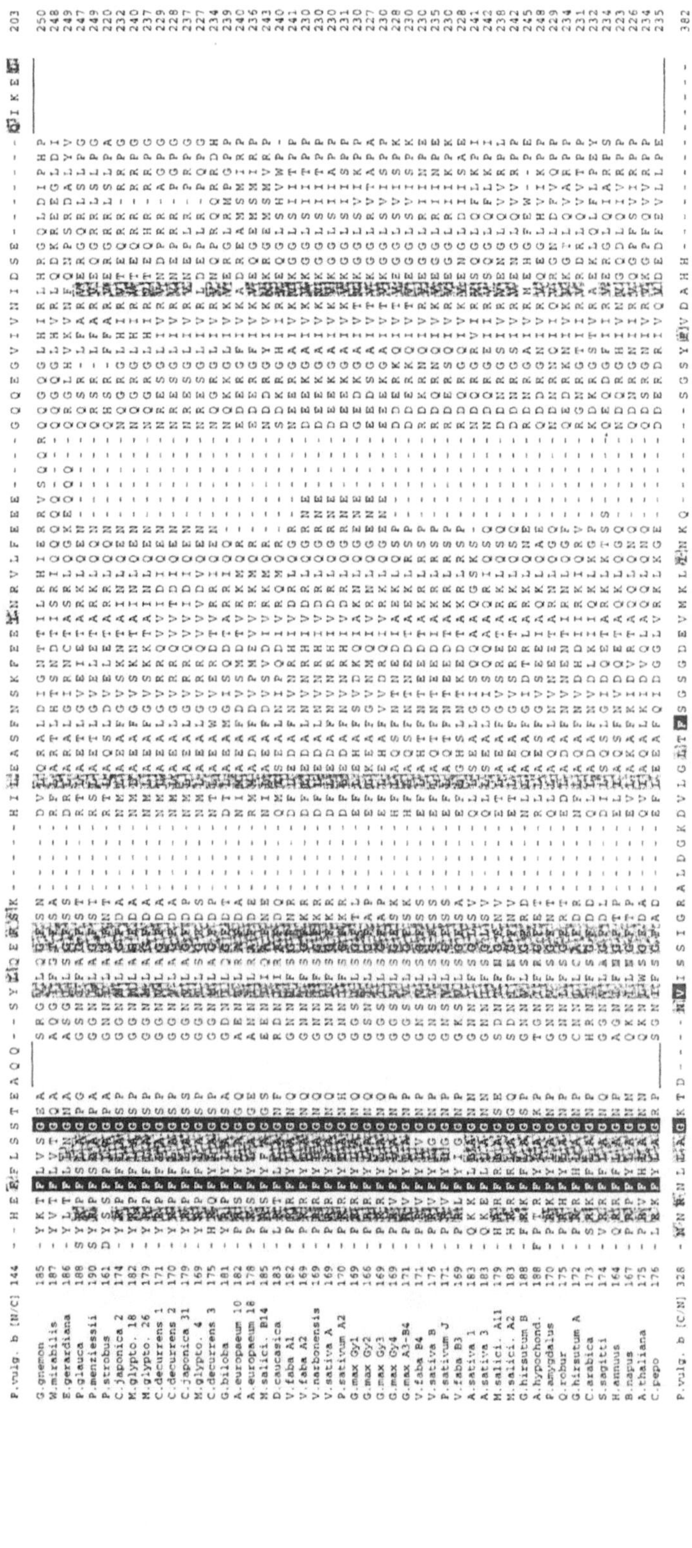

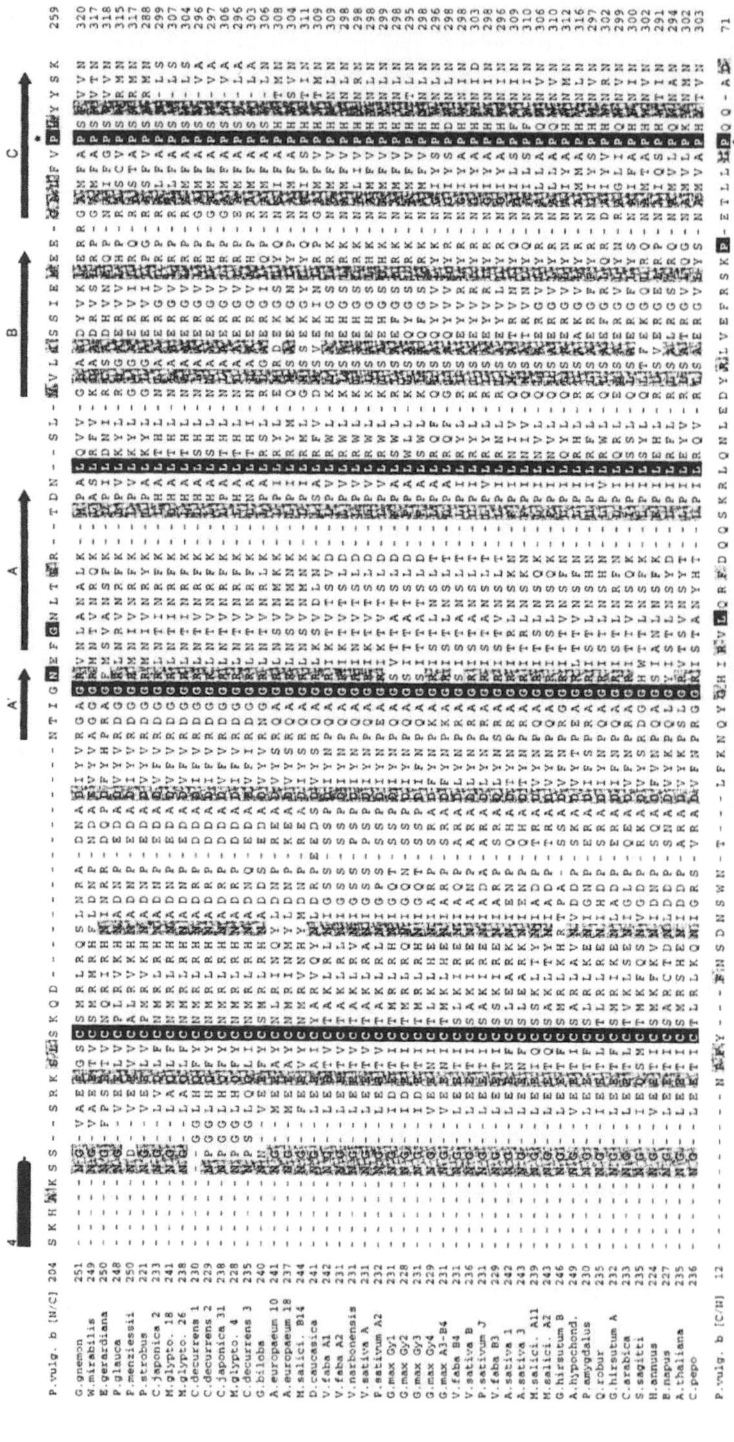

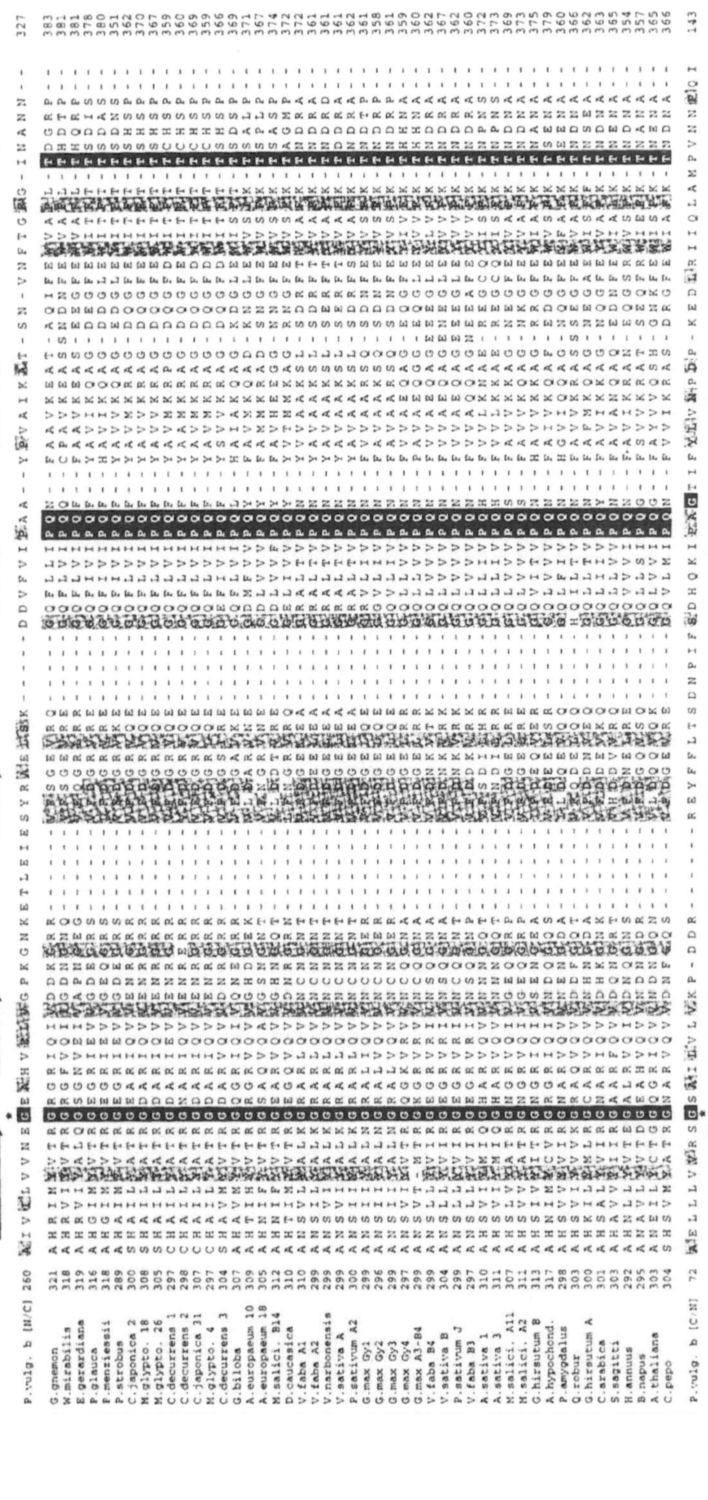

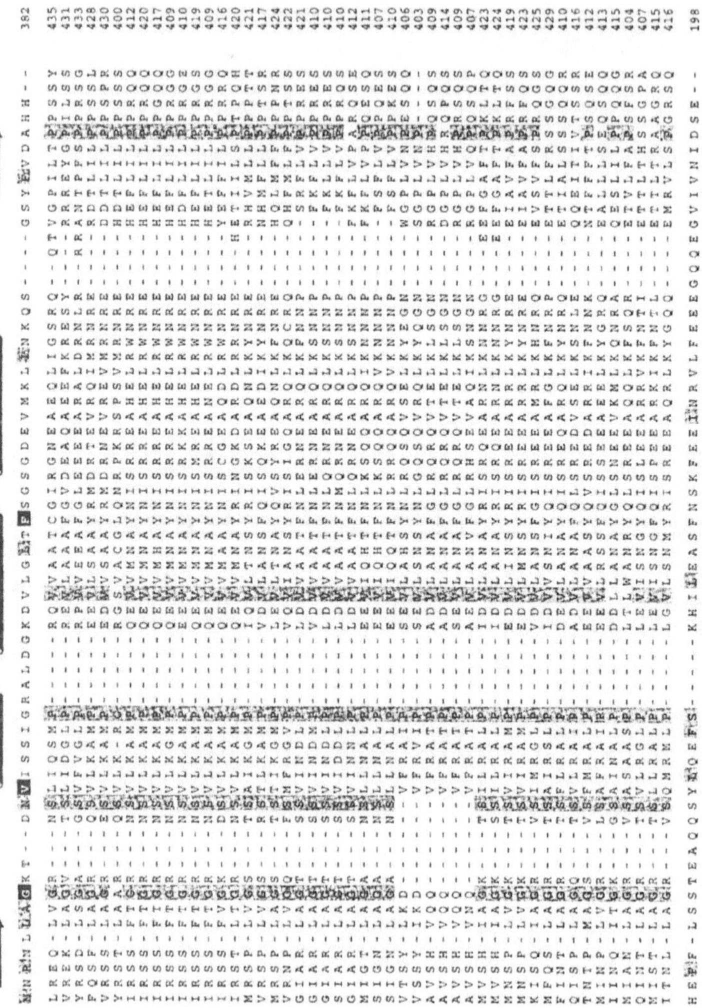

It was thus concluded that the 11S proteins shared a similar modular structure to the 7S proteins, the acidic and basic chains each forming an $\alpha+\beta$ module, and these then assembling in a manner similar to that found in the 7S proteins.

Whilst this prediction now appears readily accepted, Shutov *et al.* (1995) argued that a better alignment could be achieved by matching the 11S acidic chains with the 7S *C*-terminal polypeptide rather than with the 7S *N*-terminal polypeptide. Such a correspondence would be consistent with a gene triplication event (see Figure 1b). Table 2 represents an illustrative alignment of 11S sequences, based on that given by Lawrence *et al.* (1994) but supplemented by sequences subsequently deposited in GENBANK and by the alignment proposed by Shutov *et al.* (1995). The triplication hypothesis is an intriguing alternative, and is favoured in particular by the fact that it assigns the large inserts observed in the acidic chains (hypervariable region II) to the EF barrel loop, corresponding to a site of insertion in the 7S *C*-terminal domains (Table 1). The location of intron sites within the 7S and 11S genes is not immediately reconcilable with either the gene duplication or the gene triplication model (Shutov *et al.*, 1995); with the gene triplication model in particular one is left with the enigma that the intron positions within the 11S acidic chain correspond more closely with those of the *N*-terminal 7S polypeptide than with those of the *C*-terminal 7S polypeptide. The gene triplication model does also not explain the overwhelming preference for a $P(X_{13})G$ motif in the 7S *N*-terminal polypeptides and 11S acidic chains versus a $P(X_{14})G$ motif in the 7S *C*-terminal polypeptides and 11S basic chains. Further gene and protein sequence information would be required to distinguish convincingly between the duplication and triplication hypotheses, and indeed in a fresh investigation Shutov and Bäumlein argue against their triplication model (see Chapter 23) .

It is interesting to ask whether a single domain protein homologous to the proposed 7S/11S ancestor can still be found. Bäumlein *et al.* (1995) have demonstrated that both wheat germins and myxomycete spherulins are potential candidates, in that their sequences can be aligned well with that of the canonical single 7S/11S module. In particular the Pro/Gly motif is conserved in the germins and spherulins, as well as a further $G(X)_4IPXG$ motif located in strand G (see Table 1). There is also some correlation of location of the genomic introns. Spherulins and germins are believed to play a role in cellular desiccation or rehydration and are thus not functionally dissimilar to seed storage proteins. Such matters are considered in more detail in chapter 23.

6. LIMITATIONS OF THE 11S MODEL

Whilst sequence relationships suggest the same underlying modular structure for both the 7S and 11S structures, fundamental questions regarding the structure of the 11S proteins remain unanswered:
(1) Nothing is known about the structural role of the hypervariable regions of the 11S proteins. It is not known whether they adopt any regular secondary structure or indeed whether they play any role in the oligomerization of the 11S proteins.
(2) It is not clear (Lawrence *et al.*, 1994; Shutov *et al.*, 1995) how the disulphide bridges of the 11S structure are accommodated within the modular structure.
(3) The nature of the hexameric arrangement of 11S subunits is still uncertain. Whilst evidence points towards the 11S proteins being formed out of two 7S-like discs with 32 point-group symmetry, it is not impossible that the modules of individual disulphide-linked subunits rearrange in some fashion upon formation of the mature hexamer. The 23 symmetry observed for cucurbitin is unexplained.
(4) The sequence homology between 11S proteins and the 7S proteins appears considerably less well preserved in the helical domains than in the barrel domains. It is thus possible that the "α-helical" domain of the 11S proteins adopt a slightly different tertiary and/or secondary structure to that found in phaseolin and canavalin. However the pseudo two-fold symmetric association of neighbouring subunits found in the 7S globulins is most likely retained within each 11S trimer.

7. CONCLUSION

In summary, it appears that all known 7S globulins share a similar core folding topology to that observed in phaseolin and canavalin. The re-analysis of the 11S sequences has led to the predicition that both the acidic and basic chains of the molecules most likely adopt a similar modular structure to that found in the 7S proteins. This is an important result in that it provides a preliminary basis for for the design of mutant 11S proteins with altered composition or functionality. On the other hand, the nature of oligomerization of the 11S subunits remains uncertain and it is unlikely that much further information can be derived from gene or protein sequence analysis alone.

The remarkable conservation of 7S-like modular structure, to the extent that it can be conferred from primary sequence, is intriguing. One may speculate as to what role this structure may play in embryogenesis. In this regard little if anything is known in the case of the 7S or 11S proteins. Does the disc-like structure assist in any way with the accummulation and packaging of the protein within the storage bodies? Is it related to the crystalline nature of the storage bodies that has been found in some species? What interaction does the structure have with the processing events that occur upon germination? Does the channel through the centre of the 7S trimer play any role in transport of water, ions or metabolites in or out of the storage bodies, as suggested by Ko *et al.* (1993a,b)? These questions remain unanswered.

The most likely avenue for progress is recombinant production of the 11S protein, either in a host system with the processing machinery necessary to produce the mature protein, or in combination with *in vitro* processing to yield mature 11S hexamers (see, for example, chapters 12 and 13). It may also be possible to utilize genetic engineering to remove all or part of the 11S hypervariable regions to assist with crystallization. Modern X-ray crystallography, capable of working with smaller and more weakly diffracting crystals through the advent of cryo-cooling and synchrotron radiation sources, might thus fulfil the goal that was glimpsed more than a century ago.

ACKNOWLEDGEMENTS

I thank my colleagues at the Biomolecular Research Institute and the CSIRO Division of Molecular Science for their contribution over many years to the structural work on phaseolin, in particular Peter Colman, Bob Blagrove, Bert van Donkelaar and Peter Tulloch.

REFERENCES

Argos, P., Narayana, S.V.L. & Nielsen, N.C. (1985) Structural similarity between legumin and vicilin storage proteins from legumes. *EMBO Journal* **4**, 1111-7.

Baker, S.S., Rugh, C.L., Whitmore, F.W. & Kamalay, J.C. (1996) Genes encoding 11S-globulin-like proteins are expressed in the megagametophyte soon after fertilization in eastern white pine (*Pinus strobus* L.). *International. Journal of Plant Science* **157**, 453-61.

Barba de la Rosa, A.P., Herrera-Estrella, A., Utsumi, S. & Paredes-Lopez, O. (1996) Molecular characterization, cloning and structural analysis of a cDNA encoding an amaranth globulin. *Journal of Plant Physiology* **149**, 527-32.

Barton, G.J. (1993) ALSCRIPT a tool to format multiple sequence alignments. *Protein Engineering* **6**, 37-40.

Bassüner, R., Hai, N.V., Jung, R., Saalbach, G. & Müntz, K. (1987) The primary structure of the predominating vicilin storage protein subunit from field bean seeds (*Vicia faba* L. var. minor cv. Fribo). *Nucleic Acids Research* **15**, 9609.

Bäumlein, H., Wobus, U., Pustell, J. & Kafatoc, F.C. (1986) The legumin gene family: structure of a B type gene of *Vicia faba* and a possible legumin gene specific regulatory element. *Nucleic Acids Research* **14**, 2707-20.

Bäumlein, H., Braun, H., Kakhovskaya, I.A.& Shutov, A.D. (1995) Seed storage proteins of spermatophytes share a common ancestor with dessication proteins of fungi. *Journal of Molecular Evolution* **41**, 1070-5.

Belanger, F.C. & Kriz, A.L. (1991) Molecular basis for allellic polymorphism of the maize *Globulin-1* gene. *Genetics* **129**, 863-72.

Boulter, D. & Croy, R.R.D. (1997) The structure and biosynthesis of the legume seed storage proteins: a biological solution to the storage of nitrogen in seeds. *Advances in Botanical Research* **27**, 2-84.

Braun, H., Czihal, A., Shutov, A.D. & Bäumlein, H. (1996) A vicilin-like seed protein of cycads: similarity to sucrose-binding proteins. *Plant Molecular Biology* **31**, 35-44.

Carr, H.J., Plumb, G.W., Parker, M.L. & Lambert, N. (1990) Characterisation and crystallisation of an 11S seed storage globulin from coconut (*Cocus nucifera*). *Food Chemistry* **38**, 11-20.

Chlan, C.A., Pyle, J.B., Legocki, A.B.& Dure, L.III. (1986) Devęlopmental biochemistry of cottonseed embryogenesis and germination. XVIII. cDNA and amino acid sequences of members of the storage protein families. *Plant Molecular Biology* **7**, 475-89.

Cho, T.J & Nielsen, N.C. (1989) The glycinin Gy3 gene from soybean. *Nucleic Acids Research* **17**, 4388.

Colman, P.M., Suzuki, E. & Van Donkelaar, A. (1980) The structure of cucurbitin: subunit symmetry and organization *in situ*. *European Journal of Biochemistry* **103**, 585-8.

Danielsson, C. E. (1949) Seed globulins of the Gramineae and Leguminoseae. *Biochemistry Journal* **44**, 387-400.

De Rijk, P. & De Wachter, R. (1993) DCSE, an interactive tool for sequence alignment and secondary structure research. *Computer Appl. Biosci.* **9**, 735-40.

Devereux, , J. Haerbli, P. & Smithies, O. (1984) A comprehensive set of sequence analysis programs for the VAX. *Nucleic Acids Research* **12**, 387-95.

Dickinson, C.G., Hussein, E.H.A. & Nielsen, N.C. (1989) Role of post-translational cleavage in glycinin assembly. *Plant Cell* **1**, 459-69.

Doyle, J.J., Schuler, M.A., Godette, W.D., Zenger, V., Beachy, R.N. & Slightom, J.L. (1986) The glycosylated seed storage proteins of *Glycine max* and *Phaseolus vulgaris*. *Journal of Biological Chemistry* **261**, 9228-38.

Drenth, J. & Wiebenga, E.H. (1955) Excelsin, edestin and tobacco seed globulin crystals. *Recueil* **74**, 813-31.

Fischer, H., Haake, V., Horstmann, C. & Jensen, U. (1995) Characterization and evolutionary relationships of Magnolia legumin encoding cDNAs representing two divergent gene families. *European Journal of Biochemistry* **229**, 645-50.

Fischer, H., Chen, L. & Wallisch, S. (1996) The evolution of angiosperm seed proteins: a methionine-rich legumin subfamily present in lower angiosperm clades. *Journal of Molecular Evolution* **43**, 399-404.

Fukazawa, C., Momma, T., Hirano, H., Harada, K. & Udaka, K. (1985) Glycinin A3B4 mRNA: cloning and sequencing of a double stranded cDNA complementary to a soybean storage protein. *Journal of Biological Chemistry* **260**, 6234-9.

Galau, G.A., Wang, H. Y.-C. & Hughes, D.W. (1991) Sequence of the Gossypium hirsutum D-genome alloallele of legumin and its mRNA. *Plant Physiology* **97**, 1268-70.

Garcia-Mas, J., Messeguer, R., Arus, P. & Puigdomenech, P. (1995) Molecular characterization of cDNAs corresponding to genes expressed during *almond (Prunus amygdalus* Batsch) seed development. *Plant Molecular Biology* **27**, 205-10.

Gatehouse, J.A., Bown, D., Gilroy, J., Levasseur, M., Castleton, J. & Ellis, T.H. (1988) Two genes encoding 'minor' legumin polypeptides in pea (*Pisum sativum* L.). Characterization and complete sequence of the LegJ gene. *Biochemistry Journal* **250**, 15-24.

Gibbs, P.E.M., Strongin, K.B. & McPherson, A. (1989) Evolution of legume storage proteins: a domain common to legumins and vicilins is duplicated in vicilins. *Molecular Biology and Evolution* **6**, 614-23.

Gidamis, A.B., Mikami, B., Katsube, T., Utsumi, S. & Kito, M. (1994) Crystallization and preliminary X-ray analysis of soybean proglycinins modified by protein engineering. *Bioscience, Biotechnology & Biochemistry* **58**, 703-6.

Häger, K.-P., Braun, H., Czihal, A., Müller, B. & Bäumlein, H. (1995) Evolution of seed storage protein genes: legumin genes of *Ginkgo biloba*. *Journal of Molecular Evolution* **41**, 457-66.

Häger, K.-P. & Wind, C. (1997) Two ways of legumin precursor processing in conifers. Characterization and evolutionary relationships of *Metasequoia* cDNAs representing two legumin gene subfamilies. *European Journal of Biochemistry* **246**, 763-71.

Hara, J., Wada, K., Wakabayashi, S. & Matsubara, H. (1976) Pumpkin (*Cucurbita* sp.) seed globulin 1. Purification, characterisation and subunit structure. *Plant and Cell Physiology* **17**, 799-814.

Harada, J.J., Barker, S.T. & Goldberg, R.B. (1989) Soybean ß-conglycinin genes are clustered in several DNA regions and are regulated by transcriptional and post-transcriptional processes. *Plant Cell* **1**, 415-25.

Hasegawa, K. Murata, M. & Fujino, S. (1978) Characterisation of subunits and temperature dependent dissociation of 13S globulin of sesame seed. *Agricultural & Biological Chemistry* **42**, 2291-7.

Hayashi, M., Mori, H., Nishimura, M., Akazawa, T. & Hara-Nishimura, I.(1988) Nucleotide sequence of cloned cDNA for pumpkin 11-S globulin ß-subunit. *European Journal of Biochemistry* **172**, 627-32.

Heck, G.R., Chamberlain, A.K. & Ho, T.-H. D. (1993) Barley embryo globulin 1 gene, *Beg1*: Characterization of cDNA, chromosome mapping and regulation of expression. *Molecular & General Genetics* **239**, 209-18.

Heim,U., Baumlein,H. & Wobus,U. (1994) The legumin gene family: a reconstructed *Vicia faba* legumin gene encoding a high-molecular-weight subunit is related to type B genes. *Plant Molecular Biology* **25**, 131-5.

Higgins, T.J.V., Newbigin, E.J., Spencer, D., Llewellyn, D.J. & Craig, S. (1988) The sequence of a pea vicilin gene and its expression in transgenic tobacco. *Plant Molecular Biology* **11**, 683-95.

Kamiya, N., Sakabe, K., Sakabe, N., Sasaki, K., Sakakibara, M., & Noguchi, H. (1983) Structural properties of brazil nut 11S globulin, excelsin. *Agricultural & Biological Chemistry* **47**, 2091-8.

Ko, T.-P., Ng, J.D. & McPherson, A. (1993a) The three-dimensional structure of canavalin from jack bean (*Canavalia ensiformis*). *Plant Physiology* **101**, 729-44.

Ko, T.-P., Ng, J.D., Day, J., Greenwood, A. & McPherson, A. (1993b) Determination of three crystal structures of canavalin by molecular replacement. *Acta Crystallographica Section D* **49**, 478-89.

Lawrence, M.C., Suzuki, E., Varghese, J.N., Davis, P.C., Van Donkelaar, A., Tulloch, P.A. & Colman, P.M. (1990) The three-dimensional structure of the seed storage protein phaseolin at 3 Å resolution. *EMBO Journal* **9**, 9-15.

Lawrence, M.C., Izard, T., Beuchat, M., Blagrove, R.J. & Colman, P.M. (1994) Structure of phaseolin at 2.2 Å resolution: implications for a common vicilin/legumin structure and the genetic engineering of seed storage proteins. *Journal of Molecular Biology* **238**, 748-76.

Leal, I. & Misra, S. (1993) Molecular cloning and characterization of a legumin-like storage protein cDNA of Douglas fir seeds. *Plant Molecular Biology* **21**, 707-15.

McHenry, L. & Fritz, P.J. (1992) Comparison of the structure and nucleotide sequences of vicilin genes of cocoa and cotton raise questions about vicilin evolution. *Plant Molecular Biology* **18**, 1173-6.

Newbigin, E.J., De Lumen, B.O., Chandler, P.M., Gould, A., Blagrove, R.J., March, J.F., Kortt, A.A. & Higgins, T.J. (1990) Pea convicilin: structure and primary sequence of the protein and expression of a gene in the seeds of transgenic tobacco. *Planta* **180**,461-70.

Newton, C.H., Flinn, B.S. & Sutton, B.C.S. (1992) Vicilin-like seed storage proteins in the gymnosperm interior spruce (*Picea glauca / engelmanii*). *Plant Molecular Biology* **20**, 315-22.

Ng, J.D., Ko, T.-P. & McPherson, A. (1993) Cloning, expression, and crystallization of jack bean (*Canavalia ensiformis*) canavalin. *Plant Physiology* **101**, 713-28.

Nicholls, A.J., Sharp., K.A., & Honig, B. (1991) Protein folding and association: insights from interfacial and thermodynamic properties of hydrocarbons. *Proteins: Structure Function & Genetics* **11**, 281-96.

Nielsen, S.S., Deshpande, S.S., Hermodson, M.A. & Scott, M.P. (1988) Comparative digestibility of legume storage proteins. *Journal of Agricultural & Food Chemistry* **36**, 896-902.

Osborne, T.B. (1909) The Vegetable Proteins, Longmans, Green & Co., New York.

Osborne, T.B. (1924) The Vegetable Proteins, 2nd Edition, Longmans, Green & Co., New York.

Patel, S., Cudney, R. & McPherson, A. (1994) Crystallographic characterization and molecular symmetry of edestin, a legumin from hemp. *Journal of Molecular Biology* **235**, 361-3.

Reichelt, R., Schwenke, K.D., König, T., Pähtz, W. & Wangermann, G. (1980) Electron microscopic studies for estimation of the quaternary structure of the 11S globulin (helianthinin) from sunflower seed (*Helianthus annuus* L.). *Biochimie und Physiologie der Pflanzen*. **175**, 653-63.

Rerie, W.G., Whitecross, M.I. & Higgins, T.J. (1990) Nucleotide sequence from an A-type legumin gene from pea. *Nucleic Acids Research* **18**, 655.

Richardson, J.S. (1981) The anatomy and taxonomy of protein structure. *Advances in Protein Chemistry* **34**, 167-339.

Rossmann, M.G., Abad-Zapatero, C., Murthy, M.R.N., Liljas, L., Jones, T.A. & Strandberg, B. (1983) Structural comparisons of some small spherical plant viruses. *Journal of Molecular Biology* **165**, 711-36.

Schepman, A.M.H., Wichertjes, T. & Van Bruggen, E.F.J. (1972) Visibility of subunits in crystals of oligomeric proteins. Electron microscopy and optical diffraction of edestin and excelsin. *Biochimica et Biophysica Acta* **271**, 279-85.

Schlesier, B., Bassüner, R., Nong, V.H. & Müntz, K. (1990) The cDNA derived primary structure of two distinct legumin A subunit precursors from field bean *(V. faba* L.). *Nucleic Acids Research* **18**, 7146.

Sebastiani, F.L., Farrell, L.B., Schuler, M.A. & Beachy, R.N. (1990) Complete sequence of a cDNA of α-subunit of soybean β-conglycinin. *Plant Molecular Biology* **15**, 197-201.

Shewry, P.R. (1995) Plant storage proteins. *Biological Reviews* **70**, 375-426.

Shutov, A.D., Kakhovskaya, I.A., Braun, H., Bäumlein, H. & Müntz, K.(1995) Legumin-like and vicilin-like seed storage proteins: evidence for a common single-domain ancestral gene. *Journal of Molecular Evolution* **41**,1057-69.

Simon, A.E., Tenbarge, K.M., Scofield, S.R., Finkelstein, R.R. & Crouch, M.L.(1985) Nucleotide sequence of a cDNA clone of Brassica napus 12S storage protein shows homology with legumin from Pisum sativum. *Plant Molecular Biology* **5**, 191-201.

Sims, T.L. & Goldberg, R.B. (1989) The glycinin Gy1 gene from soybean. *Nucleic Acids Research* **17**, 4386.

Slightom, J.L., Drong, R.F., Klassy, R.C. & Hoffman, L.M. (1985) Nucleotide sequences from phaseolin cDNA clones: the major storage proteins from *Phaseolus vulgaris* are encoded by two unique gene families. *Nucleic Acids Research* **13**, 6483-98.

Tanchak, M.A., Giband, M., Potier, B., Schernthaner, J.P., Dukiandjiev, S. & Altosaar, I. (1995) Genomic clones encoding 11S globulins in oats (*Avena sativa* L.). *Genome* **38**, 627-34.

Thanh, V.H., Tumer, N.E. & Nielsen, N.C. (1989) The glycinin Gy2 gene from soybean. *Nucleic Acids Research* **17**, 4387.

Utsumi, S., Gidamis, A.B., Mikami, B. & Kito, M. (1993) Crystallization and preliminary X-ray crystallographic analysis of soybean proglycinin expressed in *E. coli*. *Journal of Molecular Biology* **233**, 177-8.

Utsumi, S., Matsumura, Y. & Mori, T. (1997) Structure-function relations of soy proteins, *in Food proteins and their applications* (eds S. Damodaran & A. Paraf), Dekker, New York, pp. 257-91.

Utsumi, S., Gidamis, A.B., Takenaka, Y., Maruyama, N., Adachi, M. & Mikami, B. (1996) Crystallization and X-ray analysis of normal and modified recombinant soybean proglycinins. Three-dimensional structure of normal proglycinin at 6 Å resolution, in *Macromolecular interactions in food technology* (eds N. Parris, A. Kato, L.K. Creamer and J. Pearce), American Chemical Society, Washington, D.C., pp. 257-70.

Von der Haar, R.A., Allen, R.D., Cohen, E.A., Nessler, C.L. & Thomas T.L. (1988) Organization of the sunflower 11S storage protein family. *Gene* **74**, 433-43.

Wind, C. & Häger, K.-P. (1996) Legumin encoding sequences from the redwood family (Taxodiaceae) reveal precursors lacking the conserved Asn-Gly processing site. *FEBS Letters* **383**, 46-50.

Wright, D.J. (1988) The seed globulins - part II, in *Developments in Food Proteins - 6* (ed. B.J.F. Hudson), Elsevier, London, pp. 119-78.

Xue, Z.T., Xu, M.L. Shen, W., Zhuang, N.L., Hu, W.M. & Shen, S.C. (1992) Characterization of a Gy4 glycinin gene from soybean *Glycine max* cv. forrest. *Plant Molecular Biology* **18**, 897-908.

23. Origin and Evolution of Seed Storage Globulins

ANDREI D. SHUTOV[1] and HELMUT BÄUMLEIN[2]
[1]*Laboratory of Protein Chemistry, State University of Moldova, MD-2009 Kishinev, Moldova.* [2]*Institut für Pflanzengenetik und Kulturpflanzenforschung D-06466, Gatersleben, Germany.*

1. INTRODUCTION

Seed storage globulins have frequently been adopted as an experimental system to study various basic aspects of plant molecular biology. Because of the importance of seed globulins for human nutrition, most data are available only from a rather limited set of crop plants. For a deep understanding of the extant structural and functional features of storage globulins, however, it would be valuable to know more about their origin and evolution. In this chapter we therefore consider: what proteins have been recruited as a progenitor molecule and what are the principle evolutionary milestones to convert this progenitor into a seed storage globulin. Some aspects of these questions have been addressed previously (Bäumlein *et al.*, 1995). In addition, there is evidence for the existence of a superfamily consisting of structurally related storage and non-storage proteins (Braun *et al.*, 1996). The recognition of several new members of this superfamily, and especially the characterisation of a fern spore-specific vicilin-like protein (Shutov *et al.*, 1998a), has essentially confirmed and extended our knowledge concerning the evolutionary pathway of seed storage globulins.

In this chapter we will review the available data, starting from a brief description of structural and evolutionary relationships between storage globulin genes, as the basis for a more detailed characterisation of the storage globulin superfamily and a separate storage globulin lineage.

P.R. Shewry and R. Casey (eds.), *Seed Proteins*, 543–561.

2. VICILIN AND LEGUMIN HAVE EVOLVED FROM A COMMON SINGLE-DOMAIN ANCESTRAL GENE

Structural similarity between vicilin and legumin subunits was first reported by Argos *et al.* (1985). These findings have been further corroborated by interfamily alignments based on vicilin tertiary structure (Lawrence *et al.*, 1994; Shutov *et al.*, 1995; and see chapter 22). A vicilin-like structural model for legumins has been proposed (Lawrence *et al.*, 1994) and generally confirmed by preliminary data on the prolegumin tertiary structure (Utsumi *et al.*, 1996). Therefore, the primary structures of the both vicilin and legumin subunits can be described in terms of higher order structures.

The subunits of both vicilin and legumin are composed of structurally equivalent *N*-terminal and *C*-terminal domains (*N*- and *C*-domains, respectively) (see also chapter 22). Each domain consists of a β-barrel of antiparallel β-strands followed by a group of α-helices (Lawrence *et al.*, 1994). The sequence region of the β-barrel is most conserved in the domains of both vicilin and legumin. Sequence regions outside the β-barrel structure are rather variable and comprise a hypervariable *N*-terminal extension of the *N*-domain of M_r 70,000 vicilin subunits as well as a hypervariable interdomain linker of legumin subunits. The loop between the β-strands E and F is always extended by a hydrophilic variable insertion I1 (Shutov *et al.*, 1995) in the *C*-domain of vicilins and the *N*-domain of legumins. Moreover, the legumin *N*-domain tolerates a hydrophilic variable insertion I2 (Shutov *et al.*, 1995) located at the junction between the β-barrel and α-helix structures. The other domains (legumin *C*-domain and vicilin *N*-domain) contain an extremely short EF loop which never tolerates insertions.

The structural resemblance of all four domains (*N*- and *C*-domains of both vicilin and legumin subunits) led to the conclusion that the domains of both storage globulins have evolved from a common single-domain ancestral gene (Shutov *et al.*, 1995). Either a duplication (Gibbs *et al.*, 1989) or triplication (Shutov *et al.*, 1995) of the putative single-domain ancestor was suggested as a way of creating either a two- or a three-domain storage globulin ancestor. The triplication scheme was based on cross-wise structural resemblance of legumin *N*-domains and vicilin *C*-domains, both containing an extended EF loop and a conserved CPE/H motif inside the inserts I1 (Shutov *et al.*, 1995).

3. A METHODOLOGICAL STRATEGY TO STUDY THE EVOLUTION OF SEED STORAGE GLOBULINS

Gene families encoding structurally related but functionally distinct proteins have been defined as a superfamily (Hood *et al.*, 1985). Non-globulin seed proteins, such as 2S albumins and the prolamins of cereals, can be traced back to enzyme inhibitors and have been united in a superfamily (Shewry and Tatham, 1990; and see chapter 24). The set of structurally similar storage and non-storage proteins given in Table 1 also implies the existence of a superfamily for seed storage globulins. Some of the members represent molecules with rather distantly related structures and a simple sequence analysis alone might be insufficient to distinguish between convergence and common ancestry. To be confident that common ancestry, rather than other causes (Doolittle, 1994), has led to the observed similarities between the members of the superfamily, the following criteria have been used: a) the existence of a common structural framework characteristic for all proteins of the superfamily; b) statistically significant similarity between sequences over the total region of this structural framework; c) the occurrence of intermediates between distantly related proteins and d) independent features such as exon/intron patterns, common promoter motives involved in gene regulation, similar tissue- and developmental specificity, etc.

Three approaches were used to follow the relationships between the proposed members of the storage globulin superfamily:

a) Choice of the sequence region. Since rather distantly related proteins were compared in this study, the choice of the most conserved sequence region, common to all members of the superfamily, is essential. Thus, based on previous results of interdomain/interfamily alignments (Lawrence *et al.*, 1994; Shutov *et al.*, 1995; Bäumlein *et al.*, 1995; and see chapter 22), a region of approximately one hundred residues covering almost the entire vicilin/legumin β-barrel structures (β-strands A to I) was used for all comparisons. The *C*-terminal border of this region corresponds to the position of the fourth intron (intron D) of gymnosperm legumin genes (Shutov *et al.*, 1998b).

b) Selection of suitable sequences. The most reliable interfamily relationships can be obtained by comparing sequences that exhibit most ancient features. Gymnosperm storage globulin sequences that meet these requirements (Braun *et al.*, 1996; Shutov *et al.*, 1998b) were used preferentially, whereas the number of angiosperm sequences was strongly limited. For the same reason we have also included a limited set of germin-like sequences that exhibit ancient features in our study.

Table 1 A brief characterisation of the members of a seed storage globulin superfamily. [a]The sequence which exhibits the highest interfamily similarity to those in the first column is indicated. The numbers refer to the statistical significance of similarity. The SD-values correspond to the C-terminal domain sequence region selected for similarity comparisons (see Fig. 1). The averaged SD-values are given when more than two sequences are included into the comparison. [b]Representative legumin and euvicilin sequences as well as spherulin sequences are specified in Fig. 1.
[c]Representative sets of eugermins (M63223, L15737, U01963, U75187, Y09915, Y09917) and germin-like proteins (D45425, M93041, U21743, U75189, U75191, U75192, U75194, U75207, U81162, U81163, U95036, X84786, X91975, X93171) were selected on the basis of multiple alignment of essentially all available germin/germin-like sequences (Shutov *et al.*, 1998a).

Protein/cluster: a brief description	Interfamily similarity[a]	
		SD's
Legumins[b]	MVP:	9.7
Euvicilins[b]	SBP:	11.7
Vicilin-like proteins		
SBP, sucrose binding protein (L06038) - *Glycine max* seeds - membrane associated - sucrose transport/immobilization?- osmotic pressure adjustment/desiccation tolerance?	euvicilins:	11.7
Lectin - *Mucuna derringiana* seeds - binds to Forssman antigen and Forssman disaccharide - homodimer of M_r 54,000 subunits - structural relation to SBP (up to 72% of identities)	SBP:	9.0
MP27/32 precursor (D29803) - *Cucurbita sp.* seeds - legumin-like processing generating separate MP27 and MP32 bound to protein body membrane - desiccation tolerance?	Gea8:	18.8
Gea8 (U62395) - *Daucus carota* - expression during somatic an zygotic embryogenesis - vicilin-like intron/exon pattern - promoter RY *cis*-element	MP27:	18.8
MVP (Z54364) - *Matteuccia struthiopteris* - spore specific - vicilin-like intron/exon pattern - storage protein? - desiccation tolerance?	MP27:	14.9
Eugermins[c]		
plant proteins - single-domain structure - homopentamer - oxalate oxidase activity - cell wall and cytoplasm component - promoter RY *cis*-element - desiccation/hydration tolerance?	spherulins:	10.0
Germin-like proteins[c]		
plant proteins - single-domain structure - multifunctional (oxalate oxidase, circadian expansion of plant cells, protection against infection, auxin and heavy metal binding)	spherulins:	9.0
Fungal spherulins[b]		
Physarum polycephalum - single-domain structure - expressed during spherulation process of myxomycetes - desiccation/hydration tolerance?	eugermins:	10.0

c) Multiple alignment based on tertiary structure. The alignment of vicilin *N*- and *C*-terminal domain sequences was manually improved according to the canavalin (Ko *et al.*, 1993) and phaseolin (Lawrence *et al.*, 1994) tertiary structures. The resulting alignment was used as a basis for comparing two- and single-domain members of the superfamily.

4. A SEED STORAGE GLOBULIN SUPERFAMILY

4.1 A brief characterisation of the members.

In this section we describe the members of the seed storage globulin superfamily. For a crude classification they are divided into legumins, euvicilins and vicilin-like proteins as well as single-domain proteins (Table 1).

i) Legumins: Legumin genes have recently been characterised in several species of lower angiosperms (Fischer *et al.*, 1996) and gymnosperms, including a number of different conifers (Häger *et al.*, 1996) and three gnetalean species (Shutov *et al.*, 1998b). Comparative analysis of essentially all legumin sequences, both at the amino acid and nucleotide levels, revealed four distinct sequence clusters (Shutov *et al.*, 1998b). They comprise legumins of gnetalean species, all other gymnosperms, a group of most ancient angiosperms described by Fischer *et al.* (1996) and a diverse group of other angiosperms representing 13 orders. The latter group covers a wide phylogenetic distance within angiosperms ranging from *Magnoliales* to *Asterales* and *Poales*. In general, the evolutionary pathway of legumin genes reconstituted by the neighbour-joining distance method follows the conventional phylogeny of spermatophytes.

ii) Euvicilins: In contrast to legumins, the set of available vicilins is rather poor. Only two gymnosperm vicilin sequences with obvious ancient features have been characterised (Newton *et al.*, 1992; Braun *et al.*, 1996). Angiosperm vicilins with rather derived sequence features represent crops from only three plant orders. However, the vicilin tertiary structure is known (Ko *et al.*, 1993; Lawrence *et al.*, 1994) and is essential for primary structure comparisons.

The structural and functional relationship of vicilin-related seed storage proteins of different species is well known (Lawrence *et al.*, 1994; Braun *et al.*, 1996). Otherwise, several non-storage proteins closely similar to vicilins have also been described (see below). To prevent any confusion it seems reasonable to distinguish both protein groups by separate designations. We therefore propose the term *euvicilin* for all vicilin seed storage proteins and *vicilin-like protein* for several other proteins of either non-storage or unknown function (see Table 1), which share obvious vicilin-like sequence features with euvicilins.

iii) Vicilin-like proteins: The sequence of a spore-specific protein from the fern *Matteuccia struthiopteris* (MVP) reveals most characteristic features of the vicilin family, including a convicilin-like *N*-terminal extension and a rather short interdomain linker. Possibly, the mature MVP is a conventional vicilin-like trimeric molecule (Shutov *et al.*, 1998a), in contrast to other vicilin-like proteins (see Table 1) such as the sucrose binding protein (SBP) from *Glycine max* (Grimes *et al.*, 1992), lectin from *Mucuna derringiana* seeds (Mo and Goldstein, 1994) and MP27/MP32, which are associated with the protein body membrane in *Cucurbita sp.* seeds (Inoue *et al.*, 1995). The Gea8 protein from *Daucus carota* is highly similar to MP27/32 precursor even in the region of the interdomain linker (10.1 SD's from random) and contains a similar potential processing site (Shutov *et al.*, 1998a).

iv) Single-domain proteins: It was shown that the sequences of the *Triticum* germin gf-2.8 and the *Physarum* spherulin 1a can be aligned with vicilin/legumin domain sequences (Bäumlein *et al.*, 1995). Recently, a number of new germin-like sequences have been deposited in the EMBL Data Base, including auxin-binding proteins (U81162 and U81163). All these germin-like sequences exhibit a high degree of intrafamily similarity within the sequence region most related to storage globulins (Bäumlein *et al.*, 1995). Nevertheless, neighbour-joining analysis of these sequences aligned with spherulin sequences revealed that a cluster of at least six germins, named here as *eugermins*, is even more closely related to spherulins than to most other *germin-like proteins* (Table 1). This interfamily similarity suggests that eugermins represent the most ancient features descending from a common prokaryotic eugermin/spherulin ancestor (Shutov *et al.*, 1998a). Both spherulins and eugermins are thought to be involved in cellular (perhaps cell-wall) responses to desiccation, hydration and osmotic stress (Lane *et al.*, 1991).

4.2 Members of the storage globulin superfamily share a common primary structure framework.

When the amino acid sequence region corresponding to the euvicilin β-barrel structure (β-strands A - I) is selected for analysis, the domain sequences of all representative members of the superfamily can be aligned with confidence (Fig. 1). The conservativity profile shown in Fig. 2 reveals the primary structure framework common to a wide set of 40 domain sequences. Thus, the vicilin-like β-barrel seems to be a core domain structure, characteristic of all the members of the superfamily.

4.3 Intron/exon patterns of legumin and vicilin genes.

Independent evidence for the common ancestry of legumins and vicilins is based on the interfamily comparison of intron/exon patterns of their genes (Shutov *et al.*, 1995). The genes of both legumin and vicilin subunits comprise introns A, B and C, D which interrupt the *N*- and *C*-domain coding region, respectively (Fig. 3). An additional intron, found only in vicilin genes, is located between introns B and C. Intron positions of legumin and vicilin genes are either identical (introns A, B and C) or similar (intron D) (Shutov *et*

Fig. 1 Multiple alignment of amino acid sequences of the members of the storage globulin superfamily. The following sequences were aligned. ***Legumins:*** *Dio*, *Dioscorea caucasica* (X95510); *Mgn*, *Magnolia salicifolia* (X82465); *Pin*, *Pinus strobus* (Z11486); *Gnk*, *Ginkgo biloba* (Z50778); *Wlw*, *Welwitschia mirabilis* (Z50780); ***Euvicilins:*** *Trt*, *Triticum aestivum* (M81719); *Can*, *Canavalia ensiformis* (X59467); *Pic*, *Picea glauca* (X63191); *Zam*, *Zamia furfuraceae* (Z50791); ***Vicilin-like proteins:*** SBP, sucrose binding protein of *Glycine max* (L06038); MVP, *Matteuccia struthiopteris* (Z54364); MP27/32 precursor of *Cucurbita sp.* (D29803), Gea8 of *Daucus carota* (U62395); Grm1, *Triticum aestivum* germin gf-2.8 (M63223), Grm2, *Hordeum vulgare* oxalate oxidase (L15737); Sph 1a and Sph 1b, *Physarum polycephalum* spherulins (M18428 and M18429). Residues forming β-strands A-I in the subunit of *Canavalia ensiformis* euvicilin are underlined. Residues of the five major conserved positions are printed in bold letters. The motif PHXXPRATEI common to single-domain proteins and MVP is also highlighted in bold typeface. Residues regarded as inserts either in Grm/Sph sequences or inside interstrand loops of two-domain proteins are shown as lower case letters. All these inserts as well as part of the EF loop labelled as (:) were omitted from the alignment for further analysis.

al., 1998b). Consistent with the concept of domain shuffling (Stoltzfus *et al.*, 1994), both introns B and D disrupt gene sequences that encode the protein region closely adjacent to the β-barrel/α-helix junction. It is possible that the β-barrel structural module of storage globulins limited by a B/D-like intron was recruited in total from a putative single-domain ancestor before its duplication. This is supported by the finding that the fourth intron D of legumin genes is shifted by only 2 bp in comparison to that of the single-domain *Physarum* spherulin gene (Bäumlein *et al.*, 1995). The introns A and

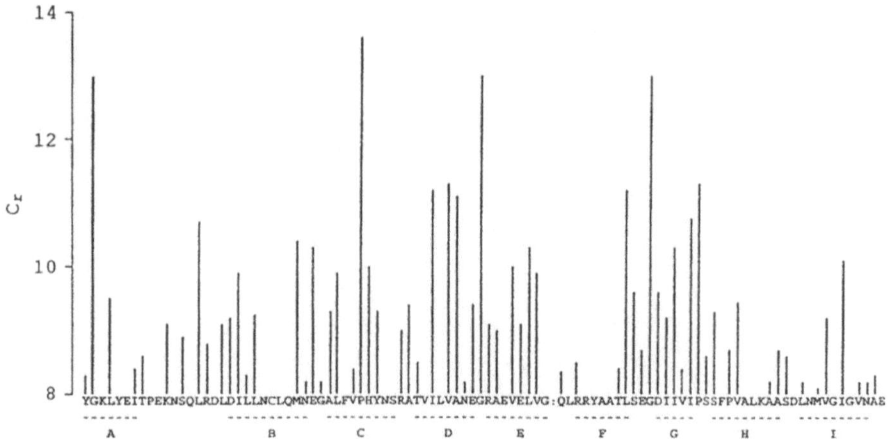

Fig. 2 Members of the storage globulin superfamily share a common primary structure framework: the profile of conservativity. The *N*- and *C*-terminal domain sequences of eight euvicilins/vicilin-like proteins and eight legumins were aligned with eight eugermins/spherulins. The complete *C*-terminal domain sequence of *Canavalia* euvicilin can be aligned with the other sequences with the single deletion of the EF loop extension designated as (:). The set of sequences selected for the construction of the profile is specified in Fig. 1 (excluding Gea8 which is highly similar to MP27/32 precursor). The additional sequences are legumins from *Cryptomeria japonica* (X95542), *Asarum europaeum* (X95508) and *Sagittaria sagittifolia* (Y09116), euvicilin from *Theobroma cacao* (A20606) and four eugermins (Y09915, Y09917, U01963, and U75187). To build the profile, each residue from a position *r* was compared with each of the other residues at the same position and the average of the corresponding $m_r(i,j)$ values (MDM-78 of Dayhoff, 1978 rescaled to positive matrix elements by the addition of the constant value of 8) was taken as a measure of the conservativity C_r of the position *r*.

C, which interrupt the gene region encoding the β-barrel structural module, are interpreted as later insertion events. Their absence in the *Welwitschia* legumin gene (Fig. 3) might reflect its ancient feature (Shutov *et al.*, 1998b). It is remarkable that the intron D, which is exclusively found in gymnosperm legumin genes, has been lost prior to, or at the very beginning of, angiosperm diversification (Häger *et al.*, 1996; Shutov *et al.*, 1998b). All the positions of introns hitherto known in the genes of the vicilin-like Gea8 protein and the MVP match precisely those of euvicilin genes. Taken together, the conservation of the intron/exon patterns provides further evidence for the common ancestry of the members of the superfamily.

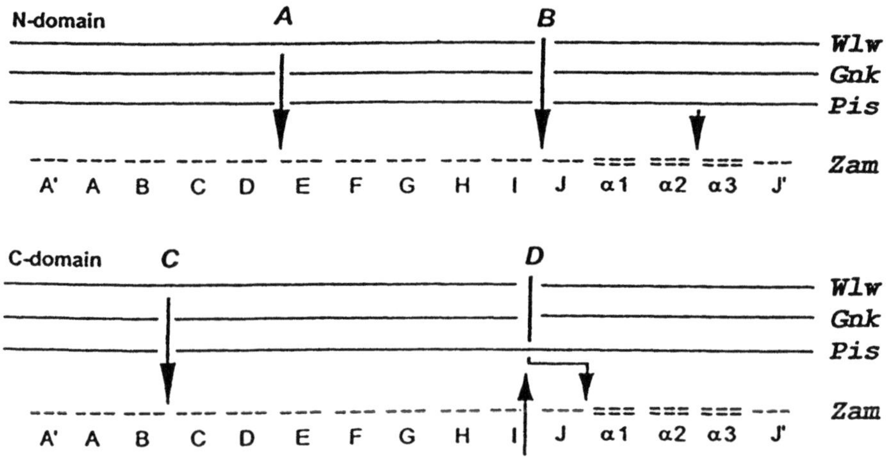

Fig. 3 Gene intron/exon structures of the members of storage globulin superfamily. The scheme is projected into aligned domain sequences (Fig. 1). Positions of introns are labelled by (↓). The secondary structures characteristic of vicilins (Ko *et al.*, 1993) are shown. The third intron of vicilins is not found in legumin genes. The positions of the introns *A*, *B* and *C* match precisely in the aligned sequences of legumins and vicilins/vicilin-like proteins. The position of the second intron in the spherulin gene (↑) is shifted by 2 bp upstream compared to the legumin gene intron *D* (Bäumlein *et al.*, 1995). The demonstrated intron/exon patterns refer to the legumin genes of *Welwitschia mirabilis* (*Wlw*, Z55780), *Ginkgo biloba* (*Gnk*, Z50778) and *Pisum sativum* (*Pis*, X00634) and vicilin gene of *Zamia furfuraceae* (*Zam*, Z50791).

4.4 *C*-domains of two-domain proteins represent the most ancient sequence features.

A similarity dendrogram based on the β-barrel sequences shown in Fig. 1 reveals three basic sequence clusters (Fig. 4). The upper two clusters N and C represent the *N*- and *C*-domain sequences of legumins and euvicilins/vicilin-like proteins. The formation of these two clusters is independent of the number of storage globulin sequences that are included in the comparative analysis. The third basic cluster S unites single-domain members of the superfamily. The formation of the clearly distinct N and C clusters supports the proposed duplication event in the evolution of seed storage globulins (Gibbs *et al.*, 1989) and argues against our previously proposed triplication model (Shutov *et al.*, 1995). Furthermore, the cluster C seems to be even more closely related to plant/fungal non-storage proteins of the cluster S, than to the cluster N. Obviously, the *C*-domains of storage globulins/vicilin-like proteins represent the most ancient sequence features and are therefore most suitable for a reliable evolutionary analysis. A recent detailed analysis of the evolutionary relationships between legumin sequences led to the same conclusion (Shutov *et al.*, 1998b).

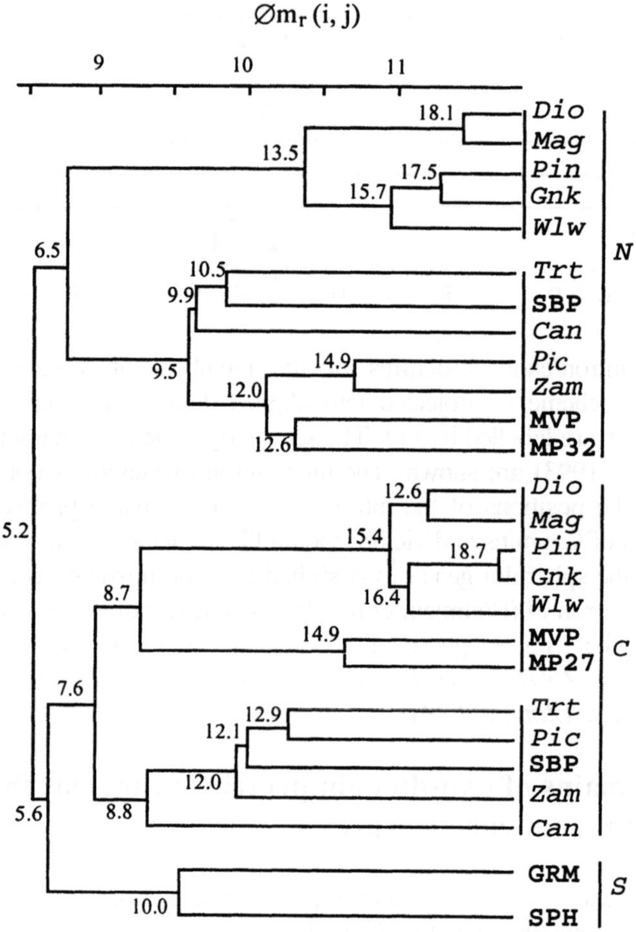

Fig. 4 Similarity relationships between members of storage globulin superfamily. The dendrogram was built on the basis of the alignment of Fig. 1. N/C, the clusters of *N-/C*-domain sequences, respectively. S, the cluster of single-domain germins and spherulins (all the four sequences specified in Fig. 1 are included in the analysis). Numbers at the branches refer to the statistical significance of sequence similarity (SD's from random). Averaged SD-values are given when more than two sequences are included into the comparison.

4.5 MVP represents an evolutionary missing link.

Sequence comparisons reveal that among all known two-domain members of the storage globulin superfamily, the MVP is most similar to single-domain germins and spherulins. Thus, contemporary germins and spherulins contain the structurally conserved decapeptide motif **PH(I/T)HPRATEI** thought to be essential for function (Lane *et al.*, 1991). A similar decapeptide **(PHWNPRATEI)** can be found at the identical position in the MVP C-terminal domain (Fig. 1). Obviously, this ancient sequence element was modified in extant euvicilins as well as in the MP27/32 precursor and the Gea8 protein.

Although the total sequence of MVP exhibits obvious vicilin-like, rather than legumin-like, features (Shutov *et al.*, 1998a) and the MVP *N*-domain sequence alone is clearly vicilin-like as well, its *C*-domain appears to be legumin-like (Fig. 4). The MVP C-domain probably represents an intermediate evolutionary step before the bifurcation of the storage globulin branch into separate legumin and vicilin families. The structural relationship of MVP to both two- and single-domain members of the superfamily can be demonstrated by the profiles shown in Fig. 5.

In conclusion, the reported similarities suggest that MVP is the best known representation of the proposed two-domain ancestor of both legumin and vicilin and could, therefore, be considered as a molecular missing link between single- and two-domain members of the superfamily.

5. EVOLUTIONARY PATHWAY OF SEED STORAGE GLOBULINS

As extensively discussed above, a quantitative evaluation of distantly related sequences has to be restricted to the analysis of the most conserved structural features as exhibited by the β-barrel region of *C*-domains limited by the intron D. Moreover, to study the evolutionary pathway of seed globulins separately from that of non-storage proteins, it is necessary to use only those sequences which obviously share a common evolutionary lineage with storage globulins. Thus, fungal spherulins, which are less similar to storage globulins than plant germins (Bäumlein *et al.*, 1995), obviously belong to a separate evolutionary branch and have to be excluded. At present, it is an open question whether or not the vicilin-like MP27/32, Gea8 protein and even SBP share a common evolutionary lineage with storage globulins. Therefore, for evolutionary analysis we selected a limited set of storage globulins with most ancient features, the MVP as a structural intermediate between storage

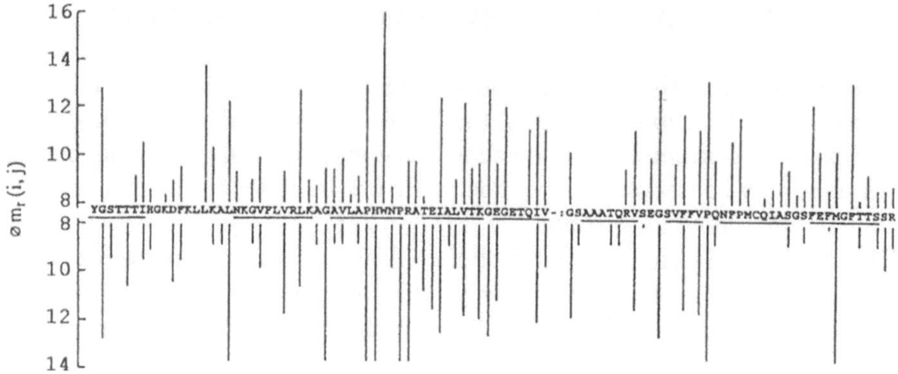

Fig. 5 *Matteuccia* vicilin-like protein (MVP) shares a common primary structural framework with both single-domain (lower profile) and two-domain (upper profile) proteins of the storage globulin superfamily. The sequence of the *C*-domain region of MVP is shown between the profiles. Positions of the β-strands from the *C*-domain of *Canavalia* euvicilin projected on the MVP sequence (see Fig. 1) are underlined. Six eugermin sequences specified in Table 1 and the *C*-domain sequences of five legumins and four euvicilins (see Fig. 4) together with SBP were used to build the lower and upper profiles, respectively. To build the profiles, a residue from a position *r* of the MVP sequence was compared with residues at the same position of other sequences involved in the comparison and the average of the corresponding $m_r(i,j)$ values is given to characterise the position *r* of the MVP sequence.

globulins and single-domain progenitor as well as an eugermin as an obvious outgroup.

In general, the gene tree constructed by the neighbour-joining distance method (Fig. 6) exhibits the same clusters as the amino acid similarity dendrogram (Fig. 4). Based on both trees we propose the following hypothetical scenario for the evolution of seed storage globulins as well as the other members of the storage globulin superfamily (Fig. 7).

A putative prokaryotic ancestor molecule, possibly involved in basic desiccation/hydration processes, has evolved into structurally and functionally similar plant eugermins and fungal spherulins. A storage globulin ancestor was recruited from a limited set of developmentally-regulated proteins specific for tissues that can tolerate desiccation, prolonged periods of storage and fast rehydration, as exemplified by spherules or spores.

Conceivably, the β-barrel of nine antiparallel β-strands A-I (see Fig. 1) provided an optimum structural basis for further development of the globulin storage function. Furthermore, this β-barrel structure was used by eugermins and germin-like proteins for a variety of different functions such as enzymatic

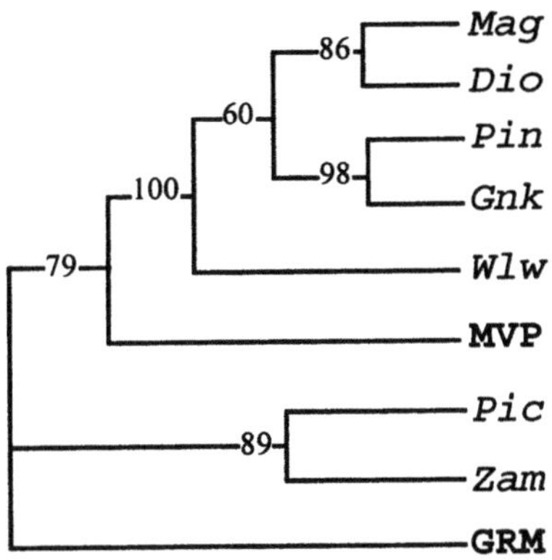

Fig. 6 Reconstruction of the putative evolutionary pathway of storage globulins by the neighbour-joining distance method. The nucleotide sequences representing the most conserved region of the *C*-domains of two-domain proteins (see Fig. 1) were analysed, omitting of third codon positions. Germin gf-2.8 (M63223) was used as an outgroup sequence. Numbers along branches refer to bootstrap values (100 replicates).

activity (as oxalate oxidase) as well as auxin- and heavy metal-binding proteins (Table 1). At the gene level the β-barrel-encoding exon is flanked by the intron D, which could be involved in the shuffling of the suitable protein module (Shutov *et al.*, 1998b).

Subsequently, an ancient single-domain eugermin gene became internally duplicated, generating a gene that encodes a two-domain protein. The principle features of this proposed two-domain ancestor might have been similar to the fern spore-specific vicilin-like protein. As discussed above, the sequence of MVP is more closely related to germins/spherulins than any other two-domain member of the storage globulin superfamily. Moreover, it exhibits both vicilin- and legumin-like features. Thus, MVP obviously retained the most ancient sequence features among the two-domain members and can be regarded as the best known representation of the proposed ancestral molecule at a stage just after duplication but before the divergence into the separate lineages of vicilin-like proteins, euvicilins and legumins.

Assuming that MVP features are ancient, the presence of an extended convicilin-like *N*-terminal sequence in MVP supports the previous hypothesis of an ancient origin of those extensions in M_r 70,000 vicilin subunits (Dure, 1990; Shutov *et al.*, 1996). Thus, the conversion of the spore-specific MVP into seed-specific vicilins includes either the loss (M_r 50,000 subunits) of the *N*-terminal extension or its further enlargement and modification (M_r 70,000 subunits) as well as the formation and extension of the insertion I1 between the β-strands E and F of the *C*-terminal domain.

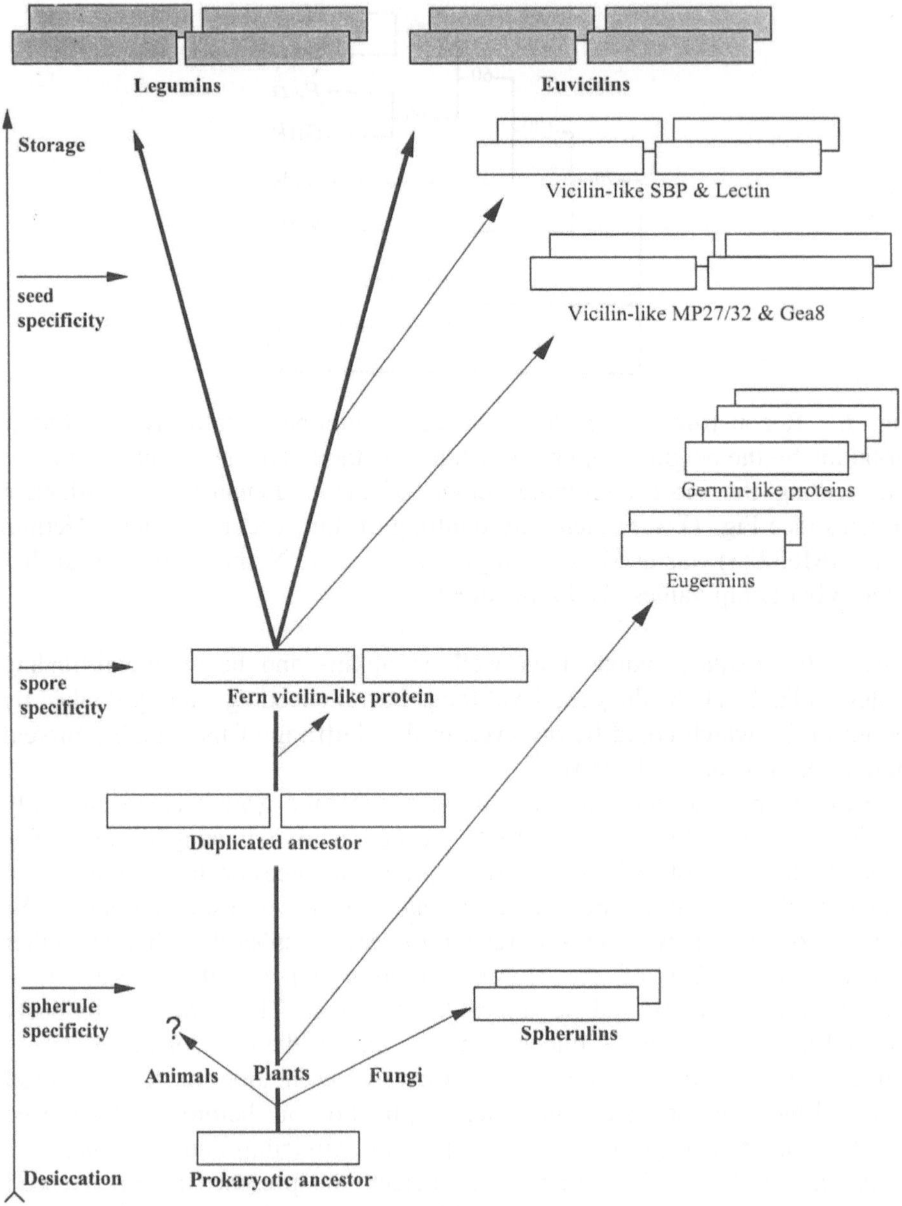

Fig. 7 A hypothetical scheme of storage globulin evolution. See also Chapter 22.

The evolutionary pathway of a separate legumin family then includes the formation of the processing site within the interdomain linker as well as the interdomain disulphide bridge. It still remains unclear whether the extended EF loop of legumin *N*-terminal domains shares a common origin with that of vicilin *C*-terminal domains as proposed previously (Shutov *et al.*, 1995; 1996).

Further diversification of both the legumin and vicilin family includes the partial loss of introns as well as the formation of subfamilies and non-functional pseudogenes.

The known structural and deduced functional diversity (Table 1) of single- and two-domain members of the storage globulin superfamily suggests the existence of more, as yet unidentified proteins, which share a common ancestry with the proposed single-domain storage globulin progenitor.

The discussion of the structural relationships between the members of the superfamily, as well as their evolutionary pathway, can be augmented by several other considerations, as follows.

Spherulins have been suggested to be involved in desiccation/hydration processes during spherulation (Bernier *et al.*, 1987; Lane *et al.*, 1991). Spherulation, which is in principle similar to seed maturation, leads to an ordered removal of water from the tissues and finally to the differentiation of a dormant cyst suitable for dispersion and able to survive harsh environmental conditions. Similarly, apoplast-associated germins have been suggested to play a role in embryo-specific desiccation/hydration processes (Lane *et al.*, 1991).

The only known function of seed storage globulins is to provide resources for the growth of the seedling. Following the outlined evolutionary pathway from desiccation-related proteins to seed storage globulins, we speculate that the storage function, and mechanisms of seed specific gene regulation, can also be traced back to functional and regulatory principles connected to basic cellular desiccation and rehydration.

Among the members of the superfamily SBP might be of special interest to seed physiologists. SBP is thought to function as a sucrose transport protein also in seeds (Grimes *et al.*, 1992; Overvoorde *et al.*, 1996). Because of the involvement of SBP in developmentally-regulated processes associated with storage and desiccation protection, we speculate that its sucrose binding activity might be important in a related physiological context. Thus, it might be that SBP is involved in the adjustment of the osmotic pressure related to the protection of seed tissues from desiccation *via* physiological mechanisms different from those of eugermins.

According to the evolutionary scheme (Fig. 7), SBP has evolved from a separate vicilin progenitor. It is highly similar to eugermins even in non-conserved regions (Braun *et al.*, 1996). *In situ* hybridisation reveals that the histological expression pattern of the *Vicia faba* SBP in the storage parenchyma of the cotyledons closely resembles that of vicilin genes (H. Weber, pers. comm.). Therefore, as an alternative view it is possible to suggest that SBP has been derived from an euvicilin subunit. Thus, it is tempting to speculate that the sucrose binding activity of SBP could be connected to the replacement of a highly conserved proline by isoleucine in the *C*-terminal domain of euvicilin. This exchange might create structural hindrances for the intersubunit contacts and prevent the trimerisation of SBP subunits (Braun *et al.*, 1996). In this connection it would be of interest to know if vicilin subunits still retain a residual sucrose binding activity. Considering the abundance of vicilin in protein bodies even a slight activity could be of physiological relevance to seed sucrose metabolism.

In histological terms, the expression of seed protein-like genes seems to be quite variable. Related genes of angiosperms have been reported to be expressed in the triploid secondary endosperm as well as in the storage cotyledons of the diploid embryo, both tissues being the result of a double fertilisation event. In gymnosperms, seed globulin genes are expressed in the haploid primary endosperm, which is homologous to the gametophytic megaprothallium of heterosporic ferns. The MVP gene demonstrates the expression of a seed globulin-like gene in fern spores. Although spores and seeds differ dramatically in their structure and ontogeny, they fulfil the same function of providing a desiccation-tolerant, dormant structure essential for the propagation and dispersion of the next generation. At present it is not known if the storage function of vicilin and legumin is restricted to spermatophytes. Possibly, the MVP also represents a functional intermediate which has acquired a storage function, like euvicilins.

The developmental pathway of late embryogenesis and seed maturation includes storage, dormancy and desiccation tolerance. These functions are controlled by the products of several regulator genes like *Vp1* in maize (McCarty, 1995) and *ABI3*, *FUS3* and *LEC1* in *Arabidopsis* (Meinke, 1995). For instance, it was shown that the activity of the legumin-like cruciferin gene of *Arabidopsis*, as well as that of the *Vicia faba* legumin B4 promoter, depend on an intact *FUS3* gene product (Bäumlein *et al.*, 1994). Moreover, a functionally important RY *cis*-element (CATGCATG) has been identified in the promoters of many genes which become active during late embryogenesis and seed maturation, including storage globulin genes (Bäumlein *et al.*, 1992; Lelievre *et al.*, 1992). Similar *cis*-elements can be found in the gene promoters of eugermins (F. Bernier, pers. comm.) and SBP of *Vicia faba* (U. Heim, pers. comm.). It would be interesting to characterise these elements and their interaction with the regulator genes in non-seed plants.

6. CONCLUSIONS

In conclusion, we have suggested that legumin and vicilin, the two-domain seed storage globulins, were derived from a common single-domain ancestor *via* a duplication event. Extant single-domain plant germins and fungal spherulins might best represent the basic features of this ancestor. Several non-storage two-domain proteins, very similar to vicilins, have been identified, suggesting the existence of a superfamily of structurally related, but functionally different proteins that share common ancestry with storage globulins. Extensive sequence comparisons and the analysis of gene structures led to the reconstruction of an evolutionary pathway for seed storage globulins. The model proposes that a prokaryotic ancestral molecule, involved in basic cellular desiccation/hydration processes, has evolved into functionally similar fungal spherulins and plant germins. A storage globulin ancestor was recruited from a limited set of developmentally-regulated proteins specific for tissues which tolerate desiccation and rapid rehydration. Among these proteins, a germin-like progenitor provides an optimum structural basis for further development of storage globulin function. This development includes a duplication event which creates a two-domain progenitor, best reflected by an extant fern spore-specific vicilin-like protein, as the common ancestor of the storage globulins vicilin and legumin as well as several related non-storage proteins

ACKNOWLEDGEMENTS

We would like to thank Prof. Dr. K. Müntz for stimulating discussions and encouragement.

REFERENCES

Argos, P., Narayana, S.V.L. and Nielsen, N.C. (1985) Structural similarity between legumin and vicilin storage proteins from legumes. *EMBO Journal* **2**, 1111-7.

Bäumlein, H., Nagy, I., Villarroel, R., Inzé, D. and Wobus, U. (1992) *Cis*-analysis of a seed protein gene promoter: The conservative RY repeat CATGCATG within the legumin box is essential for tissue-specific expression of a legumin gene. *Plant Journal* **2**, 233-9.

Bäumlein, H., Miséra, S., Luerßen, H., Kölle, K., Horstmann, C., Wobus, U. and Müller, A.J. (1994) The *FUS3* gene of *Arabidopsis thaliana* is a regulator of gene expression during late embryogenesis. *Plant Journal* **6**, 379-87.

Bäumlein, H., Braun, H., Kakhovskaya, I.A. and Shutov, A.D. (1995) Seed storage proteins of spermatophytes share a common ancestor with desiccation proteins of fungi. *Journal of Molecular Evolution* **41**, 1070-5.

Bernier, F., Lemieux, G. and Pallotta, D. (1987) Gene families encode the major encystment-specific proteins of *Physarum polycephalum* plasmodia. *Gene* **59**, 265-77.

Braun, H., Czihal, A., Shutov, A.D. and Bäumlein, H. (1996) A vicilin-like seed of cycads: Similarity to sucrose binding proteins. *Plant Molecular Biology* **31**, 35-44.

Bäumlein, H., Nagy, I., Villarroel, R., Inzé, D. and Wobus, U. (1992) *Cis*-analysis of a seed protein gene promoter: The conservative RY repeat CATGCATG within the legumin box is essential for tissue-specific expression of a legumin gene. *Plant Journal* **2**, 233-9.

Bäumlein, H., Miséra, S., Luerßen, H., Kölle, K., Horstmann, C., Wobus, U. and Müller, A.J. (1994) The *FUS3* gene of *Arabidopsis thaliana* is a regulator of gene expression during late embryogenesis. *Plant Journal* **6**, 379-87.

Bäumlein, H., Braun, H., Kakhovskaya, I.A. and Shutov, A.D. (1995) Seed storage proteins of spermatophytes share a common ancestor with desiccation proteins of fungi. *Journal of Molecular Evolution* **41**, 1070-5.

Bernier, F., Lemieux, G. and Pallotta, D. (1987) Gene families encode the major encystment-specific proteins of *Physarum polycephalum* plasmodia. *Gene* **59**, 265-77.

Braun, H., Czihal, A., Shutov, A.D. and Bäumlein, H. (1996) A vicilin-like seed of cycads: Similarity to sucrose binding proteins. *Plant Molecular Biology* **31**, 35-44.

Dayhoff, M.O. (1978) A model of evolutionary change in proteins: Matrices for detecting distant relationships, in *Atlas of protein sequences and structures* (ed. M.O.Dayhoff), Academic Press, NY, vol. 5, suppl. 3, pp. 345-58.

Doolitle, R.F. (1994) Convergent evolution: the need to be explicit. *TIBS* **19**, 15-18.

Dure, L. III (1990) An unstable domain of the vicilin genes of higher plants. *New Biologist* **2**, 487-93.

Gibbs, P.E.M., Strongin, K.B. and McPherson, A (1989) Evolution of legume seed storage proteins - a domain common to legumins and vicilins is duplicated in vicilins. *Molecular and Biological Evolution* **6**, 614-23.

Grimes, H.D., Overvoorde, P.J., Ripp, K., Franceschi, V.R. and Hitz, W.D. (1992) A 62kD sucrose binding protein is expressed and localised in tissues actively engaged in sucrose transport. *Plant Cell* **4**, 1561-74.

Fischer, H., Chen, L. and Wallisch, S. (1996) The evolution of angiosperm seed proteins: A methionine-rich legumin subfamily present in lower angiosperm clades. *Journal of Molecular Evolution* **43**, 399-4.

Häger, K.P., Müller, B., Wind, C., Erbach, S. and Fisher, H. (1996) Evolution of legumin genes: Loss of an ancestral intron at the beginning of angiosperm diversification. *FEBS Letters* **387**, 94-8.

Hood, L., Kronenberg, M. and Hunkapillar, T. (1985) T cell antigen receptors and the immunoglobulin supergene family. *Cell* **40**, 225-9.

Inoue, K., Motozaki, A., Takeuchi, Y., Nishimura, M. and Hara-Nishimura, I. (1995) Molecular characterisation of proteins in protein-body membrane that disappear most rapidly during transformation of protein bodies into vacuoles. *Plant Journal* **7**, 235-43.

Ko, T.P., Ng, J.D. and McPherson, A. (1993) The three-dimensional structure of canavalin from jack bean (*Canavalia ensiformis*). *Plant Physiology* **101**, 729-44.

Lane, B.G., Bernier, F., Dratewka-Kos, E., Shafai, R., Kennedy, T.D., Pyne, C., Munro, J.R., Vaughan, T., Walters, T. and Altomare, F. (1991) Homologies between members of the germin family in hexaploid wheat and similarities between these wheat germins and certain *Physarum* spherulins. *Journal of Biological Chemistry* **266**, 10461-9.

Lawrence, M.C., Izard, T., Beuchat, M., Blagrove, R.J. and Colman, P.M. (1994) Structure of phaseolin at 2.2 A resolution. *Journal of Molecular Biology* **238**, 748-76.

Lelievre, J.M., Oliveira, L.O. and Nielsen, N.C. (1992) 5'-CATGCAT-3' elements modulate the expression of glycinin genes. *Plant Physiology* **98**, 387-91.

McCarty, D.R. (1995) Genetic control and integration of maturation and germination pathways in seed development. *Annual Review of Plant Physiology and Plant Molecular Biology* **46**, 71-93.

Meinke, D.W. (1995) Molecular genetics of plant embryogenesis. *Annual Review of Plant Physiology and Plant Molecular Biology* **46**, 369-94.

Mo, H. and Goldstein, I.J. (1994) Isolation and characterisation of a Forssman antigen-binding lectin from velvet bean (*Mucuna derringiana*) seeds. *Glycoconjugate Journal* **11**, 424-31.

Newton, C.H., Flinn, B.S. and Sutton, B.C. (1992) Vicilin-like seed storage proteins in the gymnosperm interior spruce (*Picea glauca/engelmanii*). *Plant Molecular Biology* **20**, 315-22.

Overvoorde, P.J., Frommer, W.B. and Grimes, H.D. (1996) A soybean sucrose binding protein independently mediates nonsaturable sucrose uptake in yeast. *Plant Cell* **8**, 271-80.

Shewry, P. R. and Tatham, A. S. (1990) The prolamin storage proteins of cereal seeds: Structure and evolution. *Biochemical Journal* **267**, 1-12.

Shutov, A.D., Kakhovskaya, I.A., Braun, H., Bäumlein, H. and Müntz, K. (1995) Legumin-like and vicilin-like seed storage proteins: Evidence for a common single-domain ancestral gene. *Journal of Molecular Evolution* **41**, 1057-69.

Shutov, A.D., Kakhovskaya, I.A., Bastrygina, A.S., Bulmaga, V.P., Horstmann, C. and Müntz, K. (1996) Limited proteolysis of β-conglycinin and glycinin, the 7S and 11S storage globulins from soybean (*Glycine max* (L.) Merr.). Structural and evolutionary implications. *European Journal of Biochemistry* **241**, 221-8.

Shutov, A.D., Braun, H., Chesnokov, Yu. V. and Bäumlein, H. (1998a) A vicilin-like gene is specifically expressed in fern spores. Evolutionary pathway of seed storage globulins. *European Journal of Biochemistry* **252**, 79-89.

Shutov, A.D., Braun, H., Chesnokov, Yu. V., Horstmann, C., Kakhovskaya, I.A. and Bäumlein, H. (1998b) Sequence peculiarity of gnetalean legumin-like seed storage proteins. *Journal of Molecular Evolution* (in press).

Stoltzfus, A., Spencer, D.F., Zuker, M, Logsdon, J.M. and Doolitle, W.F. (1994) Testing the exon theory of genes: The evidence from protein structure. *Science* **265**, 202-7.

Utsumi, S., Gidamis, A.B., Takenaka, Y., Maruyama, N., Adachi, M. and Mikami, B. (1996) Crystallisation and X-ray analysis of normal and modified recombinant soybean proglycinins. Three-dimensional structure of normal proglycinin at 6 A resolution, in *Macromolecular Interactions in Food Technology* (eds N.Parris, A.Kato, L.K.Creamer and J.Pearce) ASC Symp. Ser. 650, pp.257-70.

24. The 2S Albumin Storage Proteins

PETER R. SHEWRY and MAYA J. PANDYA
IACR-Long Ashton Research Station, Department of Agricultural Sciences, University of Bristol, Long Ashton, Bristol BS41 9AF, UK

1. INTRODUCTION

The term albumin was initially applied to plant proteins which resembled hen egg albumin in being coagulable but was subsequently restricted to proteins which were soluble in water and coagulable by heat (Osborne, 1924). Osborne (1924) also recognised a problem in defining albumins which remains to this day: direct water extracts of tissues will contain low concentrations of salts which need to be removed by dialysis in order to obtain a true albumin fraction. Osborne concluded that albumins are generally present in small quantities in seeds and listed four types which had been characterized in most detail: leucosin from barley, wheat and rye, legumelin from a range of legumes (including pea, soybean and cowpea), phaselin from kidney bean and ricin from castor bean.

Subsequent studies provided detailed information on some seed albumins, but it was only in the 1980's that attention became focused on albumins as a major storage protein fraction.

The catalyst for this explosion in interest was a study by Youle and Huang (1981). They used sucrose gradient ultracentrifugation to compare salt-soluble protein fractions from a botanically diverse selection of twelve species including two legumes (peanut and lupin) and one monocot (*Yucca*, Liliaceae). The ultracentrifugation profiles showed three major peaks, with 2S and 11S components present in all species and 7S components in all except three species (Table 1). The 2S components were defined as albumins on the basis of their solubility in water and their amounts were estimated, based on the ultracentrifugation profiles, to range from between 20 and 60% of the total soluble protein. The values would clearly be lower if expressed on a total seed protein basis but the results nevertheless demonstrated that the 2S albumins form a major group of storage proteins in many species.

P.R. Shewry and R. Casey (eds.), Seed Proteins, 563–586.

Table 1 The proportions of 2S, 7S and 11S proteins in salt-soluble fractions from a range of seeds and the sulphur amino acid content of the 2S fraction. Data from Youle and Huang (1981).

Family	Species	Proteins[1]			Amino Acids[2]	
		2S	7S	11S	Cys	Met
Compositae	sunflower (*Helianthus annuus*)	62	0	38	6.59	2.84
Cruciferae	mustard (*Brassica* spp.)	62	0	38	9.01	2.60
Linaceae	linseed (*Linum usitatissimum*)	42	0	58	8.17	1.93
Rosaceae	almond (*Prunus amygdalus*)	25	15	60	nd	nd
Leguminosae	lupin (*Lupinus polyphyllus*)	38	26	36	6.15	0.66
	peanut (*Arachis hypogaea*)	20	6	74	1.81	1.84
Cucurbitaceae	cucumber (*Cucumis sativus*)	56	17	27	8.92	3.79
Lecythidaceae	Brazil nut (*Bertholletia excelsa*)	30	9	61	13.11	17.33
Betulaceae	hazelnut (*Corylus* spp.)	28	12	60	10.27	3.69
Liliaceae	yucca (*Yucca* spp.)	27	16	57	9.29	3.17
Euphorbiaceae	castor bean (*Ricinus communis*)	44	14	42	8.5	1.60
Malvaceae	cotton (*Gossypium hirsutum*)	33	35	32	7.74	1.80

[1]proportions of proteins calculated from areas under sucrose gradient ultracentrifugation curves.
[2]expressed as mol %
nd = not determined

Youle and Huang (1981) also reported amino acid compositions of the 2S albumin fractions, demonstrating that many contain high levels of cysteine. In addition, the fraction from Brazil nut contained over 17 mol% methionine (Table 1). Since legume seeds and forage legumes used as livestock feed are generally deficient in sulphur-containing amino acids these results provided a considerable stimulus to the analysis of 2S albumins as sources of improved feed quality for genetic engineering. In addition, the 2S albumins have been used to study basic processes of gene regulation and the processing and sorting of proteins within the cell. These studies have revealed the 2S albumins as a fascinating group of proteins with an intriguing range of diversity within an essentially conserved structure.

2. CHARACTERISTICS

Although the pioneering study of Youle and Huang (1981) showed the presence of 2S proteins in seeds of a wide range of plants it did not establish that these constituted a family of structurally-related proteins. Subsequent detailed characterization of examples from a range of dicotyledonous species has confirmed that the 2S albumins do indeed form a major family of homologous proteins. No examples have been characterized so far from monocotyledonous species, despite the report of Youle and Huang (1981) that they are present in *Yucca* (Liliaceae), but related proteins have been characterized from spores of the ferns *Onoclea* and *Matteuccia* (Templeman *et al.*, 1987, 1988; Templeman and De Maggio, 1990; Rödin and Rask 1990) where they are located in protein bodies and digested on germination.

As with other groups of storage proteins, the 2S albumins show a high level of polymorphism, as illustrated by the analyses of fractions from hybrid and inbred lines of sunflower in Fig. 1. Such polymorphic variants may show considerable differences in their structures and properties, with even greater differences between the 2S albumins present in different species. Nevertheless it is possible to define the structure and properties of a "typical" 2S albumin. This is a heterodimeric protein comprising subunits of about 30-40 and 60-90 residues, respectively. It is synthesised as a precursor protein which is co-translationally transported into the lumen of the endoplasmic reticulum. After the formation of four intra-chain disulphide bonds, involving eight conserved cysteine residues, the folded protein is transported to the vacuole. Here proteolytic processing occurs, yielding the two subunits which remain associated by two disulphide bonds with the loss of short linker and flanking sequences. Such "typical" 2S albumins are exemplified by napin, conglutin δ and the Brazil nut albumin in Fig. 2.

However, a range of variant types of 2S albumin also occur which differ from the "typical" 2S albumins in their structure or mode of biosynthesis. For example, the 2S albumins of sunflower are not cleaved into large and small subunits, while in both sunflower and castor bean two mature albumins may be synthesised as a single proprotein (Fig. 2). Variation in disulphide bond formation also occurs with an additional free cysteine residue being present in conglutin δ of lupin whereas the pea albumin subunits PA1a and PA1b do not appear to be associated by disulphide bonds. There is also considerable variation in the levels of cysteine and methionine present in 2S

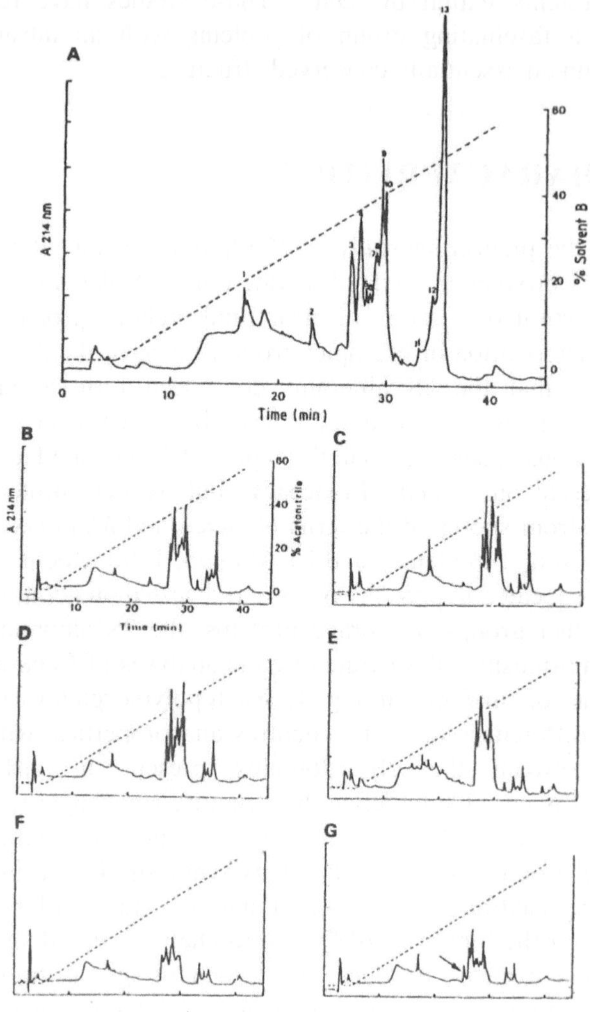

Fig. 1 Polymorphism of 2S albumin fractions from sunflower, revealed by reversed-phase HPLC. The major components in Sunbred 246 (part A) are numbered. Parts B-G show similar fractions from a range of lines: B, hybrid Asmer 4; C, VIR 111; D, VIR 130; E, VIR 143; F, VIR 169; G, VIR 369. The VIR lines are inbred lines from the collection at the N.I. Vavilov Institute of Plant Industry, St. Petersburg.Taken from Anisimova *et al.* (1995) with kind permission from Kluwer Academic Publishers.

albumin fractions (Table 1), with individual methionine-rich proteins being present in sunflower, cotton, Brazil nut and *Amaranthus* and cysteine-rich proteins in pea and quinoa (Table 2).

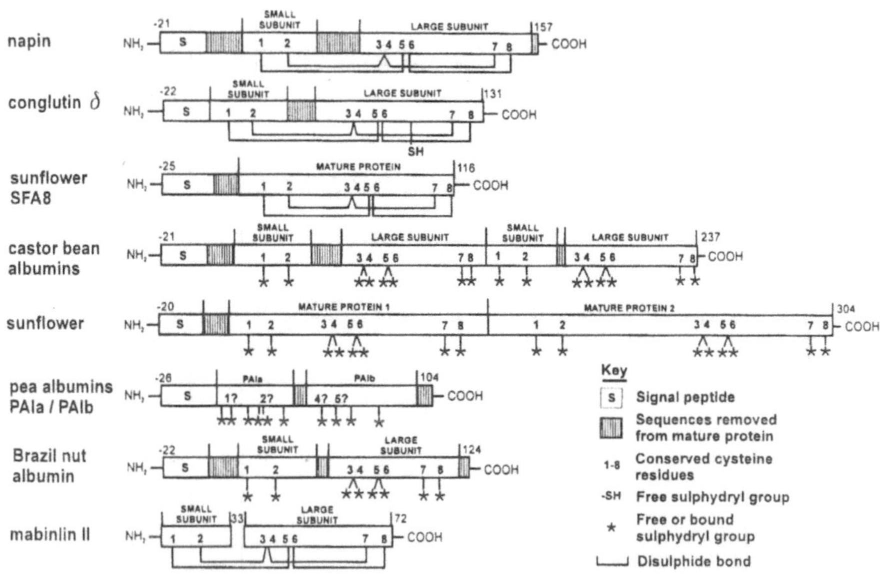

Fig. 2 Schematic depictions of various types of 2S albumin, indicating their origins from precursor proteins and their disulphide structures. Conserved cysteine residues are numbered 1-8 and free sulphydryl groups shown as -SH. Cysteine residues whose behaviour is unknown are indicated by *. The precise correspondence between the cysteines in the pea albumins PA1a and PA1b and those in the other albumins is not known and potentially conserved residues are indicated by the number and ? The origin of the two mabinlin II subunits is not known but they are presumed to derive from a single precursor protein as do the other heterodimeric albumins. The pea albumins PA1a and PA1b differ from the dimeric albumins in that the two subunits do not remain associated by inter-chain disulphide bonds. Based on data and alignments in Crouch *et al.* (1983); Ericson *et al.* (1986); Lilley and Inglis (1986); Higgins *et al.* (1986); Ampe *et al.* (1986); Altenbach *et al.* (1987); Gaylor *et al.* (1990); Irwin *et al.* (1990); Kortt *et al.* (1991); Liu *et al.* (1993); Egorov *et al.* (1996) and Thoyts *et al.* (1996). Redrawn in part from Shewry (1995) with kind permission of Cambridge University Press.

Vacuolar processing of the 2S albumin proproteins to release the mature proteins appears to occur by proteolytic cleavage at the *C*-terminal side of asparagine residues (Fig. 3) but the precise number and characteristics of the enzymes involved is still uncertain (Hara-Nishimura *et al.*, 1993; D'Hondt *et al.*, 1993).

Figure 2 summarises the sequences of a range of "typical" and variant 2S albumins, indicating their sites of proteolytic processing and disulphide

Table 2 Amino acid compositions (mol %) of S-rich and "typical" 2S albumins.

	Methionine-Rich				Cysteine-Rich		"Typical"				
	Brazil nut	Sunflower SFA8	Cotton Mat 5	Amaranthus MRP1	Quinoa 2S	Pea PA1a+b	Lupin Conglutin δ	Oilseed Rape Napin pN2	Castor bean 2S	Sunflower *N*-terminal	Sunflower *C*-terminal
Asp	1.0	2.9	3.9	} 6.7	} 5.3	7.8	3.4	0	2.0	0.8	3.6
Asn	2.0	5.8	1.0			6.7	5.1	1.7	2.0	4.1	3.0
Thr	0	1.0	3.9	1.9	0.4	1.1	0	3.3	2.0	0	1.2
Ser	6.9	4.9	5.9	4.3	2.0	8.9	8.5	5.0	7.1	5.7	6.0
Glu	14.9	9.7	8.8	} 21.2	}30.8	5.6	18.8	3.3	7.1	7.3	8.3
Gln	11.9	7.8	15.7			0	18.8	21.5	23.2	31.7	25.6
Pro	5.9	5.8	2.0	3.9	4.5	11.1	2.6	10.7	4.0	3.3	8.9
Gly	5.9	4.9	7.8	6.7	7.4	10.0	3.4	5.8	9.1	4.1	4.2
Ala	1.0	2.9	2.9	3.2	1.4	4.4	0.9	5.8	4.0	3.3	0.6
Cys	7.9	7.8	7.8	6.5	15.6	11.1	7.7	6.6	8.1	6.5	4.8
Val	0	1.9	1.0	2.2	0.7	2.2	0.9	5.0	3.0	6.5	3.6
Met	18.8	15.5	9.8	18.6	0.6	0	0	2.5	2.0	0.8	1.2
Ile	1.0	2.9	4.9	0.8	1.3	6.7	4.3	3.3	3.0	3.3	4.2
Leu	5.0	8.7	4.9	3.6	2.2	3.3	8.5	6.6	6.1	6.5	3.0
Tyr	1.0	2.9	2.0	3.9	2.9	3.3	0.9	0.8	1.0	0	6.0
Phe	0	0	0	1.3	1.2	5.6	1.7	2.5	3.0	1.6	2.4
His	2.0	2.9	2.9	1.2	7.6	2.2	2.6	3.3	1.0	2.4	0
Lys	0	3.9	7.8	6.4	0.9	5.6	2.6	7.4	2.0	1.6	0.6
Arg	14.9	6.8	6.9	7.3	15.2	3.3	9.4	4.1	10.1	9.8	12.5
Trp	0	1.0	0	0	nd	1.1	0	0.8	0	0.8	0.6
Total residues	101	103	102	nd	nd	90	117	121	99	123	168
Reference	1	1	2	3	4	5	6	7	8	9	9

References: 1, Altenbach and Simpson, 1990 based on sequences in Sun *et al.*, 1987; Altenbach *et al.*, 1987; Kortt *et al.*, 1991; 2, Galau *et al.*, 1982; 3, Segura-Nieto *et al.*, 1994; 4, Brinegar, 1997; 5, Higgins *et al.*, 1986; 6, Lilley and Inglis, 1986; 7, Crouch *et al.*, 1983; Ericson *et al.*, 1986; 8, Irwin *et al.*, 1990; 9, Thoyts *et al.*, 1996, based on the two proteins encoded by the *N*-terminal and *C*-terminal parts of the cDNA pHAO.

```
                SIGNAL PEPTIDE                  PROSEQUENCE      SMALL SUBUNIT
Pumpkin  MARLTSIIAL FAVALLV--A DAYAYRTTIT TVEVEE--NR QGRE--ERCR QMSAREE-LR   53
R. com.I MAKLIPTIAL VSVFLFIIAN ASFAYRTTIT TIEIDESKGE REGSSSQQCR QEVQRKD-LS   59
R. com.II ---------- ---------- ---------- ---RQTRTNP SQ----QGCR GQIQEQQNLR   23
B. exc.  MAKI--SVAA AALLVLMALG HATAFRATVT TTVVEEE-NQ ------EECR EQMQRQQMLS   50
A. tha.  MANKLFLVCA -ALALCFLLT NASIYR-TVV EFEEDDATNP IGP-KMRKCR KEFQKEQHLR   57
B. nap.  MANKLFLVSA -TLAFFFLLT NASIYR-TVV EFDEDDATNS AGPFRIPKCR KEFQQAQHLR   58

                                                    LARGE SUBUNIT
Pumpkin  SCEQYLRQQS R--------- ---DVLQMRG IENPWRREGG S--FDECCRE LKNVDEECRC   99
R. com.I SCERYLRQSS SRRSTG---- --EEVLRMPG DENQQQESQQ ---LQQCCNQ VKQVRDECQC  110
R. com.II QCQEYIKQQV S--GQG---- -------PRR SDNQE----- -RSLRGCCDH LKQMQSQCRC   64
B. exc.  HCRMYMRQQM E--ESP---- ----YQTM-- ---P-RRGME -PHMSECCEQ LEGMDESCRC   94
A. tha.  ACQQLMLQQA RQGRSD---- ---EFDFEDD MENPQGQQQE QQLFQQCCNE LRQEEPDCVC  110
B. nap.  ACQQWLHKQA MQSGGGPSWT LDGEFDFEDD MENPQGPQQR PPLLQQCCNE LHQEEPLCVC  118

Pumpkin  DMLEE----I AREEQRQA-- RGQE--GRQM LQKARNLPSM CGIRPQR-CD F---------  141
R. com.I EAIKY----I AEDQIQQGQL HGEE--SERV AQRAGEIVSS CGVR----C- M---------  150
R. com.II EGLRQA---I EQ-QQSQGQL QGQD--VFEA FRTAANLPSM CGVSPTE-CR F---------  108
B. exc.  EGLRMM---M MRMQQEEMQP RGEQ--MRRM MRLAENIPSR CNLSPMR-CP MGGSIAG-F-  146
A. tha.  PTLKQAAKAV RLQGQHQPMQ ------VRKI YQTAKHLPNV CDIPQVDVCP FNIPSFPSFY  164
B. nap.  PTLKGASKAV KQQIQQQGQQ QGKQQMVSRI YQTATHLPKV CNIPQVSVCP FQKTMPGPSY  178
```

Fig. 3 Comparison of the amino acid sequences and processing sites of 2S albumin precursor proteins from pumpkin (*Cucurbita* sp., Cucurbitaceae), castor bean (*R. com.* I and *R. com.* II, corresponding to the *N*- and *C*-terminal mature proteins, respectively), Brazil nut (*B. exc*), *Arabidopsis* (*A. tha* corresponding to AT2S1 of Krebbers *et al.*, 1988) and rape seed (*B. nap*). Taken from Hara-Nishimura *et al.*, (1993) incorporating sequences from Sharief and Li, 1982; Crouch *et al.*, 1983; Ampe *et al.*, 1986; Ericson *et al.*, 1986; Altenbach *et al.*, 1987, 1992; Krebbers *et al.*, 1988 and Irwin *et al.*, 1990. Complete amino acid sequences of subunits determined by direct sequencing are underlined. *N*-terminal sequences determined directly for the pumpkin subunits are underlined with arrowheads. Cysteine residues are indicated by asterisks. The closed arrowhead indicates the cleavage site of the signal peptide of the pumpkin 2S albumin precursor and the open arrowheads the two sites for vacuolar processing, adjacent to asparagine residues. Reproduced with kind permission of Blackwell Science Limited.

structures. These and other examples are discussed in detail in the following section.

3. CHARACTERIZED 2S ALBUMINS

3.1. Brassicaceae (Cruciferae)

The 2S albumins of brassicas and related species have been widely studied, particularly the napins of oilseed rape (*Brassica napus*). The first detailed study of napins was carried out by Lönnerdal and Janson (1972) who showed the presence of at least four components, with M_r between 12,000 and 14,000 and each consisting of two polypeptide chains. Ericson *et al.* (1986) sequenced a mature napin protein, showing that the large and small subunits consisted of 86 and 29 residues with M_r of about 9,000 and 4,000, respectively. Comparison of this sequence with further sequences derived from cloned cDNAs and genes (Crouch *et al.*, 1983; Josefsson *et al.*, 1987; Schofield and Crouch, 1987) indicated that this protein is probably typical of the major group of napins. Such comparisons also showed that napins are synthesised as preproproteins with the subsequent loss of *N*-terminal, *C*-terminal and linker regions in addition to the signal peptide (see Crouch *et al.*, 1983) (Fig. 2). Although most napins from *B. napus* are similar to that sequenced by Ericsson *et al.* (1986) with an M_r of about 14,500, at least two lower M_r napins (about 12,500) are also present (Monsalve and Rodriguez, 1990). Direct protein sequencing has shown that the large subunits of these proteins contain only 74-75 and 79 residues (Monsalve *et al.*, 1991). Southern blot analyses have shown the presence of about 10 (Josefsson *et al.*, 1987) and over 16 (Schofield and Crouch, 1987) napin genes per haploid genome, although it should be borne in mind that *B. napus* is an amphidiploid hybrid between *B. oleracea* and *B. rapa*. Napins contain eight conserved cysteine residues which have been shown, by NMR spectroscopy, to form the same pattern of disulphide bonds to that in conglutin δ, SFA8 and mabinlin II (see below).

2S albumins related to napins are also present in other cruciferous species including *B. rapa* (turnip), *B. oleracea* (cabbage) (Monsalve and Rodriguez, 1990), yellow mustard (*Sinapis alba*) (Menéndez-Aria *et al.*, 1988; González de la Peña *et al.*, 1993), oriental mustard (*B. juncea*) (González de la Peña *et al.*, 1993), radish (*Raphanus sativus*) (Laroche *et al.*, 1984; Raynal *et al.*, 1991), *Medicago* (Coulter and Bewley, 1990) and *Arabidopsis* (Krebbers *et al.*, 1988). In radish, Raynal *et al.* (1991) characterized three cDNAs representing two subfamilies of napins and estimated that the gene family consisted of a minimum of six members.

The exact 2S albumin gene copy number is known only in *Arabidopsis*, where Krebbers *et al.* (1988) have isolated and sequenced the whole family of four genes. However, comparison with the directly determined protein sequence indicated that only one gene was responsible for most, if not all, of the protein present in the seed. Alignment of the sequences with each other and with the sequences of napin from *B. napus* and 2S albumins from Brazil nut (*Bertholletia excelsa*) and castor bean (*Ricinus communis*) predicted similar patterns of processing with the loss of signal, *N*-terminal, *C*-terminal and linker peptides. This comparison also demonstrated the presence of conserved and variable regions within the mature proteins which were subsequently exploited in "molecular farming" studies (see below).

3.2 Castor bean

The castor bean (*Ricinus communis*, Euphorbiaceae) 2S albumin was the first member of the 2S albumin family to be sequenced, showing the presence of small and large subunits of 34 and 61 residues, respectively (Sharief and Li, 1982). Although long thought to be a "typical" 2S albumin, a more recent study by Irwin *et al.* (1990) showed an unusual mode of synthesis. They isolated a cDNA encoding two heterodimeric albumins, one of which corresponded to the component sequenced by Sharief and Li (1982). The second albumin, encoded by the 5' end of the cDNA, has since been purified and shown to be a major allergen designated Ric cIII (Machado and Silva, 1992). Release of the two heterodimers from the proprotein involves the loss of short peptides: from the *N*-terminus of the whole proprotein and from between the two subunits of each albumin (Fig. 2).

3.3 Leguminosae

The 2S albumin of lupin (*Lupinus angustifolius*), conglutin δ, represents about 20-30% of the total salt-soluble protein fraction and contains about 70% of the total sulphur present in the seed (Lilley, 1986a). It consists of two subunits with M_r determined as 9,401 and 4,597 by direct protein sequencing (Lilley and Inglis, 1986) and predicted as 9,558 and 4,520 from the sequence of a corresponding cDNA (Gayler *et al.*, 1990). These differences are consistent with the presence of multiple forms encoded by a multigene family (Lilley, 1986a; 1986b). Gayler *et al.* (1990) also demonstrated that the two subunits are synthesised as a precursor with the loss of a 13 residue linker peptide during processing (Fig. 2). This processing does not involve the loss of *N*- and *C*-terminal flanking sequences, unlike the situation with some other heterodimeric albumins.

Conglutin δ was the first 2S albumin to be disulphide mapped, establishing the pattern of four conserved disulphide bonds which is now known to be common to most 2S albumins (Lilley and Inglis, 1986) (Fig. 2). In addition to the eight conserved cysteines that form these four bonds, the protein contains an additional cysteine residue at position 45 in the large subunit; this residue appears to be present as a free sulphydryl group as it is readily blocked by *N*-ethyl maleimide (Lilley and Inglis, 1986). The high contribution of conglutin δ to the sulphur content of the whole seed results from the presence of nine cysteines (Table 1) and no methionine-rich forms appear to be present.

Pea (*Pisum sativum*) seeds contain two major albumins, PA1 and PA2, of which PA1 appears to be related to the 2S albumins, although only distantly. It has been reported to contribute about half of the total sulphur in the mature seed although representing less than 10% of the total protein (Higgins *et al.*, 1986). PA1 consists of two proteins, PA1a which has an M_r of about 6,000 and PA1b which has an M_r of about 4,000. They are synthesised as a single proprotein which is cleaved, presumably in the vacuole, to release the two mature proteins with the loss of short linker and *C*-terminal peptides (Fig. 2) (Higgins *et al.*, 1986). Comparisons of cDNA and protein sequences indicated that at least four different PA1 genes are expressed in the developing seed.

Unlike the "typical" 2S albumins, PA1a and PA1b do not appear to be associated to form dimers, being readily separated by chromatography or sedimentation equilibrium ultracentrifugation under non-denaturing conditions (Higgins *et al.*, 1986). They have limited sequence homology with the other 2S albumins, notably in the presence of conserved cysteine residues (Fig. 2), but it is not known if these conserved cysteines form intra-molecular disulphide bonds or remain unpaired.

3.4 Brazil nut

The seeds of *Bertholletia excelsa* (Lecythidaceae) have long been known as a rich source of sulphur-containing amino acids (Osborne, 1924), suitable for dietary supplementation of sulphur-poor legume seeds (Antunes and Markakis, 1977). It therefore came as no surprise when Youle and Huang (1981) demonstrated that the 2S albumin fraction from this species contained over 17 mol % methionine (Table 1). This contrasted with total 2S albumin fractions from other species, two of which (sunflower and cotton) have since been shown to contain some methionine-rich components. Sun *et al.* (1987) subsequently confirmed that the 2S albumins accounted for about 30% of the total protein fraction extracted with buffer containing salt or SDS and showed the presence of two major polypeptides of M_r about 9,000 and 3,000. Their amino acid composition is given in Table 2. Ampe *et al.* (1986) showed that the large subunit occurred in at least six forms, of which four were completely sequenced and shown to consist of 73 residues including 15 methionines. Similar values of 77 total residues and 14 methionines were reported in an independent study (Altenbach *et al.*, 1987). Only one type of small subunit appears to be present (Ampe *et al.*, 1986). The mature heterodimeric albumin is released from an M_r 18,000 precursor by at least three stepwise cleavages (Sun *et al.*, 1987; Altenbach *et al.*, 1987).

Methionine-rich 2S albumins are also present in other nuts of the Lecythidaceae, paradise nut (*Lecythis zabucajo*) and cannonball (*Couroupita quianensis*) (Zuo and Sun, 1996).

3.5 Sunflower

The 2S albumins of sunflower (*Helianthus annuus*, Compositae) account for about a third of the total salt-soluble proteins, the remainder being predominantly 11S globulins (helianthinin) (Mazhar *et al.*, 1998). They can be separated by electrophoresis and RP-HPLC into some 11 to 13 major components (Fig. 1) which vary in occurrence and amount between different genotypes (Anisimova *et al.*, 1995). Kortt and Caldwell (1990) purified and characterized nine individual components, two of which contained about 15 mol% methionine.

The major methionine-rich albumin, SFA8, was subsequently sequenced and a corresponding cDNA isolated (Kortt *et al.*, 1991). This showed that it was synthesised as a preproprotein with a signal sequence and a 13 residue prosequence which was lost from the mature protein (Fig. 2). However, unlike the "typical" 2S albumins, the mature protein was not cleaved into small and large subunits but remained as a single monomeric subunit. The high methionine content derives from sixteen residues out of 103 in the mature protein. Eight conserved cysteine residues are also present and form an identical pattern of disulphide bonds to that described for the dimeric conglutin δ (Egorov *et al.*, 1996).

The other, low-methionine, albumins have not been characterized in as much detail as SFA8. However, all appear to be monomeric proteins with eight conserved cysteines which presumably form four intra-chain disulphide bonds. In addition, most, if not all, appear to be synthesised in pairs as a preproprotein. Three cloned DNAs encoding such preproalbumins have been isolated (Allen *et al.*, 1987; Thoyts *et al.*, 1996) and comparisons with directly determined *N*-terminal sequences indicate that processing involves the loss of a short *N*-terminal prosequence but little or no loss of other sequences (Thoyts *et al.*, 1996) (Fig. 2). The predicted M_r of the albumins encoded by the cDNA of Thoyts *et al.* (1996) are 13,388 (*N*-terminal) and 20,191 (*C*-terminal) which is consistent with the M_r range of 10,000 to 18,000 determined by SDS-PAGE (Kortt and Caldwell, 1996; Anisimova *et al.*, 1995).

3.6 Cottonseed

Galau *et al.* (1992) reported the isolation of cDNAs and genes encoding two alloallelic 2S albumins encoded by the allotetraploid genome of cotton

(*Gorsypium hirsutum*, Malvaceae). The two encoded proteins differed only in three out of 139 residues and the authors estimated that the corresponding mRNAs accounted for about 2% of the total mRNA population in maturation stage embryos. Comparison of the sequences with those of Brazil nut and *Arabidopsis* albumins suggested that the mature protein comprised small and large subunits of 27 and 75 residues, respectively, including 10 methionines.

3.7 Pumpkin

A cDNA for the pumpkin (*Cucurbita* sp, Cucurbitaceae) seed 2S albumin has been isolated (Hara-Nishimura *et al.*, 1993). This appears to encode a typical heterodimeric protein which is processed from a proprotein by cleavage at specific asparagine residues catalysed by a vacuolar protease (Hara-Nishimura *et al.*, 1993) (Fig. 3).

3.8 Pseudocereals

The pseudocereals comprise three botanically unrelated species, *Amaranthus*, buckwheat and quinoa, which are grouped together on account of their small grain-like seeds (see Chapter 19).

Albumins are major storage proteins in seeds of *Amaranthus hypochondriacus*, although their amount has been reported to vary from about 10 to over 60% of the total seed proteins depending on the methods used for their extraction and separation from globulins (see Segura-Nieto *et al.*, 1994). Sedimentation analysis showed the presence of a major albumin fraction of about 1.7S which consisted of polypeptides with M_r ranging from about 4,000 to 20,000 (Segura-Nieto *et al.*, 1992; 1994). Two methionine-rich proteins (MRPs1 and 2) were purified from this fraction, containing 18.6 and 16.6 mol % methionine, respectively (Table 2). However, preliminary *N*-terminal sequence analysis showed no homology with other methionine-rich seed proteins (Segura-Nieto *et al.*, 1994).

Quinoa (*Chenopodium quinoa*) contains a 2S albumin fraction estimated to represent about 35% of the total protein (Brinegar *et al.*, 1996). SDS-PAGE under reducing conditions showed the presence of polypeptides in the M_r range 8-9,000, although the authors acknowledged that smaller components, which could correspond to the small subunits of heterodimeric albumins, may not have been resolved (Brinegar *et al.*, 1996). The amino acid composition of the 2S albumin fraction is characterized by high levels of cysteine (15.6 mol %) but low methionine (0.6 mol %) (Brinegar *et al.*, 1996; Brinegar, 1997) (Table 2).

Radovi□ *et al.* (1996) fractionated the salt-soluble proteins of buckwheat (*Fagopyrum esculentum*) by sucrose density ultracentrifugation. Components of M_r 14-23,000 accounted for about 10% of the total, corresponding to less than 5% of the total seed proteins. The work of Takumi *et al.* (1995) suggested that most of these low M_r proteins are water-soluble albumins but their relationship, if any, to the 2S albumin family of proteins was not established.

More detailed studies of buckwheat albumins were reported by Ono *et al.* (1978). Electrophoresis under non-denaturing conditions gave about 12 components but SDS-PAGE gave only a single band of M_r about 8,000 when separated under reducing conditions with a second band of M_r about 13,000 when separated without reduction. This is consistent with the presence of some heterdimeric proteins comprising large subunits of M_r about 8,000 and small subunits of M_r 4-5,000 which were not resolved by electrophoresis. However, the presence of M_r 8,000 proteins in the unreduced preparations suggests that some single chain albumins were also present.

3.9 *Capparis*

Seeds of *Capparis masaikai* (Capparidaceae) contain an unusual sweet protein, called mabinlin, which shows unusual stability to heating, retaining its sweetness even after boiling for 48 hours (Liu *et al.*, 1993). This protein comprises two chains of 72 and 33 residues and is clearly homologous with 2S albumins. This homology includes the presence of eight conserved cysteine residues which form the same disulphide bonds as in other 2 S albumins (conglutin δ, SFA8, napin) (Nirasawa *et al.*, 1993).

3.10 Oil Palm

Embryos of oil palm (*Elaeis guineensis*, *Palmae*) contain 2S albumins with high cysteine (4.9 mol %) and low methionine (0.7 mol %). Two major components are present, an acidic protein of M_r 22,000 and $pI<6$ and a basic protein of M_r 19,000 and $pI>9$ (Morcillo *et al.*, 1997). These proteins appear to comprise single subunits as in sunflower, but no sequence data are available to confirm their identity as members of the 2S albumin family of proteins.

4. BIOLOGICAL ACTIVITY OF 2S ALBUMINS

The major *in vivo* role of the 2S albumins is clearly as storage proteins, being deposited in protein bodies in the developing seed. Several types of *in vitro* biological activity have been reported but the *in vivo* significance of these is unknown.

4.1. As enzyme inhibitors

The 2S albumins are members of a much wider family of proteins called the "prolamin superfamily" (see Chapter 2). Also falling within this family, and perhaps most closely related to the 2S albumins, is a group of inhibitors of hydrolytic enzymes (α-amylases and serine proteinases) present in cereal grains (see Chapters 25 and 26). Only three members of the 2S albumin group have so far been shown to exhibit similar inhibitory activity. These are all napins, from kohlrabi (*B. napus* var *rapifera*), charlock (*Sinapis arvensis*) and black mustard (*B. nigra*). All showed inhibitory activity against trypsin and subtilisin whereas the charlock protein also inhibited α-chymotrypsin (Svendsen *et al.*, 1989; Svendsen *et al.*, 1994; Genov *et al.*, 1998).

4.2 As antifungal proteins

Terras *et al.* (1992) showed that 2S albumins (napins) from radish (*Raphanus sativus*) inhibited the growth of a number of plant pathogenic fungi, although the concentrations required to give 50% growth inhibition were higher than those of "defensin" proteins characterized in the same study. Their antifungal activity was also antagonised by the presence of cations including K^+ or Ca^{2+}. Further work showed that their antifungal activity was significantly enhanced (by x2 to x33) in the presence of thionins from barley or wheat (see Chapter 30) and that this effect occurred with Gram-positive bacteria as well as filamentous fungi but not with Gram-negative bacteria, yeast or cultured cells (Terras *et al.*, 1993). A similar synergistic effect was observed with isolated small or large subunits of the napins and scanning electron microscopy and measurements of K^+ leakage indicated that the mode of action involved membrane permeabilization.

4.3 As antagonists of calmodulin

Polya *et al.* (1993) purified a family of three calmodulin antagonists from radish (*R. sativus*), using reducing and denaturing conditions. These proteins were phosphorylated by wheat embryo Ca^{2+} - dependent protein kinase and inhibited chicken gizzard calmodulin-dependent myosin light chain kinase using a synthetic peptide substrate. Mass spectroscopy showed M_r of 4,537, 4,543 and 4,532 and *N*-terminal sequencing showed high homology with the small subunits of napins. The biological significance of this activity is not known but Polya *et al.* (1993) suggested that the albumin small chains might play a role in the regulation of processes occurring during seed germination.

4 As sweet proteins

Mabinlin II from seeds of *Capparis masaikai* (mabinlang) is one of a small number of sweet proteins of plant origin, the best known being thaumatin from arils of *Thaumatococcus danielii* (Verrips, 1992). The biological significance of the sweet property of mabinlin II is unknown. Similarly, it is not known whether this property is shared by any other 2S albumins.

4.5 As allergens

Several 2S albumins are major allergens, particularly in mustards where the Bra jIE allergen of oriental mustard (*B. juncea*) and the Sin aI allergen of yellow mustard (*Sinapis alba*) have both been cloned and shown to be members of the napin family (Menéndez-Arias *et al.*, 1988; González de la Peña *et al.*, 1991, 1993; Monsalve *et al.*, 1993). A monoclonal antibody to Sin aI was used for epitope mapping, demonstrating that both proteins contained an essential histidine residue which was absent from the non-allergenic Bn III napin from *B. napus* (Monsalve *et al.*, 1993). The Sin aI protein also binds to acid phospholipid vesicles suggesting that interactions with membranes might be involved in its allergenic activity (Oñaderra *et al.*, 1994). Other allergenic 2S albumins such as the Ric cIII allergen of castor bean (*Ricinus communis*) (Machado and Silva, 1992) and the methionine-rich albumin from Brazil nut (Melo *et al.*, 1994) have been studied in less detail.

4.6 Structures of 2S albumins

Circular dichroism spectroscopy of the napins has given variable results, with high levels of α-helix (35-50%) and low levels of β-sheet reported for napins from *Sinapis alba*, *Raphanus* spp. and *B. napus* (Menéndez-Arias *et al.*, 1987; Monsalve *et al.*, 1994; Mur9n *et al.*, 1996) but low levels of α-helix (12 and 18%) and high levels of β-sheet (45-50%) for inhibitory napins from *S. arvensis* and *B napus* var *rapifera*, respectively (Svendsen *et al.*, 1989; 1994). In fact, a low resolution NMR structure recently reported for napin from *B. napus* shows the presence of a high content of α-helix (52%), with no β-sheet.

This structure (Rico *et al.*, 1996) (Fig. 4a) shows five α-helices and a *C*-terminal loop in a right-handed spiral, stabilized by the four conserved disulphide bonds. The authors noted that the global fold was similar to the structures reported for other low M_r S-rich proteins : the soybean hydrophobic protein (Baud *et al.*, 1993) and the non-specific lipid transfer proteins from wheat (Gincel *et al.*, 1994) and maize (Shin *et al.*, 1995: Gomar *et al.*, 1996). A similar global fold has also been reported for a bifunctional α-amylase/trypsin inhibitor from ragi (Indian finger millet, *Eleusine coracana*) (Strobl *et al.*, 1995).

Using the ragi inhibitor structure as a template, Pandya *et al.* (1998) developed a structural model for the methionine-rich sunflower albumin SFA8. This model structure (Fig. 4b) comprised four α-helices connected by flexible loops and held together by the four intra-chain disulphide bonds to form a right handed fold, similar to that reported for napin by Rico *et al.* (1996). The model predicted a content of α-helix of about 55%, which agreed with data from circular dichroism and NMR spectroscopy.

SFA8 is a highly active emulsifier in oil/water mixtures, as discussed below. Analysis of the protein model surface for hydrophobic patches showed the presence of polar and apolar faces (Plate 9), the latter including contributions from methionine, leucine and isoleucine residues. The authors proposed that the large hydrophobic face could account for the emulsification properties of the protein. Commensurate with this unusual distribution of hydrophobic residues, Pandya *et al.* (1998) also showed that the folding pathway is highly atypical with initial formation of a misfolded state, or states, which must rearrange before formation of the native fold.

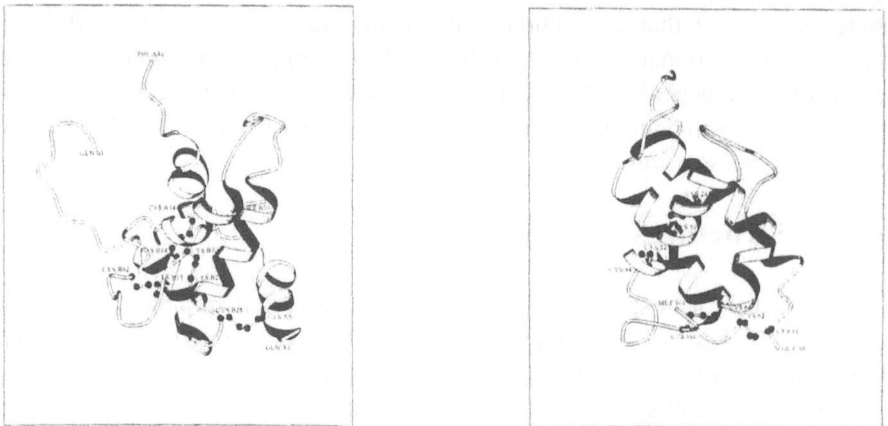

Fig. 4 Ribbon diagrams showing the 3D structures of a, the *B. napus* napin Bn1b determined by NMR spectroscopy (Rico *et al.*, 1996) and b, the sunflower albumin SFA8, modelled by Pandya *et al.* (1998). The positions of cysteine residues and the *N*- and *C*-terminal residues of the polypeptide chains are indicated, with the SFA8 model omitting residues 1-9.The figures were produced using the program MOLSCRIPT (Kraulis, 1991).Part a, reprinted with permission from Rico *et al.* (1996). Copyright 1996 American Chemical Society.

5. EXPLOITATION OF 2S ALBUMINS

5.1. As sources of sulphur amino acids

Several groups of workers have expressed the methionine-rich 2S albumin of Brazil nut in transgenic plants with the aim of enhancing the sulphur amino acid content of the seed. Expression levels, and the impact on the total methionine in the seed, have varied considerably. The most successful reports include increases in total seed methionine of 20% in *Arabidopsis* (Conceição *et al.*, 1994), 30% in tobacco (Altenbach *et al.*, 1989), 33% in oilseed rape (Altenbach *et al.*, 1992) and 300% in *Vicia narbonensis* (narbon bean) (Saalbach *et al.*, 1995). High level expression has also been achieved in soybean (R. Jung, pers. Commun.). However, the Brazil nut albumin has the disadvantage that it is strongly allergenic to humans (Melo *et al.*, 1994) and this allergenicity has also been reported when expressed in transgenic soybean (Nordlee *et al.*, 1996). This allergenicity to humans should not be a problem when soybean is used for livestock feed. However, the difficulty in segregating soybean destined for human consumption means that cultivars expressing the Brazil nut albumin are not considered suitable for commercial production. Similarly, the use of chimaeric S-rich albumins containing part of the Brazil nut protein (De Clercq *et al.*, 1990) is not commercially acceptable.

De Clercq *et al.* (1990) also reported the expression of modified 2S albumins, containing 5, 8 or 12 additional methionine residues, in seeds of tobacco, *Arabidopsis* and oilseed rape but the expression levels were not sufficient to have an impact on the total seed methionine.

The methionine-rich sunflower albumin SFA8 has not been reported to be allergenic and is therefore a more attractive target for expression in transgenic plants. This approach has been adopted by T.J. Higgins and co-workers, focusing on the improvement of forage legumes for grazing sheep (Tabe *et al.*, 1993, 1995). Addition of a C-terminal endoplasmic reticulum retention signal resulted in enhanced accumulation in leaves of transgenic alfalfa but the highest levels reported were only about 0.1% of the total soluble leaf protein. More recently, Molvig *et al.* (1997) reported the expression of SFA8 in seeds of lupin (*Lupinus angustifolius*). The transgenic seeds contained similar levels of total sulphur and nitrogen to those of control material, but less sulphate and more amino acid sulphur. The amount of methionine was increased by 94% with a small reduction (12%) in cysteine. Feeding trials with rats showed significant increases in live weight gain, true protein digestibility, biological value and net protein utilization.

5.2 As hosts for bioactive peptides

Vandekerckhove *et al.* (1989) proposed a novel and imaginative use of 2S seed storage proteins, as hosts for the production of bioactive peptides. Comparison of the sequences of the four 2S albumin proteins encoded by the *Arabidopsis* gene family with 2S albumins from Brazil nut, castor bean and oilseed rape showed the presence of a variable region between the 6th and 7th cysteine residues. This was considered likely to form a structurally independent loop and this has since been confirmed by the 3D structure reported for napin (Rico *et al.*, 1996). The sequence encoding this region of one of the *Arabidopsis* genes was therefore mutated to encode the biologically active peptide Leu-enkephalin (Tyr.Gly.Gly.Phe.Leu) flanked by lysine residues to facilitate release by cleavage of the proteins with trypsin followed by removal of the *C*-terminal lysine with carboxypeptidase B. Expression in seeds of *Arabidopsis* and oilseed rape gave yields of up to 200 nmoles of peptide per gram of seed (Vandekerckhove *et al.*, 1989) and the same approach has also been used to express 2S albumins containing a 28 residue antibacterial peptide from *Xenopus* (Krebbers *et al.*, 1993). Although Krebbers *et al.* (1993) calculated that 1Kg/hectare of a 25 residue peptide could be produced in oilseed rape based on current yields, the technology has not yet been adopted for commercial production.

5.3 As protein emulsifiers

Gueguén *et al.* (1996) compared the functional (emulsification and foaming) properties of 2S albumin fractions from sunflower. The emulsification properties varied with the cultivar of origin and this appeared to be related to the amounts and hydrophobicities of the individual 2S albumin components. In particular, the methionine-rich SFA8 component was the most effective at stabilizing emulsions, remaining in the interfacial layer when the emulsion was allowed to settle for 5 hours. In contrast, none of the albumin fractions were able to stabilize foams.

Although SFA8 could be used as an emulsifier in foods and other products (e.g. cosmetics) its commercial production may be limited by the presence of high levels of chlorogenic acid in sunflower seeds. Consequently it may be more viable in the long term to produce SFA8 in an alternative host plant. SFA8 may also prove to be a valuable model system for exploring the molecular basis of emulsification properties and designing novel emulsifiers, particularly if the hydrophobic patch model discussed above proves to be valid.

ACKNOWLEDGEMENTS

IACR receives grant-aided support from the Biotechnology and Biological Sciences Research Council of the United Kingdom.

REFERENCES

Allen, R.D., Cohen, E.A., Vonder Haar, R.A., Adams, C.A., Ma, D.P., Nessler, C.L., Thomas, T.L. (1987) Sequence and expression of a gene encoding an albumin storage protein in sunflower. *Molecular and General Genetics* **210**, 211-218.

Altenbach, S.B., Pearson, K.W., Leung, F.W., Sun, S.S.M. (1987) Cloning and sequence analysis of a cDNA encoding a Brazil nut protein exceptionally rich in methionine. *Plant Molecular Biology* **8**, 239-250.

Altenbach, S.B., Pearson, K.W., Meecker, G., Staraci, L.C., Sun, S.S.M. (1989) Enhancement of the methionine content of seed proteins by the expression of a chimeric gene encoding a methionine-rich protein in transgenic plants. *Plant Molecular Biology* **13**, 513-522.

Altenbach, S.B., Simpson, R.B. (1990) Manipulation of methionine-rich protein genes in plant seeds. *Trends in Biotechnology* **8**, 156-160.

Altenbach, S.B., Kuo, C.-C., Staraci, L.C., Pearson, K.W., Wainwright, C., Georgescu, A. Townsend J. (1992) Accumulation of a Brazil nut albumin in seeds of transgenic canola results in enhanced levels of seed protein methionine. *Plant Molecular Biology* **18**, 235-245.

Ampe, C., van Damme, J., de Castro, L.A.B., Sampaio, M.J.A.M., van Montagu, M., Vanderkerckhove, J. (1986) The amino-acid sequence of the 2S sulfur-rich proteins from seeds of Brazil nut (*Bertholletia excelsa* H.B.K.) *European Journal of Biochemistry* **159**, 597-604.

Anisimova, I.N., Fido, R.J., Tatham, A.S., Shewry, P.R. (1995) Genotypic variation and polymorphism of 2S albumins of sunflower. *Euphytica* **83**, 15-23.

Antunes, A.J., Markakis, P.J. (1977) Protein supplementation of navy beans with Brazil nuts. *Journal of Agricultural and Food Chemistry* **25**, 1096-1098.

Brinegar, C. (1997) The seed storage proteins of quinoa. In: "Food Proteins and Lipids" S. Damodaran (ed) Plenum Press, New York, pp. 109-115.

Brinegar, C., Sine, B., Nwokocha, L. (1996) High-cysteine 2S seed storage proteins from quinoa (*Chenopodium quinoa*). *Journal of Agricultural and Food Chemistry* **44**, 1621-1623.

Conceição, A.de S., Van Vliet, A., Krebbers, E. (1994) Unexpectedly high expression levels of a chimeric 2S albumin seed protein transgene from a tandem array construct. *Plant Molecular Biology* **26**, 1001-1005.

Coulter, K.M., Bewley, J.D. (1990) Characterization of a small sulphur-rich storage albumin in seeds of alfalfa (*Medicago sativa* L.) *Journal of Experimental Botany* **41**, 1541-1547.

Crouch, M.L., Tenbarge, K.M., Simon, A.E., Ferl, R. (1983) cDNA clones for *Brassica napus* seed storage proteins: evidence from nucleotide sequence analysis that both subunits of napin are cleaved from a precursor polypeptide. *Journal of Molecular and Applied Genetics* **2**, 273-283.

De Castro, L.A.B., Lacerada, A., Aramayo, R.A., Sampaio, M.J.A.M., Gander, E.S. (1987) Evidence for a precursor molecule of Brazil nut 2S seed proteins from biosynthesis and cDNA analysis. *Molecular and General Genetics* **206**, 338-343.

De Clercq, A., Vandewiele, M., Van Damme, J., Guerche, P., Van Montagu, M., Vanderkerckhove, J., Krebbers, E. (1990) Stable accumulation of modified 2S albumin seed storage proteins with higher methionine contents in transgenic plants. *Plant Physiology* **94**, 970-979.

D'Hondt, K., Bosch, D., Van Damme, J., Goethals, M., Vandekerckhove, J. and Krebbers, E. (1993) An aspartic proteinase present in seeds cleaves *Arabidopsis* 2S albumin precursors *in vitro*. *Journal of Biological Chemistry* **268**, 20884-20891.

Egorov, T.A., Odintsova, T.I., Musolyamov, A.Kh., Fido, R., Tatham, A.S., Shewry, P.R. (1996) Disulphide structure of a sunflower seed albumin: conserved and variant disulphide bonds in the cereal prolamin superfamily. *FEBS Letters* **396**, 285-288.

Ericson, M.L., Rödin, J., Lenman, M., Glimelius, K., Josefsson, L-G., Rask, L. (1986) Structure of the rapeseed 1.7 S storage protein, napin, and its precursor. *Journal of Biological Chemistry* **261**, 14576-14581.

Galau, G.A., Wang, H. Y-C., Hughes, D.W. (1992) Cotton *Mat5-A* (C164) Gene and *Mat5-D* cDNAs encoding methionine-rich 2S albumin storage proteins. *Plant Physiology* **99**, 779-782.

Gayler, K.R., Kolivas, S., Macfarlane, A.J., Lilley, G.G., Baldi, M., Blagrove, R.J., Johnson, E.D. (1990) Biosynthesis, cDNA and amino acid sequences of a precursor of conglutin δ, a sulphur-rich protein from *Lupinus angustifolius*. *Plant Molecular Biology* **15**, 879-893.

Genov, N., Goshev, I., Nikolova, D., Georgieva, D.N., Filippi, B., Svendsen, I. (1998) A novel thermostable inhibitor of trypsin and subtilisin from the seeds of *Brassica nigra*: amino acid sequence, inhibitory and spectroscopic properties and thermostability. *Biochimica et Biophysica Acta* In press.

Gincel, E., Simorre, J-P., Caille, A., Marion, D., Ptak, M., Vovelle, F. (1994) Three-dimensional structure in solution of a wheat lipid-transfer protein from multidimensional ^{1}H-NMR data. A new folding for lipid carriers. *European Journal of Biochemistry* **226**, 413-422.

Gomar, J., Petit, M.C., Sodano, P., Sy, D., Marion, D., Kader, J.C., Vovelle, F., Ptak, M. (1996) Solution structure and lipid binding of a nonspecific lipid transfer protein extracted from maize seeds. *Protein Science* **5**, 565-577.

González de la Peña, M.A., Menéndez-Arias, L., Monsalve, R.I., Rodríguez, R. (1991) Isolation and characterization of a major allergen from oriental mustard seeds, *Bra j I*. *International Archives of Allergy and Applied Immunology* **96**, 263-270.

González de la Peña, M.A., Villalba, M., García-López, J.L., Rodríguez, R. (1993) Cloning and expression of the major allergen from yellow mustard seeds, *Sin a I*. *Biochemical and Biophysical Research Communications* **190**, 648-653.

Guéguen, J., Popineau, Y., Anisimova, I.N., Fido, R.J., Shewry, P.R., Tatham, A.S. (1996) Functionality of the 2S albumin seed storage proteins from sunflower (*Helianthus annuus* L.). *Journal of Agricultural and Food Chemistry* **44**, 1184-1189.

Hara-Hishimura, I., Takeuchi, Y., Inoue, I., Nishimura, M. (1993) Vesicle transport and processing of the precursor to 2S albumin in pumpkin. *The Plant Journal* **4**, 793-800.

Higgins, T.J.V., Chandler, P.M., Randall, P.J., Spencer, D., Beach, L.R., Blagrove, R.J., Kortt, A.A., Inglis, A.S. (1986) Gene structure, protein structure, and regulation of the synthesis of a sulfur-rich protein in pea seeds. *Journal of Biological Chemistry* **261**, 11124-11130.

Irwin, S.D., Keen, J.N., Findlay, J.B.C., Lord, J.M. (1990) The *Ricinus communis* 2S albumin precursor: a single preprotein may be processed into two different heterodimeric storage proteins. *Molecular and General Genetics* **222**, 400-408.

Josefsson, L-G., Lenman, M., Ericson, M.L., Rask, L. (1987) Structure of a gene encoding the 1.7 S storage protein, napin, from *Brassica napus*. *Journal of Biological Chemistry* **262**, 12196-12201.

Kortt, A.A., Caldwell, J.B. (1990) Low molecular weight albumins from sunflower seed: identification of a methionine-rich albumin. *Phytochemistry* **29**, 2805-2810.

Kortt, A.A., Caldwell, J.B., Lilley, G.G., Higgins, T.J.V. (1991) Amino acid and cDNA sequences of a methionine-rich 2S protein from sunflower seed (*Helianthus annuus* L.). *European Journal of Biochemistry* **195**, 329-334.

Kraulis, P.J. (1991) MOLSCRIPT - a program to produce both detailed and schematic plots of protein structure. *Journal of Applied Crystallography* **24**, 946-960.

Krebbers, E., Herdies, L., De Clercq, A., Seurinch, J., Leemans, J., Van Damme, J., Segura, M., Gheysen, G., Van Montagu, M., Vanderkerckhove, J. (1988) Determination of the processing sites of an *Arabidopsis* 2S albumin and characterization of the complete gene family. *Plant Physiology* **87**, 859-66.

Krebbers, E., da Silva Conceição, A., Denis, M., D'Hondt, K., Vanderkerckhove, J. (1993) Modification of plant seed storage proteins. In: "Seed Storage Compounds: Biosynthesis, Interactions, and Manipulation" P.R. Shewry and K. Stobart (eds) Oxford University Press, Oxford pp. 317-324.

Laroche, M., Aspart, L., Delseny, M., Penon, P. (1984) Characterization of radish (*Raphanus sativus*) storage proteins. *Plant Physiology* **74**, 487-93.

Lilley, G.G. (1986a) Isolation of conglutin δ, a sulphur-rich protein from the seeds of *Lupinus angustifolius* L. *Journal of the Science of Food and Agriculture* **37**, 20-30.

Lilley, G.G. (1986b) The subunit structure and stability of conglutin δ, a sulphur-rich protein from the seeds of *Lupinus angustifolius* L. *Journal of the Science of Food and Agriculture* **37**, 895-907.

Lilley, G.G., Inglis, A.S. (1986) Amino acid sequence of conglutin δ, a sulfur-rich seed protein of *Lupinus angustifolius* L. Sequence homology with the C-III α-amylase inhibitor from wheat. *FEBS Letters* **195**, 235-241

Liu, X., Maeda, S., Hu, Z., Aiuchi, T., Nakaya, K., Kurihara, Y. (1993) Purification, complete amino acid sequence and structural characterization of the heat-stable sweet protein, mabinlin II. *European Journal of Biochemistry* **211**, 281-287.

Lonnerdal, B., Janson, J.C. (1972) Studies on *Brassica* seed proteins. I. The low molecular weight proteins in rape-seed. Isolation and characterization. *Biochimica et Biophysica Acta* **278**, 175-83.

Machado, O.L.T., Silva, J.G. Jr. (1992) An allergenic 2S storage protein from *Ricinus communis* seeds which is a part of the 2S albumin precursor predicted by c-DNA data. *Brazilian Journal of Medical and Biological Research* **25**, 567-582.

Mazhar, H., Quayle, R., Fido, R., Stobart, A.K., Napier, J.A., Shewry, P.R. (1998) Synthesis of storage reserves in developing seeds of sunflower. *Phytochemistry* **48**, 429-432.

Melo, V.M.M., Xavier-Filho, J., Lima, M.S., Prouvost-Danon, A. (1994) Allergenicity and tolerance to proteins from brazil nut (*Bertholletia excelsa* H.B.K.) *Food and Agricultural Immunology* **6**, 185-195.

Menéndez-Arias, L., Monsalve, R.I., Gavilanes, J.G., Rodríguez, R. (1987) Molecular and spectroscopic characterization of a low molecular weight seed storage protein from yellow mustard (*Sinapis alba* L.). *International Journal of Biochemistry* **19** , 899-907.

Menéndez-Arias, L., Moneo, I., Dominguez, J., Rodríquez, R. (1988) Primary structure of the major allergen of yellow mustard (*Sinapis alba* L.) seed, *Sin al. European Journal of Biochemistry* **177**, 159-66.

Molvig, L., Tabe, L.M., Eggum, B.O., Moore, A.M., Graig, S., Spencer, D., Higgins, T.J.V. (1997) Enhanced methionine levels and increased nutritive value of seeds of transgenic lupins (*Lupinus angustifolius* L.) expressing a sunflower seed albumin gene. *Proceedings of the National Academy of Science* **94**, 8393-8398.

Monsalve, R.I., Rodríguez, R. (1990) Purification and characterization of proteins from the 2S fraction from seeds of the Brassicaceae family. *Journal of Experimental Botany* **41**, 89-94.

Monsalve, R.I., Lopez-Otin, C., Villalba, M., Rodríguez, R. (1991) The new distinct group of 2S albumins from rapeseed. Amino acid sequence of two low molecular weight napins. *FEBS Letters* **295**, 207-210.

Monsalve, R.I., González de la Peña, M.A., Menéndez-Arias, L., Lopez-Otin, C., Villalba, M., Rodríguez, R. (1993) Characterization of a new oriental-mustard (*Brassica juncea*) allergen, *Bra j IE*: detection of an allergenic epitope. *Biochemical Journal* **293**, 625-632.

Monsalve, R.I., Menéndez-Arias, L., González de la Peña, M.A., Batanero, E., Villalba, M., Rodríguez, R. (1994) Purification and characterization of napin-like proteins from radish. *Journal of Experimental Botany* **45**, 1169-1176.

Morcillo, F., Aberlenc-Bertossi, F., Trouslot, P., Hamon, S., Duval, Y. (1997) Characterization of 2S and 7S storage proteins in embryos of oil palm. *Plant Science* **122**, 141-151.

Murén, E., Ek, B., Björk, I., Rask, L. (1996) Structural comparison of the precursor and the mature form of napin, the 2S storage protein in *Brassica napus*. *European Journal of Biochemistry* **242**, 214-219.

Nirasawa, S., Liu, X., Nishino, T., Kurihara, Y. (1993) Disulfide bridge structure of the heat-stable sweet protein mabinlin II. *Biochimica et Biophysica Acta* **1202**, 277-280.

Nordlee, J.A., Taylor, S.L., Townsend, J.A., Thomas, L.A., Bush, R.K. (1996) Identification of Brazil nut allergen in transgenic soybeans. *The New England Journal of Medicine* **334**, 688-692.

Oñaderra, M., Monsalve, R.I., Mancheño, J.M., Villalba, M., Martinez del Pozo, A., Gavilanes J.G., Rodriquez, R. (1994) Food mustard allergen interaction with phospholipid vesicles. *European Journal of Biochemistry* **225**, 609-615.

Ono, T., Sato, T., Odagiri, S. (1978) Albumins in buckwheat seed. *Agricultural and Biological Chemistry* **42**, 1779-1780.

Osborne, T.B. (1924) The vegetable proteins. Longmans, Green & Co., London.

Pandya, M.J., Sessions, R.B., Williams, P.B., Dempsey, C.E., Tatham, A.S., Shewry, P.R. Clarke, A.R. (1998) Direct kinetic evidence for misfolding of an emulsifying protein. *Submitted.*

Polya, G.M., Chandra, S., Condron, R. (1993) Purification and sequencing of radish seed calmodulin antagonists phosphorylated by calcium-dependent protein kinase. *Plant Physiology* **101**, 545-551.

Radovi□, S.R., Maksimovi□, V.S., Varkonji-G|si□, E.I. (1996) Characterization of buckwheat seed storage proteins. *Journal of Agricultural and Food Chemistry* **44**, 972-974.

Raynal, M., Depigny, D., Grellet, F., Delseny, M. (1991) Characterization and evolution of napin-encoding genes in radish and related crucifers. *Gene* **99**, 77-86.

Rico, M., Bruix, M., González, C., Monsalve, R.I., Rodríguez, R. (1996) ^{1}H NMR assignment and global fold of napin BnIb, a representative 2S albumin seed protein. *Biochemistry* **35**, 15672-15682.

Rödin, J., Rask, L. (1990) Characterization of matteuccin, the 2.2S storage protein of the ostrich fern. Evolutionary relationship to angiosperm seed storage proteins. *European Journal of Biochemistry* **192**, 101-107.

Saalbach, I., Waddell, D., Pickardt, T., Schieder, O., Müntz, K. (1995) Stable expression of the sulphur-rich 2S albumin gene in transgenic *Vicia narbonensis* increases the methionine content of seeds. *Journal of Plant Physiology* **145**, 674-681.

Scofield, S.T., Crouch, M.L. (1987) Nucleotide sequence of a member of the napin storage protein family from *Brassica napus*. *Journal of Biological Chemistry* **262**, 12202-12208.

Segura-Nieto, M., Vázquez-Sánchez, N., Rubio-Velázquez, H., Olguín-Martínez, L.R., Rodíquez-Nester, C., Herrera-Estrella, L. (1992) Characterization of amaranth (*Amaranthus hypochondriacus* L.) seed proteins. *Journal of Agricultural and Food Chemistry* **40**, 1553-1558.

Segura-Nieto, M., Barba de la Rosa, A.P., Paredes-López, O. (1994) Biochemistry of amaranth proteins. In: "Amaranth: Biology, Chemistry and Technology" O. Paredes-López, (ed) Boca Raton: CRC Press pp. 75-106.

Sharief, F.S., Li, S.-L. (1982) Amino acid sequence of small and large subunits of seed storage protein from *Ricinus communis*. *Journal of Biological Chemistry* **257**, 14753-14759.

Shewry, P.R. (1995) Plant storage proteins. *Biological Reviews* **70**, 375-426

Shin, D.H., Lee, J.Y., Hwang, K.Y., Kim, K.K., Suh, S.W. (1995) High resolution of the non-specific lipid-transfer protein from maize seedlings. *Structure* **3**, 189-199.

Strobl, S., Mühlhahn, P., Bernstein, R., Wilscheck, R., Maskos, K., Wunderlich, M., Huber, R., Glockhuber, R., Holak, T.A. (1995) Determination of the three-dimensional structure of the bifunctional α-amylase/trypsin inhibitor from ragi seeds by NMR spectroscopy. *Biochemistry* **34**, 8281-8293.

Sun, S.S.M., Altenbach, S.B., Leung, F.W. (1987) Properties, biosynthesis and processing of a sulfur-rich protein in Brazil nut (*Bertholletia excelsa* H.B.K.). *European Journal of Biochemistry* **162**, 477-483.

Sun, S.M.S., Leung, F.W., Tomic, J.C. (1987) Brazil nut (*Bertholletia excelsa* H.B.K.) proteins: fractionation, composition, and identification of a sulfur-rich protein. *Journal of Agricultural and Food Chemistry* **35**, 232-235.

Svendsen, I., Nicolova, D., Goshev, I., Genov, N. (1989) Isolation and characterization of a trypsin inhibitor from the seeds of kohlrabi (*Brassica napus* var. *Rapifera*) belonging to the napin family of storage proteins. *Carlsberg Research Communications* **54**, 231-239.

Svendsen, I., Nicolova, D., Goshev, I., Genov, N. (1994) Primary structure, spectroscopic and inhibitory properties of a two-chain trypsin inhibitor from the seeds of charlock (*Sinapis arvensis* L), a member of the napin protein family. *International Journal of Peptide & Protein Research* **43**, 425-430.

Tabe, L.M., Higgins, C.M., NcNabb, W.C., Higgins, T.J.V. (1993) Genetic engineering of grain and pasture legumes for improved nutritive value. *Genetica* **90**, 181-200.

Tabe, L.M., Wardley-Richardson, T., Ceriotti, A., Aryan, A., McNabb, W., Moore, A., Higgins, T.J.V. (1995) A biotechnological approach to improving the nutritive value of alfalfa. *Journal of Animal Science* **73**, 2752-2759.

Takumi, K., Koga, T., Kanoh, M., Udaka, J., Tsuji, H., Manabe, S. (1995) Immunochemical crossreactivity between globulins from buckwheat and indigo seeds. *Bioscience, Biotechnology and Biochemistry* **59**, 1971-1972.

Templeman, T.S., DeMaggio, A.E., Stetler, D.A. (1987) Biochemistry of fern spore germination: globulin storage proteins in *Matteuccia struthiopteris* L. *Plant Physiology* **85**, 343-349.

Templeman, T.S., Stein, D.B., DeMaggio, A.E. (1988) A fern spore storage protein is genetically similar to the 1.7S seed storage protein of *Brassica napus. Biochemical Genetics* **26**, 595-603.

Templeman, T.S., DeMaggio, A.E. (1990) Biochemistry of fern spore proteins: globulin storage proteins in *Onoclea sensibilis* and *Osmunda cinnamomea. American Journal of Botany* **77**, 284-287.

Terras, F.R.G., Schoofs, H.M.E., De Bolle, M.F.C., Van Leuven, F., Rees, S.B., Vanderleyden, J., Cammuie, B.P.A., Broekaert, W.F. (1992) Analysis of two novel classes of plant antifungal proteins from radish (*Raphanus sativus* L.) seeds. *Journal of Biological Chemistry* **267**, 15301-15309

Terras, F.R.G., Schoofs, H.M.E., Thevissen, K., Osborn, R.W., Vanderleyden, J., Cammue, B.P.A., Broekaert, W.F. (1993) Synergistic enhancement of the antifungal activity of wheat and barley thionins by radish and oilseed rape 2S albumins and by barley trypsin inhibitors. *Plant Physiology* **103**, 1311-1319.

Thoyts, P.J.E., Napier, J.A., Millichip, M., Stobart, A.K., Griffiths, W.T., Tatham, A.S., Shewry, P.R. (1996) Characterization of a sunflower seed albumin which associates with oil bodies. *Plant Science* **118**, 119-125.

Vanderkerckhove, J., Van Damme, J., Van Lijsebettens, M., Botterman, J., De Block, M., Vandewiele, M., De Clercq, A., Leemans, J., Van Montagu, M., Krebbers, E. (1989) Enkephalins produced in transgenic plants using modified 2S seed storage proteins. *Bio/Technology* **7**, 929-932.

Verrips, C.T. (1992) Structure and protein engineering of thaumatin and other sweet proteins. In: "Plant Protein Engineering" P.R. Shewry and S. Gutteridge (eds) Cambridge University Press, Cambridge, pp. 219-234.

Youle, R.J., Huang, A.H.C. (1981) Occurrence of low molecular weight and high cysteine containing albumin storage proteins in oilseeds of diverse species. *American Journal of Botany* **68**, 44-48.

Zuo, W-N., Sun, S.S.M. (1996) Purification and characterization of the methionine-rich 2S seed proteins from the Brazil nut family (*Lecythidaceae*). *Journal of Agricultural and Food Chemistry* **44**, 1206-1210.

25. Enzyme Inhibitors of Seeds: Types and Properties

PETER R. SHEWRY
IACR-Long Ashton Research Station, Department of Agricultural Sciences, University of Bristol, Long Ashton, Bristol BS41 9AF, UK

1. INTRODUCTION

Plant seeds, and other storage organs such as tubers, are rich sources of proteins which may confer resistance to pests and pathogens. Of these the most abundant, widely distributed and intensively studied are the enzyme inhibitors, and Richardson estimated in 1991 that over 20,000 articles had been published on this group of proteins alone (Richardson, 1991). Other "defence-related" seed proteins are endohydrolases (notably chitinases and glucanases), chitin-binding proteins (hevein and lectins), thionins, defensins, non-specific phospholipid transfer proteins and ribosome-inactivating proteins. Many of these proteins are considered in detail in following chapters (28-31) while a recent review focuses on their biological role in resistance and defence (Shewry and Lucas, 1997). These proteins are usually expressed "constitutively" in the seed tissues, but related proteins may be induced by infection or tissue damage in vegetative tissues, forming part of the well-characterized PR (pathogenesis-related) response (Van Loon, 1985).

The present chapter aims to provide a broad introduction to enzyme inhibitors present in seed tissues including their classification, properties and distribution. More detailed accounts of the cereal inhibitor superfamily and of the major groups of inhibitors present in legume seeds are given in chapters 26 and 27, respectively. The account is not intended to be exhaustive and the reader is referred to several excellent review articles for further information (Richardson, 1977; 1991; Xavier-Filho and Campos, 1989; Ryan, 1990; Norton, 1991; Brzin and Kidric, 1995; Wilson, 1997).

P.R. Shewry and R. Casey (eds.), Seed Proteins, 587–615.

2. CLASSIFICATION AND DISTRIBUTION

Enzyme inhibitors were first isolated from seeds in the 1940s, with the characterization of Kunitz and Bowman-Birk type proteinase inhibitors from soybean (Kunitz, 1945; Bowman, 1946) and of an α-amylase inhibitor from cereal grains (Kneen and Sandstedt, 1943). Since then an immense number of inhibitors have been identified and characterized from a wide range of organs and species. Seeds are by far the richest source and it is probable that all seeds contain one or more inhibitors, although the levels vary greatly and their detection is dependent on the selection of the appropriate target enzymes and assay conditions. The vast majority of the inhibitors characterized to date are active against endoproteinases, with a smaller number of amylase inhibitors, some of which are bifunctional (i.e. inhibit both amylases and proteinases). Inhibitors of other enzymes appear to occur more rarely, although this may partly reflect the lack of systematic searches. Target enzymes for seed inhibitors so far identified include catalase in maize scutella (Tsaftaris *et al.*, 1980), urease in melon (Makkar *et al.*, 1980) and lipase in a range of species including sunflower, soybean and cereals (Lairon *et al.*, 1985; Chapman, 1987; Gargouri *et al.*, 1984; Wang and Huang, 1984; Huang and Wang, 1992).

3. FAMILIES OF INHIBITORS

Comparison of the amino acid sequences of proteinase inhibitors allows them to be classified into a number of families, some of which also contain related or bifunctional α-amylase inhibitors. These families are summarised in Table 1. In addition, at least four seed α-amylase inhibitors fall outside these groups and may represent separate families.

The structures and properties of these inhibitors will now be discussed, with the exception of the potato II, carboxypeptidase and arrowhead inhibitor families which have not so far been identified in seeds.

3.1 The Bowman-Birk family

The Bowman-Birk inhibitors are one of the most widely distributed and intensively studied groups, being particularly abundant in legume seeds. They are low M_r proteins (≅ 8-9,000) with seven intra-chain disulphide bonds, and most are "double-headed" with two reactive sites which are inhibitory to the same or different serine proteinases (usually trypsin and chymotrypsin). These inhibitory sites are present in regions of internal homology and the proteins are thought to have evolved from the duplication of an ancestral single domain protein. This is illustrated by the schematic structure of the soybean inhibitor shown in Fig. 1.

Table 1 Proteinase inhibitor families of plants

Family	Monomer M_r x10^{-3}	Monomer Cys	Enzymes inhibited	Distribution	Weak affinities with other families
1. Bowman-Birk	8-9(14)	14(18)	Trypsin Chymotrypsin Elastase	Leguminosae Gramineae	6, 7
2. Kunitz	21-22	4	Trypsin Chymotrypsin Subtilisin Kallikrein Amylase	Leguminosae Gramineae Araceae Alismataceae	8
3. Potato I	8-9	0-2	Chymotrypsin Trypsin Subtilisin	Solanaceae Gramineae Leguminosae Polygonaceae Cucurbitaceae	
4. Potato II	6(12)	8	Trypsin Chymotrypsin	Solanaceae	7, 9
5. Squash	3	6	Trypsin Hageman factor	Cucurbitaceae	
6. Cereal family	12-13	10	Amylase Trypsin Hageman factor	Gramineae Cruciferae Euphorbiaceae Lecythidaceae Leguminosae	1
7. Thaumatin-related	22-23	16	Amylase Trypsin	Gramineae Solanaceae	2, 7
8. Carboxy-peptidase	4	6	Carboxy-peptidase	Solanaceae	4
9. Cystatins	12	0	Cysteine-proteinases	Gramineae	
10. Mustard trypsin inhibitor (MTI-2)	7	8	Trypsin	Cruciferae	1
11. Serpins	43	2	Trypsin Chymotrypsin Cathepsin G	Gramineae	
12. Arrowhead inhibitors	16.5	6	Trypsin Chymotrypsin Kallikrein	Alismataceae	

Reproduced with modifications from Richardson (1991), with permission of the author and publisher. Data on additional families is from Yang *et al.* (1992), Menegatti *et al.* (1993), Brandt *et al.* (1990) and Rosenkrands *et al.* (1994).

A large number of amino acid sequences of Bowman-Birk inhibitors are available, some of which are aligned in Fig. 2. Most of these sequences correspond to legume seed inhibitors, which are discussed in detail in chapter 27. However, the sequences of four cereal seed inhibitors are also included:

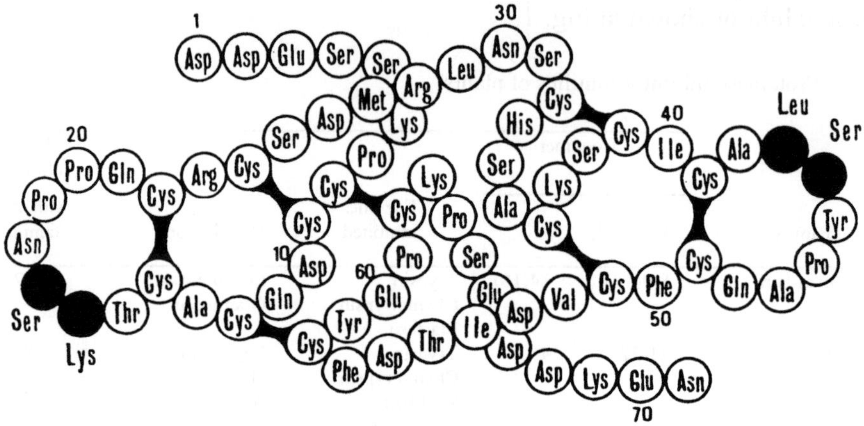

Fig. 1 Schematic structure of the soybean Bowman-Birk proteinase inhibitor, showing the two active sites (Lys Ser and Leu Ser). See chapter 27 (Figure 5) for further details. Reproduced from Odani and Ikenaka (1973) with permission of the author.

from rice bran (i.e. mainly aleurone), wheat germ and *Coix* seeds. It is of interest that the rice bran inhibitor and the wheat germ inhibitor I-2b both have M_r of about 14,000 and appear to have arisen by duplication of a typical M_r 7,000 Bowman-Birk inhibitor (Odani *et al.*, 1986; Tashiro *et al.*, 1987).

Three-dimensional structures of several Bowman-Birk inhibitors have been reported, from peanut (Suzuki *et al.*, 1987) from soybean (Werner and Wemmer, 1992) and the complex between the adzuki bean inhibitor and trypsin (Tsunogae *et al.*, 1986). The former shows that the protein has an extended shape with approximate dimensions of 45 x 15 x 15Å. It consists of two similar domains, containing independent protease-binding sites which protrude from the molecule on opposite sides.

3.2 The Kunitz family

The Kunitz inhibitors are the second major family of inhibitors which are widely distributed, and often very abundant, in seeds of leguminous plants, but also occur in other groups of plants including cereal seeds. The "typical" legume proteins are trypsin inhibitors of M_r about 21,000 with four cysteine residues that form two intra-chain disulphide bonds. However, in members of the legume sub-family Mimosoideae a proteolytic cleavage occurs between the third and fourth cysteine residues, resulting in a heterodimeric protein comprising chains of M_r about 5,000 and 16,000 linked by a single disulphide

```
                                                     1
1 SOYBEAN CII                                        S D H S S S D D E
2 PEANUT AII                                               E A S S S S
3 ADZUKI BEAN                                S G H H E D T T D E P S E
4 MACROTYLOMA DE3                                D H H H S T D E P S E
5 VICIA ANGUSTIFOLIA                                             G D D
6 MUNG BEAN                                  S S H H H D S S D E P S E
7 ALFALFA LEAF
8 RICE BRAN                                                        M E
9 WHEAT GERM II-4                                                  A T
10 WHEAT GERM I-2b                                             A A K K
11 COIX                                                        G D E K

           10                  20        ↓                        30
1          - S S K P C C D L C M C T A S M - - P P - - Q C H C A D I R L N
2          - D D N V C C N G C L C D R R A - - P P Y F E C V C Y D T F D H
3          - S S K P C C D Q C - C T K S M - - P P - - K C R C S D I R L D
4          - S S K P C C D E C A C T K S I - - P P - - Q C R C T D V R L N
5          - V K S A C C D T C L C T R S Q - - P P - - T C R C V D V G E R
6          - S S E P C C D S C R C T K S I - - P P - - Q C H C A D I R L N
7              T T A C C N F C P C T R S I - - P P - - Q C R C T D I G E T
8          R P W K - C C D N I K R L P T K P D P P - - Q W R C N D E L E P
8 *        R P W G D C C D K A F C N K M N - - P P - - T C R C M D E V K -
9          R P W K - C C D R A I C T K S F - - P P - - M C R C M D M V E -
10         R P W K - C C D Q A V C T R S I - - P P - - I C R C M D Q V F E
11         R P W E - C C D I A M C T R S I - - P P - - I C R C V D K V D R
                       10                       20                       30
                                         ↓
                40                       50                       60
1          - S C H S A C D R C A C T R S M - - P - - - G C R C L D T T D F
2          - - C P A S C N S C V C T R S N - - P P - - Q C R C T D K T Q G
3          - S C H S A C K S C A C T Y S I - - P A - - K C F C T D I N D F
4          - S C H S A C S S C V C T F S I - - P A - - Q C V C V D M K D F
5          - - C H S A C N H C V C N Y S N - - P P - - Q C Q C F D T H K F
6          - S C H S A C K S C M C T R S M - - P G - - K C K C L D T D D F
7          - - C H S A C K T C L C T K S I - - P P - - Q C R C T D I T N F
8          S Q C T A A C K S C R E A P G P F G K L I - - C E D I Y W G A D
8 *        - E C A D A C K D C Q R V E S S E - P P R Y V C K D K F T G H P
9          - Q C A A T C K K C G P A T S D S S R R - - V C E D X Y
10         - - C P S T C K A C G P S V G D P S R R - - V C Q D Q Y V
11         - - C S D A C K D C E E T E D - - N R H - - V C F D T Y I G D P
                                 40                      50

                           70
1          - C Y K P - C K S S D E D D D
2          R C P V T E C R S
3          - C Y E P - C K S S R D D D W D N
4          - C Y A P - C K S S H D D
5          - C Y K A - C H S S E K E E V I K N
6          - C Y K P - C E S M D K D D D
7          - C Y P K - C N
8          P - - G P F C T P
8 *        - - - G P V C K P R
11         - - - G P T C H D D
                     60
```

Fig. 2 Comparison of the amino acid sequences of typical Bowman-Birk inhibitors from legume seeds with those from cereal seeds, including the M_r 14,000 inhibitors of rice bran and wheat germ (I-2b). (1) soybean *Glycine max* inhibitor CII (Odani and Ikenada, 1977); (2) peanut *Arachis hypogaea* inhibitor AII (Norioka and Ikenaka, 1983); (3) adzuki bean *Phaseolus angularis* (Yoshikawa *et al.*, 1979); (4) *Macrotyloma axillare* (Joubert *et al.*, 1979); (5) *Vicia angustifolia* (Shimokawa *et al.*, 1984); (6) mung bean *Phaseolus radiata* (Wilson and Chen, 1983); (7) wounded alfalfa leaf *Medicago sativa* (Brown *et al.*, 1985); (8) rice bran *Oryza sativa* residues 1-69, residues 70-133 are an internal duplication homologous with residues 3-

69 and are shown as (8*) (Tashiro *et al.*, 1987); (9) wheat germ *Triticum aestivum* inhibitor II-4; partial sequence (Odani *et al.*, 1986); (10) wheat inhibitor I-2b; incomplete sequence (Odani *et al.*, 1986); (11) Job's tears *Coix lachryma-jobi*, inhibitor TI1 (Ary *et al.*, 1988). Only the numbering of the soybean (top line) and the *Coix* (bottom line) inhibitor sequences are shown. The sequences are aligned for maximum homology, resulting in some gaps (-) which may represent insertions or deletions. The arrows indicate the position of the reactive sites determined for the legume inhibitors. The cysteine residues are boxed to facilitate comparisons. Reproduced from Richardson (1991) with permission of the author and publishers.

bond (see Richardson, 1991 for details). The legume Kunitz-type inhibitors are discussed in detail in chapter 27.

However, not all Kunitz-related proteins of legume seeds are proteinase inhibitors. The winged bean albumin-1 storage protein from *Psophocarpus tetragonolobus* accounts for about 15% of the total seed protein. It comprises 175 amino acid residues with an M_r of 19,333 and contains a single disulphide bond. It shows 38% and 28% sequence similarity with Kunitz inhibitors from soybean and winged bean, respectively, but has no inhibitory activity (Kortt *et al.*, 1989).

Also related to the Kunitz inhibitors are a group of bifunctional α-amylase/subtilisin inhibitors from cereals (Fig. 3). The best characterized member of this family is the barley α-amylase/subtilisin inhibitor, usually called BASI. BASI consists of 181 residues with M_r of 19,865 (Svendsen *et al.*, 1986). It inhibits barley α-amylase-2 with a stoichiometric ratio of 2:1 (Halayko *et al.*, 1986) and subtilisin with a ratio of 1:1 (Yoshikawa *et al.*, 1979) and appears to be "double headed" with separate inhibitory sites for the two enzymes.

Weselake *et al.* (1985) used an antibody to the barley inhibitor to demonstrate the presence of related proteins in wheat, rye and triticale and this has been confirmed by subsequent studies. The wheat α-amylase inhibitor, WASI, has a similar spectrum of activity to BASI, inhibiting α-amylases from wheat, barley, rye and oats but not those from sorghum, rice, larvae of the coleopteran *Telebrio molitor* (yellow mean worm), hog pancreas, human saliva, *Aspergillus oryzae* or *Bacillus subtilis* (Mundy *et al.*, 1984). It has a high level of sequence identity to BASI (Fig. 3) (Maeda, 1986). A related protein is also present in rye (Weselake *et al.*, 1985; Sadowski *et al.*, 1986) where it is enriched in the bran fraction (Täufel *et al.*, 1991). This protein also inhibits endogenous α-amylase. In contrast, the homologous inhibitor from rice (RASI) (Fig. 3) inhibits subtilisin and α-amylase from *T. molitor* larvae, but not barley α-amylase, even when present in a 10-fold excess (Ohtsubo and Richardson, 1992).

```
          1       10        20        30        40        50        60
RASI      APPPVYDTEGHELSADGSYYVLPASPGHGGGLTMAPRRVIPCPLLVAQETDERRKGFPVR
WASI      DPPPVHDTDGNELRADANYYVLPANRAHGGGLTMAPGHGRRCPLFVSQEADGQRDGLPVR
BASI     ADPPPVHDTDGHELRADANYYVLSANRAHGGGLTMAPGHGRHCPLFVSQDPNGQHDGFPVR
           **** ** * ** **  **** *   *********     *** * *       * * *

                  70        80        90         100       110      120
RASI      FTPWGGAASPRTLRVSTDVRIRFNAATLCVQSTEWHVGDEPLTGARRVVTGPLIGPSPSGR
WASI      IAPHGGAPSDKIIRLSTDVRISFRAYTTCVQSTEWHIDSELVSGRRHVITGPVRDPSPSGR
BASI      ITPYGVAPSDKIIRLSTDVRISFRAYTTCLQSTEWHIDSELAAGRRHVITGPVKDPSPSGR
            * * * *    * ****** * * * * ******   *   * * * ***   ******

                 130            140       150        160       170
RASI      ENAFRVEKYGGG----YKLVSCRDSCQDLGVSRDGAGAAW-IGASQPPHVVVFKKARPSP
WASI      ENAFRIEKYSGAEVHEYKLMACGDSCQDLGVFRDLKGGAWFLGATEPYHVVVFKKAPPA
BASI      ENAFRIEKYHGAEVSEYKLMSCGDWCQDLGVFRDLKGGAWFLGATEPYHVVVFKKAPPA
          ***** *** *     ***  * * ****** **  * **  **  * **********
```

Fig. 3 Comparison of the amino acid sequences of the rice bifunctional inhibitor of subtilisin and α-amylase (RASI) with WASI, the subtilisin/α-amylase inhibitor from wheat (Maeda, 1986) and with BASI, the barley subtilisin/α-amylase inhibitor (Svendsen *et al.*, 1986). The numbering of residues refers only to RASI. -, indicates deletions inserted in sequences to maximize homology. *, indicates residues which are invariant in all sequences shown. Reproduced from Ohtsubo and Richardson (1992) with permission of the author and publisher.

BASI, and its homologues from wheat and rye, are the only plant protein inhibitors which have been shown to inhibit endogenous α-amylases. They are therefore thought to play a physiological role in the regulation of starch degradation *in vivo* and this is supported by the fact that the synthesis of BASI is regulated by abscisic acid and dehydration stress in embryos (Robertson *et al.*, 1989; Liu and Hill, 1995) and by gibberellic acid and abscisic acid in developing and mature aleurone cells (Leah and Mundy, 1989). The biological significance of their inhibition of subtilisin is not known but this could relate to protection against pests and pathogens.

Three-dimensional structures of Kunitz inhibitors from soybean (Sweet *et al.*, 1974) and *Erythrina* (Onesti *et al.*, 1991), the winged bean albumin-1 (McCoy and Kortt, 1997) and WASI (Zemke *et al.*, 1990) have been determined. The proteins have similar structures comprising 12 β-strands, six of which form an anti-parallel β-barrel. The structure of the winged bean albumin-1 shows that it contains an active site residue (lysine 64) and that its lack of activity is probably related to an unfavourable conformation of the inhibitory loop in which this residue is present.

3.3 The potato I inhibitor family

This family includes the chymotrypsin/subtilisin inhibitor I from potato (Melville and Ryan, 1972) and the wound-inducible tomato inhibitor (Graham *et al.*, 1985) but also includes chymotrypsin/subtilisin inhibitors from barley seeds (see below) and a trypsin inhibitor from *Momordica charantia* (Cucurbitaceae) (Zeng *et al.*, 1988). In addition, a related inhibitor, eglin, has been characterized from the medicinal leech (*Hirudo medicinalis*) (See Muller *et al.*, 1977; McPhalen *et al.*, 1985a). All are low M_r proteins, comprising about 60-80 amino acid residues, with either one (potato and tomato inhibitors) or no (barley, *Momordica*, eglin) disulphide bonds.

The barley chymotrypsin inhibitors CI-1 and CI-2 were initially identified as lysine-rich proteins, contributing to the high lysine phenotype of the line Hiproly (Hejgaard and Boisen, 1980). Subsequent studies showed that both proteins comprised two isoforms with the major form of CI-2 (CI-2A) comprising 84 residues with an M_r of 9,380 and the two isoforms of CI-1 (CI-1A and CI-1B) comprising 83 residues with M_r of 8,790 and 8,960, respectively (Svendsen *et al.*, 1980a, b; 1982; Williamson *et al.*, 1987, 1988). The individual isoforms of the two proteins are highly similar while the sequence identity between CI-1 and CI-2 is about 42% (Fig. 4).

The small size of CI-2 and its lack of disulphide bonds have made it an attractive target for structural, folding and protein engineering studies. Its 3-D structure has been determined by X-ray crystallography and NMR spectroscopy (see McPhalen *et al.*, 1985b; McPhalen and James, 1987; Clore *et al.*, 1987a, b). This shows a wedge-shaped disc of about 28 x 27 x 19Å, with a single α-helix of 3.6 turns and four strands of β-sheet with a characteristic left-handed twist (Fig. 5). The reactive (inhibitory) site is present in a loop region (residues Gly^{54} - Tyr^{61}) which has been mutated in protein engineering studies to alter the rate of dissociation of the complex with chymotrypsin and the proteinase specificity (to a trypsin inhibitor) (Campbell, 1992).

3.4 The squash inhibitor family

This is a family of highly homologous inhibitors which have been isolated only from seeds of squash and other members of the Cucurbitaceae and have no known sequence homology with any other groups of proteins. Over 40 members of the family have been sequenced (Fig. 6), showing that all consist of 27-33 amino acid residues with a single reactive site between residues 5 (Lys, Arg or Leu) and 6 (Ile) (Otlewski and Krowarsch, 1996). They are the smallest known protein inhibitors of serine proteinases, and most are powerful inhibitors of bovine trypsin. However, other serine proteinases may also be

Amino Acid Sequences of CI-1 and CI-2 of Barley

CI - 1A	M	S	S	M	E	G	S	V	L	K	Y	P	E	P	T	E	G	S	I	G
1B	M	R	S	M	E	G	S	V	P	K	Y	P	E	P	T	E	G	S	I	G
CI - 2A	M	S	S	V	E	-	-	-	-	K	K	P	E	G	V	N	T	G	A	G
2B																*	D	C	L	C
CI - 1A	A	S	S	A	-	K	T	S	W	P	E	V	V	G	M	S	A	E	K	A
1B	A	S	G	A	-	K	R	S	W	P	E	V	V	G	M	S	A	E	K	A
CI - 2A	D	R	H	N	L	K	T	E	W	P	E	L	V	G	K	S	V	E	E	A
2B	D	C	Q	N	Q	K	T	E	W	P	E	L	V	E	K	S	V	E	E	A
CI - 1A	K	E	I	I	L	R	D	K	P	N	A	Q	V	E	V	I	P	V	D	A
1B	K	E	I	I	L	R	D	K	P	D	A	Q	I	E	V	I	P	V	D	A
CI - 2A	K	K	V	I	L	Q	D	K	P	E	A	Q	I	I	V	L	P	V	G	T
2B	K	K	V	I	L	Q	D	K	P	E	A	Q	I	I	V	L	P	V	G	T
CI - 1A	M	V	H	L	N	F	D	P	N	R	V	F	V	L	V	-	-	-	-	A
1B	M	V	P	L	D	F	N	P	N	R	I	F	I	L	V	-	-	-	-	A
CI - 2A	I	V	T	M	E	Y	R	I	D	R	V	R	L	F	V	D	K	L	D	N
2B	I	V	T	M	E	Y	R	I	D	R	V	R	L	F	V	D	R	L	D	N
CI - 1A	V	A	R	T	P	T	V	G												
1B	V	A	R	T	P	T	V	G												
CI - 2A	I	A	Q	V	P	R	V	G												
2B	I	A	Q	V	P	R	V	G												

* sequence incomplete

Fig. 4 Amino acid sequences of chymotryptic inhibitors CI-1 and CI-2 of barley, based on cDNA sequences reported by Williamson *et al.* (1987, 1988). Reproduced from Shewry (1993) with permission of the publisher.

inhibited: kallikrein, thrombin, subtilisin, elastase and chymotrypsin (see Otlewski and Krowarsch, 1996). They are among the most potent inhibitors of serine proteinases that are known, with high association constants.

Comparison of the amino acid sequences of the inhibitors show that the major differences lie in the presence in some of *N*-terminal extensions of one to five residues, with two inhibitors also differing in the presence of one or two additional *C*-terminal residues (Fig. 6) (Otlewski and Krowarsch, 1996). Comparison of the directly determined sequence of the towel gourd (*Luffa cylindrica*) trypsin inhibitor with the sequences of corresponding cDNA and genomic DNA sequences showed the presence of *N*-terminal pre and pro sequences of 21 and 13 residues, respectively (Ling *et al.*, 1993), indicating that the inhibitor is translocated into the endoplasmic reticulum and then

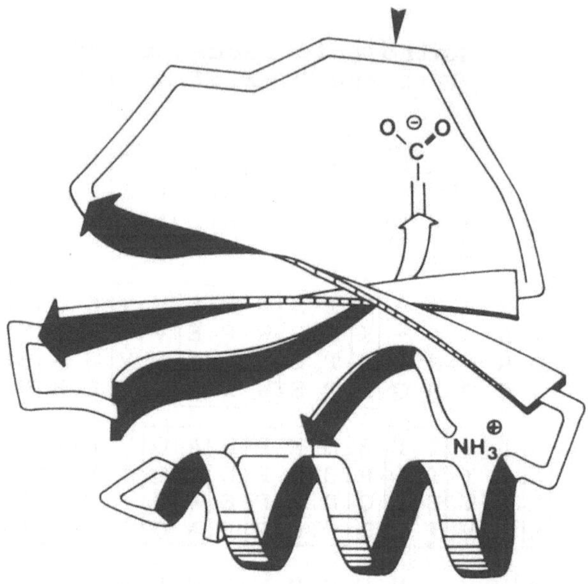

Fig. 5 The three-dimensional structure of chymotrypsin inhibitor CI-2. Large arrows denote β-strands; wide ribbons are α-helix; and narrow ribbons are turns or unclassified structures. The small arrow indicates the reactive (inhibitory) site. Reproduced from McPhalen *et al.* (1985b) with the author's permission.

processed proteolytically to remove the pro sequence, probably within the vacuole. All the inhibitors contain six cysteine residues which form three intra-chain disulphide bonds.

The small size, compactness and rigidity of the inhibitors has made them attractive subjects for 3-D structure analysis, resulting in the determination of the structures of several inhibitors and their complexes with proteinases by NMR spectroscopy and X-ray crystallography (Holak *et al.*, 1989a, b; 1991; Heitz *et al.*, 1989; Likos, 1989; Bode *et al.*, 1989; Krishnamoorthi *et al.*, 1992; Nilges *et al.*, 1991; Huang *et al.*, 1992). These studies show that the inhibitory site is present on a hydrophobic loop region which appears to form a short fragment of anti-parallel β-sheet with the proteinase (Fig. 7).

3.5 The cereal inhibitor family

This group of inhibitors is present only in seeds of cereals, occurring in the Triticeae (wheat, barley, rye and relatives), rice, maize and finger millet. It includes inhibitors of trypsin, exogenous α-amylases, bifunctional inhibitors and forms which apparently lack biological activity. The subunit M_r varies

	1 ↓5 6 10 20 25A 29	Ref
CMTI I	R V C P R I L M E C K K D S D C L A E C V C L E H - G Y C G	[7]
CMTI III	R V C P R I L M K C K K D S D C L A E C V C L E H - G Y C G	[7]
CMTI IV	H E E R V C P R I L M K C K K D S D C L A E C V C L E H - G Y C G	[13]
CPGTI I	R V C P K I L M K C K K D S D C L A E C I C L E H - G Y C G	[13]
CPTI II	R V C P K I L M K C K K D S D C L A E C I C L E H - G Y C G	[13]
CPTI III	H E E R V C P K I L M K C K K D S D C L A E C I C L E H - G Y C G	[13]
CSTI IIb	M V C P K I L M K C K H D S D C L L D C V C L E D I G Y C G V S	[13]
CSTI IV	M M C P R I L M K C K H D S D C L P G C V C L E H I E Y C G	[13]
CVTI I	G R R C P R I Y M E C K R D A D C L A D C V C L Q H - G I C G	[15]
BDTI II	R G C P R I L M R C K R D S D C L A G C V C Q K N - G Y C G	[15]
MR I	G I C P R I L M E C K R D S D C L A Q C V C K R Q - G Y C G	[16]
EETI II	G C P R I L M R C K Q D S D C L A G C V C G P N - G F C G	[22]
MCTI I	<E R R C P R I L K Q C K R D S D C P G E C I C M A H - G F C G	[17]
MCTI II	R I C P R I W M E C K R D S D C M A Q C I C V D - - G H C G	[17]
MCTI III	<E R G C P R I L K Q C K Q D S D C P G E C I C M A H - G F C G	[80]
MCTI A	R S C P R I W M E C T R D S D C M A K C I C V A - - G H C G	[21]
MCEI I	R I C P L I W M E C K R D S D C L A Q C I C V D - - G H C G	[17]
MCEI II	E R I C P L I W M E C K R D S D C L A Q C I C V D - - G H C G	[80]
MCEI III	E E R I C P L I W M E C K R D S D C L A Q C I C V D - - G H C G	[80]
MCEI IV	E E E R I C P L I W M E C K R D S D C L A Q C I C V D - - G H C G	[80]
LCTI 1	R I C P R I L M E C S S D S D C L A E C I C L E Q - G F C G	[19]
LCTI 2	R I C P R I L M E C S S D S D C L A E C I C L E Q D G F C G	[19]
LCTI II	R I C P R I L M E C S Y D S D C F G E C I C L P S - G Y C G	[75]
LCTI III	R I C P R I L M E C S S D S D C L A E C I C L E N - G F C G	[75]
LLDTI I	<E R R C P R I Y M E C K H D S D C L A D C V C L E H - G I C G	[18]
LLDTI II	R R C P R I Y M E C K H D S D C L A D C V C L E H - G I C G	[18]
LLTI I	<E R R C P R I Y M E C K H D S D C L A D C V C L E H - G I C G G	[20]
LLTI II	R R C P R I Y M E C K H D S D C L A D C V C L E H - G I C G	[20]
LLTI III	E R R C P R I Y M E C K H D S D C L A D C V C L E H - G I C G	[20]
BHTI I	R C P R I Y M E C K H D S D C L A D C V C L P Q - G I C G	[18]
BHTI III	R R C P R I Y M E C K H D S D C L A D C V C L P Q - G I C G	[18]
TTI I	C P R I L M P C K V N D D C L R G C K C L S N - G Y C G	[68]
TTI II	C P R I L M P C Q V N D D C L R G C K C L S N - G Y C G	[68]
HMTI	V G C P R I L M K C K T D D D C L L G C K C L S N - G Y C G	[23]
TGTI I	I C P R I L M P C S S D S D C L A E C I C L E N - G F C G	[23]
TGTI II	G I C P R I L M P C K T D D D C M L D C R C L S N - G Y C G	[23]
CMeTI A	R M C P K I L M K C K Q D S D C L L D C V C L K E - G F C G	[81]
CMeTI B	V G C P R I L M K C K T D R D C L T G C T C K R N - G Y C G	[81]
CMCTI I	M C P K I L N K C K Q D S D C L L D C V C L K E - G F C G	[79]
CMCTI II	R M C P K I L N K C K Q D S D C L L D C V C L K E - G F C G	[79]
CMCTI III	<E R M C P K I L N K C K Q D S D C L L D C V C L K E - G F C G	[79]

Fig. 6 Amino-acid sequences of squash inhibitors of serine proteinases. An arrow indicates the reactive site peptide bond. The inhibitor nomenclature is based on the name of the plant species (e.g. CM for *Cucurbita maxima*) and the target enzyme (TI, trypsin inhibitor; EI, elastase inhibitor). The species are: CM, *Cucurbita maxima*; CPG, *Cucurbita pepo* var. *Giromontia*; CP, *Cucurbita pepo*; CS, *Cucumis sativus*; CV, *Citrulus vulgaris*; BD, *Bryonia dioica*; MR, *Momordica repens*; EE, *Ecballium elaterium*; MC, *Momordica charantia*; LC, *Luffa cylindrica*; LLD, *Lagenaria leucantha* var. *Depressa*; LL, *Lagenaria leucantha* var. *Gourda*; BH, *Benincasa hispida*; T, *Trichosanthes*; HM, Hami melon; TG, towel gourd, CMe, *Cucumis melo*; CMC, *Cucumis melo* var. *Conomon*. Reproduced with permission from Otlewski and Krowarsch (1996) which gives full references to individual sequences.

from about 12-16,000 with monomeric, dimeric and tetrameric forms. These proteins are discussed in detail in chapter 26 while their relationships to the cereal prolamin superfamily are discussed in chapter 2.

The cereal inhibitor superfamily also includes the 2S albumin storage proteins discussed in chapter 24. Although most of these proteins appear to lack

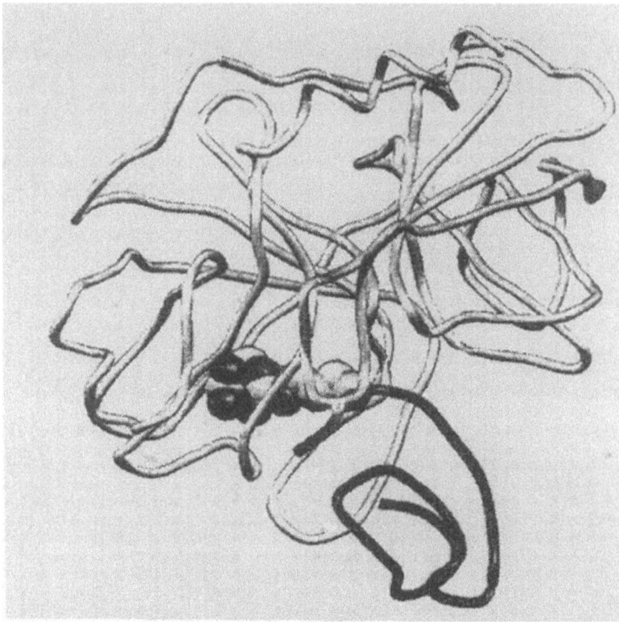

Fig. 7 General view of the CMTII-bovine β-trypsin complex, according to the 2.0Å crystal structure of the complex. The Asp189-Arg5 interaction is shown as space filling atoms. The main chain of the inhibitor is shaded. Reproduced from Otlewski and Krowarsch (1996) with permission of the publishers.

biological activity, three napin-type albumins from brassicas have been shown to have activity as trypsin inhibitors (see chapter 24).

3.6 The thaumatin-related family

This extensive family of proteins is named after the first member to be characterized. This is thaumatin, an intensely sweet protein from fruits of the West African shrub *Thaumatococcus danielii*, which does not exhibit any inhibitory activity. However, in 1987 Richardson *et al.*, described an M_r 22,000 protein from maize grain which inhibited bovine trypsin and α-amylase from *Tribolium castaneum* (red flour beetle) *in vitro* and had over 50% sequence identity with thaumatin. This protein, called zeamatin, has since been shown to belong to a group of antifungal proteins from seeds which have been called permeatins (Vigers *et al.*, 1991). Zeamatin, and related permeatins from other cereals (barley, oats, wheat, sorghum) and flax, inhibit the growth of a range of fungi (*Candida albicans*, *Neurospora crassa*, *Trichoderma reesii*, *Fusarium oxysporum*, *Alternaria solani*), probably by permeabilization of the hyphal membrane, resulting in leakage and rupture (Roberts and Selitrennikoff, 1986; 1990; Hejgaard *et al.*, 1991; Vigers *et al.*, 1991; Huynh *et al.*, 1992; Malehorn *et al.*, 1994). However, these studies have failed to show any inhibitory activity for purified zeamatin or permeatin from barley (Hejgaard *et al.*, 1991; Vigers *et al.*, 1991), casting doubt on the original report by Richardson *et al.* (1987).

In addition to the permeatins, the thaumatin-related family includes two other groups of proteins (Dudler *et al.*, 1994). These are the PR-5 group of pathogenesis-related proteins which are induced in tobacco and a range of other plants by infection with viruses or pathogenic fungi and the osmotins which are induced in a range of tissues in response to osmotic stress (see Cornelissen *et al.*, 1986; Dudler *et al.*, 1994; Shewry and Lucas, 1997). The PR-5 proteins and osmotins resemble the permeatins in having antifungal activity but neither group has been shown to contain enzyme inhibitors.

3.7 The cystatins

Cystatins are inhibitors of papain and other cysteine endoproteinases. Although initially characterized from egg white (Barrett, 1981), a large number of related proteins have since been characterized from animals (mainly vertebrates) and have been classified into three types (Barrett, 1987). In plants, cysteine proteinase inhibitors have been characterized from seeds of a number of species including rice (Abe and Arai, 1985), maize (Abe *et al.*, 1980; 1992; 1994), cowpea (Fernandes *et al.*, 1993) and *Wisteria* (Hirashiki *et al.*, 1990). They have also been reported in potato tubers (Hildmann *et al.*, 1992) and pineapple stems (Reddy *et al.*, 1975). Although this distribution is consistent with a protective role, the levels are generally lower and less variable than those of serine proteinase inhibitors, which is consistent with a metabolic rather than a protective role in most species (Fernandes *et al.*, 1991).

With the exception of the pineapple inhibitor, all the characterized plant cysteine proteinase inhibitors are related to the animal cystatins and comprise about 100-110 residues. They resemble the type I animal cystatins in the absence of disulphide bonds, but have stronger sequence similarity to the type 2 cystatins. It has, therefore, been suggested that they form a fourth type (Kondo *et al.*, 1990). The most intensively studied plant cystatin is oryzacystatin of rice, which consists of two forms called oryzacystatins I and II (Abe *et al.*, 1987a, b; 1988; Kondo *et al.*, 1990; Arai *et al.*, 1991). Whereas the oryzacystatins and the cowpea cystatin lack a signal sequence, one appears to be present on the maize inhibitor suggesting that it passes through the secretory pathway. The amino acid sequences of a range of plant cystatins are aligned in Fig. 8.

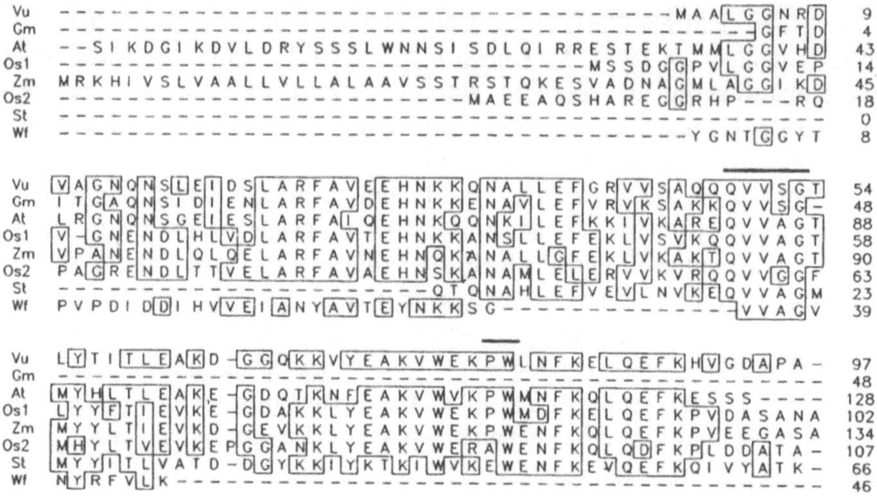

Fig. 8 Alignment of the cowpea cysteine protease inhibitor sequence (Vu) with those of other inhibitors of cysteine proteinase from different plant sources: soybean (Gm) (Brzin *et al.*, 1990); *Arabidopsis* (At; EMBL accession number Z176618); rice oryzacystatin 1 (Os1) (Abe *et al.*, 1987a); maize (Zm; EMBL accession number D10622); rice oryzacystatin 2 (Os2) (Kondo *et al.*, 1990); potato (St) (Hildmann *et al.*, 1992); and *Wisteria* (Wf) (Hirashiki *et al.*, 1990). The soybean, *Arabidopsis*, potato, and *Wisteria* sequences are incomplete at the *N*-terminus. Identical residues are boxed. The overlined pentapeptide sequence QVVSG is characteristic of inhibitors of cysteine proteinases, and the dipeptide sequence PW of the Type 2 cystatins (see Barratt, 1987). Reproduced from Fernandes *et al.* (1993) by kind permission of Kluwer Academic Publishers.

Although the *in vivo* role of the cystatins is still not established, there is clear evidence from feeding trials and transgenic plants that they can be exploited to confer resistance to a range of invertebrate pests which use cysteine endoproteinases as digestive enzymes. These include larvae of coleopteran insects (beetles), with a range of species having cysteine proteinases which are inhibited either by the chemical inhibitor E64 or cystatins from cowpea, rice or soybean (Gatehouse *et al.*, 1985; Wolfson and Murdock, 1987; Murdock *et al.*, 1987, 1988; Campos *et al.*, 1989; Lemos *et al.*, 1990; Thie and Houseman, 1990a, b; Hines *et al.*, 1991; Liang *et al.*, 1991; Silva and Xavier-Filho, 1991; Chen *et al.*, 1992; Edmonds *et al.*, 1996). Extracts of potato leaves expressing oryzacystatin I (Benchekroun *et al.*, 1995) and rice seeds expressing corn cystatin (Irie *et al.*, 1996) have also been shown to inhibit digestive proteinases of colorado beetle (*Leptinotarsa decemlineata*) and maize weevil (*Sitophilus zeamais*), respectively, but feeding trials with such transgenic material have not so far been reported. Lepidopteran larvae do not generally use cysteine proteinases for digestion so cystatins are unlikely to provide effective protection against insect pests of this group.

Although insect pests have been studied most widely with respect to protection by cystatins, the most spectacular success has been achieved with an unrelated invertebrate pest, the potato cyst nematode (*Globodera pallida*). Adult females of this species contain high levels of cysteine endoproteinases (Koritsas and Atkinson, 1994) which are inhibited by oryzacystatin I expressed in *E.coli* (Urwin *et al.*, 1995). In addition, the growth and development of the nematode were inhibited when the cystatin was expressed in hairy roots of tobacco, with a mutant form of the protein proving more effective than the wild-type protein (Urwin *et al.*, 1995). Further work showed that feeding of *Heterodera schachtii* (beet cyst nematode) and *Melidogyne incognita* (root knot nematode) on transgenic *Arabidopsis* expressing th. mutant form of oryzacystatin resulted in reduced growth and fecundity of females which was correlated with loss of cysteine proteinase activity in the intestine (Urwin *et al.*, 1997).

Work in our laboratory has also demonstrated that the major endoproteinase in the gut of the slug *Deroceras reticulatum*, a major pest of arable crops in Western Europe, is a cysteine proteinase which is inhibited by E64 and plant cystatins (Walker *et al.*, 1998). In addition, the growth of juvenile slugs was reduced and their mortality increased when fed leaves of transgenic *Arabidopsis* expressing the oryzacystatin I mutant of Urwin *et al.* (1995) (Walker, 1997).

These studies suggest that plant cystatins can be used to provide protection against a range of invertebrate pests, with protein engineering providing an opportunity to optimise their activity.

3.8 Mustard trypsin inhibitors

Menegatti *et al.* (1985) isolated a serine proteinase inhibitor (MTI-1) from seeds of white mustard (*Sinapis alba*) which inhibited bovine β-trypsin but had little activity against α-chymotrypsin. The protein had an M_r of about 18,000 and was highly labile to deactivation on heating to 80°C. This thermal lability was exploited to facilitate the purification of a second, and thermostable, inhibitor (MTI-2) which showed activity against β-trypsin and α-chymotrypsin (Menegatti *et al.*, 1992). Amino acid sequencing of MTI-2 showed the presence of 63 amino acid residues ($M_r \cong 7{,}000$) with eight cysteines which appeared to form four intra-chain disulphide bonds. The sequence showed no clear homology with those of other proteinase inhibitors, except for a short region of similarity to part of the rice bran Bowman-Birk inhibitor (residues 51-77 of the sequence in Table 1).

3.9 Serpins

The serpin superfamily of serine proteinase inhibitors is widely distributed in animal tissues, the best studied example being the human plasma α_1-antitrypsin (Carrell *et al.*, 1987).

The only plant serpin which has been studied in detail is a barley seed protein called protein Z. This is a major grain albumin, accounting for about 5% of the total fraction (Hejgaard, 1982). Like β-amylase and the chymotrypsin inhibitors CI-1 and CI-2, it is present at higher levels in the high lysine genotype Hiproly, where its relatively high lysine content (≅ 5.3 mol%) may contribute to the elevated lysine content of the whole grain (Hejgaard 1982). It is very resistant to degradation and denaturation during barley malting and brewing and is the only well-defined protein of barley origin present in beer, where it is called antigen-1 and may contribute to foam stability and/or haze formation (Hejgaard, 1977; Hejgaard and Kaersgaard, 1983). Protein Z also resembles β-amylase in being present in the mature grain in free and bound (thiol or papain-extractable) forms, with the free form apparently including some heterodimers with β-amylase (Hejgaard, 1977; Hejgaard and Carlsen, 1977).

Protein Z purified from mature barley grain consists of four antigenically-related forms of M_r about 40,000 and pI 5.56 to 5.8 (Hejgaard, 1982). These appear to be encoded by genes on two separate chromosomes, with the major form (Z_4) encoded by genes on chromosome 4 and a less abundant form (Z_7) by genes on chromosome 7 (Hejgaard, 1984). The complete amino acid sequence of Z_4, determined from partial protein and cDNA sequences and a full length genomic clone, shows it to consist of 399 amino acid residues with an M_r of 43,128 (Hejgaard *et al.*, 1985; Brandt *et al.*, 1990). It contains only two cysteine residues and its presence in thiol-extractable forms (see above) suggest that these do not form an intra-chain disulphide bond. It does not appear to be synthesised with an *N*-terminal signal peptide which is consistent with the lack of disulphide bond formation.

Early studies of purified protein Z_4 failed to show any inhibitory activity to serine proteinases which are inhibited by other serpins (Hejgaard *et al.*, 1985). However, it now appears that this probably resulted from technical problems (e.g. proteolysis, aggregation, instability at pH6) as heterologous expression of a protein Z_4 cDNA clone in *E.coli* gives protein which inhibits trypsin, chymotrypsin, cathepsin G, thrombin, kallikrein and other coagulation factors (Dahl *et al.*, 1996a,b). The protein does not inhibit trypsin from *Fusarium* and is rapidly cleaved as a substrate by subtilisin

(Dahl *et al.*, 1996b). Inhibitory activity to α-chymotrypsin has also been demonstrated for purified Z_7 (Lundgard and Svensson, 1989) while Dahl *et al.* (1996a) showed that a third barley protein Z gene, BSZ_x (Rasmussen, 1993), could be expressed in *E.coli* to give inhibitory protein, although the gene product was not detectable in the grain. Similarly, a related inhibitory serpin (Z_1) has been purified from wheat grain (Rosenkrands *et al.*, 1994) and reactive protein expressed in *E.coli* (Rasmussen *et al.*, 1996). Comparison of the inhibitory properties of the Z_4 and Z_x proteins of barley and the Z_1 protein of wheat expressed in *E.coli* showed differences related to their reactive loop sequences (Dahl *et al.*, 1996a)

3.10 Other α-Amylase Inhibitors

Four α-amylase inhibitors have been identified which do not belong to the proteinase inhibitor families discussed above.

3.10.1 The Ragi I-2 Inhibitor

Campos and Richardson (1984) isolated and sequenced a 95 residue protein from seeds of ragi (Indian finger millet, *Eleusine coracana*) which inhibited α-amylases from animals (human saliva, human and pig pancreas) (Shivaraj and Pattabiraman, 1980). A related cDNA clone was then isolated from barley and called "probable amylase/protease inhibitor" (PAPI) (Mundy and Rogers, 1986), as no biological activity had been identified. However, subsequent studies showed that both PAPI and the ragi I-2 inhibitor had sequence homology with the non-specific phospholipid transfer proteins (LTPs) (Bernhard and Somerville, 1989) and that PAPI was capable of transferring phosphatidyl choline from liposomes to potato mitochondria *in vitro* (Breu *et al.*, 1989). A range of LTPs have been characterized (see Kader, 1997) and current evidence indicates that they play an *in vivo* role in plant defence rather than lipid transport, as discussed in chapter 31. The ragi I-2 inhibitor remains the only member of this family which has been demonstrated to have biological activity as an enzyme inhibitor.

3.10.2 The bean inhibitor

Phaseolus vulgaris contains a lectin-like α-amylase inhibitor which is discussed in detail in chapter 27.

3.10.3 The *Coix* bifunctional α-amylase inhibitor/endochitinase

Seeds of Jobs tears (*Coix lachryma-jobi*), a minor panicoid cereal, contain a protein which inhibits α-amylase from locust gut and exhibits activity as an endochitinase (Ary *et al.*, 1989). The sequence of this protein shows strong homology with those of other plant endochitinases, some of which also display activity as lysozyme (see Collinge *et al.*, 1993). The bifunctional endochitinase/lysozyme activity is consistent with the marked similarity noted between the three-dimensional structures of barley endochitinase and human lysozyme (Holm and Sander, 1994). However, the *Coix* protein was not assayed for lysozyme activity and it is possible that this protein is in fact trifunctional.

3.10.4 The *Amaranthus* "knottin"-like inhibitor

Amaranthus hypochondriacus seeds contain a 32 residue peptide which has three intra-chain disulphide bonds and inhibits α-amylases from larvae of the insects *Tribolium castaneum* (red flour beetle) and *Prostephanus truncatus* (grain borer) (Chagolla-Lopez *et al.*, 1994). It shows some homology with the *Mirabilis jalapa* antimicrobial peptides (see chapter 31), with the squash family of proteinase inhibitors (see above) and with the "knottins"; a group of proteins which includes the ω-conotoxins of snake venoms and the cellulose binding domains of cellobiohydrolase (see Chagolla-Lopez *et al.*, 1994).

4. GENERAL PROPERTIES OF INHIBITORS

The proteinase and α-amylase inhibitors present in seeds can be classified into a number of families which vary widely in their structures and properties. Although there are weak structural homologues between some of these groups, as indicated in Table 1, many are related most closely to other protein groups which do not share their inhibitory properties. Most notably, the cereal inhibitors discussed in detail in chapter 26 belong to a major superfamily of seed proteins which include the prolamins and 2S albumin storage proteins (chapter 2). In some cases inhibitors of α-amylase and proteinases are present in the same family, with at least two families containing bifunctional inhibitors (The Kunitz and cereal families).

Although the various families of inhibitors vary widely in their structures, most are low M_r (≅ 3,000 to 20,000) proteins stabilized by extensive intra-chain disulphide bonds. However, exceptions do occur, for example, CI-1 and CI-2 lack disulphide bonds while the serpins are high M_r proteins which appear to lack intra-chain bonds. In addition, most appear to be active as single monomeric subunits, although dimeric and tetrameric forms of the cereal inhibitors occur. However, the latter are stabilized by non-covalent interactions and the individual subunits may exhibit biological activity, although at a lower level. Finally, most inhibitors are synthesised with signal peptides which direct them into the lumen of the endoplasmic reticulum and the secretory pathway. However, exceptions again occur, notably the serpins and some cystatins which appear to be synthesised in the cytoplasm.

The vast majority of the proteinase inhibitors are active against serine proteinases, notably trypsin but also other serine proteinases including chymotrypsin and subtilisin. However, a significant number of inhibitors of cysteine proteinases have been described over the past decade and there is little doubt that systematic searches will identify inhibitors of other proteinase groups, although these are probably present in smaller amounts than the major families of inhibitors discussed here.

It is also of interest that many seeds contain complex mixtures of inhibitors belonging to the same or different families. The most well known example of this is the Kunitz and Bowman-Birk type inhibitors which are present in significant amounts in seeds of many legumes, as discussed in chapter 27. Other examples occur in the major temperate cereals, wheat and barley. One or both of these species have been shown to contain active inhibitors of the Kunitz, Bowman-Birk, Potato I and cereal families together with apparently non-inhibitory members of the ragi I-2 and thaumatin-related families.

5. BIOLOGICAL ROLE OF INHIBITORS

The biological role of the inhibitors has been the subject of much speculation with no firm conclusions drawn. In some cases, their properties are consistent with a role in regulating metabolism within the plant. For example, the Kunitz-type bifunctional α-amylase/subtilisin inhibitors of cereals are inhibitory to endogenous α-amylase-2 and may regulate starch degradation *in vivo*. Similarly, the low and fairly constant levels of cysteine proteinase inhibitors present in many plant tissues has been interpreted as being consistent with a role in regulating metabolism (Fernandes *et al.*, 1991).

However, most of the inhibitors which accumulate in seeds are thought to play a role in providing broad spectrum protection against pests and pathogens, most notably to seed-eating insects. There are several lines of evidence for this. Firstly, many of the inhibitors are inactive against endogenous enzymes but show clear activity against digestive enzymes of insect pests. Secondly, they accumulate in high amounts in storage tissues which are particularly vulnerable to insect infestation. Thirdly, a range of inhibitor types may be present within a single tissue, widening the spectrum of activity. Similarly, the same seeds may also contain a range of other antifungal and protective proteins, as discussed in chapter 31 and by Shewry and Lucas (1997). Thus, the proteinase and α-amylase inhibitors discussed here may be components of a highly evolved system for providing broad protection against a range of specialised and opportunist pests and pathogens.

It is, of course, more difficult to demonstrate such a role *in vivo* than to propose a hypothesis based on circumstantial and correlative evidence. Nevertheless, supporting evidence is accumulating from genetic studies (e.g. comparison of null and normal isolines), from *in vitro* feeding tests and from *in planta* expression in transgenic plants (see Shewry and Lucas, 1997). Although no transgenic crops with improved resistance to pests and pathogens based on expression of inhibitors are currently grown commercially, there is little doubt that enzyme inhibitor genes will be successfully exploited for this purpose over the next decade.

ACKNOWLEDGEMENTS

IACR receives grant-aided support from the Biotechnology and Biological Sciences Research Council of the United Kingdom. The author is particularly grateful to Dr. Mike Richardson (Durham) for his advice and critical comments on the text.

REFERENCES

Abe, K. and Arai, S. (1985) Purification of a cysteine protease inhibitor from rice, *Oryza sativa* L. *japonica*. *Agricultural and Biological Chemistry* **49**, 3349-3350.

Abe, K., Arai, S., Kato, H. and Fujimaki, M. (1980) Thiol protease inhibitors occurring in endosperm of corn. *Agricultural and Biological Chemistry* **44**, 685-686.

Abe, K., Emori, Y., Kondo, H., Arai, S. and Suzuki, K. (1988) The NH_2-terminal 21 amino acid residues are not essential for the papain-inhibitory activity of oryzacystatin, a member of the cystatin superfamily. *Journal of Biological Chemistry* **263**, 7655-7659.

Abe, K., Emori, Y., Kondo, H., Suzuki, K. and Arai, S. (1987a) Molecular cloning of a cysteine proteinase inhibitor of rice (oryzacystatin). *Journal of Biological Chemistry* **262**, 16793-16797.

Abe, K., Kondo, H. and Arai, S. (1987b) Purification and characterisation of a rice cysteine protease inhibitor. *Agricultural and Biological Chemistry* **51**, 2763-2768.

Abe, M., Abe, K., Iwabuchi, K., Domoto, C. and Arai, S. (1994) Corn cystatin I expressed in *Escherichia coli*: investigation of its inhibitory profile and occurrence in corn kernels. *Journal of Biochemistry* **116**, 488-492.

Abe, M., Abe, K., Kuroda, M. and Arai, S. (1992) Corn kernal cysteine protease inhibitor as a novel cystatin superfamily member of plant origin. *European Journal of Biochemistry* **209**, 933-937.

Arai, S., Watanabe, H., Kondo, H., Emori, Y. and Abe, K. (1991) Papain-inhibitory activity of oryzacystatin, a rice seed cysteine protease inhibitor, depends on the central Gln-Val-Val-Ala-Gly region conserved among cystatin superfamily members. *Journal of Biochemistry* **109**, 294-298.

Ary, M.B., Shewry, P.R. and Richardson, M. (1988) The amino acid sequence of a cereal Bowman-Birk type trypsin inhibitor from seeds of Jobs tears (*Coix lachryma-jobi* L.). *FEBS Letters* **229**, 111-118.

Ary, M.B., Richardson, M. and Shewry, P.R. (1989) Purification and characterization of an insect α-amylase inhibitor/endochitinase from seeds of Jobs tears (*Coix lachryma-jobi*). *Biochimica et Biophysica Acta* **993**, 260-266.

Barrett, A.J. (1981) Cystatin, the egg white inhibitor of cysteine proteinases. *Methods in Enzymology* **80**, 771-778.

Barrett, A.J. (1987) The cystatins: a new class of peptidase inhibitors. *Trends in Biochemical Science* **12**, 193-196.

Benchekroun, A., Michaud, D., Nguyen-Quoc, B., Overney, S., Desjardins, Y. and Yelle, S. (1995) Synthesis of active oryzacystatin I in transgenic potato plants. *Plant Cell Reports* **14**, 585-588.

Bernhard, W.R. and Somerville, C.R. (1989) Coidentity of putative amylase inhibitors from barley and finger millet with phospholipid transfer proteins inferred from amino acid sequence homology. *Archives of Biochemistry and Biophysics* **269**, 695-697.

Bode, W., Greyling, H.J., Huber, R., Otlewski, J. and Wilusz, T. (1989) The refined 2.0 Å X-ray crystal structure of the complex formed between bovine β-trypsin and CMTI-I, a trypsin inhibitor from squash seeds (*Cucurbita maxima*). *FEBS Letters* **242**, 285-292.

Bowman, D.E. (1946) Differentiation of soybean antitrypsin factors. *Proceedings of the Society for Experimental Biology and Medicine* **63**, 547-550.

Brandt, A., Svendsen, I. and Hejgaard, J. (1990) A plant serpin gene. Structure, organization and expression of the gene encoding barley protein Z_4. *European Journal of Biochemistry* **194**, 499-505.

Breu, V., Guerbette, F., Kader, J.-C., Kannagara, C.G., Svensson, B. and von Wettstein-Knowles, P. (1989) A 10 kDa barley basic protein transfers phosphatidylcholine from liposomes to mitochondria. *Carlsberg Research Communications* **54**, 81-84.

Brown, W.E., Takio, K., Titani, K. and Ryan, C.A. (1985) Wound-induced trypsin inhibitor in alfalfa leaves: Identity as a member of the Bowman-Birk inhibitor family. *Biochemistry* **24**, 2105-2108.

Brzin, J., Ritonja, A., Popovic, T. and Turk, V. (1990) Low molecular mass protein inhibitor of cysteine proteinases from soybean. *Biologistito Chimie Hoppé-Seyler Supplement* **371**, 167-170.

Brzin, J. and Kidric, M. (1995) Proteinases and their inhibitors in plants: Role in normal growth and in response to various stress conditions. *Biotechnology and Genetic Engineering Reviews* **13**, 421-467.

Campbell, A.F. (1992) Protein engineering of the barley chymotrypsin inhibitor 2. In: Plant Protein Engineering, eds. Shewry, P.R. and Gutteridge, S. Edward Arnold, Sevenoaks, U.K. pp.257-268.

Campos, F.A.P. and Richardson, M. (1984) The complete amino acid sequence of the α-amylase inhibitor 1-2 from seeds of ragi (Indian finger millet, *Eleusine coracana* Gaern.). *FEBS Letters* **167**, 221-225.

Campos, F.A.P., Xavier-Filho, J., Silva, C.P. and Ary, M.B. (1989) Resolution and partial characterization of proteinases and α-amylases from midguts of larvae of the bruchid beetle *Callosobruchus maculatus* (F.). *Comparative Biochemistry and Physiology* **92B**, 51-57.

Carrell, R.W., Pemberton, P.A. and Boswell, D.R. (1987) The serpins: Evolution and adaptation in a family of protease inhibitors. *Cold Spring Harbor Symposia on Quantitative Biology* **52**, 527-535.

Chagolla-Lopez, A., Blanco-Labra, A., Patthy, A., Sánchez, R. and Pongor, S. (1994) A novel α-amylase inhibitor from amaranth (*Amaranthus hypocondriacus*) seeds. *Journal of Biological Chemistry* **269**, 23675-23680.

Chapman, G.W. (1987) A proteinaceous competitive inhibitor of lipase isolated from *Helianthus annuus* seeds. *Phytochemistry* **26**, 3127-3131.

Chen, M.-S., Johnson, B., Wen, L., Muthukrishnan, S., Kramer, K.J., Morgan, T.D. and Reeck, G.R. (1992) Rice cystatin: bacterial expression, purification, cysteine proteinase inhibitory activity, and insect growth suppressing activity of a truncated form of the protein. *Protein Expression and Purification* **3**, 41-49.

Clore, G.M., Gronenborn, A.M., Kjaer, M. and Poulsen, F.M. (1987a) The determination of the three-dimensional structure of barley serine proteinase inhibitory 2 by nuclear magnetic resonance, distance geometry and restrained molecular dynamics. *Protein Engineering* **1**, 305-311.

Clore, G.M., Gronenborn, A.M., James, M.N.G., Kjaer, M., McPhalen, C.A. and Poulsen, F.M. (1987b) Comparison of the solution and X-ray structures of barley serine proteinase inhibitor 2. *Protein Engineering* **1**, 313-318.

Collinge, D.B., Kragh, K.M., Mikkelsen, J.D., Nielsen, K.K., Rasmussen, U. and Vad, K. (1993) Plant chitinases. *The Plant Journal* **3**, 31-40.

Cornelissen, B.J.C., Hooft van Huijsduijnen, R.A.M. and Bol, J.F. (1986) A tobacco mosaic virus-induced tobacco protein is homologous to the sweet-tasting protein thaumatin. *Nature* **321**, 531-532.

Dahl, S.W., Rasmussen, S.K. and Hejgaard, J. (1996a) Heterologous expression of three plant serpins with distinct inhibitory specificities. *Journal of Biological Chemistry* **271**, 25083-25088.

Dahl, S.W., Rasmussen, S.K., Petersen, L.C. and Hejgaard, J. (1996b) Inhibition of coagulation factors by recombinant barley serpin BSZx. *FEBS Letters* **394**, 165-168.

Dudler, R., Mauch, F. and Reimmann, C. (1994) Thaumatin-like proteins. In "Thaumatin" (M. Witty and J.D. Higginbotham, eds), pp.193-199. CRC Press Inc., Boca Raton FL.

Edmonds, H.S., Gatehouse, L.N., Hilder, V.A. and Gatehouse, J.A. (1996) The inhibitory effects of the cysteine protease inhibitor, oryzacystatin, on digestive proteases and on larval survival and development of the southern corn rootworm (*Diabrotica undecimpunctata howard*). *Entomologia Experimentalis et Applicata* **78**, 83-94.

Fernandes, K.V.S., Campos, F.A.P., Do Val, R.R. and Xavier-Filho, J. (1991) The expression of papain inhibitors during development of cowpea seeds. *Plant Science* **74**, 179-184.

Fernandes, K.V.S., Sabelli, P.A., Barratt, D.H.P., Richardson, M., Xavier-Filho, J. and Shewry, P.R. (1993) The resistance of cowpea seeds to bruchid beetles is not related to levels of cysteine proteinase inhibitors. *Plant Molecular Biology* **23**, 215-219.

Gargouri, Y., Julien, R., Pieroni, G., Verger, R. and Sarda, L. (1984) Studies on the inhibition of pancreatic and microbial lipases by soybean proteins. *Journal of Lipid Research* **25**, 1214-1221.

Gatehouse, A.M.R., Butler, K.J., Fenton, K.A. and Gatehouse, J.A. (1985) Presence and partial characterisation of a major proteolytic enzyme in the larval gut of *Callosobruchus maculatus*. *Entomologia Experimentalis et Applicata* **39**, 279-286.

Graham, J.S., Pearce, G., Merryweather, J., Titani, K., Ericsson, L. and Ryan, C.A. (1985) Wound-induced proteinase inhibitors from tomato leaves. I. The cDNA-deduced primary structure of pre-inhibitor I and its posttranslational processing. *Journal of Biological Chemistry* **260**, 6555-6560.

Halayko, A.J., Hill, R.D. and Svensson, B. 1986. Characterisation of the interaction of barley α-amylase II with an endogenous α-amylase inhibitor from barley kernels. *Biochimica et Biophysica Acta* **873**, 91-101.

Heitz, A., Chiche, L., Le-Nguyen, D. and Castro, B. (1989) 1H 2D NMR and distance geometry study of the folding of *Ecballium elaterium* trypsin inhibitor, a member of the squash inhibitors family. *Biochemistry* **28**, 2392-2398.

Hejgaard, J. (1977) Origin of a dominant beer protein. Immunochemical identity with a β-amylase-associated protein from barley. *Journal of the Institute of Brewing* **83**, 94-96.

Hejgaard, J. (1982) Purification and properties of protein Z - A major albumin of barley endosperm. *Physiologia Plantarum* **54**, 174-182.

Hejgaard, J. (1984) Gene products of barley chromosomes 4 and 7 are precursors of the major antigenetic beer protein. *Journal of the Institute of Brewing* **90**, 75-87.

Hejgaard, J. and Carlsen, S. (1977) Immunoelectrophoretic identification of a heterodimer β-amylase in extracts of barley grain. *Journal of the Science of Food and Agriculture* **28**, 900-904.

Hejgaard, J. and Boisen, S. (1980) High lysine proteins in Hiproly barley breeding: identification, nutritional significance and new screening methods. *Hereditas* **93**, 311-320.

Hejgaard, J. and Kaersgaard, P. (1983) Purification and properties of the major antigenic beer protein of barley origin. *Journal of the Institute of Brewing* **89**, 402-410.

Hejgaard, J., Rasmussen, S.K., Brandt, A. and Svendsen, I. (1985) Sequence homology between barley endosperm protein Z and protease inhibitors of the α_1-antitrypsin family. *FEBS Letters* **180**, 89-94.

Hejgaard, J., Jacobsen, S. and Svendsen, I. (1991) Two antifungal thaumatin-like proteins from barley grain. *FEBS Letters* **291**, 127-131.

Hildmann, T., Ebneth, M., Peña-Cortés, H., Sánchez-Serrano, J.J., Willmitzer, L. and Prat, S. (1992) General roles of abscisic and jasmonic acids in gene activation as a result of mechanical wounding. *Plant Cell* **4**, 1157-1170.

Hines, M.E., Osuala, C.I. and Nielsen, S.S. (1991) Isolation and partial characterization of a soybean cystatin cysteine proteinase inhibitor of Coleopteran digestive proteolytic activity. *Journal of Agricultural and Food Chemistry* **39**, 1515-1520.

Hirashiki, I., Ogata, F., Yoshida, N., Makisumi, S. and Ito, A. (1990) Purification and complex formation analysis of a cysteine proteinase inhibitor (cystatin) from seeds of *Wisteria floribunda*. *Journal of Biochemistry* **108**, 604-608.

Holak, T.A., Bode, W., Huber, R., Otlewski, J. and Wilusz, T. (1989a) Nuclear magnetic resonance and X-ray structures of squash trypsin inhibitor exhibit the same conformation of the proteinase binding loop. *Journal of Molecular Biology* **210**, 649-654.

Holak, T.A., Gondol, D., Otlewski, J. and Wilusz, T. (1989b) Determination of the complete three-dimensional structure of the trypsin inhibitor from squash seeds in aqueous solution by nuclear magnetic resonance and a combination of distance geometry and dynamic simulated annealing. *Journal of Molecular Biology* **210**, 635-648.

Holak, T.A., Habazettl, J., Oschkinat, H. and Otlewski, J. (1991) Structure of proteins in solution derived from homonuclear three-dimensional NOE-NOE nuclear magnetic resonance spectroscopy. High resolution structure of squash trypsin inhibitor. *Journal of the American Chemical Society* **113**, 3196-3198.

Holm, L. and Sander, C. (1994) Structural similarity of plant chitinase and lysozymes from animals and phage. *FEBS Letters* **340**, 129-132.

Huang, Q., Liu, S. and Tang, Y. (1992) Refined 1.6 Å resolution crystal structure of the complex formed between porcine β-trypsin and MCTI-A, a trypsin inhibitor of the squash family. *Journal of Molecular Biology* **229**, 1022-1036.

Huang, A.H.C. and Wang, S.M. (1992) Proteinaceous inhibitors of lipase activities in soybean and other oil seeds. In: Seed Analysis, eds. Linskens, H.-F. and Jackson, J.F. Springer-Verlag, Berlin. pp.263-271

Huynh, Q.K., Borgmeyer, J.R. and Zobel, J.F. (1992) Isolation and characterization of a 22 kDa protein with antifungal properties from maize seeds. *Biochemical and Biophysical Research Communications* **182**, 1-5.

Irie, K., Hosoyama, H., Takeuchi, T., Iwabuchi, K., Watanabe, H., Abe, M., Abe, K. and Aria, S. (1996) Transgenic rice established to express corn cystatin exhibits strong inhibitory activity against insect gut proteinases. *Plant Molecular Biology* **30**, 149-157.

Joubert, F.J., Kruger, H., Townshend, G.S. and Botes, D.P. (1979) Purification, some properties and the complete primary structures of two protease inhibitors (DE-3 and DE-4) from *Macrotyloma axillare* seed. *European Journal of Biochemistry* **97**, 85-91.

Kader, J.-C. (1997) Lipid-transfer proteins: a puzzling family of plant proteins. *Trends in Plant Science* **2**, 66-70.

Kneen, E. and Sandstedt, R.M. (1943) An amylase inhibitor from certain cereals. *Journal of the American Chemical Society* **65**, 1247.

Kondo, H., Abe, K., Nishimura, I., Watanabe, H., Emori, Y. and Arai, S. (1990) Two distinct cystatin species in rice seeds with different specificities against cysteine proteinases. *Journal of Biological Chemistry* **26**, 15832-15837.

Koritsas, V.M. and Atkinson, H.J. (1994) Proteinases of females of the phytoparasite *Globodera pallida* (potato cyst nematode). *Parasitology* **109**, 357-365.

Kortt, A.A., Strike, P.M., and De Jersey, J. (1989) Amino acid sequence of a crystalline seed albumin (winged bean albumin-1) from *Psophocarpus tetragonolobus* (L.) DC. Sequence similarity with Kunitz-type seed inhibitors and 7S storage globulins. *European Journal of Biochemistry* **181**, 403-408.

Krishnamoorthi, R., Lin, C.-L.S., Gong, Y.X., VanderVelde, D., and Hahn, K. (1992) Proton NMR studies of *Cucurbita maxima* trypsin inhibitors: evidence for pH-dependent conformational change and His25-Tyr27 interaction. *Biochemistry* **31**, 905-910.

Kunitz, M. (1945) Crystallization of a trypsin inhibitor from soybean. *Science* **101**, 668-9.

Lairon, D., Borel, P., Termine, E., Grataroli, R., Chabert, C., and Hauton, J.C. (1985) Evidence for a proteinic inhibitor of pancreatic lipase in cereals, wheat bran and wheat germ. *Nutrition Reports International* **32**, 1107-1113.

Leah, R. and Mundy, J. 1989. The bifunctional α-amylase/subtilisin inhibitor of barley: Nucleotide sequence and patterns of seed-specific expression. *Plant Molecular Biology* **12**, 673-682.

Lemos, F.J.A., Campos, F.A.P., Silva, C.P. and Xavier-Filho, J. (1990) Proteinases and amylases of larval midgut of *Zabrotes subfasciatus* reared on cowpea (*Vigna unguiculata*) seeds. *Entomologia Experimentalis et Applicata* **56**, 219-227.

Liang, C., Brookhart, G., Feng, G.H., Reeck, G.R. and Kramer, K.J. (1991) Inhibition of digestive proteinases of stored grain coleoptera by oryzacystatin, a cysteine proteinase inhibitor from rice seed. *FEBS Letters* **278**, 139-142.

Likos, J.J. (1989) ^{1}H NMR studies of squash seed trypsin inhibitor. *International Journal of Peptide Protein Research* **34**, 381-386.

Ling, M.-H., Qi, H. and Chi, C. (1993) Protein, cDNA and genomic DNA sequences of the towel gourd trypsin inhibitor. *Journal of Biological Chemistry* **268**, 810-814.

Liu, J.-H. and Hill, R.D. (1995) Post-transcriptional regulation of bifunctional α-amylase/subtilisin inhibitor expression in barley embryos by abscisic acid. *Plant Molecular Biology* **29**, 1087-1091.

Lundgard R. and Svensson, B. (1989) A 39 kD barley seed protein of the serpin superfamily inhibits α_1-chymotrypsin. *Carlsberg Research Communications* **54**, 173-180.

Maeda, K. (1986) The complete amino-acid sequence of the endogenous α-amylase inhibitor in wheat. *Biochimica et Biophysica Acta* **871**, 250-256.

Makkar, H.P.S., Sharma, O.P., Pal, R.N. and Negi, S.S. (1980) In vitro inhibition of rumen urease by melon (*Cucumis melo*) seed urease inhibitor. *Journal of Dairy Science* **63**, 785-788.

Malehorn, D.E., Borgmeyer, J.R., Smith, C.E. and Shah, D.M. (1994) Characterization and expression of an antifungal zeamatin-like protein (ZIp) gene from *Zea mays*. *Plant Physiology* **106**, 1471-1481

McCoy, A.J. and Kortt, A.A. (1997) The 1.8 Å crystal structure of winged bean albumin 1, the major albumin from *Psophocarpus tetragonolobus* (L.) DC. *Journal of Molecular Biology* **269**, 881-891.

McPhalen, C.A. and James, M.N.G. (1987) Crystal and molecular structure of the serine proteinase inhibitor C1-2 from barley seeds. *Biochemistry* **26**, 261-269.

McPhalen, C.A., Schnebli, H.P. and James, M.N.G. (1985a) Crystal and molecular structure of the inhibitor eglin from leeches in complex with subtilisin Carlsberg. *FEBS Letters* **188**, 55-58.

McPhalen, C.A., Svendsen, I., Jonassen, I. and James, M.N.G. (1985b) Crystal and molecular structure of chymotrypsin inhibitor 2 from barley seeds in complex with subtilisin Novo. *Proceedings of the National Academy of Sciences USA*, **82**, 7242-7246.

Melville, J.C. and Ryan, C.A. (1972) Chymotrypsin inhibitor I from potatoes: large scale preparation and the characterization of its subunit components. *Journal of Biological Chemistry* **247**, 3445-3453.

Menegatti, E., Palmieri, S., Walde, P. and Luisi, P.L. (1985) Isolation and characterization of a trypsin inhibitor from white mustard (*Sinapis alba* L.). *Journal of Agricultural and Food Chemistry* **33**, 784-789.

Menegatti, E., Tedeschi, G., Ronchi, S., Bortolotti, F., Ascenzi, P., Thomas, R.M., Bolognesi, M. and Palmieri, S. (1992) Purification, inhibitory properties and amino acid sequence of a new serine proteinase inhibitor from white mustard (*Sinapis alba* L.) seed. *FEBS Letters* **301**, 10-14.

Mundy, J., Hejgaard, J., and Svendsen, I. (1984) Characterization of a bifunctional wheat inhibitor of endogenous α-amylase and subtilisin. *FEBS Letters* **167**, 210-214.

Mundy, J. and Rogers, J.C. (1986) Selective expression of a probable amylase/protease inhibitor in barley aleurone cells: comparison to the barley amylase/subtilisin inhibitor. *Planta* **169**, 51-63.

Murdock, L.L., Brookhart, G., Dunn, P.E., Foard, D.E., Kelley, S., Kitch, L., Shade, R.E., Shukle, R.H. and Wolfson, J.L. (1987) Cysteine digestive proteinases in Coleoptera. *Comparative Biochemistry and Physiology* **87**, 783-787.

Murdock, L.L., Shade, R.E. and Pomeroy, M.A. (1988) Effects of E-64, a cysteine proteinase inhibitor, on cowpea weevil growth, development, and fecundity. *Environmental Entomology* **17**, 467-469.

Nilges, M., Habazettl, J., Brünger, A.T., Holak, T.A. (1991) Relaxation matrix refinement of the solution structure of squash trypsin inhibitor. *Journal of Molecular Biology* **219**, 499-510.

Norioka, S. and Ikenaka, T. (1983) Amino acid sequences of trypsin-chymotrypsin inhibitors (A-I, A-II, and B-II) from peanut (*Arachis hypogaea*): A discussion on the molecular evolution of legume Bowman-Birk type inhibitors. *Journal of Biochemistry* **94**, 589-599.

Norton, G. (1991) Proteinase Inhibitors. In: D'Mello, J.P.F. *et al.,* Toxic substances in Crop Plants. pp.68-106. ISBN-0-85186-863-0. Royal Society of Chemistry: Cambridge, England.

Odani, S. and Ikenaka, T. (1973) Studies on soybean trypsin inhibitors VIII. Disulfide bridges in soybean Bowman-Birk proteinase inhibitor. *Journal of Biochemistry* **74,** 697-715.

Odani, S. and Ikenaka, T. (1977) Studies on soybean trypsin inhibitors. XI. Complete amino acid sequence of a soybean trypsin-chymotrypsin-elastase inhibitor, C-II. *Journal of Biochemistry* **82**, 1523-1531.

Odani, S., Koide, T. and Ono, T. (1986) Wheat germ trypsin inhibitors. Isolation and structural characterization of single-headed and double-headed inhibitors of the Bowman-Birk type. *Journal of Biochemistry* **100**, 975-983.

Ohtsubo, K.-I., and Richardson, M. (1992) The amino acid sequence of a 20 kDa bifunctional subtilisin/ α-amylase inhibitor from brain of rice (*Oryza sativa* L.) seeds. *FEBS Letters* **309**, 68-72.

Onesti, S., Brick, P. and Blow, D.M. (1991) Crystal structure of a Kunitz-type trypsin inhibitor from *Erythrina caffra* seeds. *Journal of Molecular Biology* **217**, 153-176.

Otlewski, J. and Krowarsch, D. (1996) Squash inhibitor family of serine proteinases. *Acta Biochimica Polonica* **43**, 431-444.

Rasmussen, S.K. (1993) A gene coding for a new plant serpin. *Biochimica et Biophysica Acta* **1172**, 151-154.

Rasmussen, S.K., Dahl, S.W., Nørgård, A. and Hejgaard, J. (1996) A recombinant wheat serpin with inhibitory activity. *Plant Molecular Biology* **30**, 673-677.

Reddy, M.N., Keim, P.S., Heinrikson, R.L. and Kezdy, F.J. (1975) Primary structural analysis of sulphydryl protease inhibitors from pineapple stem. *Journal of Biological Chemistry* **250**, 1741-1750.

Richardson, M. (1977) The proteinase inhibitors of plants and micro-organisms. *Phytochemistry* **16**, 159-169.

Richardson, M. (1991) Seed storage proteins: the enzyme inhibitors. *Methods in Plant Biochemistry* **5,** 259-305.

Richardson, M., Valdes-Rodriquez, S. and Blanco-Labra, A. (1987) A possible function for thaumatin and a TMV-induced protein suggested by homology to a maize inhibitor. *Nature* **327**, 432-434.

Roberts, W.K. and Selitrennikoff, C.P. (1986) Isolation and partial characterization of two antifungal proteins from barley. *Biochimica et Biophysica Acta* **880**, 161-170.

Roberts, W.K. and Selitrennikoff, C.P. (1990) Zeamatin, an antifungal protein from maize with membrane-permeabilizing activity. *Journal of General Microbiology* **136**, 1771.

Robertson, M., Walker-Simmonds, M., Munro, D. and Hill, R.D. (1989) Induction of α-amylase inhibitor synthesis in barley embryos and young seedlings by abscisic acid and dehydration stress. *Plant Physiology* **91**, 415-420.

Rosenkrands, I., Hejgaard, J., Rasmussen, S.K. and Bjørn, S.E. (1994) Serpins from wheat grain. *FEBS Letters* **343**, 75-80.

Ryan, C.A. (1990) Protease inhibitors in plants: genes for improving defenses against insects and pathogens. *Annual Review of Phytopathology* **28**, 425-449.

Sadowski, J., MacGregor, A. and Daussant, J. (1986) Alpha-amylase inhibitor in cereals: comparison of the protein in different rye, wheat and triticale seeds by using immunoblotting. *Electrophoresis* **7**, 176-179.

Seemuller, U., Meier, M., Ohlsson, K., Muller, H.-P. and Fritz, H. (1977) Isolation and characterisation of a low molecular weight inhibitor (of chymotrypsin and human granulocytic elastase and cathepsin G) from leeches. *Hoppe-Seyler's Zeitschrift für Physiologische Chemie* **358**, 1105-17.

Shewry, P.R. (1993) Barley seed proteins. In: Barley: Chemistry and Technology (Eds. MacGregor, J. and Bhatty, R.) AACC St. Paul Minnesota, U.S.A. pp. 131-197.

Shewry, P.R. and Lucas, J.A. (1997) Plant proteins that confer resistance to pests and pathogens. *Advances in Botanical Research* **26**, 135-192.

Shimokawa, Y., Kuromizu, K., Araki, T., Ohata, J. and Abe, O. (1984) Primary structure of *Vicia angustifolia* proteinase inhibitor. *European Journal of Biochemistry* **143**, 677-684.

Shivaraj, B. and Pattabiraman, T.N. (1980) Natural plant enzyme inhibitors: Part VIII - Purification and properties of two α-amylase inhibitors from ragi (*Eleusine coracana*) grains. *Indian Journal of Biochemistry and Biophysics* **17**, 181-185.

Silva, C.P. and Xavier-Filho, J. (1991) Comparison between the levels of aspartic and cysteine proteinases of the larval midguts of *Callosobruchus maculatus* (F.) and *Zabrotes subfasciatus* (BOH.) (Coleoptera: Bruchidae). *Comparative Biochemistry and Physiology* 99, 529-533.

Suzuki, A., Tsunogae, Y., Tanaka, I., Yamane, T., Ashida, T., Norioka, S., Hara, S. and Ikenaka, T. (1987) The structure of Bowman-Birk type protease inhibitor A-II from peanut (*Arachis hypogaea*) at 3.3 Å resolution. *Journal of Biochemistry* **101**, 267-274.

Svendsen, I. Boisen, S. and Hejgaard, J. (1982) Amino acid sequence of serine protease inhibitor CI-1 from barley. Homology with barley inhibitor C1-2, potato inhibitor 1 and leech eglin. *Carlsberg Research Communications* **47**, 45-53.

Svendsen, I., Martin, B. and Jonassen, I. (1980a) Characteristics of Hiproly barley III. Amino acid sequences of two lysine rich proteins. *Carlsberg Research Communications* **45**, 79-85.

Svendsen, I., Jonassen, I., Hejgaard, J. and Boisen, S. (1980b) Amino acid sequence homology between a serine protease inhibitor from barley and potato inhibitor I. *Carlsberg Research Communications* **45**, 389-395.

Svendsen, I.B., Hejgaard, J. and Mundy, J. (1986) Complete amino acid sequence of the α-amylase/subtilisin inhibitor from barley. *Carlsberg Research Communications* **51**, 43-50.

Sweet, R.M., Wright, H.T., Janin, J., Chothia, C.H., and Blow, D.M. (1974) Crystal structure of the complex of porcine trypsin with soybean trypsin inhibitor (Kunitz) at 2.6 - ? resolution. *Biochemistry* **20**, 4212-4228.

Tashiro, M., Hashino, K., Shiozaki, M., Ibuki, F. and Maki, Z. (1987) The complete amino acid sequence of rice bran trypsin inhibitor. *Journal of Biochemistry* **102**, 297-306.

Täufel, A., Behnke, U., Emmer, I. and Gabor, R. (1991) Studies on the germination specific α-amylase and its inhibitor of rye (*Secale cereale*). *Zeitschrift für Lebensmittel-untersuchung und - Forchung* **193**, 9-14.

Thie, N.M.R. and Houseman, J.G. (1990a) Identification of cathepsin B,D and H in the larval midgut of colorado potato beetle, *Leptinotarsa decemlineata* say (Coleoptera: Chrysomelidae). *Insect Biochemistry* **20**, 313-318.

Thie, N.M.R. and Houseman, J.G. (1990b) Cysteine and serine proteolytic activities in larval midgut of yellow mealworm, *Tenebrio molitor* L. (Coleoptera: Tenebrionidae). *Insect Biochemistry* **20**, 741-744.

Tsaftaris, A.S., Sorenson, J.C. and Scandalios J.G. (1980) Glycosylation of catalase inhibitor necessary for activity. *Biochemical and Biophysical Research Communications* **92**, 889-895.

Tsunogae, Y., Tanaka, I., Yamane, T., Kikkawa, J., Ashida, T., Ishikawa, C., Watanabe, K., Nakamura, S. and Takahashi, K. (1986) Structure of the trypsin-binding domain of Bowman-Birk type protease inhibitor and its interaction with trypsin. *Journal of Biochemistry* **100**, 1637-1646.

Urwin, P.E., Atkinson, H.J., Waller, D.A. and McPherson, M.J. (1995) Engineered oryzacystatin-I expressed in transgenic hairy roots confers resistance to *Globodera pallida*. *The Plant Journal* **8**, 121-131.

Urwin, P.E., Lilley, C.J., McPherson, M.J. and Atkinson, H.J. (1997) Resistance to both cyst and root-knot nematodes conferred by transgenic *Arabidopsis* expressing a modified plant cystatin. *The Plant Journal* **12**, 455-461.

Van Loon, L.C. (1985) Pathogenesis-related proteins. *Plant Molecular Biology* **4**, 111-116.

Vigers, A.J., Roberts, W.K. and Selitrennikoff, C.P. (1991) A new family of plant antifungal proteins. *Molecular Plant-Microbe Interactions* **4**, 315-323.

Walker, A.J. (1997) Characterisation of the proteolytic activity in the digestive system of the field slug *Deroceras reticulatum* as a target for novel methods of control. *PhD Thesis, University of Bristol.*

Walker, A.J., Glen, D.M. and Shewry, P.R. (1998) Purification and characterisation of a digestive cysteine proteinase from the field slug (*Deroceras reticulatum*): A potential target for slug control. *Journal of Agricultural and Food Chemistry.* In press

Wang, S-M. and Huang, A.H.C. (1984) Inhibitors of lipase activities in soybean and other oil seeds. *Plant Physiology* **76**, 929-934.

Werner, M.H. and Wemmer, D.E. (1992) Three-dimensional structure of soybean trypsin/chymotrypsin Bowman-Birk inhibitor in solution. *Biochemistry* **31**, 999-1010.

Weselake, R.J., MacGregor, A.W. and Hill, R.D. (1985) Endogenous alpha-amylase inhibitor in various cereals. *Cereal Chemistry* **62**, 120-123.

Williamson, M.S., Forde, J., Buxton, B. and Kreis, M. (1987) Nucleotide sequence of barley chymotrypsin inhibitor-2 (C1-2) and its expression in normal and high-lysine barley. *European Journal of Biochemistry* **165**, 99-106.

Williamson, M.S., Forde, J. and Kreis, M. (1988) Molecular cloning of two isoinhibitor forms of chymotrypsin inhibitor 1 (Cl-1) from barley endosperm and their expression in normal and mutant barleys. *Plant Molecular Biology* **10**, 521-535.

Wilson, K.A. (1997) The protease inhibitors of seeds. In: Cellular and Molecular Biology of Plant Seed Development, eds. Larkins, B.A. and Vasil, I.K. Kluwer Academic Publishers, Dordrecht. pp.331-374.

Wilson, K.A. and Chen, J.C. (1983) Amino acid sequence of mung bean trypsin inhibitor and its modified forms appearing during germination. *Plant Physiology* **71**, 341-349.

Wolfson, J.L. and Murdock, L.L. (1987) Suppression of larval Colorado potato beetle growth and development by digestive proteinase inhibitors. *Entomologia Experimentalis et Applicata* **44**, 235-240.

Xavier-Filho, J. and Campos, F.A.P. (1989) Proteinase inhibitors. In "Toxicants of Plant Origin" (P.R. Cheeke, Ed.), Vol. III: Proteins and Amino Acids, pp.1-27. CRS Press Inc., Boca Raton, FL.

Yang, H-L., Luo R-S., Wang, L-X., Zhu, D-X. and Chi, C-W. (1992) Primary structure and disulfide bridge location of arrowhead double-headed proteinase inhibitors. *Journal of Biochemistry* **111**, 537-545.

Yoshikawa, M., Iwasaki, T., Fujii, M. and Oogaki, M. (1976) Isolation and some properties of a subtilisin inhibitor from barley. *Journal of Biochemistry* **79**, 765-773.

Yoshikawa, M., Kiyohara, T., Iswasaki, T., Ishii, Y. and Kimura, N. (1979) Amino acid sequences of proteinase inhibitors II and II' from adzuki beans. *Agricultural and Biological Chemistry* **43**, 787-796.

Zemke, K.J., Müller-Fahrnow, A., Jany, K.-D., Pal, G.P. and Saenger, W. (1990) The three -dimensional structure of the bifunctional proteinase K/ α-amylase inhibitor from wheat (PKI3) at 2.5 Å resolution. *FEBS Letters* **279,** 240-242.

Zeng, F.-Y., Qian, R.-Q. and Wang, Y. (1988) The amino acid sequence of a trypsin inhibitor from the seeds of *Momordica charantia* Linn. Cucurbitaceae. *FEBS Letters* **234**, 35-38.

26. A Multigene Family of Trypsin/α-Amylase Inhibitors from Cereals

PILAR CARBONERO and FRANCISCO GARCÍA-OLMEDO
Laboratorio de Bioquímica y Biología Molecular, Departamento de Biotecnología - UPM, E.T.S. Ingenieros Agrónomos, E-28040 Madrid, Spain

1. CHARACTERISTICS.

1.1 Distribution and solubility.

Plant proteinaceous inhibitors of hydrolases from heterologous systems (fungi, insects, vertebrates, etc.) have been extensively studied. Recent advances in the molecular biology of these inhibitors have greatly increased our knowledge of their structure and *in vitro* properties, and have allowed previously unsuspected relationships between functionally different proteins to be established.

In wheat and barley, a substantial fraction of the total endosperm protein content is represented by toxins and inhibitors that are active towards heterologous systems. In these cereals, a single protein family of trypsin/α-amylase inhibitors is prominently represented among albumins and globulins from endosperm. More than 20 different members from this family have been characterized (reviewed by Carbonero *et al.*, 1993; García-Olmedo *et al.*, 1987; 1992). Their apparent molecular weights are in the 12-16,000 range, and some of them can be selectively extracted with chloroform/methanol mixtures and have therefore been designated CM-proteins. The α-amylase inhibitors can be classified according to their degree of aggregation into monomeric, dimeric, and tetrameric forms. The trypsin inhibitors are monomeric. No members of this family have been found in tissues other than endosperm, although this aspect deserve further investigation. During kernel development, their synthesis precedes that of the main storage proteins and they are rapidly degraded upon germination. These inhibitors have been found not only in wheat and barley but also in other species of the Poaceae (Gramineae) such as rye, rice, maize and finger-millet (Carbonero *et al.*, 1993; García-Olmedo *et al.*, 1987; 1992).

P.R. Shewry and R. Casey (eds.), Seed Proteins, 617–633.

1.2 Amino acid sequences

Amino acid sequences of members of this inhibitor family from wheat, barley and other cereals - directly determined or deduced from nucleotide sequences of cDNA clones - have been aligned in Figure 1. The sequences have been organized into three domains (A, B, C) and grouped into ten subfamilies according to sequence similarity and *in vitro* activity. In addition to the homology relationships summarized in Figure 1, weaker and more elusive relationships have been proposed between this family and the 2S storage proteins from dicots, with dispersed sulphur-rich domains from cereal prolamins and even with the Kazal secretory trypsin inhibitor from bovine pancreas (reviewed in García-Olmedo *et al.*,1992).

Barley trypsin inhibitor BTI-CMe, one of the best characterized members of the family (Odani *et al.*, 1983; Rodriguez-Palenzuela *et al.*, 1989; Royo *et al.*, 1996; Shewry *et al.*, 1984), belongs to the same subfamily as the trypsin inhibitors from rye (RTI), maize (MTI) and ragi (RBI) (Campos and Richardson, 1983; Lyons *et al.*, 1987; Mahoney *et al.*, 1984; Wen *et al.*, 1992). The wheat homologue has not yet been isolated, although a related cDNA (pCMx) has been characterized (Sanchez de la Hoz *et al.*, 1994). The reactive site of BTI-CMe is the motif glycine-proline-arginine-leucine (GPRL) that is located at the right-hand border of the A domain, a region that is quite variable throughout the family. This same reactive site appears in RTI, MTI and RBI. Two related members, represented by cDNA clones pUP23 from barley (Lazaro *et al.*, 1988a) and its wheat homoeologue pUP-88 have the variant sequence proline-serine-arginine-proline in the same position (Carbonero *et al.*, 1993). However, *in vitro* inhibition tests of the native proteins would be needed to confirm that these proteins are not trypsin inhibitors.

The monomeric α-amylase inhibitor from barley, BMAI-1, whose amino acid sequence has been deduced from its cDNA, is glycosylated, a feature that is related to its allergenic properties (Barber *et al.*, 1989; Mena *et al.*, 1992), as will be discussed below. An homologous protein from rice, RAP, is also a major allergen and its cDNA and genomic DNA have been characterized (Adachi *et al.*, 1993; Izumi *et al.*, 1992). This protein has an extension of six amino acids (aspartate-histidine-histidine-glutamine-valine-tyrosine) with respect to the *N*-terminal serine of the barley BMAI-1 inhibitor (Izumi *et al.*, 1992; Mena *et al.*, 1992). Two homologous monomeric α-amylase inhibitors have been purified from wheat and the cDNA from one of them (WMAI-1) has been characterized (García-Maroto *et al.*, 1991; Gomez *et al.*, 1991; Kashlan and Richardson, 1981).

The amino acid sequences of the homodimeric α-amylase inhibitors encoded by genes from the B and D genomes of wheat are close to each other, but that of the barley homodimeric inhibitor, BDAI-1, is closer to those of the wheat monomeric inhibitors than to those of either the barley monomeric or the wheat dimeric inhibitors (Carbonero *et al.*, 1993; Lazaro *et al.*, 1988b; Maeda *et al.*, 1985).

Fig. 1. Alignment of amino acid sequences of cereal inhibitors of α-amylase/trypsin. Sequences are divided into A, B and C domains and have been grouped according to similarity and *in vitro* activities. The reactive sites of trypsin inhibitors (proline-arginine-leucine) at the end of the A domain are underlined indicated with a wavy line and putative *N*-glycosylation sites (asparragine-X-serine/threonine) in pUP23, BMAI-1, BDA-1, WTAI-CM16, WTAI-CM17 and BTAI-CMb are indicated with a double underline. Only partial *N*-terminal sequences are available for WMAI-2 and WDAI-3.

A

BTI-CME	--FGDSCAPGDALPH	NPLRACRTYVVSQIC	HQGPRLLTSD-----	-----------
RTI	-SVGGQCVPGLAMPH	NPLGACRTYVVSQIC	HVGPRLFTWD-----	-----------
RBI	-SVGTSCIPGMAIPH	NPLDSCRWYVSTRTC	GVGPRLATQE-----	-----------
MTI	-SAGTSCVPGWAIPH	NPLPSCRWYVTSRTC	GIGPRLPWPEGRLE-	-----------
PUP-23	-SVKDECQLGVDFPH	NPLATCHTYVIKRVC	GRGPSRPMLV-----	-----------
PUP-88	-SVEDECQPGVAFPH	NALATCHTYVIKRVC	GRGPSRPMLV-----	-----------
BTI-CMc	-TSIYTCYEGMGLPV	NPLQGCRFYVASQTC	GAVPLLPIEV-----	-----------
BMAI-1	-SPGEWCWPGMGYPV	YPFPRCRALVKSQ-C	AG-GQVVESIQ----	-----------
RAP	-SPGEQCRPGISYPT	YSLPQCRTLVRRQ-C	VGRGASAADEQV---	-----------
BDAI-1	SGPWMWCDPEMGHKV	SPLTRCRALVKLE-C	VG-----NRVPEDVL	-----------
WMAI-1	SGPWSWCNPATGYKV	SALTGCRAMVKLQ-C	VGSQVPEAVL-----	-----------
WMAI-2	SGPWMWCDPAMGYRV	SPLTGCRAMVKLQ-C	VGSQVPEA-------	-----------
WDAI-1	SGPWM-CYPGQAFQV	PALPGCRPLLKLQ-C	NGSQVPEAVL-----	-----------
WDAI-2	SGPWM-CYPGQAFQV	PALPACRPLLRLQ-C	NGSQVPEAVL-----	-----------
WDAI-3	SGPWM-CYPGYAFKV	PALPGCRPVLLLQ-C	NGSQVPEAVL-----	-----------
WTAI-CM1	--TGPYCYAGMGLPI	NPLEGCREYVASQTC	GIS-ISGSAVSTEPG	NT---------
WTAI-CM2	--TGPYCYPGMGLPS	NPLEGCREYVAQQTC	GVGIIVGSPVSTEPG	NT---------
BTA1-CMa	--TGQYCYAGMGLPS	NPLEGCREYVAQQTC	GVT-IAGSPVSSEPG	DT---------
WTAI-CM16	-IGNEDCTPWMSTLI	TPLPSCRDYVEQQAC	RIETPGS--------	-----------
WTAI-CM17	---NEDCTPWTSTLI	TPLPSCRNYVEEQAC	RIEMPGPPYL-----	-----------
BTAI-CMb	-VGSEDCTPWTATPI	TPLPSCRDYVEQQAC	RIETPGPPYL-----	-----------

WTAI-CM3	---SGSCVPGVAFRT	NLLPHCRDYVLQQTC	TFTPGSKLPEWMTSA	S-IYSPGKPYL
BTAI-CMd	AAAATDCSPGVAFPT	NLLGHCRDYVLQQTC	AVLTPGSKLPEWMTS	AELNYPGQPYL

B

BTI-CME	MKRRCCDELSAIP-	AYCRCEALRIIMQGV	VTWQGA--------F	EGAYFK----
RTI	MKRRCCDELLAIP-	AYCRCEALRILMDGV	VTQQGV--------F	EGGYLK----
RBI	MKARCCRQLEAIP-	AYCRCEAVRILMDGV	VTSSGQ--------H	EGRLLQ----
MTI	LKRRCCRELADIP-	AYCRCTALSILMDGA	IPP-GP---------	DAQLE-----
PUP-23	-KERCCRELAAVP-	DHCRCEALRILMDGV	RTPEG-RWEGRLG--	----------
PUP-88	-KERCCRELAVVP-	DYCRCEALRVLMDGV	RAEEGHVVEGRLG--	----------
BTI-CMc	MKDWCCRELAGISS	N-CRCEGLRVFIDRA	FPPSQSQ--GAPPQL	PPL-------
BMAI-1	--KDCCRQIAAIGD	EWCICGALGSMRGSM	YKELGVA--------	LADDKATVAE
RAP	-WQDCCRQLAAVDD	GWCRCGALDHMLSGI	YRELGAT--------	EAGHPMAE--
BDAI-1	--RDCCQEVANISN	EWCRCGDLGSMLRSV	YAALGVG--------	GGPEE-----
WMAI-1	--RDCCQQLADINN	EWCRCGDS-SMLRSV	YQELGVR--------	EGKE------
WDAI-1	--RDCCQQLADIS-	EWPRCGALYSMLDSM	YKEHGVS--------	EGQAGTG---
WDAI-2	--RDCCQQLAHIS-	EWCRCGALYSMLDSM	YKEHGAQ--------	EGQAGTG---
WDAI-3	--RD-CQQ------	---------------	---------------	----------
WTAI-CM1	PRDRCCKELYDAS-	QHCRCEAVRYFIGR-	-RSDPN---------	SGVLK-----
WTAI-CM2	PRDRCCKELYDAS-	QHCRCEAVRYFIGR-	-TSDPN---------	SGVLK-----
WTA1-CMa	PKDRCCQELDEAP-	QHCRCEAVRYFIGR-	-RSHPD---------	WSVLK-----
WTAI-CM16	AKQQCCGELANIP-	QQCRCQALRYFMGP-	-KSRPD--------Q	SGLM------
BTAI-CMb	AKQQCCGELANIP-	QQCRCQALRFFMGR-	-KSRPD--------Q	SGLM------
WTAI-CM17	AKQECCEQLANIP-	QQCRCQALRYFMGP-	-KSRPD--------Q	SGLM------
WTAI-CM3	AKLYCCQELAEIS-	QQCRCEALRYFIALP	VPSQPVDPRSGNVGE	SGLI------
BTAI-CMd	AKLYCCQELAKIP-	QQCRCEALRYFMALP	VPSQPVDPSTGNVGQ	SGLM------

C

BTI-CME	DSPNCPRERQTSYAA	NLVTPQECNLGTIHG	S-----AYCPELQPG	YGVVL
RTI	DMPNCPRVTQRSYAA	TLVAPQECNLPTIHG	S-----PYCPTLQAG	Y
RBI	DLPGCPRQVQRAFAP	KLVTEVECNLATIHG	G-----PFCLSLLGA	GE
MTI	DLPGCPREVQRGFAA	TLVTEAECNLATISG	V-----AECPWILGG	GTMPSK
PUP-23	DRRDCPREEQPAFAA	TLVTAAECNLSSVQE	P-----GVRLVLLAD	G

```
PUP-88      DRRDCPREAQREFAA TLVTAAECNLPTVS- ------GVGSTLGAT GRWMTIELPK

BTI-CMc     AT-ECPAEVKRDFAR TLALPGQCNLPAIHG G-----AYCVFP

BMAI-1      VFPGCRTEV--MDRA VASLPAVCNQYIPNT NGT--DGVCY--WLS YYQPPRQMSSR
RAP         VFPGCRRGD--LERA AASLPAFCNVDIPNG PGG---VVCY--WLG YPRTPRTGH

BDAI-1      VFPGCQKDV--MKLL VAGVPALCNVPIPNE A-AGTRGVCY--WSA STDT

WMAI-1      VLPGCRKEV--MKLT AASVPEVCKVPIPNP SGD-RAGVCYGDWAA YPDV

WDAI-1      AFPSCRREV--VKLT AASITAVCRLPIVVD ASGDGAYVCK-DVAA YPDA
WDAI-2      AFPRCRREV--VKLT AASITAVCRLPIVVD ASGDGAYVCK-DVAA YPDA

WTAI-CM1    DLPGCPREPQRDFAK VLVTSGHCNVMTVHN A-----PYCLGLDI
WTAI-CM2    DLPGCPREPQRDFAK VLVTPGHCNVMTVHN T-----PYCLGLDI
BTA1-CMa    DLPGCPKEPQRDFAK VLVTPGQCNVLTVHN A-----PYCLGLDI

WTAI-CM16   ELPGCPREVQMDFVR ILVTPGYCNLTTVHN T-----PYCLAMEES QWS
WTAI-CM17   ELPGCPREVQMNFVP ILVTPGYCNLTTVHN T-----PYCLGMEES QWS
BTAI-CMb    ELPGCPREVQMDFVR ILVTPGFCNLTTVHN T-----PYCLAMDEW QWNRQFCSS

WTAI-CM3    DLPGCPREMQWDFVR LLVAPGQCNLATIHN V-----RYCPAVEQP LWI
BTAI-CMd    DLPGCPREMQRAFVR LLVAPGQCNLATIHN V-----RYCPAVEQP LWI
```

A barley tetrameric α-amylase inhibitor has been characterized and its subunits have been identified as the previously described proteins CMa, CMb and CMd (Sanchez-Monge *et al*., 1986). The CMd subunit, the most hydrophobic of the three, is present in two copies in the tetramer. The complete sequences of WTAI-CMa, WTAI-CMb and WTAI-CMd have been deduced from the corresponding cDNAs (Halford *et al*., 1988; Medina *et al*., 1993; Paz-Ares, *et al*., 1986; Rasmussen and Johansson, 1992). Tetrameric inhibitors have also been characterized in hexaploid wheat *Triticum aestivum* (genomes AABBDD), as well as in tetraploid *Triticum turgidum* (AABB) and in diploid *Triticum tauschii* (DD) and cDNAs corresponding to the three types of subunits have been cloned (García-Maroto *et al*., 1990).

2. GENETICS.

2.1 Chromosomal locations of inhibitor genes.

The trypsin/α-amylase inhibitors are encoded by a multi-gene family dispersed over several chromosomes both in wheat and in barley. Present knowledge of the genomic organization of this family is summarized in Table 1. Five out of the seven homoeologous chromosome groups carry genes coding for these inhibitors. Genes in the A genome of hexaploid wheats appear to be silenced (pseudogenes) (García-Maroto *et al.*, 1990).

2.2 Gene expression

In all the cases where a complete cDNA has been characterized, the mature protein is preceded by a typical signal peptide of aproximately 30 amino acids, which is in agreement with the observation that the synthesis of these inhibitors takes place in membrane-bound polysomes as pre-proteins that are co-translationally processed (Paz-Ares *et al.*, 1983).

The effects of high-lysine mutations on the expression of different genes from this family in barley have been investigated. The most remarkable effect concerns the gene for trypsin inhibitor BTI-CMe, which is regulated in *trans* by the *Lys3a* locus. The CMe protein is present in the mature endosperm of the mutant Risϕ 1508 at less than 2-3% of the wild type level, and the steady state level of the CMe-mRNA is about 1% (Lazaro *et al.*, 1985; Rodriguez-Palenzuela *et al.*, 1989; Royo *et al.*, 1996; Salcedo *et al.*, 1984). Southern blot analysis of wheat-barley addition lines has shown that chromosome 3H of barley carries the gene for CMe. One or two copies of the CMe gene per haploid genome have been estimated both in the wild type and in the mutant and DNA restriction patterns are identical in both stocks, so neither a change in copy number nor a major rearrangement of the structural gene account for the markedly decreased expression. The mutation at the *Lys3a* locus in Risϕ 1508 has been previously mapped in chromosome 5H. A single dose of the wild type allele at this locus (*Lys3a*) restores the expression of gene CMe in chromosome 3H to normal levels (Rodriguez-Palenzuela *et al.*, 1989).

The gene *Itr1*, encoding trypsin inhibitor BTI-CMe from barley, has been functionally analysed by transient expression in protoplasts derived from different barley tissues showing that, under these conditions, its promoter retains both its endosperm specificity and its regulation in *trans* by the *Lys3a* gene (Royo *et al.*, 1996). The proximal promoter extending 343 bp upstream of the initiation codon is sufficient to confer endosperm-specific expression in wild-type protoplasts, whereas expression in protoplasts from the Risϕ1508 *lys3a* mutant was less than 5 % of that in wild type protoplasts (barley cv.

Bomi). Nuclear proteins extracted from the two types of endosperm gave differential patterns in gel retardation experiments. Several transcription factors belonging to the bZIP class have been isolated from barley and their possible involvement in the regulation of the *Itr1* gene is currently being investigated (Vicente-Carbajosa *et al.*, 1998; Oñate *et al.*, in preparation).

Table 1 Chromosomal Location of Genes Encoding Trypsin/α-Amylase Inhibitors

Inhibitory activity against	Aggregation	Protein	Gene	Chromosome, genome, arm(*)
TRYPSIN	monomeric	BTI-CMe	*Itr1*	3HS
		RTI	*Itr-R1*	3R
		BTI-CMc	*Itr2*	7HS
α -AMYLASE	monomeric	BMAI-1	*Iam1*	2H
		WMAI-1 (syn. 0.28)	*Imha-D1*	6DS
		WMAI-2	*Imha-B1*	6BS
α -AMYLASE	homodimeric	BDAI-1	*Iad1*	6H
		WDAI-1 (syn. 0.53)	*IdhaB1.1*	3BS
		WDAI-2 (syn. 0.19)	*IdhaD1.1*	3DS
		WDAI-3	*IdhaB1.2*	3BS
α -AMYLASE	tetrameric			
1st SUBUNIT		BTAI-CMa	*Iat1*	7HS
		WTAI-CM1	*IthaD1*	7DS
		WTAI-CM2	*IthaB1*	7BS
2nd SUBUNIT		BTAI-CMb	*Iat2*	4HL
		WTAI-CM16	*IthaB2*	4BS
		WTAI-CM17	*IthaD2*	4DS
3rd SUBUNIT (2 copies)		BTAI-CMd	*Iat3*	4HL
		WTAI-CM3B	*IthaB3*	4BS
		WTAI-CM3D	*IthD3*	4D
UNKNOWN				
clone pUP23	-	-	-	6HL
clone pUP88	-	-	-	6AL
				6BL
				6DL

(*)S= short, L= long chromosome arms; B,D= wheat (genomes AABBDD); H= barley (genome HH); R= rye (genome RR).

3. BIOLOGICAL PROPERTIES.

3.1 Inhibition of insect enzymes

Barley BTI-CMe and its maize homologue (MTI) are not only active against trypsin but also against Hageman factor XII-a of the blood-clotting cascade, and BTI-CMe is also active against Kallikrein (Chong and Reeck, 1987). BTI-CMe is inactive against chymotrypsin, papain, pepsin, bacterial and fungal proteases and the endogenous barley proteases (Mikola and Soulinna, 1969), as well as against α-amylases (Barber *et al.*, 1986b), which is in contrast with the bifunctional activity of its homologues from maize and ragi (Chen *et al.*, 1992; Shivaraj and Pattabiraman, 1981).

The monomeric, dimeric and tetrameric α-amylase inhibitors from wheat and barley differentially inhibit α-amylases from different origins. In general, the amylase inhibitors purified from barley are less effective than those from wheat. The wheat dimeric class is more active towards the α-amylase from human saliva than against those from insect pests (see Carbonero *et al.*, 1993). The enzymes from both *Tenebrio molitor* (Coleoptera) and *Ephestia kuehniella* (Lepidoptera) are significantly more sensitive than human salivary α-amylase to the monomeric wheat inhibitor WMAI-1. A given inhibitor class may also discriminate whithin an insect group: the *Leptinotarsa decemlineata* enzyme is more affected by the homodimeric inhibitors than by the monomeric ones, while the opposite is true for the α-amylase of *Tenebrio molitor*, both coleopterous insects (Gutierrez *et al.*, 1990). The Lepidoptera seem to be more susceptible to the tetrameric inhibitors than to the monomeric or dimeric ones (Gutierrez *et al.*, 1993).

Although no *in vivo* function is known for the cereal trypsin/α-amylase inhibitor family, the following aspects can be noted:

i) Different inhibitors show different specificities towards enzymes from different insects (Gutierrez *et al.*, 1990; 1993).

ii) Considerable intra- and interspecific variation in inhibitor levels has been observed (Gomez *et al.*, 1989; Kirsi and Ahokas, 1983).

iii) Gene silencing has occurred in some cases (García-Maroto *et al.*, 1990).

These characteristics, together with the low genetic variability at given loci, suggest that these proteins are involved in plant defense, probably as components of the non-host resistance mechanism rather than in relation to more specific interactions. More direct evidence for a defense role stems from experiments with insect pests. Thus, insects which are able to feed on wheat endosperm have unusually high levels of α-amylase (Gutierrez *et al.*, 1990). High inhibitor concentrations in an artifitial diet were required to affect the development of larvae of *Tribolium confusum*, a storage pest of wheat products, while quite low concentrations were effective against *Callosobruchus maculatus*, a pest of legume seeds (Gatehouse *et al.*, 1986). More recently, transgenic tobacco plants expressing the inhibitors BTI-CMe from barley or WMAI-1 (syn. 0.28) from wheat, under the control of the 35S promoter, have been found to be lethal in leaf-disc assays to larvae from Lepidoptera, such as *Agrotis ipsilon* and *Spodoptera littoralis* (Table2).

Table 2 Insect feeding assays in transgenic (Ro) tobacco leaves

Transgenic plant expressing	Plant no	% Mortality (L_1 to L_3)	
		Agrotis ipsilon	*Spodoptera littoralis*
BTI-CMe	2	22.6	-
	3	62.6*	37.1
	4	52.5*	32.8
	6	30.0	-
WMAI-1	6	72.7*	23.0
	8	17.4	-
	9	38.3	-
	12	65.7*	39.2
Untransformed tobacco	-	15.0	10.5
Artificial diet	-	14.5	12.0

3.2 Allergenic properties

The allergic asthma of workers with occupational exposure to cereal flour is due in part to the allergenic properties of some members of this protein family. Most of these inhibitors isolated from wheat, barley and rice are recognized by specific IgE when tested with sera from allergic patients (Barber *et al.*, 1989; García-Casado *et al.*, 1995; 1996; Gomez *et al.*, 1990; Izumi *et al.*, 1992; Sanchez-Monge *et al.*, 1992; 1996a,b). However, their IgE-binding capacities *in vitro* are very different (Sanchez-Monge *et al.*, 1992). In wheat and barley the glycosylated forms of WTAI-CM16, BTAI-CMb and BMAI-1 have been found to be the most prominent allergens, both *in vitro* and *in vivo* (García-Casado *et al.*, 1995; Sanchez-Monge *et al.*, 1992). cDNA clones encoding these three proteins have also been isolated (García-Maroto *et al.*, 1990; Medina *et al.*,1993; Mena *et al.*, 1992).

4. STRUCTURE

4.1 The reactive site of α-amylase inhibitors

Although, as already indicated, the reactive sites of trypsin inhibitors have been known for some time, it had been speculated that the ability to inhibit α-amylases was mediated by the carbohydrate moieties of the glycosylated members of this family. This question has been finally clarified through mutagenesis. Inhibitor WMAI-1 produced in *Escherichia coli* using the PT7-7 expression vector had the same specific activity towards the α-amylase from the insect *Tenebrio molitor* as the native WMAI-1 purified from wheat. This confirms that the native inhibitor, although presenting a putative *N*-glycosylation site (asparagine-proline-serine), is not glycosylated and contradicts the previous claim that a glycosyl moiety was essential for inhibition (Silano *et al.*, 1977). Site-directed mutagenesis of different regions of the inhibitor (Fig. 2) has shown that modifications of the highly conserved *N*-terminal sequence (serine-glycine-proline-tryptophan) increased the pre-incubation time required for maximum activity, while insertions in the middle of the B domain (position 58) led to inactivation (García-Maroto *et al.*, 1991). When the disulphide-bridge structure of this inhibitor was subsequently established (Poerio *et al.*, 1991), position 58 was found to be close to the *N*-terminus (Fig. 2), which suggests that both regions are part of the reactive site.

4.2 Disulphide bridges and 3D structure determination.

The location of the disulphide bridges within these cysteine-rich molecules has been investigated to a limited extent. The four-bridge structure (9 cysteines) of the wheat dimeric amylase inhibitor WDAI-1 (syn. 0.53), represented in Fig.3, was described by Maeda *et al.* (1983). The presence of one additional cysteine in the sequence of the wheat monomeric amylase inhibitor WMAI-1 (syn. 0.28) was found by Poerio *et al.* (1991) to imply not only the formation of a fifth disulphide bridge, but a general rearrangement of the disulphide structure (Fig. 3). More recently, the 3D structure of the RBI bifunctional α-amylase inhibitor from ragi has been reported (Strobl *et al.*, 1995). This inhibitor has the same disulphide structure as WMAI-1 (Fig. 3). The RBI inhibitor consists of a globular four-helix motif with a simple "up-and-down" topology (Strobl *et al.*, 1995) and there is an antiparallel β-sheet motif between the 3rd and the 4th helices (Plate 10). A location of the putative α-amylase binding site on the face of the molecule opposite to the trypsin-binding loop has been postulated (Strobl *et al.*,1995). Barley trypsin inhibitor

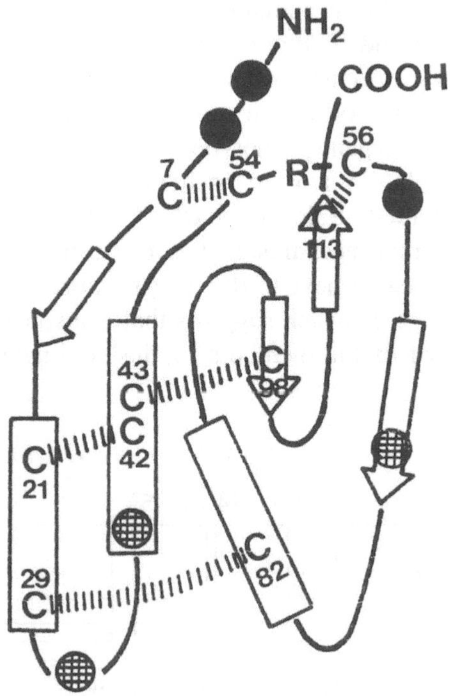

Fig. 2. Schematic view of wheat monomeric α-amylase inhibitor WMAI-1 (syn. 0.28), indicating mutagenised points. Sites where mutagenesis leads to inactivation or diminished affinity towards the α-amylase of *Tenebrio molitor* are depicted as black circles and the mutagenised sites not leading to inactivation as reticulate circles. Beta-sheets (⇒) and alpha-helixes (□) in the predicted secondary structure are indicated. Disulphide bridges are presented as (lllllll).

BTI-CMe, which has ten cysteines and 55% coincident (69% similar) residues with RBI, is likely to have the same disulphide structure.

5. EVOLUTION

The evolution of this dispersed multigene family raises a number of interesting issues that have been only investigated in a preliminary way:

i) Dispersal of the gene family over several chromosomes of a given genome must have involved both intra-chromosomal duplications and inter-chromosomal translocations. Although some degree of synteny is observed in the gene locations among species, there are significant variations between closely-related species, such as wheat and barley (Fig. 1; Table 1). A well-studied case of intra-chromosomal duplication is that affecting the *Itr1* locus in *Hordeum vulgare* and *H. spontaneum*. A single active *Itr1* gene is present in most *H. vulgare* cultivars, whereas a duplication of this locus exists in some *H. vulgare* cultivars and in most *H. spontaneum* accessions, and the duplication shows divergent phenotypes within the latter species leading both to active trypsin inhibitors or to pseudogenes (Molina-Cano *et al.*, 1987; Royo *et al.*, submitted; Salcedo *et al.*, 1984;). As to the possible inter-chromosomal dispersal mechanism, it is to be noted that the *Itr1* gene has been shown to be located next to the long terminal repeat of the "copia-like" retro-transposon *Bare-1*, which suggests that transposition may have played a role in the dispersal of the members of this multigene family (Royo *et al.*, 1996).

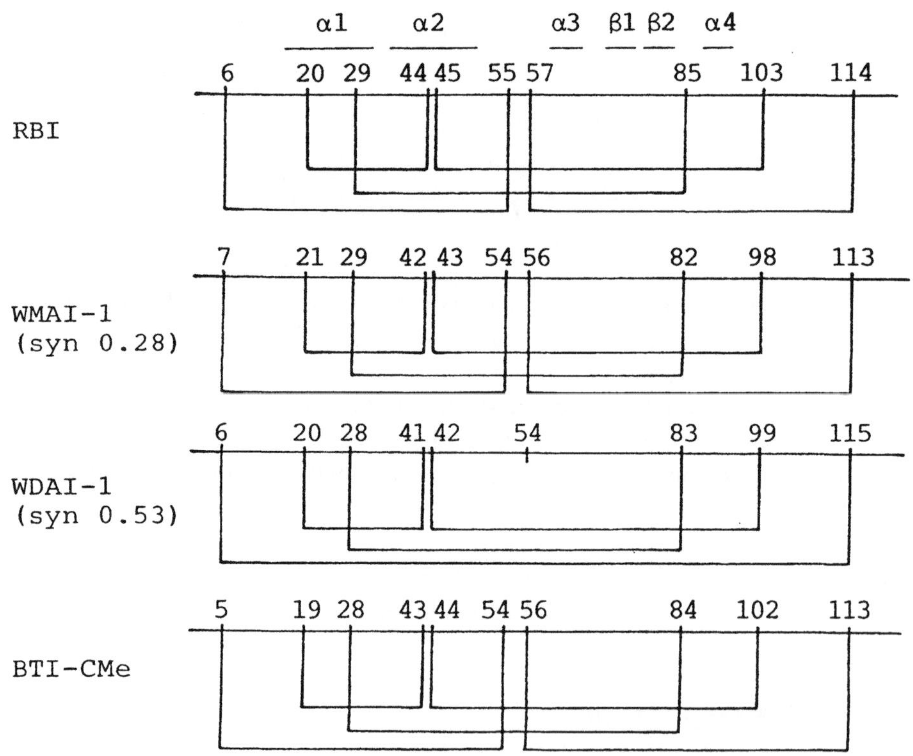

Fig. 3. Schematic representation of the disulphide bond patterns of RBI, WMAI-1 (syn. 0.28) and WDAI-1 (syn. 0.53) and postulated disulphide pattern for barley trypsin inhibitor BTI-CMe.

ii) The possible coevolution of the plant inhibitors and the insect enzymes is suggested indirectly by the available evidence. Little intraspecific variability is observed in the inhibitor pattern, as judged by two-dimensional electrophoresis. However, there are sharp pattern differences between closely-related species. This is consistent with the idea that each inhibitor pattern represents a specific response to the main pests of the area of distribution of each species. There is also evidence of specificity changes in closely-related inhibitors. Thus, barley inhibitor BTI-CMc, which is a weaker trypsin inhibitor than BTI-CMe, is more closely related in its *N*-terminal sequence (21 out of 29 coincident amino acids) to one of the subunits of the tetrameric inhibitor of α-amylase (BTAI-CMa) than to any of the cereal trypsin inhibitors described so far (Barber *et al.*, 1986a; García-Olmedo *et al.*, 1992; Paz-Ares *et al.*, 1983; Rodriguez-Palenzuela *et al.*, 1989). However the residues deduced from its cDNA at the reactive site place are valine-proline-leucine-leucine, and the rest of the sequence differs sharply from that of BTI-CMe (Figure 1). Also relevant in this context is the above described ability of different α-amylase inhibitors of this family to discriminate among the enzymes from different insect taxa.

iii) The evolution of the aggregative properties of inhibitor subunits also deserves attention. It has been recently reported that two very similar subunits with identical *N*-terminal amino acid sequences (23 residues) have strikingly different properties: one is a subunit of a tetrameric inhibitor that is active against α-amylases from the insect *Tenebrio molitor* (Coleoptera), but not towards α-amylases from other sources, such as *Ephestia kuehniella* (Lepidoptera) or human saliva, whereas the other aggregates as a homodimer and is only active against the human enzyme (García-Casado *et al.*, 1996). A second example is that of the barley dimeric inhibitor BDAI-1, whose chromosomal location and amino acid sequence suggest that this molecule is evolutively closer to the wheat monomeric than to the wheat dimeric inhibitors (Figure 1; Table 1). This would imply that BDAI-1 is a diverged form of the wheat monomeric inhibitors that has acquired the ability to self-associate (Mena *et al.*, 1992).

iv) The existence of hetero-tetrameric inhibitors suggests a molecular model for intergenome heterosis in alloploids. For example, single tetrameric species were observed in *T. tauschii* (subunits CM1, CM3D, CM17) and in *T. turgidum* (CM2, CM3D, CM16), while multiple tetrameric species were observed in *T. aestivum*, resulting from combinations of the subunits contributed by its two parental species (Gomez *et al.*, 1989). The three types of subunits were required for significant activity although binary mixtures involving subunit WTAI-CM1 (or the corresponding barley BTAI-CMa) also had some activity. Additional combinations of subunits were also reconstituted and their inhibitory activities ranged from 144% (CM1, CM3B, CM17) to 33% (CM2, CM3D, CM17) compared to the activity of the reconstituted inhibitor from *T. tauschii*. This, together with the established chromosomal locations of these genes, fit a model of alloploid heterosis at the molecular level (García-Maroto *et al.*, 1990; Gomez *et al.*, 1989).

ACKNOWLEDGEMENTS

This work was financed by grants Bio96-2303 and PB92-0325 from the Comision Interministerial de Ciencia y Tecnología (CICYT) Spain. The technical assistance of L. Lamoneda is acknowledged.

REFERENCES

Adachi, T., Izumi, H., Yamada, T., Tanaka, K., Takeuchi, S., Nakamura, R. and Matsuda, T. (1993) Gene structure and expression of rice seed allergenic proteins belonging to the α-amylase/trypsin inhibitor family. *Plant Molecular Biology* **21**, 239-248.

Barber, D., Sanchez-Monge, R., García-Olmedo, F., Salcedo, G. and Mendez, E. (1986a) Evolutionary implications of sequential homologies among members of the trypsin/α-amylase inhibitor family (CM-proteins) in wheat and barley. *Biochimica et Biophysica Acta*. **873**, 147-151.

Barber, D., Sanchez-Monge, R., Mendez, E., Lazaro, A., García-Olmedo, F. and Salcedo, G. (1986b) New α-amylase and trypsin inhibitors among the CM-proteins of barley (Hordeum vulgare). *Biochimica et Biophysica Acta*. **869**, 115-118.

Barber, D., Sanchez-Monge, R., Gomez, L., Carpizo, J., Armentia, A., Lopez-Otín, C., Juan, F. and Salcedo, G. (1989) A barley flour inhibitor of insect α-amylase is a major allergen associated with baker's asthma disease. *FEBS Letters* **248**, 119-122.

Carbonero, P., Salcedo, G., Sanchez-Monge, R., García-Maroto, F., Royo, J., Gomez, L., Mena, M., Medina, J. and Diaz, I. (1993) A multigene family from cereals which encodes inhibitors of trypsin and heterologous α-amylases. In: "Innovations in Proteases and their Inhitbitors" (F.X. Avilés, ed.) pp. 333-348. Walter de Gruyter, Berlin and New York.

Campos, F.A.P. and Richardson, M. (1983) The complete amino acid sequence of the bifunctional α-amylase/trypsin inhibitor from seeds of ragi (Indian finger millet: Eleusine coracana, Goertn). *FEBS Letters* **152**, 300-304.

Chen, M.S., Feng, G.H., Zen, K.C., Richardson, M., Valdesrodriguez, S., Reeck, G.R. and Kramer, K.J. (1992) α-Amylases from three species of stored grain Coleoptera and their inhibition by wheat and corn proteinaceous inhibitors. Insect *Biochemistry and Molecular Biology* **22**, 261-268.

Chong, G.L. and Reeck, G.R. (1987) Interaction of trypsin, β-factor XIIa and plasma kallikrein with a trypsin inhibitor isolated from barley seeds: a comparison with the corn inhibitor of activated Hageman factor. *Thrombosis Research* **48**, 211-221.

García-Casado, G., Armentia, A., Sanchez-Monge, R., Sanchez, L.M., Lopez-Otín, C. and Salcedo, G. (1995) A major baker's asthma allergen from rye flour is considerably more active than its barley counterpart. *FEBS Letters* **364**, 36-40.

García-Casado, G., Sanchez-Monge, R., Puente, X.S. and Salcedo, G. (1996) Divergence in properties of two closely related α-amylase inhibitors of barley. *Physiologia Plantarum* **98**, 523-528.

García-Maroto, F., Carbonero, P. and García-Olmedo, F. (1991) Site-directed mutagenesis and expression in Escherichia coli of WMAI-1, a wheat monomeric inhibitor of insect α-amylase. *Plant Molecular Biology* **17**, 1005-1011.

García-Maroto, F., Maraña, C., Mena, M., García-Olmedo, F. and Carbonero, P. (1990) Cloning of cDNA and chromosomal location of genes encoding the three types of subunits of the wheat tetrameric inhibitor of insec α-amylase. *Plant Molecular Biology* **14**, 845-853.

García-Olmedo, F., Salcedo, G., Sánchez-Monge, R., Gomez, L., Royo, J. and Carbonero, P. (1987) Plant proteinaceous inhibitors of proteinases and α-amylases. In: Oxford Surveys of Plant Molecular and Cell Biology. Vol. 4. (B. Miflin, ed.) pp. 275-334. Oxford University Press-ISPMB, New York.

García-Olmedo, F., Salcedo, G., Sanchez-Monge, R., Hernandez-Lucas, C., Carmona, M.J., Lopez-Fando, J.J., Fernandez, J.A., Gomez, L., Royo, J., García-Maroto, F., Castagnaro, A. and Carbonero, P. (1992) Trypsin/α-amylase inhibitors and thionins: possible defence proteins from barley. In: "Barley: Genetics, Biochemistry, Molecular Biology and Biotechnology" (P.R. Shewry, ed.) pp. 335-350. CAB International. Wallingford U.K.

Gatehouse, A.M.R., Fenton, K.A., Jepson, I. and Pavey, D.J. (1986) The effects of α-amylase inhibitors on insect storage pests: inhibition of α-amylase *in vitro* and effects on development in vivo. *Journal of the Science of Food and Agriculture* **37**, 727-734.

Gomez, L., Martin, E., Hernandez, D., Sanchez-Monge, R., Barber, D., Pozo, V., Andres, B., Armentia, A., Lahoz, C., Salcedo, G. and Palomino, P. (1990) Members of the α-amylase inhibitors family from wheat endosperm are major allergens associated with baker's asthma. *FEBS Letters* **261**, 85-88.

Gomez, L., Sanchez-Monge, R., García-Olmedo, F. and Salcedo, G. (1989) Wheat tetrameric inhibitors of insect α-amylase: alloploid heterosis at the molecular level. *Proceedings of the National Academy of Sciences USA*. **86**, 3242-3246.

Gomez, L., Sanchez-Monge, R., Lopez-Otín, C. and Salcedo, G. (1991) Wheat inhibitors of heterologous α-amylases. *Plant Physiology* **96**, 768-774.

Gutierrez, C., Sanchez-Monge, R., Gomez, L., Ruiz-Tapiador, M., Castañera, P. and Salcedo, G. (1990) α-Amylase activities of agricultural insect pests are specifically affected by different inhibitor preparations from wheat and barley endosperm. *Plant Science* **72**, 37-44.

Gutierrez, C., García-Casado, G., Sanchez-Monge, R., Gomez, L., Castañera, P. and Salcedo, G. (1993) Three inhibitor types from wheat endosperm are differentially active against α-amylases of Lepidoptera pests. *Entomology Experimental and Applicata* **66**, 47-52.

Halford, N.G., Morris, N.A., Urwin, P., Williamson, M.S., Kasarda, D.D., Lew, E.J.L., Kreis, M. and Shewry, P. (1988) Molecular cloning of the barley seed protein CMd: a variant member of the α-amylase/trypsin inhibitor family of cereals. *Biochimica et Biophysica Acta* **950**, 435-440.

Izumi, H., Adachi, T., Fujii, N., Matsuda, T., Nakamura, R., Tanaka, K., Urisu, A. and Y. Kurosawa. (1992) Nucleotide sequence of a cDNA clone encoding a major allergenic protein in rice seeds: homology of the deduced aminoacid sequence with members of α-amylase/trypsin inhibitor family. *FEBS Letters* **302**, 213-216.

Kashlan, N. and Richardson, M. (1981) The complete amino acid sequence of a major wheat protein inhibitor of α-amylase. *Phytochemistry* **20**, 1781-1784.

Kirsi, M. and Ahokas, H. (1983) Trypsin inhibitor activities in the wild progenitor of barley. *Phytochemistry* **22**, 2739-2740.

Lazaro, A., Barber, D., Salcedo, G., Mendez, E. and García-Olmedo, F. (1985) Differential effects of high-lysine mutations on the accumulation of individual members of a group of proteins encoded by a disperse multigene family in the endosperm of barley (Hordeum vulgare L.). *European Journal of Biochemistry* **149**, 617-623.

Lazaro, A., Rodriguez-Palenzuela, P., Maraña, C., Carbonero, P. and García-Olmedo, F. (1988a) Signal peptide homology between the sweet protein thaumatin II and unrelated cereal α-amylase/trypsin inhibitors. *FEBS Letters* **1**, 147-150.

Lazaro, A., Sanchez-Monge, R., Salcedo, G., Paz-Ares, J., Carbonero, P. and García-Olmedo, F. (1988b) A dimeric inhibitor of insect α-amylase from barley. *European Journal of Biochemistry* **172**, 129-134.

Lyons, A., Richardson, M., Tatham, A.S. and Shewry, P.R. (1987) Characterization of homologous inhibitors of trypsin and α-amylase from seeds of rye (Secale cereale L.) *Biochimica et Biophysica Acta.* **915**, 305-313.

Maeda, K., Wakabayashi, S. and Matsubara, H. (1983) Disulfide bridges in an α-amylase inhibitor from wheat kernel. *Journal of Biochemistry* **94**, 865-870.

Maeda, K., Wakabayashi, S. and Matsubara, H. (1985) Complete amino acid sequence of an α-amylase inhibitor in wheat kernel (0.19 inhibitor). *Biochimica et Biophysica Acta* **828**, 213-221.

Mahoney, W.C., Hermodson, M.A., Jones, B., Powers, D.D., Corfman, R.S. and Reeck, G.R. (1984) Amino acid sequence and secondary structural analysis of the corn inhibitor of trypsin and activated Hageman factor. *Journal of Biological Chemistry* **259**, 8412-8416.

Medina, J., Hueros, G. and Carbonero, P. (1993) Cloning of cDNA, expression and chromosomal location of genes encoding of three types of subunits of the barley tetrameric inhibitor of insect α-amylase. *Plant Molecular Biology* **23**, 535-542.

Mena, M., Sanchez-Monge, R., Gomez, L., Salcedo, G. and Carbonero, P. (1992) A major barley allergen associated with baker's asthma disease is a glycosylated monomeric inhibitor of insect α-amylase: cDNA cloning and chromosomal location of the gene. *Plant Molecular Biology* **20**, 451-458.

Mikola, J. and Soulinna, E.M. (1969) Purification and properties of a trypsin inhibitor from barley. *European Journal of Biochemistry* **9**, 555-560.

Molina-Cano, J.L., Fra-Mon, P., Salcedo, G., Aragoncillo, C., Roca de Tagores, F. and García-Olmedo, F. (1987) Morocco as a possible domestication center for barley: biochemical and agromorphological evidence. *Theoretical and Applied Genetics* **73**, 531-536.

Odani, S., Koide, T. and Ono, T. (1983) The complete amino acid sequence of barley trypsin inhibitor. *Journal of Biological Chemistry* **258**, 7998-8003.

Paz-Ares, J., Ponz, F., Aragoncillo, C., Hernandez-Lucas, C., Salcedo, G., Carbonero, P. and García-Olmedo, F. (1983) In vivo and *in vitro* synthesis of CM-proteins (A-hordeins) from barley (Hordeum vulgare L.). *Planta.* **157**, 74-80.

Paz-Ares, J., Ponz, F., Rodriguez-Palenzuela, P., Lazaro, A., Hernandez-Lucas, C., García-Olmedo, F. and Carbonero, P. (1986) Characterization of cDNA clones of the family of trypsin/α-amylase inhibitors (CM-proteins) in barley (Hordeum vulgare L.). *Theoretical and Applied Genetics* **71**, 842-846.

Poerio, E., Caporale, C., Carrano, L., Pucci, P. and Buonocore, V. (1991) Assignment of the five disulfide bridges in an α-amylase inhibitor from wheat kernel by fast-atom bombardment mass spectrometry and Edman degradation. *European Journal of Biochemistry* **199**, 595-600.

Rasmussen, S.K. and Johansson, A. (1992) Nucleotide sequence of a cDNA encoding the barley seed protein CMa: an inhibitor of insect α-amylase. *Plant Molecular Biology* **18**, 423-427.

Rodriguez-Palenzuela, P., Royo, J., Gomez, L., Sanchez-Monge, R., Salcedo, G., Molina-Cano, J.L., García-Olmedo, F. and Carbonero, P. (1989) The gene for trypsin inhibitor CMe is regulated in trans by the *Lys3a* locus in the endosperm of barley (Hordeum vulgare L.). *Molecular and General Genetics* **219**, 474-479.

Royo, J., Diaz, I., Rodriguez-Palenzuela, P. and Carbonero, P. (1996) Isolation and promoter characterization of the barley gene Itr1 encoding trypsin inhibitor BTI-CMe: differential activity in wild type and mutant *Lys3a* endosperm. *Plant Molecular Biology* **31**, 1051-1059.

Royo, J., Gaddour, K., Vicente-Carbajosa, J., Diaz, I and Carbonero, P. (1997) Gene duplication and evolution of the Itr1 locus in cultivated and wild Hordeum spp. (submitted).

Salcedo, G., Fra-Mon, P., Molina-Cano, J.L., Aragoncillo, C. and García-Olmedo, F. (1984) Genetics of CM-proteins (A-hordeins) in barley. *Theoretical and Applied Genetics* **68**, 53-59.

Sanchez de la Hoz, P., Castagnaro, A. and Carbonero, P. (1994) Sharp divergence between wheat and barley at loci encoding novel members of the trypsin/α-amylase inhibitors family. *Plant Molecular Biology* **26**, 1231-1236.

Sanchez-Monge, R., García-Casado, G., Barber, D. and Salcedo, G. (1996a) Interaction of allergens from house-dust mite and from cereal flours: Dermatophagoides pteronyssinus α-amylase (Der p 4) and wheat and rye α-amylase inhibitor. *Allergy* **51**, 176-180.

Sanchez-Monge, R., García-Casado, G., Malpica, J.M. and Salcedo, G. (1996b) Inhibitory activities against heterologous α-amylases and *in vitro* allergenic reactivity of Einkorn wheats. *Theoretical and Applied Genetics* **93**, 745-750

Sánchez-Monge, R., Gomez, L., García-Olmedo, F. and Salcedo, G. (1986). A tetrameric inhibitor of insect α-amylase from barley. *FEBS Letters* **207**, 105-109.

Sanchez-Monge, R., Gomez, L., Barber, D., Lopez-Otín, C., Armentia, A. and Salcedo, G. (1992) Wheat and barley allergens associated with baker's asthma. Glycosylated subunits of the α-amylase-inhibitor family have enhanced IgE-binding capacity. *Biochemical Journal* **281**, 401-405.

Shewry, P.R., Lafiandra, D., Salcedo, G., Aragoncillo, C., García-Olmedo, F., Lew, E.J.-L., Dietler, M.D. and Kasarda, D.D. (1984) *N*-terminal amino acid sequences of chloroform-methanol-soluble proteins and albumins from the endosperm of wheat, barley, and related species. *FEBS Letters* **175**, 359-363.

Shivaraj, B. and Pattabiraman, T.N. (1981) Natural plant enzyme inhibitors. Characterization of an unusual α-amylase/trypsin inhibitor from ragi (Eleusine coracana Geartn). *Biochemical Journal* **193**, 29-36.

Silano, V., Poerio, E. and Buonocore, V. (1977) A model for the interaction of wheat monomeric and dimeric protein inhibitors with α-amylase. *Molecular and Cell Biochemistry* **18**, 87-91.

Strobl, S., Mühlhahn, P., Bernstein, R., Witscheck, R., Maskos, K., Wunderlich, M., Huber, R., Glockshuber, R. and Holak, T.A. (1995) Determination of the three-dimensional structure of the bifunctional α-amylase/trypsin inhibitor from ragi seeds by NMR spectroscopy. *Biochemistry* **34**, 8281-8293.

Vicente-Carbajosa, J., Oñate, L., Lara, P., Diaz, I. and Carbonero, P. (1998) Barley BLZ1: a bZIP transcriptional activator that interacts with endosperm-specific gene promoters. *The Plant Journal* **13**, 629-640.

Wen, L., Huang, J-K., Zen, K.C., Johnson, B.H., Muthukrishnan, S., Mackay, V., Manney, T.R., Manney, M. and Reeck, G.R. (1992) Nucleotide sequence of a cDNA clone that encodes the maize inhibitor of trypsin and activated Hageman factor. *Plant Molecular Biology* **18**, 813-814.

27. Inhibitors of Legume Seeds

C. DOMONEY
John Innes Centre, Colney Lane, Norwich NR4 7UH, England

1. INTRODUCTION

Enzyme inhibitors are prevalent among many plant species and have been detected in many different plant organs. Plant proteins have been identified that inhibit many diverse enzymes, including animal digestive proteases and amylases; other animal enzymes, including elastase, thrombin, plasmin and kallikrein; bacterial enzymes, such as subtilisin; fungal enzymes; endogenous plant proteases and amylases; and insect digestive enzymes (Garcia-Olmedo *et al.*, 1987; Richardson, 1991; Hilder *et al.*, 1990; Belozersky *et al.*, 1995). Classification of inhibitors according to either the type of enzyme inhibited or the source of inhibitor is not meaningful. The source of a particular enzyme can determine its degree of interaction with an inhibitor, so that proteins that are inhibitory towards a particular enzyme from one group of animals may have little or no effect on the equivalent enzyme from a different source. This distinction among enzymes is most striking in the case of α-amylase inhibitors that do not inhibit some α-amylases (Section 2) but even a small number of amino acid substitutions in, for example, bovine, compared with porcine, trypsin may be significant in affecting complex formation with inhibitors (Sweet *et al.*, 1974), an effect that ultimately influences apparent inhibitory activity values (Le Guen and Birk, 1993). Furthermore many inhibitors can inhibit more than one enzyme at distinct, or the same, sites on the inhibitor protein and it has been speculated that some inhibitors could act on enzymes other than those tested and reported. In addition, DNA sequences have been isolated that encode proteins with homology to inhibitors but for which an inhibitory function has not been demonstrated for the corresponding protein, giving rise to nomenclature such as 'probable amylase/protease inhibitor' (PAPI) in barley and rice and inclusion of these and other deduced protein sequences, having low similarity to known inhibitors, in amino acid sequence alignments of inhibitors (Richardson, 1991). Nonetheless, a classification of enzyme inhibitors based on sequence homologies has proved useful and up to

P.R. Shewry and R. Casey (eds.), Seed Proteins, 635–655.

twelve distinct classes of plant inhibitors are generally recognized (Chapter 25).

The importance of many inhibitor proteins, both in terms of their perceived negative influence on animal digestion, as components of feed and food, and in terms of their physiological roles and beneficial functions in plants, as agents of defence, has been the subject of much debate and present the plant breeder and genetic engineer with diametrically opposed objectives, represented by both desired decreases and increases in inhibitor concentrations for particular purposes. It has been suggested that a very large number of plant inhibitors are as yet unidentified and unexplored as tools for improving the responses of plants both to disease and to attack by insects (Ryan, 1990).

Enzyme inhibitors are present at relatively high concentrations in legume seeds when compared with their concentrations in other organs and in other plant families. Nonetheless, legume seed inhibitors are quantitatively minor seed components in comparison to the major seed proteins, although they receive much attention as major "antinutritional compounds", as a result of their influence on the action of animal digestive enzymes and the consequences of this for the nutritional quality of legume seed products. Legume seed enzyme inhibitors can be divided into two groups: those that inhibit enzymes involved in starch digestion, principally the α-amylase inhibitors, and those that inhibit enzymes involved in protein digestion, of which the trypsin and chymotrypsin inhibitors are the most prevalent. Both groups have received attention as antinutritional and as nutritional components of seeds. In addition, some inhibitors are desirable seed proteins from the viewpoint of their favourable amino acid composition, in particular, their contents of the sulphur-containing amino acids that are lacking or present in low amounts in the majority of legume seed proteins.

Antinutritional effects have been demonstrated both through *in vitro* enzyme assays and in animal studies, usually based on rat feeding trials, but extrapolation of observed results to whole animals in the first instance, or to larger animals in the second, must be treated with caution. It is clear, for example, that the ratio of pancreas to total body weight is a crucial factor in determining the extent to which hypersecretion of digestive enzymes, in response to the action of inhibitors, results in pancreatic hypertrophy (Le Guen and Birk, 1993). Both groups of legume enzyme inhibitors have been implicated in positive contributions to human health. Inhibitors of α-amylase were exploited as "starch blockers" for the control of diabetes mellitus and obesity in humans before the realization that impurities in the commercial preparations used meant that the treatment was not merely useless but also harmful (Liener *et al*., 1984). Purified α-amylase inhibitor preparations still have potential application in medical treatment (Boivin *et al*., 1988). Legume seed protease inhibitors have also been implicated in positive health responses, with several being effective chemoprotectors against carcinogenesis (Le Guen and Birk, 1993; Fernandes and Banerji, 1995). However, it is the antinutritional consequences of legume enzyme inhibitors that have received most attention and this situation is likely to continue. Since the ability of both amylase and protease inhibitors to inhibit insect digestive enzymes is being exploited now for improved protection of plants, using transgenesis, it is not surprising that the expression of such genes at high concentrations in seeds of transgenic plants raises concerns as to the utility of these plants for intensive animal production. In some cases, these transgenic plants cannot be recommended for such use (Pusztai *et al.*, 1995).

2. INHIBITORS OF α-AMYLASE

Protein inhibitors of α-amylase occur widely in plants (Buonocore and Silano, 1986) with that present in seeds of *Phaseolus vulgaris*, the common bean, receiving most attention in recent years for a number of reasons.

There has been some considerable confusion over the identity of bean seed proteins having insecticidal activity, this activity having been attributed originally to phytohemagglutinin (PHA), a seed lectin (Gatehouse *et al.*, 1984). However, it was later shown that the toxic effects of PHA to cowpea weevil were attributable to an α-amylase inhibitor contaminant present in earlier and in commercial PHA preparations (Huesing *et al.*, 1991). Moreover, since studies of the bean α-amylase inhibitor (αAI) gene indicated that this protein was unrelated to any of the previously characterized families of enzyme inhibitor, it thus became the first member of a new family of plant enzyme inhibitors (Moreno and Chrispeels, 1989). The gene encoding αAI is, however, closely related to, though distinct from, the genes encoding phytohemagglutinin and a common ancestral gene is likely to have existed for the two proteins (Moreno and Chrispeels, 1989). The product of the αAI gene has also been referred to, therefore, as lectin-like-protein (Altabella and Chrispeels, 1990).

These studies have highlighted the difficulties associated with attributing biological effects to proteins that have not been completely purified. Ultimate proof that αAI, rather than PHA, is responsible for the resistance of bean seeds to cowpea weevil has come from studies with transformed plants. Expression of a chimeric gene, containing the αAI coding sequence, in tobacco plants gave rise to a seed-expressed protein that inhibited the α-amylase activity extracted from the midgut of a mealworm, chosen as a representative of coleopteran insects (Altabella and Chrispeels, 1990). Expression of the αAI chimeric gene in seeds of transgenic *Pisum sativum* (pea) plants has demonstrated the considerable negative impact that this protein has on major insect pests of stored legume grain, including *Callosobruchus maculatus* (cowpea weevil) and *Callosobruchus chinensis* (azuki bean weevil) (Shade *et al.*, 1994) (Fig. 1), as well as on the growth and development of insects that attack the growing pea crop in the field, such as *Bruchus pisorum* (pea weevil) (Schroeder *et al.*, 1995). These studies demonstrate the power of inhibitor proteins in crop protection but also serve to highlight the untapped array of such tools that can be exploited for plant genetic manipulation. For example, natural resistance to pea weevil exists in *Pisum fulvum* accessions that is not attributable to αAI protein (Schroeder *et al.*, 1995). By combining the source of this natural resistance, that may be a hitherto unidentified inhibitor, with expression of the αAI protein, the development of insect resistance may be delayed.

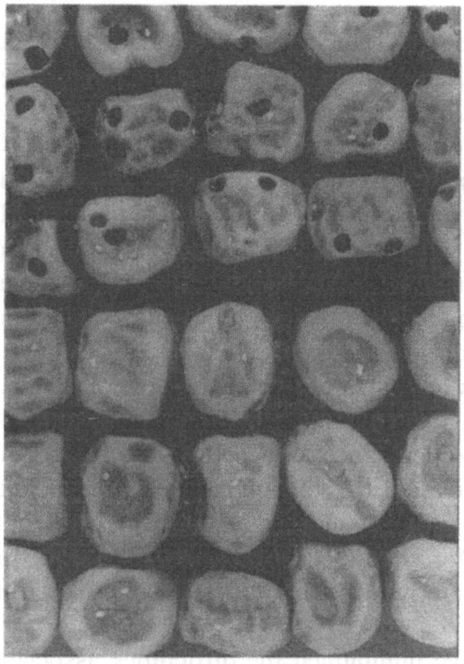

Fig. 1 Transgenic peas (left) expressing bean αAI show resistance to attack by larvae of seed-feeding weevils. Control peas (right) show cavities in the seeds where adult insects emerged. Reproduced from Shade *et al.* (1994), with permission.

αAI is normally present at 1-2% of the soluble protein in bean seeds but is expressed at a somewhat higher concentration in transgenic peas, being up to 3% of the soluble protein (Schroeder *et al.*, 1995). The oligomeric αAI protein is a thermostable glycoprotein (M_r 45000) (Moreno and Chrispeels, 1989), composed of subunits of M_r 14000 - 19000, that are derived from a preproprotein of M_r 35000 that lacks biological activity (Pueyo *et al.*, 1993). Processing is therefore essential for activity and varying degrees of glycosylation give rise to differences in the products observed in beans compared with transgenic peas (Schroeder *et al.*, 1995). Since it has been reported that αAI complexes at a 1:1 molar ratio with insect and mammalian (including human) α-amylases and some fungal amylo-glucosidases, but is inactive against bacterial and higher plant enzymes (Moreno and Chrispeels, 1989), the effects of this protein on energy metabolism in mammals warrant full investigation to allay fears of antinutritional effects. Although αAI can be inactivated by proper cooking, its presence in inadequately cooked food and in animal feed is predicted to be problematical (Pusztai *et al.*, 1995).

In view of the successes achieved with crop protection resulting from expression of αAI in transgenic plants, other members of this gene family have been sought in beans and their evolutionary relatedness examined (Mirkov *et al.*, 1994). As mentioned earlier, the genes for αAI and PHA are likely to have derived from a common ancestral gene. The same is likely to be true for the gene encoding arcelin (Osborn *et al.*, 1988). Genes for all three proteins are present at a single genetic locus in the bean genome (Nodari *et al.*, 1993) and multiple genes for all three proteins have been predicted either from DNA sequences or protein analyses (Mirkov *et al.*, 1994). Comparisons of derived amino acid sequences indicate that one or more of the four amino acid residues that are essential for carbohydrate binding, in functional lectins, are missing from arcelin and αAI sequences (Mirkov *et al.*, 1994). A unique feature of the αAI sequences is a processing that cleaves between conserved asparagine and serine residues, a step that is necessary for activation of αAI. This processing site is common among many legume seed proteins but is also present in some arcelins that are not processed, indicating that, among these very similar protein sequences, an asparagine-serine motif alone is not sufficient to ensure processing (Mirkov *et al.*, 1994). Comparisons of deduced sequences from sixteen genes in the bean lectin-arcelin-αAI gene family show that the arcelin and PHA proteins are more closely related to each other than to αAI proteins, the last having one or more gaps relative to the other sequences. However, when αAI-like sequences are included in these comparisons, the question arises once again whether or not the proteins encoded by these sequences are actually functional inhibitors. Proof that they are is likely to come in each case only through their expression in a system allowing for their correct post-translational modification *and* through identification of the appropriate source of enzyme that interacts with each inhibitor (Mirkov *et al.*, 1994).

This strategy has been adopted in the characterization of an αAI-like protein from wild accessions of *Phaseolus vulgaris* that contain arcelin as the major seed protein instead of phaseolin. This variant αAI, termed αAI-2, does not inhibit mammalian α-amylases or the insect (cowpea weevil, azuki bean weevil) α-amylases that are inhibited by αAI-1 of the cultivated bean. However, αAI-2 has been expressed in tobacco seeds and seed extracts from transformed plants have been shown to inhibit the α-amylase of *Zabrotes subfasciatus*, the Mexican bean weevil, an enzyme that is not inhibited by αAI-1 (Grossi de Sa *et al.*, 1997). Despite these distinct activities, the two αAI proteins share 78% identity, highlighting two areas of difference as specificity determinants and allowing for some elegant experiments to determine the basis of specificity. When two amino acids of a tripeptide in one of the two areas of difference were 'swapped' between the two αAI proteins, these changes were insufficient to alter inhibitor specificities (Grossi de Sa *et al.*, 1997), although replacement of this tripeptide with the equivalent sequence from PHA resulted in an inactive inhibitor (Mirkov *et al.*, 1995). Three-dimensional structural analysis of the αAI-1/porcine pancreatic α-amylase complex has highlighted the importance of the two areas mentioned above in interactions between αAI-1 and the substrate cleft of the enzyme, as well as the involvement of amino acids near the proteolytic processing site of αAI-1 as contact points with the enzyme (Bompard-Gilles *et al.*, 1996). Structural analyses such as this, in combination with data derived from specificities of variant inhibitors expressed in transgenic plants, widen the horizons for the rational design of inhibitors, for extending the specificities of inhibition to include multiple pest targets and for minimizing inhibition of mammalian enzymes.

3. INHIBITORS OF PROTEASES

Legume seeds contain two major classes of protease inhibitor, the Kunitz and the Bowman-Birk classes, so-called after the isolation of their original members from soybean seeds (*Glycine max*) by Kunitz (1945) and Bowman (1946), respectively. In the latter case, characterization was carried out by Birk *et al.* (1963), hence its present designation. Additional members of these inhibitor classes have been isolated from other leguminous species, but it is now clear that not all legume seeds contain representatives of both classes. It is also clear that these proteins are confined neither to seeds nor to legumes. Indeed, there is now a wealth of information available for Kunitz inhibitors that are synthesized in situations other than during seed development and this information can be expected to further our understanding of the physiological roles of these proteins in plants and to give clues to their evolution as seed proteins.

3.1 Kunitz protease inhibitor class

The Kunitz trypsin inhibitor (KTi) is the most prevalent soybean seed protease inhibitor, a protein of M_r 21000 that inhibits trypsin through interaction with a single site on the inhibitor and that is encoded by the *KTi3* gene (Jofuku and Goldberg, 1989). It has been estimated that *KTi3* belongs to a gene class containing at least ten members, many of which are linked in tandem pairs, and two additional genes (*KTi1* and *KTi2*) from this class have been characterized in some detail (Jofuku and Goldberg, 1989). Specificity for trypsin inhibition is determined by the two amino acid residues, arginine and isoleucine, at the active site of the KTi protein; these amino acids are considered essential for inhibitor function, although arginine and serine are the active site residues in other inhibitors (see below). However, substitution of the active site amino acids by histidine and alanine residues, together with substitution of several of the other amino acids that form possible contact points in KTi with porcine trypsin (Sweet *et al.*, 1974), in the proteins deduced from *KTi1* and *KTi2* means that the last two genes probably do not encode proteins having inhibitory activity (Jofuku and Goldberg, 1989). KTi1/2 mRNA has been detected in developing soybean embryos as well as in leaves, roots and stems; in embryos, the accumulation of KTi1 and KTi2 mRNAs is quantitatively, temporally and spatially distinct from that of KTi3 mRNA (Perez-Grau and Goldberg, 1989; Jofuku and Goldberg, 1989).

The organization of the tandemly linked *KTi1* and *KTi2* genes is very striking in that the two genes are approximately 0.5kb apart and have > 95% nucleotide sequence similarity in their 5', coding, and 3' regions, suggesting that these genes probably reflect a relatively recent gene duplication event. The *KTi3* gene is also expressed in leaf mRNA and the expression patterns of all three genes have been reproduced in the respective organs of transgenic tobacco, apart from in tobacco roots (see later), albeit at lower levels generally than in soybean (Jofuku and Goldberg, 1989). The function of the products of any, or all, of the ten *KTi* genes remains a matter for speculation. It may be assumed, however, that the KTi protein product of *KTi3* is not essential for normal plant and seed growth and development, since null lines lacking KTi in their seed have been identified (Orf and Hymowitz, 1979) and the molecular basis for such a null mutation has been demonstrated to be a frameshift in coding sequence, preventing correct translation of KTi3 mRNA (Jofuku *et al.*, 1989).

Seed-expressed genes of the Kunitz class have been less well-studied in other legumes. However, it is clear that legume inhibitors exist that are homologous to KTi, but with specificity for chymotrypsin, rather than trypsin; specificity for either enzyme can be strong or weak in the different inhibitors (Garcia-Olmedo *et al.*, 1987). The major inhibitors of winged bean (*Psophocarpus tetragonolobus*), a species with a high content of protease inhibitors, are members of the Kunitz class and include inhibitors that are active against either trypsin or chymotrypsin, with the activities of the latter inhibitors being much higher both in seeds and in tuberous roots (Habu *et al.*, 1992). The structure and expression of winged bean chymotrypsin inhibitor (WCI) genes have been analyzed and both functional genes and pseudogenes identified, with the major seed inhibitors being encoded by the near-identical genes, *WCI-3a* and *WCI-3b* (Habu *et al.*, 1992), whose transcribed regions are completely identical and whose promoters are 99% homologous (Habu *et al.*, 1993). The expression of some *WCI* genes in other plant organs has also been reported (Habu *et al.*, 1992).

A winged bean protein with homology to Kunitz protease inhibitors has been implicated in the control of proteolytic activity in senescent nodules (Manen *et al.*, 1991). The deduced protein sequence of a cDNA, isolated from a library constructed using nodule mRNA, showed clear homology to soybean KTi and to winged bean seed trypsin inhibitor. However, it was evident that, although the nodule sequence retained the four cysteine residues implicated in disulphide bridge formation, the amino acids (Arg,Pro) deduced for the region of the nodule protein corresponding to the active sites of the seed inhibitors (Arg,Ile and Arg,Ser at positions 63 and 64 in soybean and winged bean, respectively) were unlikely to effect inhibition of trypsin (Manen *et al.*, 1991). These authors speculated that the residues predicted for positions 57 and 58 (Lys,Ser) could represent the active site for trypsin inhibition, a suggestion that is incompatible with the chain folding model of soybean KTi (Fig. 2), based on the crystal structure of the KTi complex with porcine trypsin, that implicated residues 63 and 64 as an external loop of electron density which enters the active site of trypsin (Sweet *et al.*, 1974). Nonetheless, a nodule protein of M_r 21000 was demonstrated to bind to trypsin affinity columns and to inhibit the proteolytic activity, in *in vitro* assays, both of trypsin and of nodule extracts from which the M_r 21000 protein had been removed (Manen *et al.*, 1991). It is possible that another closely related gene encodes the nodule protein and that the cDNA isolated was the most abundant but did not correspond to a functional inhibitor. The observation that KTi-related proteins are involved in nodulation processes offers a plausible explanation for the lack of expression of soybean *KTi1* and *KTi2* in roots of transgenic tobacco plants (Jofuku and Goldberg, 1989), referred to above; it has been speculated that infection by *Rhizobium* of leguminous roots provides the *trans*-acting regulator of transcription for such genes (Manen *et al.*, 1991), although low-level expression of *KTi1* and *KTi2* in embryos would require an alternative control mechanism. Clearly, elucidation of function of seed-expressed genes from these multi-gene families, where expression of closely related genes is regulated organ-specifically and developmentally, presents many difficulties and it is noteworthy that, as observed earlier for the α-amylase inhibitor family, expression in appropriate transgenic plants may represent the only means of clarification of function in many cases.

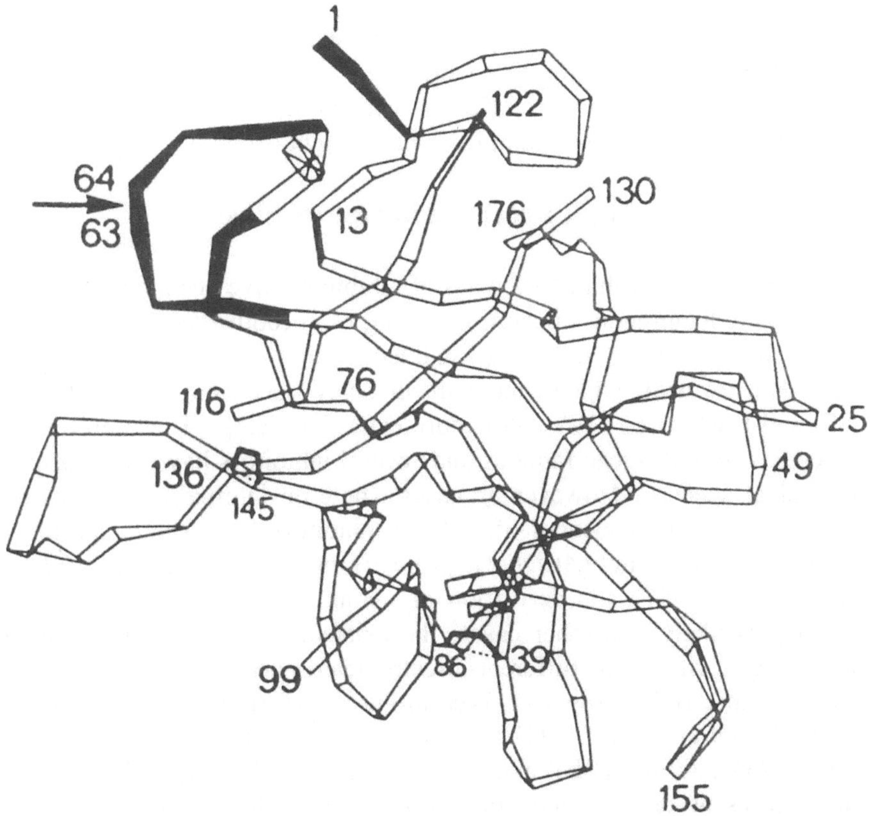

Fig. 2 The folding model of soybean seed Kunitz trypsin inhibitor, based on the crystal structure of a trypsin-trypsin inhibitor complex (Sweet *et al.*, 1974). Residues in contact with trypsin are shown in black. The arrow highlights the position of the peptide bond between Arg-63 and Ile-64 that is reversibly split by trypsin. Modified from Sweet *et al.* (1974), with permission.

Kunitz inhibitors have also been identified in seeds from many other legumes, including butterfly pea (*Clitoria ternatea*) (Macedo and Xavier-Filho, 1992), mesquite (*Prosopsis juliflora*) (Monte Negreiros *et al.*, 1991) and kafferboom or lucky bean tree (*Erythrina caffra)* (Onesti *et al.*, 1991). Although Kunitz-type inhibitors are found mainly in seeds of the Mimosideae subfamily of legumes, some members of the Papilionoideae subfamily (that includes soybean and winged bean) clearly also contain these inhibitors. The existence of homologous proteins in seeds from cereals, where the inhibitors are bifunctional and inhibit both α-amylase and proteases (Richardson, 1991), and from cocoa (*Theobroma cacao*) (Spencer and Hodge, 1991), in tubers of solanaceous species (Yeh *et al*., 1997; Yamagishi *et al*., 1993) and as stress-responsive proteins in leaves of rapeseed (*Brassica napus*) (Lee Downing *et al*., 1992) and poplar (*Populus* hybrid) (Bradshaw *et al*., 1989) indicates an early evolutionary origin for these proteins. In fact, a striking structural similarity between a legume seed Kunitz-type inhibitor and human interleukin-1β (Onesti *et al*., 1991) may point to a very ancient origin.

3.2 Bowman-Birk protease inhibitor class

Proteins of the Bowman-Birk class of inhibitors have been identified in the seeds of many leguminous species, including bean (both *Phaseolus* and *Vicia* species), chickpea (*Cicer arietinum*) and lentil (*Lens culinaris*), as well as the species that are dealt with in more detail below [see Garcia-Olmedo *et al.* (1987) and Richardson (1991) for more extensive lists]. These proteins are the products of multi-gene families and several inhibitors have been identified in individual species. However, in cases where no internal differences exist between determined sequences of isolated inhibitors from single species, as, for example, with peanut (*Arachis hypogaea*) inhibitors PI1 and PI2 (Garcia-Olmedo *et al*., 1987), the derivation of variant forms by post-translational processing is likely. In other cases, the existence of at least three distinct genes that give rise to the different directly determined and deduced inhibitor sequences expressed in cowpea (*Vigna unguiculata*) (Hilder *et al*., 1989) and soybean (Joudrier *et al*., 1987) can be inferred. In the latter case, the Bowman-Birk inhibitor (BBI) and the closely related PI IV and C-II inhibitors clearly represent distinct gene products but it is likely that a family of inhibitors PI I-III are derived from PI IV (also called D-II) through amino-terminal cleavage of a number of amino acids (Joudrier *et al*., 1987). In pea (*Pisum sativum*) seeds, multiple inhibitors of the Bowman-Birk class have been identified that are the products of two closely homologous genes; processing at both amino- and carboxy-terminal ends of these proteins has been implicated in the derivation of multiple isoforms (Domoney *et al*., 1993, 1995; Domoney and Welham, 1998) (Fig. 3).

The inhibitors of this class are generally small, having M_r 8000-9000, and are termed double-headed inhibitors, due to their ability to simultaneously inhibit two, either the same or different, enzymes. Thus trypsin/trypsin (T/T), trypsin/chymotrypsin (T/C) and trypsin/elastase (T/E) inhibitors have been identified among proteins from this class with at least one inhibitor (soybean C-II) being capable of inhibition of three different proteases, two of them at the same site (Joudrier *et al*., 1987; Hilder *et al*., 1989). The major pea seed inhibitors are of the T/C type (Domoney *et al*., 1995).

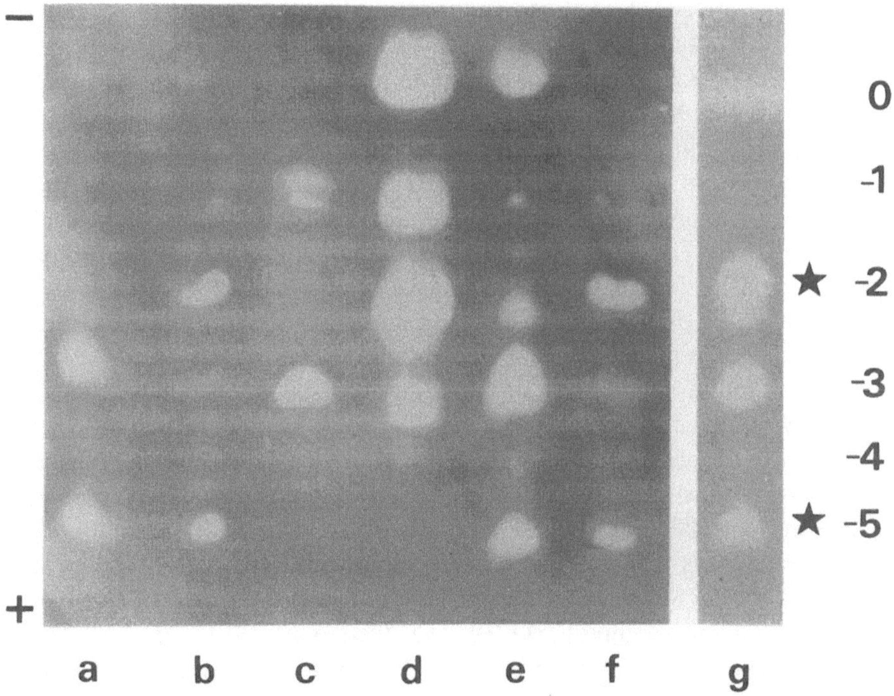

Fig. 3 Variation in pea seed trypsin inhibitor isoform patterns, as detected on non-denaturing polyacrylamide gels that have been incubated, following electrophoresis, in trypsin and stained for trypsin activity, using the substrate *N*-acetyl-DL-phenylalanine β-naphthyl ester (Domoney *et al.*, 1993). White bands show where trypsin activity has been inhibited. Whole seed extracts were analyzed in tracks a - f (seeds from the John Innes germplasm accession numbers JI 15, JI 61, JI 126, JI 281, JI 399 and JI 430, respectively); an extract from axes of cv. Birte is shown in track g. The derivation of the different isoforms of cv. Birte is deduced as follows: the two primary mature isoforms with deduced overall charges of -5 and -2 are indicated (*). Other major isoforms result from *C*-terminal processing, with concomitant overall loss of two negative charges (-3 and 0, respectively). More minor isoforms are likely to result from *N*-terminal processing of the primary mature proteins with a loss of one negative charge (-4 and -1, respectively).

Genetical analyses of the major pea seed inhibitors, utilizing variation in the proteins (Fig. 3) and gene polymorphisms, have indicated that a single genetic locus, *Tri*, on linkage group 5, contains the structural genes and that this locus also controls quantitative variation in expression of the genes (Domoney *et al.*, 1994). The identification of wild species of soybean lacking BBI has been reported (Domagalski *et al.*, 1992) but it is not clear whether or not the related inhibitors, PI I-IV and C-II, are also absent from such lines or if the BBI-less phenotype results from mutation of the gene itself or a distant regulatory gene.

The physiological functions of inhibitors of the Bowman-Birk class are no clearer than those of other classes. The view that these seed inhibitors are involved in protection of stored reserves from predation by pests is supported by the experiments demonstrating protection of transgenic tobacco plants 'constitutively' expressing a trypsin inhibitor that is normally expressed in cowpea seeds (CpTI). The leaves of such transgenic plants were protected from predation by lepidopteran insect larvae (Hilder *et al.*, 1987), which is consistent with the presence of trypsin-like and chymotrypsin-like activities as the major digestive enzymes of these insects (Johnston *et al.*, 1995). However, although early work on CpTI was based on the belief that this protein was involved in protection of cowpea seeds against attack by cowpea weevil (*Callosobruchus maculatus*), a view that led to its use in transgenic experiments, several lines of evidence now indicate that resistance to cowpea weevil is not correlated either with amounts of CpTI or with the inhibitors of cysteine proteinases that are also present in cowpea seeds (see 'The cowpea resistance story' in Shewry and Lucas, 1997, for a full account of this work). Nonetheless a gene encoding CpTI has now been exploited for fruit protection in apple (*Malus pumila*) (James *et al.*, 1992) and in strawberry (*Fragaria* hybrid) (Graham *et al.*, 1997). In the latter case, resistance to root attack by vine weevil (*Otiorhynchus sulcatus*) is dramatic (Fig. 4), although restriction of expression of CpTI to those plant parts on which the insect, rather than the intended consumer, feeds is the ultimate aim (Graham *et al.*, 1997). Inhibitors such as CpTI may also be involved in nematode resistance in plants, an effect of CpTI on the male/female ratio of the cyst nematode (*Globodera rostochiensis*) having been reported for transgenic potato (Sijmons, 1993).

Most of the evidence for a protective role for seed inhibitors comes from studies utilizing a single CpTI gene; it has been suggested that the T/T inhibitors have greatest effectiveness as protective proteins but that this could be attributable to variable regions of the molecule which might affect fine aspects of specificity, binding constant or stability (Hilder *et al.*, 1989). The importance of a *C*-terminal nonapeptide in determining the affinity of pea seed inhibitors for a digestive enzyme has been demonstrated; removal of this nonapeptide results in inhibitors that have a higher affinity for trypsin and this processing correlates with the increases in inhibitor activity that have been documented during seed desiccation (Domoney *et al.*, 1995). Whether or not such processed proteins also show increased inhibition of insect enzymes remains to be established. Nonetheless, despite the demonstrations of inhibitor proteins as protective agents, progress in their introduction into commercial crops has been slow. Although concerns have been raised as to

Fig. 4 Insect bioassay of transgenic (left) and control (right) strawberry plants, showing the improved root system in plants expressing CpTI, after plants had been inoculated with eggs of the vine weevil. Reproduced from Graham *et al*. (1997), with permission.

the development of resistance by insects to individual inhibitors (Jongsma *et al*., 1996), it is clear that, in combination with other genes encoding protective proteins, the potential for CpTI and related genes is very promising. CpTI and lectin have additive protective effects and, in combination with the δ-endotoxins of *Bacillus thuringiensis* (Bt toxins), various inhibitors can enhance the effects of such toxins (Shewry and Lucas, 1997), possibly through preventing their proteolysis. Pyramiding resistance genes (Ryan, 1990; Shewry and Lucas, 1997) clearly offers the possibilities of better protection and longer-term prevention of resistance development.

Elucidating the probable functions of seed inhibitors requires an understanding of their mode of synthesis and knowledge of their subcellular compartmentation and the organs where they are expressed. It is evident that the soybean, cowpea and pea inhibitors of the Bowman-Birk class are all synthesized as precursor proteins (Joudrier *et al*., 1987; Hilder *et al*., 1989; Domoney *et al*., 1995). However, in the case of pea and cowpea inhibitors, multiple methionine residues exist in the 'leader' peptides of potentially 42 - 66 amino acids and their apparent length, and other differences between the cowpea precursor 'leaders' and traditional 'signal sequences', led Hilder *et al*. (1989) to suggest that cleavage of these precursors differed from that of storage proteins. The subcellular location of these proteins has been cited as cytosolic (Hilder *et al*., 1989) but the existence of 'leader' peptides and the possible involvement of asparaginyl endopeptidases in *C*-terminal cleavage of the pea inhibitors would support their compartmentation, probably in protein bodies (Domoney and Welham, 1998). Characterization of the organs in which seed inhibitors of this class are synthesized is confounded by the existence of multiple genes. For example, wound-induced leaf expression of a Bowman-Birk-type inhibitor was first demonstrated in alfalfa (*Medicago sativa*) (Brown *et al.*, 1985). More recently, genes encoding inhibitors of this class have been shown to be expressed in a variety of organs of alfalfa in response to both wounding and the presence of microorganisms in the soil, but it is not clear if these vegetatively expressed genes are the same as, or distinct from, those that are expressed in alfalfa seeds (McGurl *et al.*, 1995). Although expression of the two genes encoding pea seed inhibitors was found to be seed-specific under normal growing conditions, one appeared to be expressed in roots of plants that had been subjected to a drought treatment, a response that may relate to a protective role under such conditions and may explain the early detection of these proteins in root-cap cells of immature, developing radicles (Welham *et al.*, 1998). Such inhibitors may thus be classified as LEA (late embryogenesis abundant)-like proteins in showing responses to water-stress that are similar to those documented for LEA proteins.

The potential of inhibitors of the Bowman-Birk class as protective agents has not been systematically explored. As transformation of economically important crop plants becomes more routine, the large number of unexplored BBI-type inhibitors, together with candidates from other classes, should receive more attention. For example, it is by no means clear whether the wound-induced vegetative BBI-type inhibitors (for example, in alfalfa) are more tailored to anti-pest functions than the inhibitors (for example, BBI and CpTI) that are expressed as part of the seed developmental programme. In pea, two classes of vegetatively-expressed BBI-type inhibitors have been identified that are distinct from each other and from seed inhibitors in terms of sequence, patterns of expression and processing (Domoney and Welham, 1998 and unpublished). It is clear that the inhibitors that have received most attention are those that are most abundant and, subsequently, those related to the most abundant ones. However, minor inhibitors that belong to other classes probably exist in seeds and other organs from most species; in broad bean (*Vicia faba*) seeds, for example, a subtilisin inhibitor (Svendsen *et al.*, 1984) has been placed in the potato inhibitor I class (Garcia-Olmedo *et al.*, 1987), although Bowman-Birk-type (T/C) inhibitors have also been identified in such seeds, as well as in the related vetch, *Vicia angustifolia* (Asao *et al.*, 1991; Gárcia-Olmedo *et al.*, 1987).

The utility of the Bowman-Birk class of inhibitors as anti-pest proteins has to be weighed against the opposing view of their presence in crop plants, namely, that their amounts should be decreased due to their antinutritional effects in animals, following consumption. The extraordinary heat-stability of these proteins is attributable to their seven disulphide bonds which renders them resistant to denaturation by boiling. Targeted and specific genetic manipulation, using antisense technology, may achieve a drastic reduction in inhibitor amounts in seeds, without affecting expression of related genes in vegetative organs. Alternatively, the same technology may be applied to interfere with the processing that affects inhibitor activity (Domoney *et al.*, 1995; Domoney and Welham, 1998). A third and indirect approach is suggested by experiments in soybean, where expression of a heterologous sulphur amino acid-rich protein, an albumin from Brazil nut (*Bertholletia excelsa*), had the unpredicted secondary effect of drastically reducing the amount of BBI that was synthesized, probably as a consequence of limited availability of sulphur-containing amino acids (Jung *et al.*, 1997), combined with the effects of developmental timing programmes. In pea, inhibitors are synthesized late during seed development and may thus be more amenable to disruptions in their synthesis by earlier expression of foreign proteins.

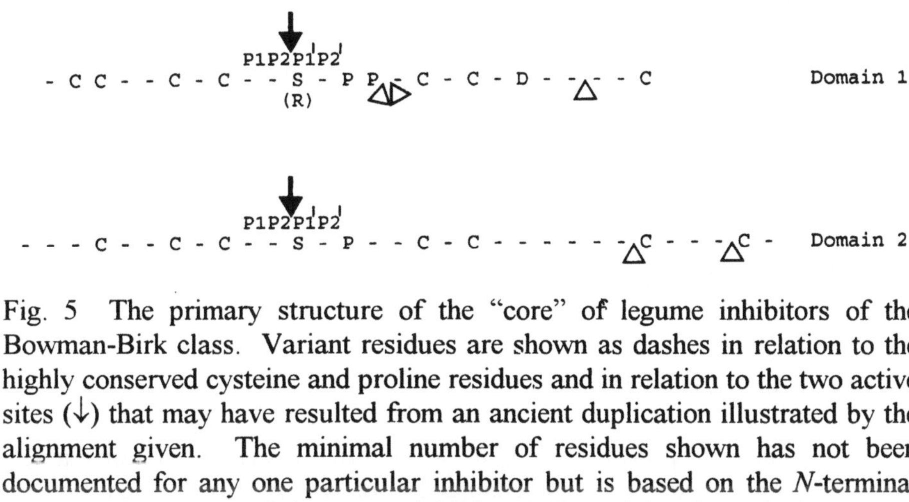

Fig. 5 The primary structure of the "core" of legume inhibitors of the Bowman-Birk class. Variant residues are shown as dashes in relation to the highly conserved cysteine and proline residues and in relation to the two active sites (↓) that may have resulted from an ancient duplication illustrated by the alignment given. The minimal number of residues shown has not been documented for any one particular inhibitor but is based on the *N*-terminal amino acid sequence of the peanut inhibitor, PI2, and the *C*-terminal sequence of the peanut inhibitor, PI3, that have the lowest number of amino acids before the first, and after the last, cysteine residues, respectively (Garcia-Olmedo *et al.*, 1987). It is likely, however, that PI2 is derived from PI1, the latter having nine additional *N*-terminal residues. "Core" length variants arise as a result of one or two amino acid insertions at the positions shown (Δ), peanut inhibitors being the main variants in this respect, since they have two extra residues at the position shown for domain 1 and an extra residue at each of the positions shown for domain 2. Peanut inhibitors also contain arginine at the P1 position of domain 1 instead of the more usual serine. Aspartic acid has been documented for almost all of the inhibitors at position 20 of the "core" shown and so this is included in the "core".

The evolution of BBI-type double-headed inhibitors by duplication of an ancestral single-headed inhibitor gene has been considered previously by other authors, including Hilder *et al.* (1989). Comparisons of 'core' regions, covering the invariant cysteine residues and active site serines (Fig. 5) have indicated very high rates of amino acid substitution, considering that 18 of the residues may be obligatorily invariant for function as BBI-type inhibitors (Hilder *et al.*, 1989). The BBI-type inhibitors that are present in cereals, for example wheat-germ (*Triticum aestivum*) and barley (*Hordeum vulgare*), may include more ancestral types of inhibitor (single-headed) and types in which a further duplication has given rise to a four-domain structure, as well as the two-domain types that are typical among legumes (Ary *et al.*, 1988; Nagasue *et al.*, 1988). BBI-type genes have also been identified in maize (*Zea mays*), in one case where a response to wounding has been documented (Rohrmeier and Lehle, 1993; Eckelkamp *et al.*, 1993) and in another, as genes that may have a solute transfer function in basal endosperm cells (Hueros *et al.*, 1997; G. Hueros and R. Thompson, pers. comm.). In neither case has an inhibitory function been demonstrated for the corresponding proteins, however, and in the first, it is not clear whether the deduced protein has one (Eckelkamp *et al.*, 1993) or two (Rohrmeier and Lehle, 1993) active sites. Speculation on the evolutionary relationships of these proteins to legume proteins of the BBI class must await further work. The same is also true of a possible relationship between BBI-type inhibitors and the small cysteine-rich proteins that have, for example, been implicated in fungal disease responses in pea (Chiang and Hadwiger, 1991).

4. CONCLUSIONS

The legume seed proteins belonging to the three enzyme inhibitor classes that have been considered in this chapter are of significance to seed biology through their nutritional and antinutritional properties and their contributions to plant defence. It is evident that the more abundant inhibitors have received most attention in all three areas of interest, although there are likely to be many as yet undiscovered or unexamined representatives of the same and other inhibitor classes. For example, a minor subtilisin inhibitor has been identified in broad bean (see earlier) and inhibitors of cysteine proteinases have not been studied to any great extent in legumes, although such inhibitors are likely to be ubiquitous in nature (Ryan, 1990). The latter have received some attention in investigations of cowpea insect resistance (Shewry and Lucas, 1997) and are likely to be of significance in defence strategies against hemipteran and coleopteran insects, where cysteine proteinases play important roles in digestion (see Chapter 24).

Elucidating the functions of individual inhibitors is confounded by the multiplicity of genes in single species and their modes of expression. Despite the estimation that over 20,000 papers have been published on plant enzyme inhibitors (Richardson, 1991), it is only through transgenesis that the functions of individual inhibitors can be really established; progress on bean α-amylase inhibitors is noteworthy in this respect.

Genetic manipulation of active sites and the engineering of affinities, and binding constants, for particular target enzymes offer many possibilities for the increase or decrease of inhibitor activity. There may be an association between potency and size for legume (pea) inhibitors of the Bowman-Birk class (see earlier). It is noteworthy in this respect that the plant protease inhibitors with the largest documented enzyme association constants are the very small squash inhibitors, comprising approximately 30 amino acids (see Chapter 25) (Richardson, 1991). It is clear, however, that multiple small inhibitors in at least some non-legume species are derived by proteolytic processing of multidomain precursors (Heath *et al.*, 1995) and the synthesis of such precursors may offer an efficiency advantage for their synthesis. The engineering of precursors having several distinct activities may additionally offer maximum efficiency for the targeted introduction into plants of genes that will result in long-lasting protection from pest attack in the organs of interest.

ACKNOWLEDGEMENTS

I acknowledge the support of the Ministry of Agriculture, Fisheries and Food and the Biotechnology and Biological Sciences Research Council for research carried out in the John Innes Centre and cited in this chapter. I am very grateful to Dr. Julie Graham of the Scottish Crop Research Institute for providing me with reprints and information related to Fig. 4. I thank Tarn Dalzell and Nicola Cooke for their patience in typing and processing this chapter.

REFERENCES

Altabella, T. and Chrispeels, M.J. (1990) Tobacco plants transformed with the bean αai gene express an inhibitor of insect α-amylase in their seeds. *Plant Physiology*, **93**, 805-10.

Ary, M.B., Shewry, P.R. and Richardson, M. (1988) The amino acid sequence of a cereal Bowman-Birk type trypsin inhibitor from seeds of Jobs' tears (*Coix lachryma-jobi* L.). *FEBS Letters*, **229**, 111-8.

Asao, T., Imai, F., Tsuji, I., Tashiro, M., Iwami, K. and Ibuki, F. (1991) The amino acid sequence of a Bowman-Birk type proteinase inhibitor from faba beans (*Vicia faba* L). *Journal of Biochemistry*, **110**, 951-5.

Belozersky, M.A., Dunaevsky, Y.E., Musolyamov, A.X. and Egorov, T.A. (1995) Complete amino acid sequence of the protease inhibitor from buckwheat seeds. *FEBS Letters*, **371**, 264-6.

Birk, Y., Gertler, A. and Khalef, S. (1963) A pure trypsin inhibitor from soya beans. *Biochemical Journal*, **87**, 281-4.

Boivin, M., Flourie, B., Rizza, R.A., Go, V.L.W. and DiMagno, E.P. (1988) Gastrointestinal and metabolic effects of amylase inhibition in diabetics. *Gastroenterology*, **94**, 387-94.

Bompard-Gilles, C., Rousseau, P., Rougé, P. and Payan, F. (1996) Substrate mimicry in the active center of a mammalian α-amylase: structural analysis of an enzyme-inhibitor complex. *Structure*, **4**, 1441-52.

Bowman, D.E. (1946) Differentiation of soybean antitrypsin factors. *Proceedings of the Society for Experimental Biology and Medicine*, **63**, 547-50.

Bradshaw, H.D. Jr., Hollick, J.B., Parsons, T.J., Clarke, H.R.G. and Gordon, M.P. (1989) Systemically wound-responsive genes in poplar trees encode proteins similar to sweet potato sporamins and legume Kunitz trypsin inhibitors. *Plant Molecular Biology*, **14**, 51-9.

Brown, W.E., Takio, K., Titani, K. and Ryan, C.A. (1985) Wound-induced trypsin inhibitor in alfalfa leaves: identity as a member of the Bowman-Birk inhibitor family. *Biochemistry*, **24**, 2105-8.

Buonocore, V. and Silano, V. (1986) Biochemical, nutritional and toxicological aspects of alpha-amylase inhibitors from plant foods. *Advances in Experimental Medicine and Biology*, **199**, 483-507.

Chiang, C.C. and Hadwiger, L.A. (1991) The *Fusarium solani*-induced expression of a pea gene family encoding high cysteine content proteins. *Molecular Plant-Microbe Interactions*, **4**, 324-31.

Domagalski, J.M., Kollipara, K.P., Bates, A.H., Brandon, D.L., Friedman, M. and Hymowitz, T. (1992) Nulls for the major soybean Bowman-Birk protease inhibitor in the genus *Glycine*. *Crop Science*, **32**, 1502-5.

Domoney, C., Welham, T. and Sidebottom, C. (1993) Purification and characterization of *Pisum* seed trypsin inhibitors. *Journal of Experimental Botany*, **44**, 701-9.

Domoney, C., Welham, T., Ellis, N. and Hellens, R. (1994) Inheritance of qualitative and quantitative trypsin inhibitor variants in *Pisum*. *Theoretical and Applied Genetics*, **89**, 387-91.

Domoney, C., Welham, T., Sidebottom, C. and Firmin, J.L. (1995) Multiple isoforms of *Pisum* trypsin inhibitors result from modification of two primary gene products. *FEBS Letters*, **360**, 15-20.

Domoney, C. and Welham, T. (1998) Limited proteolysis of enzyme inhibitor proteins during seed desiccation in *Pisum*. *Journal of Plant Physiology*, **152**, 692-5.

Eckelkamp, C., Ehmann, B. and Schopfer, P. (1993) Wound-induced systemic accumulation of a transcript coding for a Bowman-Birk trypsin inhibitor-related protein in maize (*Zea mays*) seedlings. *FEBS Letters*, **323**, 73-6.

Fernandes, A. O. and Banerji, A. P. (1995) Inhibition of benzopyrene-induced forestomach tumors by field bean protease inhibitor(s). *Carcinogenesis*, **16**, 1843-6.

Garcia-Olmedo, F., Salcedo, G., Sanchez-Monge, R., Gomez, L., Royo, J. and Carbonero, P. (1987) Plant proteinaceous inhibitors of proteinases and α-amylases. *Oxford Surveys of Plant Molecular and Cell Biology*, **4**, 275-334.

Gatehouse, A.M.R., Dewey, F.M., Dove, J., Fenton, K.A. and Pusztai, A. (1984) Effect of seed lectins from *Phaseolus vulgaris* on the development of larvae of *Callosobruchus maculatus*; mechanism of toxicity. *Journal of the Science of Food and Agriculture*, **35**, 373-80.

Graham, J., Gordon, S.C. and McNicol, R.J. (1997) The effect of the CpTi gene in strawberry against attack by vine weevil (*Otiorhynchus sulcatus* F. Coleoptera: Curculionidae). *Annals of Applied Biology*, **131**, 133-9.

Grossi de Sa, M.F., Mirkov, T.E., Ishimoto, M., Colucci, G., Bateman, K.S. and Chrispeels, M.J. (1997) Molecular characterization of a bean α-amylase inhibitor that inhibits the α-amylase of the Mexican bean weevil *Zabrotes subfasciatus*. *Planta*, **203**, 295-303.

Habu, Y., Peyachoknagul, S., Umemoto, K., Sakata, Y. and Ohno, T. (1992) Structure and regulated expression of Kunitz chymotrypsin inhibitor genes in winged bean [*Psophocarpus tetragonolobus* (L.) DC.] *Journal of Biochemistry*, **111**, 249-58.

Habu, Y., Sakata, Y., Fukasawa, K. and Ohno, T. (1993) Ubiquitous nuclear proteins bind to 5' upstream region of major Kunitz chymotrypsin inhibitor gene in winged bean. *Plant Molecular Biology*, **23**, 1139-50.

Heath, R.L., Barton, P.A., Simpson, R.J., Reid, G.E., Lim, G. and Anderson, M.A. (1995) Characterization of the protease processing sites in a multidomain proteinase inhibitor precursor from *Nicotiana alata*. *European Journal of Biochemistry*, **230**, 250-7.

Hilder, V.A., Gatehouse, A.M.R., Sheerman, S.E., Barker, R.F. and Boulter, D. (1987) A novel mechanism of insect resistance engineered into tobacco. *Nature*, **330**, 160-3.

Hilder, V.A., Barker, R.F., Samour, R.A., Gatehouse, A.M.R., Gatehouse, J.A. and Boulter, D. (1989) Protein and cDNA sequences of Bowman-Birk protease inhibitors from the cowpea (*Vigna unguiculata* Walp.). *Plant Molecular Biology*, **13**, 701-10.

Hilder, V.A., Gatehouse, A.M.R. and Boulter, D. (1990) Genetic engineering of crops for insect resistance using genes of plant origin, in *Genetic engineering of crop plants*, (eds D. Grierson and G. Lycett), Butterworths, pp. 51-66.

Hueros, G., Dues, G., Varotto, S., Yan, G., Müller, M., Maitz, M., Heyl, A., Salamini, F. and Thompson, R.D. (1997) Regulation of nutrient assimilation in the developing cereal grain, in *Abstracts of the Seventh International Seed Protein Symposium*, Gatersleben, Germany.

Huesing, J.E., Shade, R.E., Chrispeels, M.J. and Murdock, L.L. (1991) α-amylase inhibitor, not phytohemagglutinin, explains resistance of common bean seeds to cowpea weevil. *Plant Physiology*, **96**, 993-6.

James, D.J., Passey, A.J., Easterbrook, M.A., Solomon, M.G. and Barbara, D.J. (1992) Progress in the introduction of transgenes for pest resistance in apples and strawberries. *Phytoparasitica*, **20**, 83-7.

Jofuku, K.D. and Goldberg, R.B. (1989) Kunitz trypsin inhibitor genes are differentially expressed during the soybean life cycle and in transformed plants. *The Plant Cell*, **1**, 1079-93.

Jofuku, K.D., Schipper, R.D. and Goldberg, R.B. (1989) A frameshift mutation prevents Kunitz trypsin inhibitor mRNA accumulation in soybean embryos. *The Plant Cell*, **1**, 427-35.

Johnston, K.A., Lee, M.J., Brough, C., Hilder, V.A., Gatehouse, A.M.R. and Gatehouse, J.A. (1995) Protease activities in the larval midgut of *Heliothis virescens*: evidence for trypsin and chymotrypsin-like enzymes. *Insect Biochemistry and Molecular Biology*, **25**, 375-83.

Jongsma, M.A., Stiekema, W.J. and Bosch, D. (1996) Combatting inhibitor-insensitive proteases of insect pests. *TIBTECH*, **14**, 331-3.

Joudrier, P.E., Foard, D.E., Floener, L.A. and Larkins, B.A. (1987) Isolation and sequence of cDNA encoding the soybean protease inhibitors PI IV and C-II. *Plant Molecular Biology*, **10**, 35-42.

Jung, R., Martino-Catt, S., Townsend, J. and Beach, L. (1997) Expression of a sulfur-rich protein in soybean seeds causes an altered seed protein composition, in *Abstracts of the Seventh International Seed Protein Symposium,* Gatersleben, Germany.

Kunitz, M. (1945) Crystallization of a trypsin inhibitor from soybean. *Science*, **101**, 668-9.

Lee Downing, W., Mauxion, F., Fauvarque, M.-O., Reviron, M.-P., de Vienne, D., Vartanian, N. and Giraudat, J. (1992) A *Brassica napus* transcript encoding a protein related to the Kunitz protease inhibitor family accumulates upon water stress in leaves, not in seeds. *The Plant Journal*, **2**, 685-93.

Le Guen, M.P. and Birk, Y. (1993) Protein protease inhibitors from legume seeds: nutritional effects, mode of action and structure-function relationship, in *Recent advances of research in antinutritional factors in legume seeds*, (eds A.F.B. van der Poel, J. Huisman and H.S. Saini), Wageningen Pers, Wageningen, The Netherlands, pp. 157-71.

Liener, I.E., Donatucci, D.A. and Tarcza, J.C. (1984) Starch blockers: a potential source of trypsin inhibitors and lectins. *American Journal of Clinical Nutrition*, **39**, 196-200.

Macedo, M.L.R. and Xavier-Filho, J. (1992) Purification and partial characterization of trypsin inhibitors from seeds of *Clitoria ternatea*. *Journal of the Science of Food and Agriculture*, **58**, 55-8.

Manen, J.-F., Simon, P., van Slooten, J.-C., Østerås, M., Frutiger, S. and Hughes, G.J. (1991) A nodulin specifically expressed in senescent nodules of winged bean is a protease inhibitor. *The Plant Cell*, **3**, 259-70.

McGurl, B., Mukherjee, S., Kahn, M. and Ryan, C.A. (1995) Characterization of two proteinase inhibitor (ATI) cDNAs from alfalfa leaves (*Medicago sativa* var. Vernema): the expression of ATI genes in response to wounding and soil microorganisms. *Plant Molecular Biology*, **27**, 995-1001.

Mirkov, T.E., Wahlstrom, J.M., Hagiwara, K., Finardi-Filho, F., Kjemtrup, S. and Chrispeels, M.J. (1994) Evolutionary relationships among proteins in the phytohemagglutinin-arcelin-α-amylase inhibitor family of the common bean and its relatives. *Plant Molecular Biology*, **26**, 1103-13.

Mirkov, T.E., Evan, S.V., Wahlstrom, J., Gomez, L., Young, N.M. and Chrispeels, M.J. (1995) Location of the active site of the bean α-amylase inhibitor and involvement of a Trp, Arg, Tyr triad. *Glycobiology*, **5**, 45-50.

Moreno, J. and Chrispeels, M.J. (1989) A lectin gene encodes the α-amylase inhibitor of the common bean. *Proceedings of the National Academy of Sciences of the United States of America*, **86**, 7885-9.

Monte-Negreiros, A.N., Carvalho, M.M., Xavier Filho, J., Blanco-Labra, A., Shewry, P.R. and Richardson, M. (1991) The complete amino acid sequence of the major Kunitz trypsin inhibitor from the seeds of *Prosopsis juliflora*. *Phytochemistry*, **30**, 2829-33.

Nagasue, A., Fukamchi, H., Ikenaga, H. and Funatsu, G. (1988) The amino acid sequence of the barley rootlet trypsin inhibitor. *Agricultural and Biological Chemistry*, **52**, 1505-14.

Nodari, R.O., Tsai, S.M., Gilbertson, R.L. and Gepts, P. (1993) Towards an integrated linkage map of common bean 2. Development of an RFLP-based linkage map. *Theoretical and Applied Genetics*, **85**, 513-20.

Onesti, S., Brick, P. and Blow, D.M. (1991) Crystal structure of a Kunitz-type trypsin inhibitor from *Erythrina caffra* seeds. *Journal of Molecular Biology*, **217**, 153-76.

Orf, J.H. and Hymowitz, T. (1979) Inheritance of the absence of Kunitz trypsin inhibitor in seed proteins of soybeans. *Crop Science*, **19**, 107-9.

Osborn, T.C., Alexander, D.C., Sun, S.S.M., Cardona, C. and Bliss, F.A. (1988) Insecticidal activity and lectin homology of arcelin seed protein. *Science*, **240**, 207-10.

Perez-Grau, L. and Goldberg, R.B. (1989) Soybean seed protein genes are regulated spatially during embryogenesis. *The Plant Cell*, **1**, 1095-109.

Pueyo, J.J., Hunt, D.C. and Chrispeels, M.J. (1993) Activation of bean α-amylase inhibitor requires proteolytic processing of the pro-protein. *Plant Physiology*, **101**, 1341-8.

Pusztai, A., Grant, G., Duguid, T., Brown, D.S., Peumans, W.J., Van Damme, E.J.M. and Bardocz, S. (1995) Inhibition of starch digestion by α-amylase inhibitor reduces the efficiency of utilization of dietary proteins and lipids and retards the growth of rats. *Journal of Nutrition*, **125**, 1554-62.

Richardson, M. (1991) Seed storage proteins: the enzyme inhibitors. *Methods in Plant Biochemistry*, **5**, 259-305.

Rohrmeier, T. and Lehle, L..(1993) *WIP1*, a wound-inducible gene from maize with homology to Bowman-Birk proteinase inhibitors. *Plant Molecular Biology*, **22**, 783-92.

Ryan, C.A. (1990) Protease inhibitors in plants: genes for improving defenses against insects and pathogens. *Annual Review of Phytopathology*, **28**, 425-49.

Schroeder, H.E., Gollasch, S., Moore, A., Tabe, L.M., Craig, S., Hardie, D.C., Chrispeels, M.J., Spencer, D. and Higgins, T.J.V. (1995) Bean α-amylase inhibitor confers resistance to the pea weevil (*Bruchus pisorum*) in transgenic peas (*Pisum sativum* L.). *Plant Physiology*, **107**, 1233-9.

Shade, R.E., Schroeder, H.E., Pueyo, J.J., Tabe, L.M., Murdock, L.L., Higgins, T.J.V. and Chrispeels, M.J. (1994) Transgenic pea seeds expressing the α-amylase inhibitor of the common bean are resistant to bruchid beetles. Bio/technology, **12**, 793-6.

Shewry, P.R. and Lucas, J.A. (1997) Plant proteins that confer resistance to pests and pathogens. *Advances in Botanical Research*, **26**, 135-92.

Sijmons, P.C. (1993) Plant-nematode interactions. *Plant Molecular Biology*, **23**, 917-31.

Spencer, M.E. and Hodge, R. (1991) Cloning and sequencing of the cDNA encoding the major albumin of *Theobroma cacao*. Identification of the protein as a member of the Kunitz protease inhibitor family. *Planta*, **183**, 528-35.

Sweet, R.M., Wright, H.T., Janin, J., Chothia, C.H. and Blow, D.M. (1974) Crystal structure of the complex of porcine trypsin with soybean trypsin inhibitor (Kunitz) at 2.6-Å resolution. *Biochemistry*, **13**, 4212-28.

Svendsen, I., Hejgaard, J. and Chavan, J.K. (1984) Subtilisin inhibitor from seeds of broad bean (*Vicia faba*); purification, amino acid sequence and specificity of inhibition. *Carlsberg Research Communications*, **49**, 493-502.

Welham, T., O'Neill, M., Johnson, S., Wang, T.L. and Domoney, C. (1998) Expression patterns of genes encoding seed trypsin inhibitors in *Pisum sativum*. *Plant Science*, **131**, 13-24.

Yamagishi, K., Mitsumori, C., Takahashi, K., Fujino, K., Koda, Y. and Kikuta, Y. (1993) Jasmonic acid-inducible gene expression of a Kunitz-type proteinase inhibitor in potato tuber disks. *Plant Molecular Biology*, **21**, 539-41.

Yeh, K.-W., Chen, J.-C., Lin, M.-I., Chen, Y.-M. and Lin, C.-Y. (1997) Functional activity of sporamin from sweet potato (*Ipomoea batatas* Lam.): a tuber storage protein with trypsin inhibitory activity. *Plant Molecular Biology*, **33**, 565-70.

28. Seed Lectins

WILLY J. PEUMANS and ELS J.M. VAN DAMME
Katholieke Universiteit Leuven, Laboratory for Phytopathology and Plant Protection, Willem de Croylaan 42, 3001 Leuven, Belgium.

1. HISTORICAL BACKGROUND

Lectins, also called agglutinins or phytohemagglutinins, have been known since 1888 when Stillmark observed that extracts from castor beans (*Ricinus communis*) contain a proteinaceous hemagglutinating factor (Stillmark, 1888). Even though at that time the basis for the erythrocyte clumping activity was not understood, Stillmark's finding was a milestone in plant biochemistry because ricin was the first protein to which a well-defined biological activity could be ascribed. Moreover, the pioneering work of Stillmark is also considered as the onset of 'lectinology', a scientific discipline covering broad interdisciplinary research on different aspects of lectins. Soon after the discovery of ricin, agglutinating factors were also found in several other plants and the term 'hemagglutinin' was introduced as a general name for all substances with hemagglutinating activity. Due to the self-explanatory character of the term agglutinin no other names have been introduced until it turned out that some hemagglutinins exhibited a clear preference towards erythrocytes of a particular human blood group within the ABO system. At that time the term 'lectin' (from the Latin verb 'legere', which means 'to select') was introduced to emphasize the selective agglutination behavior of some hemagglutinins. Later on, the term lectin was not exclusively used for blood group specific hemagglutinins but was applied to all proteins showing a more general agglutination behavior. Therefore, the term 'hemagglutinin' is more appropriate as it refers to the capability of the carbohydrate-binding proteins to agglutinate erythrocytes. However, since most hemagglutinins also agglutinate other cells, the term 'agglutinin' should be preferred. The use of three different names for the same group of proteins is not only confusing but also incorrect since the terms 'lectin', 'hemagglutinin' and 'agglutinin' have different meanings. In the absence of a consensus the term 'lectin' is now commonly used. One should realize, however, that the original meaning

P.R. Shewry and R. Casey (eds.), Seed Proteins, 657–683.

of the term 'lectin' has been lost. The current ambiguity in the naming of lectins is also due to the fact that the underlying mechanism of their macroscopically visible agglutination activity was not understood until it was demonstrated in 1952 that the agglutination properties of lectins are based on a specific sugar binding activity (Watkins and Morgan, 1952). Once it was established that recognition and binding to carbohydrates is essential for activity, lectins could be distinguished from other proteins on the basis of a well-defined functional criterion. As a result lectins are now considered in the first place as carbohydrate-binding proteins rather than (hem)agglutinins.

2. INTRODUCTION

2.1 Plant lectins: definition and subdivision

The introduction of modern biochemistry in lectin research and especially the development of powerful affinity-purification techniques facilitated the isolation and characterization of numerous plant lectins. Comparative analyses soon revealed that lectins are a heterogeneous group of proteins, which have little in common apart from their capability to bind specific carbohydrates and agglutinate cells. Moreover, it also became evident that some carbohydrate-binding proteins do not agglutinate cells. In the absence of clear biochemical and/or structural requirements the question arose what criteria had to be met by a protein to be regarded as a lectin. According to the first definition, which was based primarily on the sugar specificity and inhibition of the agglutination reaction, *lectins are carbohydrate-binding proteins of non-immune origin which agglutinate cells or precipitate glycoconjugates* (Goldstein *et al.*, 1980). Because this definition excluded some poorly agglutinating toxins with genuine lectin subunits (like ricin, abrin, modeccin, etc.), it was later extended to cover these toxins (Kocourek and Horejsi, 1983). After it was observed that some lectins contain a second type of binding site that interacts with non-carbohydrate ligands, lectins were redefined as carbohydrate-binding proteins other than antibodies or enzymes (Barondes, 1988). This definition also needed to be updated after the molecular cloning of lectins and lectin-related proteins generated novel insights with respect to the structure and function of different types of lectins. The finding that some plant enzymes are fusion proteins of a carbohydrate-binding domain tandemly arrayed with a catalytic domain, and the discovery of several carbohydrate-binding proteins possessing a single binding site, argued for a definition based on the presence of carbohydrate-binding domain(s). Moreover, the discovery of proteins which are evolutionary related to lectins but lack active carbohydrate-binding domains necessitated the inclusion of functionality as a major criterion. To meet both requirements the presence of at least one non-catalytic domain, which binds reversibly to a specific carbohydrate, is now considered as the only criterion for a protein to be called a lectin. Accordingly, plant lectins are now defined as *'All plant proteins possessing at least one non-catalytic domain, which binds reversibly to a specific mono- or oligosaccharide'* (Peumans and Van Damme, 1995). Since a broad range of proteins with different agglutination and/or glycoconjugate precipitation properties falls within the limits set by the new definition, lectins are further subdivided according to the overall structure of the primary translation products of the lectin genes. Basically, lectins are subdivided in 'merolectins', 'hololectins', 'chimerolectins' and 'superlectins'. *Merolectins* consist exclusively of a single carbohydrate-binding domain. They are monovalent and hence do not precipitate glycoconjugates or agglutinate cells. *Hololectins* are also exclusively composed of carbohydrate-binding domains but contain at least two such domains which are either identical or very homologous and bind either the same or structurally similar sugar(s). Due to the presence of multiple binding sites hololectins agglutinate cells and/or precipitate glycoconjugates. Most of the classical plant lectins belong to the subgroup of hololectins. *Chimerolectins* are fusion proteins comprising one or more carbohydrate-binding domain(s) tandemly arrayed to an unrelated domain. This unrelated domain can have a well defined enzymatic (or biological) activity but acts independently of the carbohydrate-binding domain. Depending on the number of carbohydrate-binding sites chimerolectins behave as merolectins or as hololectins. *Superlectins* are also fusion proteins but they are composed of two tandemly arrayed carbohydrate-binding domains which are structurally different and recognize structurally unrelated sugars.

2.2 Occurrence and distribution of plant lectins

Although several decades of intensive research yielded a fairly good idea of the occurrence and distribution of lectins in plants two important misconceptions still persist. The first concerns the general occurrence of plant lectins. There is no doubt that lectins occur in many species of different taxonomic groups but they certainly cannot be regarded as an ubiquitous group of plant proteins. Lectins have been identified in less than 500 plant species. Moreover, most of these species belong to a relatively small number of plant families. Even if one assumes that (i) all relatives of these 500 species contain agglutinins and that (ii) novel lectins will be discovered in the future, the occurrence of lectins will still be limited to a very small fraction of the over one million plant species. Hence, the classical (i.e. agglutinating) lectins are rather scarce in the plant kingdom. It should be mentioned, however, that class I chitinases, which according to the updated definition are classified as (chimero)lectins, are probably present in most if not all plant species.

A second misconception concerns the location of lectins in the plant. The initial discovery of lectins in seeds, the large impact of a handful of seed lectins on early lectinology and the coincidence that the most common legume lectins are strictly confined to seeds left many scientists with the wrong impression that lectins are typical seed proteins. At present, the occurrence of lectins in various types of vegetative tissues such as leaves, bark, roots, stems, rhizomes, bulbs, tubers etc. is well documented. Moreover, non-seed lectins are at least as widespread as seed lectins and some plants contain extremely high levels of lectins, especially in vegetative storage tissues.

Seed lectins markedly differ from each other with respect to their abundance and location in the seed. Most seed lectins are predominantly located in the storage parenchyma of the cotyledons (e.g. legume seed lectins) or the endosperm (e.g. castor bean lectins) but some are strictly confined to the primary axis (e.g. wheat germ agglutinin). On average seed lectins account for 0.1-5% of the total protein. However, maximal levels can exceed 50% of the total protein (e.g. in some *Phaseolus* species), whereas other seed lectins are barely detectable.

2.3 Lectins are a heterogeneous group of proteins comprising several families of evolutionary and structurally related proteins.

The results of the early studies generated the idea that plant lectins are a heterogeneous group of proteins because they differ from each other with respect to their biochemical/physicochemical properties, molecular structure, carbohydrate-binding specificity and biological activities. Recent advances in the structural analysis of lectins and the molecular cloning of lectin genes yielded detailed information about several hundred different plant lectins. Although the novel data do not abolish the apparent heterogeneity of lectins at the protein level, the availability of detailed sequence information has allowed us to trace the evolutionary relationships and classify the lectins accordingly. At present seven distinct lectin families have been identified on the basis of protein and cDNA or genomic DNA sequences. They are further referred to as legume lectins, monocot mannose-binding lectins, chitin-binding lectins, type 2 RIP, jacalins, amaranthins and Cucurbitaceae phloem lectins. Since these seven families comprise the vast majority of all known plant lectins they are briefly described below.

2.3.1 Legume lectins

Legume lectins are an extended lectin family which have only been found in the plant family Fabaceae (Leguminosae) (Sharon and Lis, 1990). All legume lectins are composed of two or four protomers of M_r around 30,000. In some species the protomers are (partly) cleaved into two smaller polypeptides forming the so-called two chain legume lectins. All legume lectins share high sequence similarity but differ markedly with respect to their carbohydrate-binding specificity. Legume lectins contain a single carbohydrate-binding site per protomer. In addition to the carbohydrate-binding sites, legume lectins also contain metal binding sites. Depending on the number of protomers, legume lectins are either di- or tetravalent. Legume lectins have been found in seeds as well as vegetative (storage) tissues.

2.3.2 Monocot mannose-binding lectins

The monocot mannose-binding lectins are an extended superfamily of proteins which hitherto have been found in representatives of the monocot families Amaryllidaceae, Alliaceae, Araceae, Orchidaceae, Liliaceae and Bromeliaceae (Van Damme *et al.*, 1998). Monocot mannose-binding lectins are built up of one, two or four subunits of M_r around 12,000 and exhibit an exclusive specificity towards mannose. All monocot mannose-binding lectins share high sequence similarity and have a very similar overall structure (Barre *et al.*, 1996). The number of active binding sites varies between one and three per subunit. Depending on the number of subunits, native monocot mannose-binding lectins possess between one and twelve mannose-binding sites per molecule. Monocot mannose-binding lectins are present in various vegetative tissues but until now have not been detected in seeds.

2.3.3 Chitin-binding lectins

Chitin-binding lectins are a large family of proteins containing at least one hevein domain (Raikhel *et al.*, 1993). The latter domain refers to a chitin-binding polypeptide of about 40 amino acid residues that has sequence similarity with hevein, a small 43 amino-acid (chitin-binding) protein from the latex of the rubber tree (*Hevea brasiliensis*). It should be mentioned that there are also chitin-binding lectins without hevein domain(s) (e.g. some legume lectins and the Cucurbitaceae phloem lectins). However, these lectins have no sequence similarity with the chitin-binding lectins containing hevein domain(s), and hence belong to other lectin families. The family of chitin-binding lectins comprises merolectins, hololectins and chimerolectins. Hevein and the chitin-binding antimicrobial peptides from *Amaranthus caudatus* (Broekaert *et al.*, 1992) for instance contain a single (truncated) hevein domain and hence are merolectins. Several chitin-binding lectins are hololectins composed of polypeptides containing two, three or four tandemly arrayed hevein domains. Some of these lectins consist of a single polypeptide (e.g. the nettle lectin) whereas others are homodimers (e.g. wheat lectin). As a result the number of binding sites varies between two (e.g. the nettle lectin) and eight (e.g. wheat lectin). The family of chitin-binding lectins also comprises at least two different types of chimerolectins. Protomers of the Solanaceae lectins are fusion proteins of a chitin-binding domain (built up of three tandemly arrayed hevein domains) and an *O*-glycosylated serine-hydroxyproline-rich domain resembling the cell wall protein extensin (Kieliszewski *et al*., 1994; Allen *et al*., 1996). Class I chitinases are chimerolectins built up of a single hevein domain linked to a chitinase domain (Collinge *et al*., 1993). It is also worth mentioning that several chitin-binding lectins are processing products of chimeric precursors. Both hevein and the nettle lectin are cleaved from a large precursor containing an extended *C*-terminal domain with a high sequence similarity to the catalytic domain of plant chitinases (Lerner and Raikhel, 1992; Lee *et al*., 1991). Hitherto agglutinating chitin-binding lectins have been found in the families Gramineae, Solanaceae, Phytolaccaceae, Urticaceae, Papaveraceae and Viscaceae (Raikhel *et al*., 1993; Konami *et al*., 1995; Peumans *et al*., 1996) indicating that they occur both in monocots and dicots. Chitin-binding lectins occur in seeds as well as in various vegetative tissues.

2.3.4 Type 2 ribosome-inactivating proteins

Ribosome-inactivating proteins (RIP) are commonly known as proteins that catalytically inactivate eukaryotic ribosomes through the enzymatic removal of a specific adenosine residue from the large rRNA (Barbieri *et al*., 1993). Since recent experiments clearly demonstrated that RIP can use different types of polynucleotides as a substrate it has been proposed to introduce the term polynucleotide:adenosine glycosidase (PAG) (Barbieri *et al*., 1997). Whereas type 1 RIP are composed of a single polypeptide (M_r around 30,000) that exhibits PAG-activity, type 2 RIP are chimeric proteins composed of a PAG-domain (the so-called A chain) linked to a carbohydrate-binding domain (the so-called B chain). Due to the presence of the latter domain type 2 RIP are chimerolectins. Type 2 RIP are composed of one, two or four identical protomers whereby each protomer consists of disulphide-linked A and B chains. Both chains are derived from a single precursor by the excision of a linker between the *N*-terminal A domain and the *C*-terminal B domain. Type 2 RIP share a high degree of sequence similarity and have similar overall folding and three-dimensional structures, but differ from each other with respect to their catalytic activity, toxicity and carbohydrate-binding activity. Hitherto, type 2 RIP have been found within the Euphorbiaceae,

Fabaceae, Sambucaceae, Viscaceae, Ranunculaceae, Lauraceae, Passifloraceae, Iridaceae and Liliaceae which implies that they occur in monocots and dicots. Type 2 RIP occur both in seeds and in vegetative tissues. Some of them are very abundant proteins whereas others are present in minute amounts.

2.3.5 Jacalins

The jacalins have only recently been recognized as a separate lectin family. Jackfruit (*Artocarpus integrifolia*) as well as a few other *Artocarpus* species and osage orange (*Maclura pomifera*) (Young *et al.*, 1989) contain galactose-specific lectins thought to be characteristic for the Moraceae family. All Moraceae lectins are built up of four identical protomers. Each protomer contains a small and a large subunit, which are both derived from a single precursor by complex proteolytic processing. Recently, a mannose/maltose binding lectin which shows sequence similarity with jacalin has been discovered in rhizomes of *Calystegia sepium* (hedge bindweed, family Convolvulaceae) (Van Damme *et al.*, 1996). The occurrence of similar lectins in Moraceae and Convolvulaceae species indicates that the jacalin family occurs in several taxonomic groups. Jacalins not only exhibit relatively low sequence similarity but also differ with respect to their carbohydrate-binding specificity. They occur in seeds as well as in vegetative tissues.

2.3.6 Cucurbitaceae phloem lectins

The phloem exudates of many Cucurbitaceae species contain high concentrations of lectins with a high affinity for oligomers of *N*-acetylglucosamine (GlcNAc) (Wang *et al.*, 1994). All currently known Cucurbitaceae phloem lectins are composed of two identical subunits of M_r around 24,000. They have high sequence similarity but have no apparent similarity with any other protein. Cucurbitaceae phloem lectins have only been found in phloem exudates.

2.3.7 Amaranthaceae lectins

Amaranthins are a small family of lectins occurring in seeds of *Amaranthus* species. All amaranthins are homodimeric proteins composed of two identical subunits of M_r around 30,000 and exhibit specificity towards *N*-acetylgalactosamine (GalNAc). Amaranthins have no sequence similarity with any other plant lectin and have a unique three-dimensional structure (Raina and Datta, 1992; Transue *et al.*, 1997).

2.3.8 Other lectins

Most of the currently known plant lectins belong to one of the above described lectin families. Due to the lack of sequence information several lectins cannot be classified yet. Once protein or cDNA sequences have been determined it will be clear whether these lectins belong to the established or to novel lectin families.

2.4 Lectins exhibit marked differences in carbohydrate binding specificity

Plant lectins exhibit marked differences in sugar binding properties (Goldstein and Portez, 1986). Since one can reasonably assume that the physiological role of lectins is primarily determined by their sugar binding activity this issue needs special attention. It is important to realize that lectins display a broad range of specificities covering monosaccharides as well as complex oligosaccharides. Usually, the specificity of lectins is given in terms of preferential binding to one or more monosaccharides. According to this criterion, mannose-, mannose/glucose-, mannose/maltose-, Gal/GalNAc-, GlcNAc/$(GlcNAc)_n$-, fucose- and sialic acid-binding lectins are distinguished (Van Damme *et al.*, 1998). Lectins which bind exclusively complex oligosaccharides are classified as having a 'complex' specificity. The distinction between simple and complex specificity is merely arbitrary because lectins which bind simple sugars usually have a much higher affinity for more complex oligosaccharides. For example, GlcNAc is about 4500-fold less effective than $(GlcNAc)_4$ in the inhibition of the hemagglutination of wheat germ agglutinin.

Structural analyses demonstrated that the overall three-dimensional structure of the carbohydrate-binding sites are well conserved within a given lectin family. In spite of the obvious conservation of the binding sites different members of a lectin family may differ in fine specificity (e.g. the monocot mannose-binding lectins) or even in overall specificity (e.g. legume lectins). As far as can be concluded from the currently available data the binding sites of each lectin family have a unique overall structure. Nevertheless, members of different lectin families can have a very similar specificity. For example, mannose-binding lectins are found within the legume lectins, the monocot mannose-binding lectins and the jacalin family.

The preferential binding of plant lectins to complex glycans (rather than to mono- or disaccharides) indicates that their natural receptors are most probably oligosaccharides. In addition, the fact that many lectins have an exclusive specificity towards carbohydrates which are absent in plants but are common in animals (e.g. sialic acid) or insects (e.g. chitin) suggests that most lectins are predestined to interact with foreign glycans. The latter observation is of great relevance whenever the physiological role of lectins comes into discussion (Peumans and Van Damme, 1995).

3. OVERVIEW OF SEED LECTINS

3.1 Legume seed lectins

Legume seed lectins are an important group of proteins for several reasons. They not only played a central role in the development of lectinology but also served as a model for the study of various aspects of plant proteins. Moreover, the presence of lectins in seeds of several economically important legume crops is a major issue in food science and technology because most legume lectins are antinutritional.

Lectins have been found in seeds of several hundred species of legumes. Less than 100 of these lectins have been purified and characterized in some detail with respect to their molecular structure, carbohydrate-binding specificity and biological activities (summarized in Table 1). Many seed lectins have been completely sequenced by protein sequencing techniques. In addition, sixteen legume seed lectins have been cloned at present. Biochemical analyses and molecular cloning revealed that many legumes contain two or more different seed lectins, which are encoded by separate genes. In some legumes the distinct seed lectins closely resemble each other with respect to their molecular structure and specificity whereas in others the individual lectins strongly differ in their structure, specificity and/or biological activities (Table 1).

Table 1 Legume seed lectins: Molecular structure and specificity

Species	Structure*	Specificity	Sequence available
Amphicarpaea bracteata	$[P32,30,28,27]_4$	GalNAc	
Arachis hypogaea	$[P27]_4$	Gal	protein, cDNA
Bauhinia purpurea	$[P32]_4$	GalNAc	protein, cDNA
Bowringia milbraedii	$[P(13+12)]_2$	Mannose/Glucose	protein
Butea monosperma	$[P32, 34]_2$	GalNAc	
Canavalia ensiformis	$[P26]_4$	Mannose/Glucose	protein, cDNA
Caragana arborescens	I $[P30]_4$ II $[P30]_2$	GalNAc Gal	
Cicer arietinum	$[P26]_2$	Complex	
Crotalaria juncea	$[P30]_4$	Gal>GalNAc	
Crotalaria striata	$[P30]_4$	GalNAc>Gal	
Cytisus multiflorus	I $[P42]_2$ II $[P42]_2$	GlcNAc oligomers Gal	
Cytisus scoparius	I $[P30]_4$	Gal/GalNAc	

	II [P30]$_4$	GalNAc>Gal	protein
Cytisus sessilifolius	I [P27]$_4$	GlcNAc oligomers	protein
	II [P34]$_2$	Gal	protein
Dioclea grandiflora	[P25]$_4$	Mannose/Glucose	protein
Dolichos biflorus	[P28]$_4$	GalNAc	cDNA
Erythrina corallodendron	[P30]$_2$	GalNAc>Gal	protein, cDNA
Erythrina cristagalli	[P26, 28]$_2$	GalNAc	
Falcata japonica	[P34]$_4$	GalNAc	
Galactia tashiroi	[P24]$_4$	GalNAc	
Glycine max	[P30]$_4$	GalNAc>Gal	cDNA
Griffonia simplicifolia	I-A4 [P32]$_4$**	GalNAc	
	I-B4 [P33]$_4$**	Gal	
	II [P30]$_4$	GlcNAc	
	IV [P29 + P27]	Fucose	
Lablab purpureus	[P(15 + 12)]$_2$	Mannose/Glucose	protein
Laburnum alpinum	I [P32]$_4$	GlcNAc	protein
	II [P36]$_2$	Gal	
Lathyrus ochrus	[P(6 + 20)]$_2$	Mannose/Glucose	protein
Lens culinaris	[P(6 + 18)]$_2$	Mannose/Glucose*	protein
Lotononis bainesii	[P(32 + 35)]$_2$	Gal	
Lotus tetragonolobus	[P27]$_{2-4}$	Fucose	protein
Maackia amurensis	MAH [P33]$_4$	NANA	protein, cDNA
	MAL [P37]$_4$	NANA	cDNA
Macrotyloma axillare	[P27]$_4$	GalNAc	
Onobrychis viciifolia	[P26]$_2$	Mannose/Glucose	protein
Phaseolus acutifolius	[P29]$_4$	Complex	
Phaseolus coccineus	[P35]$_4$	Complex	
Phaseolus lunatus	[P30]$_{2-4}$	GalNAc	cDNA
Phaseolus vulgaris	E4 [P31]$_4$**	Complex	cDNA
	L4 [P31]$_4$**	Complex	cDNA
	Pinto [P28]$_2$	Complex	
Pisum sativum	[P(17 + 6)]$_2$	Mannose/Glucose	protein, cDNA
Psophocarpus tetragonolobus	I [P29]$_2$	GalNAc	protein, cDNA
	II [P29]$_2$	GalNAc	
Robinia pseudoacacia	I [P34]$_4$	Complex	cDNA
	II [P29]$_4$	Complex	cDNA
Sophora japonica	[P30]$_4$	GalNAc>Gal	cDNA
Ulex europaeus	I [P29, 31]$_4$	Fucose	protein
	II [P25]$_4$	GlcNAc	protein
	III [P34]$_4$	Gal	
Vicia cracca	I [P33]$_4$	GalNAc	
	II [P(17 + 6)]$_2$	Mannose/Glucose	
Vicia faba	[P(6 + 21)]$_2$	Mannose/Glucose	protein
Vicia graminea	[P25]$_4$	GalNAc	
Wisteria floribunda	WFA [P28]$_4$	GalNAc	
	WFM [P32]$_2$	Mannose/Glucose	

* [PX] stands for protomer with a M_r of Xx10^{-3}. [P(Y+Z)] indicates that the protomer is cleaved in two polypeptides of M_r Yx10^{-3} and Zx10^{-3}.

**The lectin is a mixture of isolectins composed of two different subunits which associate randomly

3.1.1 Molecular structure, biosynthesis and processing of legume lectins

The legume lectin family nicely illustrates how proteins with a different molecular structure are formed starting from similar primary translation products (Van Damme and Peumans, 1996). All legume seed lectins are synthesized, indeed, on the endoplasmic reticulum (ER) as precursors comprising a signal peptide followed by a sequence of about 250 amino acid residues. Co-translational removal of the signal peptide yields a lectin protomer of M_r 25,000-30,000. For some lectins the polypeptide backbone of the protomers is not modified and the protomers associate by non-covalent interactions into dimers or tetramers. In other cases the lectin protomers undergo further proteolytic processing. Proteolytic removal of a short *C*-terminal peptide (as for example in the lectins from *Dolichos biflorus* and *Erythrina corallodendron*) leaves a lectin subunit, which closely resembles the original protomer. Cleavage within the protomer (which may or may not be accompanied by the excision of a small internal peptide) yields two fragments. Depending on the cleavage site the resulting fragments are similar (e.g. *Bowringia milbraedii* lectin) or dissimilar (e.g. *Pisum sativum* agglutinin) in size. Since the two fragments of the cleaved protomers are held together by non-covalent interactions they behave as uncleaved protomers and associate into dimers or tetramers. Legume lectins which are processed by an internal proteolytic cleavage are usually referred to as two chain legume lectins to distinguish them from the so-called one chain lectins. In addition to proteolytic processing legume lectins may also undergo other modifications. Many legume lectins are synthesized as glycosylated protomers carrying *N*-linked oligosaccharide side chains. In some cases, the glycan chains are lost during post-translational processing. Usually, the oligosaccharide side chains remain attached to the lectin polypeptides but are, to a varying extent, subjected to modifications. As a result, legume lectins carry both high mannose and modified *N*-glycans.

The above described processing scheme holds true for all legume lectins except for Concanavalin A (ConA), the lectin from *Canavalia ensiformis*, and the lectins from other species belonging to the tribe Diocleae, which undergo a unique but very complex processing. Basically, the conversion of proConA into mature ConA involves the simultaneous elimination of a short internal glycosylated peptide and the reassociation of the remaining peptides by transposition and ligation (Carrington *et al.*, 1985).

3.1.2 Topogenesis, intracellular location, location in the seed and physiological role of legume seed lectins

According to the available data the biosynthesis of all legume lectins follows the secretory pathway. The primary translation products of the lectin mRNAs are synthesized on the ER. In the ER the signal peptide is cotranslationally removed. Glycan chains may also be added via the dolichol pathway to asparagine residues of putative glycosylation sites and undergo the first modifications. Lectin protomers already associate in the ER into dimers or tetramers before they are translocated to the Golgi apparatus where the oligosaccharide side chains are further trimmed and modified. From the Golgi the lectins are transported to storage vacuoles where a final proteolytic processing may take place (Van Driessche, 1988).

Legume seed lectins are mainly located in the storage parenchyma cells of the cotyledons but are also present in small quantities in the cells of the primary axis. There are striking differences in the abundance of legume seed lectins. In some species, the seed lectins are barely detectable (e.g. lupin) whereas in others they are the predominant seed proteins representing up to 50 % of the total protein (e.g. in some wild *Phaseolus* species). On average, lectins account for 0.1-1% of the total seed protein.

Studies of the temporal and spatial regulation of legume seed lectins revealed that they behave as typical storage proteins. These findings, in combination with the relatively high abundance of many legume seed lectins, argue for a storage role, which, however, does not preclude that they fulfil a second more important function. It is worth mentioning, indeed, that at least some legume lectins are believed to play a role in the plant's defence mechanism(s) against insects and/or predating animals.

3.1.3 Legume lectins are evolutionary conserved

All legume lectins, irrespective of the molecular structure of the mature protein, share a marked sequence identity and have very similar folding and three-dimensional structures (Sharon and Lis, 1990; Rougé *et al.*, 1991). In particular, the amino acids involved in metal (Mn^{2+} and Ca^{2+}) binding are highly conserved. The amino acids which constitute the sugar binding sites are less conserved but are often very similar (Young and Oomen, 1992; Sharma and Surolia, 1997). This explains why legume lectins, in contrast to the other lectin families, have a broad range of carbohydrate-binding specificities.

Some legumes contain seed proteins which are clearly related to the legume lectins but are devoid of carbohydrate-binding activity. Well known examples of such proteins are the *Phaseolus vulgaris* α-amylase inhibitors and arcelins. Both proteins exhibit high sequence similarity with the bean lectin and have similar overall folding and three-dimensional structures but do not possess functional carbohydrate-binding sites (Mirkov *et al.*, 1994).

3.1.4 Agronomic and economic aspects of legume seed lectins

Many legumes are important crop plants. Grain legumes are used on a large scale in food and feed industry as sources of protein, oil and starch. In addition, legumes such as beans, peas, broad beans and lentils are also grown as vegetables. Nutritional studies performed with whole legume seed meal and purified or partially purified protein fractions indicated that several legume lectins are toxins/antinutrients especially for monogastric animals (Pusztai and Bardocz, 1996). Due to the presumed adverse effects of lectins the presence of such noxious proteins in legume crops is of great concern to the food and feed industry. To ensure food safety, products derived from lectin-containing legumes (e.g. soybean, bean) must be subjected to a thorough (heat) processing in order to inactivate the lectins (and other antinutrients such as protease inhibitors).

3.2 Chitin-binding seed lectins

In contrast to legume lectins, the occurrence of chitin-binding lectins is not confined to a single plant family. So far, seed lectins composed of hevein domains have been found in seeds of the Gramineae, Solanaceae, Amaranthaceae and Papaveraceae (Table 2). Chimerolectins composed of a hevein domain linked to a chitinase domain (the so-called class I chitinases) are not discussed here because they fall beyond the scope of this review. For more details we refer to some recent reviews on chitinases (Meins *et al.*, 1992; Collinge *et al.*, 1993).

3.2.1 Gramineae lectins

Gramineae lectins are the best characterized chitin-binding seed lectins. They have been found in the seeds of all species of the tribe Triticeae (comprising the genera *Aegilops*, *Agropyrum*, *Eremopyrum*, *Haynaldia*, *Heteranthelium*, *Secale*, *Triticosecale*, *Triticum*, *Elymus*, *Hordeum*, *Psathyrostachys*), and the genera *Brachypodium* and *Oryza*. All Gramineae seed lectins exhibit a similar carbohydrate-binding specificity towards GlcNAc and GlcNAc-oligomers, the inhibitory potency of which increases with chain length up to 4 monosaccharide units. Besides GlcNAc, Gramineae lectins also bind sialic acid.

a) Molecular structure, biosynthesis and processing of Gramineae seed lectins

Table 2 Chitin-binding seed lectins: Molecular structure and specificity

Species	Structure*	Specificity	Sequence available
Gramineae			
Agropyrum repens, A. sp.	$[P18]_2$	GlcNAc	
Brachypodium sylvaticum, B. sp.	$[P18]_2$	GlcNAc	
Hordeum vulgare, H. sp.	$[P18]_2$	GlcNAc	cDNA
Oryza sativa	$[P18]_2$, $[P(8+10)]_2$	GlcNAc	cDNA
Oryza alta	$[P18]_2$	GlcNAc	
Oryza minuta	$[P18]_2$, $[P(8+10)]_2$	GlcNAc	
Secale cereale, S. sp.	$[P18]_2$	GlcNAc	
Triticum aestivum, T. sp.	$[P18]_2$	GlcNAc	cDNA
Others			
Amaranthus caudatus	[P4]	GlcNAc oligomers	protein, cDNA
Chelidonium majus	$[P10+P12]_2$	GlcNAc oligomers	
Datura stramonium	$[P28]_2$, $[P32]_2$, [P28+P32]	GlcNAc oligomers	
Urtica dioica	[P9]	GlcNAc oligomers	protein, cDNA

* See footnote Table 1

All Gramineae seed lectins are dimeric proteins composed of two identical or very similar protomers of about 170 amino acids. Interestingly, the mature protomers are composed of four repetitive domains of 42-43 amino acids which have a high sequence similarity with hevein. Molecular cloning demonstrated that the primary translation products of the lectin genes from wheat, barley and rice have a similar structure (Raikhel *et al.*, 1993). They all contain an *N*-terminal signal peptide followed by a sequence with four repetitive domains corresponding to the mature lectin protomer, and a *C*-terminal propeptide with a putative *N*-glycosylation site. *In vivo* and *in vitro* studies demonstrated that the biosynthesis of the Gramineae lectins follows the secretory pathway. In the ER the signal peptide is cotranslationally removed and at the same time the *C*-terminal propeptide is glycosylated. The glycosylated prolectins possess carbohydrate-binding activity and associate into dimers in the ER. From the ER, the precursors are transported via the Golgi apparatus to the vacuoles. Removal of the glycosylated propeptide occurs during transport or upon arrival of the precursor to its final destination. In some rice species the lectins undergo (most probably in the vacuoles) a second post-translational proteolytic cleavage whereby the protomer is cleaved into two fragments of M_r 10,000 and 8,000 which are held together by intramolecular disulphide bridges. For example in *Oryza sativa* and all other *Oryza* species carrying the A or F genomes about 90% of the protomers are cleaved. In contrast, species with the C, D and E genomes contain exclusively uncleaved protomers.

In (allo)polyploid Gramineae species each genome encodes its own lectin subunit. Due to the high sequence similarity between the subunits encoded by the individual genomes, both homodimers and heterodimers are formed. As a result tetra-, hexa- and octaploid species contain three, six and ten isoforms, respectively (Stinissen and Peumans, 1985).

b) Physiology and regulation of Gramineae seed lectins

Localization studies demonstrated that Gramineae seed lectins are strictly confined to the embryo. No lectin has been detected in the endosperm or the aleurone. Within the embryo the lectins are only located in the peripheral tissues, namely the embryonic root cap, the coleorhiza, and in the outer cell layers of the coleoptile, the radicle and the scutellum. It appears, therefore, that only organs that come in contact with the soil during germination contain lectin and that the lectin is confined to the periphery of these organs. It should be mentioned here that root caps of wheat, barley and rice seedlings also contain lectins, which are indistinguishable from the seed lectins but are definitely not derived from the embryos (Raikhel *et al.*, 1993).

Gramineae lectins are present in low quantities. The total lectin content amounts to about 0.2-0.5 μg per embryo representing 0.2-0.5% of the soluble embryo protein. On a seed base the lectin accounts for less than 0.02% of the total protein, which indicates that the Gramineae seed lectins are certainly not storage proteins. However, taking into consideration that the lectin is confined to a small part of the embryo the lectin containing cells probably contain up to 10% lectin.

Under natural conditions Gramineae lectins are typically synthesized during seed formation. Some lectin mRNAs survive dessication of the seeds and enable germinating embryos to synthesize some lectin *de novo*. As soon as the preformed mRNAs disappear lectin synthesis normally stops. The synthesis of seed lectins is apparently under the control of abscisic acid (ABA). Exogenous ABA stimulates the synthesis of lectin in developing embryos and prolongs the *de novo* synthesis in germinating wheat and rice embryos (Raikhel *et al.*, 1993).

3.2.2 Solanaceae seed lectins

Chitin-binding lectins occur in several genera of the nightshade family. So far, Solanaceae lectins have been found in several *Solanum* species, in *Lycoperiscum esculentum* (tomato), in *Cyphomandra betacea* (tamarillo) and in *Datura stramonium* (thorn apple) and *D. innoxia*. Most of the currently known Solanaceae lectins have been isolated from vegetative tissues. Only the *Datura stramonium* and *D. innoxia* lectins are typical seed proteins.

Both *Datura* seed lectins are dimeric proteins composed of two identical or slightly different subunits of M_r 30,000-35,000. Based on analogy with the well-characterized potato lectin one can reasonably assume that the *Datura* seed lectins are chimerolectins composed of a chitin-binding part comprising three hevein domains linked to an extensin-like *O*-glycosylated hydroxyproline-serine rich domain.

The *Datura* lectins exhibit specificity towards GlcNAc-oligomers but are not inhibited by the monomer. *Datura* lectins are abundant seed proteins representing up to 20% of the total seed protein. Electron microscope localization studies suggested that at least the *D. stramonium* lectin is an extracellular protein. Immunofluorescent staining of seed sections further indicated that the lectin is predominantly located in the seed coat and the epidermis. This particular location explains why the *D. stramonium* lectin rapidly elutes upon imbibition of the seeds. Taken together with the fact that the lectin blocks the movements of normally motile bacteria at the air-water interface, the latter observation may suggest that the thorn apple lectin plays a role in the seed's defence against potentially harmful microbes.

3.2.3 Other chitin-binding seed lectins

a) The Urtica dioica *seed lectin*

Rhizomes, roots and seeds of stinging nettle (family Urticaceae) contain a small chitin-binding lectin built up of a single polypeptide chain of M_r around 8,500. Amino acid sequence analysis showed that the so-called *Urtica dioica* agglutinin (UDA) contains two hevein domains separated by a short hinge region. Molecular cloning revealed that the primary translation product of the UDA gene contains a signal peptide followed by a peptide of 89 amino acid residues corresponding to the mature UDA, and a *C*-terminal extension of 171 amino acids (Lerner and Raikhel, 1992). The *C*-terminal extension has high sequence similarity with the catalytic domains of class I and class II chitinases. Since proUDA expressed in *Escherichia coli* definitely has chitinase activity one can reasonably assume that the *in planta* synthesized precursor is also a chitinase. The question remains to be answered whether proUDA is a stable protein or an intermediate in the biosynthesis of UDA.

UDA exhibits carbohydrate-binding specificity towards GlcNAc-oligomers, the inhibitory potency of which increases with chain length up to 4 monosaccharide units. The monomer has a very low inhibitory activity on the agglutination activity of the lectin. Due to its small size UDA has very low specific agglutination activity.

UDA inhibits the growth and development of fungi (Broekaert *et al*., 1989). The lectin does not affect the metabolism of the fungus but disturbs the synthesis and/or deposition of chitin in the cell wall. As a result the morphology of the mycelium is thoroughly changed (Van Parijs *et al*., 1992).

Since UDA has no fungicidal activity it is questionable whether the lectin is involved in the nettle's defence against pathogenic microbes. The particular spatial expression pattern of UDA (namely in rhizomes/roots and seeds) and the profound effect on the hyphal structure suggest that this chitin-binding lectin helps to control the colonization of rhizomes and roots by endomycorrhiza.

b) The Chelidonium majus *seed lectin*

Seeds of the greater celandine (family Papaveraceae) contain a chitin-binding lectin which, according to its origin, has been called *Chelidonium majus* agglutinin or CMA. The lectin is a heterodimer composed of subunits of M_r 10,000 and 12,000 (Peumans *et al.*, 1985). No sequence data are available for CMA but its but amino acid composition closely resembles that of the above described chitin-binding lectins. CMA exhibits specificity towards oligomers of GlcNAc but is not inhibited by the monosaccharide. The greater celandine lectin is only found in the seeds. Until now, no similar lectins have been reported in other representatives of the Papaveraceae.

c) The chitin-binding antifungal peptides from Amaranthus caudatus

Seeds of the grain amaranth (*Amaranthus caudatus*) contain, in addition to a GalNAc-specific lectin called amaranthin (see below), two small chitin-binding peptides with strong antimicrobial activity (Broekaert *et al.*, 1992). The two peptides, called Ac-AMP1 and Ac-AMP2, are 29 and 30 amino acids in length, respectively. Ac-AMP1 has the same sequence as Ac-AMP2 except that it lacks the *C*-terminal arginine residue. The primary translation product of the Ac-AMP gene contains an *N*-terminal signal peptide, a domain corresponding to the mature Ac-AMP, and a *C*-terminal extension. The synthesis of Ac-AMP probably follows the secretory pathway.

The amaranth peptides are very potent antimicrobial agents. The concentration required for 50% inhibition of fungal growth ranges between 1 and 10 μg/ml. In addition, the Ac-AMP are also active against Gram-positive but not Gram-negative bacteria. Although the Ac-AMPs are only minor seed proteins their concentrations are high enough to protect young seedlings against potentially harmful microorganisms.

3.3 Type 2 ribosome-inactivating proteins from seeds

Type 2 RIP occur in many plant families belonging to different taxonomic groupings. So far, well-characterized seed-specific type 2 RIP have only been documented in castor bean (*Ricinus communis*), jequirity bean (*Abrus precatorius*), elderberry (*Sambucus nigra*), bitter gourd (*Momordica charantia*) and camphor tree (*Cinnamonum camphora*) (Table 3).

All ribosome-inactivating proteins are capable of catalytically inactivating eukaryotic ribosomes *in vitro* by the removal of a specific adenine residue from the 28S rRNA. However, most RIP are not cytotoxic because they cannot get access to the ribosomes. Type 1 RIP, for example, are not capable of penetrating into intact cells. At least some type 2 RIP can enter intact cells by endocytosis after binding of the B chain to glycan receptors on the cell surface. Once endocytosed the A chain may be delivered to the cytosol and inactivate the ribosomes resulting in cell death. Type 2 RIP show marked differences in their cytotoxicity. Some type 2 RIP are virtually non-toxic, even at concentrations as high as a few mg/ml, whereas others cause 50% cell death at concentrations below 1 ng/ml.

Table 3 Type 2 ribosome-inactivating proteins from seeds: Molecular structure and specificity

Species	**Structure***	**Specificity**	**Sequence available**
Abrus precatorius	Abrin [P(34 + 32)]	Gal>GalNAc	protein, cDNA
	APA [P(33 + 29)]$_2$	Gal	
Cinnamonum camphora	[P(30 + 33)]$_?$	Unknown	
Momordica charantia	[P(28 + 30)]$_2$	Gal>GalNAc	
Ricinus communis	Ricin [P(32 + 34)]	Gal>GalNAc	protein, cDNA
	RCA [P(32 + 36)]$_2$	Gal>>GalNAc	protein, cDNA
Sambucus nigra	SNA-V [P(26 + 32)]$_2$	GalNAc	cDNA
	SNA-III [P30]$_2$	GalNAc	

* See footnote Table 1

3.3.1 The castor bean type 2 RIP

a) Molecular structure, biosynthesis, processing and topogenesis

Castor beans contain a mixture of type 2 RIP. In addition to the extremely toxic ricin(s), a far less toxic hemagglutinating type 2 RIP called *Ricinus communis* agglutinin (RCA) has been isolated (Lord *et al*., 1994). Ricin is composed of a single [A-s-s-B]-pair of which both the A (M_r 32,000) and the B (M_r 34,000) chains are glycosylated. Ricin is a Gal/GalNAc-specific lectin with a clear preference for Gal over GalNAc. Ricin has a low specific agglutination activity but is extremely toxic. In addition to the classic ricin (also called ricin D) some castor bean varieties contain a variant toxin called ricin E. Ricin E has the same A chain as ricin D but contains a B chain which contains the first domain of the B chain of ricin D and the second domain of the B chain of the castor bean agglutinin.

The castor bean agglutinin RCA closely resembles ricin except that is built up of two [A-s-s-B]-pairs. The A and B chains of RCA have approximately the same M_r as the A and B chains of ricin and are also glycosylated. RCA is far less toxic than ricin but is a potent Gal-specific agglutinin.

Most of the information on the biosynthesis, processing and topogenesis of type 2 RIP comes from biochemical and molecular biological studies of ricin. The synthesis of ricin follows the secretory pathway. Ricin is synthesized on the ER as a prepropotein which undergoes cotranslational cleavage of a signal peptide and *N*-glycosylation at four sites. The resulting proricin is transported from the ER via the Golgi complex to the storage protein vacuoles. During this transport the oligosaccharide side chains of proricin are modified. Final modifications of the proricin polypeptide backbone take place in the protein bodies where an endopeptidase excises a 12-residue linker peptide between the A and B domains. In addition to the excision of the linker, a short peptide is also removed from the *N*-terminus of the proricin (after the signal peptide). Since proricin has no polynucleotide:adenosine glycosidase activity it appears that the toxin is transported from the ER in a catalytically inactive form and becomes proteolytically activated only upon arrival in the protein bodies. This mechanism possibly prevents the accidental miscompartimentalization of ricin into the cytosol causing cell death (Lord *et al.*, 1994).

b) Physiology, regulation and role

Castor beans contain high levels of type 2 RIP, which are apparently strictly seed-specific. Ricin and RCA accumulate during seed formation in the endosperm cells (and to a lesser extent also in the primary axes and cotyledons) where they are targeted to the protein bodies. During germination of the seeds both the toxin and the lectin are rapidly degraded (Lord *et al.*, 1994). This specific pattern of accumulation during seed formation and degradation upon early germination, together with the high concentration of the type 2 RIP and their preferential location in a highly specialized storage tissue, indicate that ricin and RCA are typical storage proteins. However, the fact that ricin and RCA are storage proteins does not preclude that they fulfil another function. Taking into consideration the extremely high toxicity of ricin it seems likely that this RIP is somehow involved in the plant's defence against possible predators.

3.3.2 The jequirity bean type 2 RIP

Seeds of *Abrus precatorius* contain a complex mixture of type 2 RIP, which comprises several isoforms of abrin and the *Abrus precatorius* agglutinin (APA). Abrins are composed of a single [A-s-s-B]-pair. The molecular weights of the A and B chains are approximately 34,000 and 32,000, respectively. The agglutinins consist of two [A-s-s-B]-pairs with A and B chains of M_r 33,000 and 29,000, respectively. All abrins and agglutinins are glycoproteins. Molecular cloning demonstrated that the different isoforms of abrin are encoded by highly similar though distinct genes.

Both abrin and APA are Gal-specific lectins. There are, however, slight differences in fine specificity between the toxins and the lectins. The jequirity bean type 2 RIP also differ from each other with respect to their biological activities. Abrin is an extremely potent toxin but has a very low agglutinating activity. In contrast, APA has a low cytotoxicity but is a strong agglutinin.

The complex mixture of toxins/agglutinins of the jequirity bean shows striking similarities with the castor bean. Both species, although not closely related taxonomically, have apparently developed a very similar system of toxic and less toxic type 2 RIP with distinct molecular structures and slightly different carbohydrate-binding specificities.

3.3.3 The elderberry seed-specific type 2 RIP

Elderberry (*Sambucus nigra*) trees express complex mixtures of type 2 RIP and related proteins in bark, leaves, fruits and also in seeds. Protein purification and molecular cloning yielded detailed information about the type 2 RIP and related proteins from bark, leaves and fruits. In contrast, the seed proteins have not been characterized in as much detail and their corresponding genes have not yet been cloned. We recently isolated and characterized a typical type 2 RIP from elderberry seeds. The elderberry seed RIP is a GalNAc-specific lectin composed of two [A-s-s-B]-pairs. *N*-terminal sequencing of the A (M_r 30,000) and B (M_r 35,000) chains indicated that they are virtually identical to these of the previously described GalNAc-specific type 2 RIP SNA-V from the bark. In addition to the type 2 RIP, elderberry seeds also contain a GalNAc-specific lectin SNA-III built up of two identical subunits of M_r 35,000. A reinvestigation of SNA-III revealed that the *N*-terminal sequence of the seed lectin is highly similar to that of the subunits of a fruit-specific lectin called SNA-IV. Molecular cloning demonstrated that SNA-IV is encoded by a type 2 RIP gene with a major deletion in the A domain (Van Damme *et al.*, 1997). It is possible, therefore, that the seed lectin SNA-III is also encoded by a truncated type 2 RIP gene.

3.3.4 Other seed-specific type 2 RIP

In addition to the above described seed-specific type 2 RIP, similar proteins have been isolated from seeds of the camphor tree (*Cinnamonum camphora*, Lauraceae family) and bitter gourd (*Momordica charantia*, Cucurbitaceae family). In addition, type 2 RIP also occur in seeds of mistletoe (*Viscum album*, Viscaceae family).

The *Momordica charantia* agglutinin (MCA) is a typical type 2 RIP composed of two glycosylated [A-s-s-B]-pairs. The A and B chains have molecular weights of 28,000 and 30,000, respectively. MCA is a Gal-specific type 2 RIP with a clear preference for Gal over GalNAc.

Cinnamomin, the type 2 RIP from *Cinnamonum camphora*, is built up of one or more glycosylated [A-s-s-B]-pair(s) and clearly exhibits RIP activity. Interestingly, the seeds of the camphor tree also contain a type 1 RIP called camphorin.

Mistletoe (*Viscum album*) expresses a complex mixture of Gal/GalNAc-specific type 2 RIP (the so-called mistletoe lectins ML-I, ML-II and ML-III) in leaves, stems and fruits. Similar lectins are also found in the seeds of this half parasite. Since the seed lectins have not yet been characterized in detail their exact relationship to the classical mistletoe type 2 RIP remains to be established.

3.4 Seed jacalins

A small group of Gal-specific lectins has been found in seeds of Moraceae species belonging to the genera *Artocarpus* and *Maclura*. The first member of this lectin family was isolated from jackfruit (*Artocarpus integrifolia*) and has been named jacalin. Jacalin has been studied in detail at both the protein and the molecular levels (Young *et al.*, 1989). Mature jacalin is composed of four α-chains of 133 amino acid residues and four ß-chains of 20 amino acid residues. cDNA cloning demonstrated that jacalin is synthesized as a preproprotein which contains a signal peptide followed by a propeptide of 39 residues, the ß-peptide of 20 residues, a linker of four amino acids, and the α-peptide of 133 residues. Most probably, jacalin is synthesized on the ER, where the signal peptide is cotranslationally removed, and subsequently transported via the ER to the protein bodies. During or after transport, the propeptide is cleaved at three different sites resulting in the removal of the *N*-terminal propeptide and the excision of the linker between the ß-chain and α-chain. The chains remain associated by non-covalent bonds and form the protomer of the jacalin. Four (or possibly two) such protomers associate to form the mature lectin.

Jacalin exhibits specificity towards galactose and has high reactivity with terminal non-reducing α-D-galactosyl residues. Although jacalin is usually considered as a seed-specific lectin it is not known whether the lectin also occurs in other tissues.

In addition to jacalin and other *Artocarpus* lectins, a similar agglutinin has also been isolated from seeds of *Maclura pomifera* (osage orange). The *Maclura pomifera* agglutinin (MPA) consists of four (or possibly two) protomers composed of non-covalently linked ß- and α-chains of 133 and 20 amino acid residues, respectively. MPA shares marked sequence similarity with jacalin and is probably synthesized in a similar way. MPA displays a high specificity towards terminal non-reducing α-D-galactosyl residue.

However, unlike jacalin the *Maclura* lectin has a higher affinity for GalNAc than Gal.

3.5 Amaranthins

A small family of GalNAc-specific lectins occurs in seeds of the genus *Amaranthus* (family Amaranthaceae). Amaranthin, the lectin from *Amaranthus caudatus* (tassel flower, Inca wheat) has been studied in great detail. Native amaranthin is a dimeric protein composed of two identical subunits of M_r 33,000. The protein has been sequenced completely and analyzed by X-ray diffraction (Transue *et al.*, 1997). Amaranthin has not been cloned but its amino acid sequence shares high similarity with a previously described seed-specific protein from *A. leucocarpus* (Raina and Datta, 1992).

Amaranthin exhibits carbohydrate-binding specificity towards GalNAc and has a preferential affinity for T- and Tn-antigen. Amaranthin is an abundant seed-specific protein representing 3-4% of the total seed protein. The biosynthesis and processing of the *Amaranthus* lectins has not been studied but, based on the deduced amino acid sequence of the cDNA clones encoding the *A. leucocarpus* protein, it can be concluded that they are synthesized as preproteins on the ER. The lectin is probably cotranslationally processed in the ER and subsequently transported to the protein bodies.

3.6 Other seed lectins

In addition to the lectins described above, a number of yet unclassified seed agglutinins have been described. In the absence of any clear criterion to classify these lectins, they are grouped according to the plant family in which they have been found.

3.6.1 Celastraceae seed lectins

Arils from the seeds of *Euonymus europaeus* (spindle tree) are a rich source of a lectin, which, according to its origin, is usually referred to as *Euonymus europaeus* agglutinin or EEA. The molecular structure of EEA is still unclear. Molecular masses of 120,000 and 166,000 have been reported for the native protein, which apparently consists of polypeptides of M_r 17,000 and 35,000. Since EEA is located in the arils it can not be considered as a true seed lectin. It is not known whether the *Euonymus* lectin is related to other plant lectins.

3.6.2 Euphorbiaceae seed lectins

In addition to the type 2 RIP from castor bean, Euphorbiaceae seed lectins have also been isolated from species belonging to the genera *Croton*, *Euphorbia*, *Hura* and *Tetracarpidium*.

A lectin called CTA has been isolated from seeds of *Croton tiglium*. CTA is a glycoprotein of M_r 120,000 consisting of two identical subunits of M_r 56,000 which are not disulphide-linked. The lectin is not inhibited by any simple sugar suggesting that it has a complex specificity.

A seed-specific lectin EHA has been purified from *Euphorbia heterophylla*. This lectin is a glycoprotein of M_r 65,000 composed of two identical subunits of M_r 32,000. EHA is a GalNAc-specific lectin. Since no sequence information is availablc, possiblc rclationships bctwccn EHA and other Euphorbiaceae lectins cannot be traced. In contrast to the castor bean lectins, which are typical endosperm proteins, the *Euphorbia heterophylla* lectin is predominantly located in the primary axes and cotyledons of the seed.

Hura crepitans (sand-box tree) also expresses a seed lectin. This lectin is a tetrameric protein composed of four identical glycosylated subunits of M_r 30,000, and exhibits specificity towards GalNAc. It is not clear whether the *Hura crepitans* seed lectin is related to the latex lectins from the same species.

Seeds of *Tetracarpidium conophorum* (Nigerian walnut) contain two distinct lectins called TCA-I and TCA-II. TCA-I is a glycoprotein consisting of disulphide-linked subunits of M_r 34,000 and exhibits specificity towards GalNAc. TCA-II has been described as a monomeric lactose-specific lectin composed of a single polypeptide of M_r 34,000. Both TCA-I and TCA-II show sequence similarity with the B-chain of type 2 RIP. It is not clear, however, whether the *Tetracarpidium* lectins are derived from type 2 RIP.

3.6.3 Cucurbitaceae seed lectins

Within the Cucurbitaceae family seed lectins have been found in some *Trichosanthes* and *Telfaira* species.

Seeds of *Trichosanthes kirilowii* (serpent cucumber) contain a glycosylated lectin (M_r 57,000) composed of disulphide-linked subunits of M_r 37,000 and M_r 25,000. The lectin preferentially binds galactose. Although both the molecular structure and the amino acid composition of the *Trichosanthes* lectin are similar to that of typical type 2 RIP, it cannot be classified as type 2 RIP until conclusive sequence information is available.

Telfaira occidentalis (fluted gourd) possesses a seed lectin that has been described as a hexamer (M_r 180,000) with three pairs of disulphide-linked subunits of M_r 30,000. The lectin is glycosylated and exhibits specificity towards galactose. Although the occurrence of disulphide-linked subunits is reminiscent of type 2 RIP, the *Telfaira* lectin cannot be classified as a type 2 RIP until sequence information is available.

3.6.4 Labiatae lectins

Within the Labiatae family lectins have been found in seeds of a few species of the genera *Moluccella* and *Salvia*.

Moluccella laevis seeds contain a mixture of two isolectins, a major one of M_r 130,000 and a minor one of M_r 250,000. Both lectins consist of three subunits: a M_r 67,000 subunit composed of two disulphide-bound polypeptides of M_r 28,000 and M_r 46,000 and two non-covalently linked subunits of M_r 46,000 and M_r 42,000. Both isolectins are GalNAc-specific.

A lectin has been isolated from seeds of *Salvia sclarea* (fetid clary sage). It is a glycoprotein composed of two identical disulphide-linked subunits of M_r 35,000. The lectin is specific for terminal GalNAc residues. A similar lectin has also been isolated from seeds of *Salvia horminum* (salvia).

Since no sequence information is available it is not clear whether the *Salvia* and *Moluccella* lectins share sequence similarity. For the same reasons, possible evolutionary relationships between the Labiatae lectins and other plant lectins can not be traced.

3.6.5 Fabaceae family

Mung beans (*Vigna radiata*) contain a seed lectin which does not belong to the legume lectin family. This lectin has been described as a glycosylated protein (M_r 160,000) composed of four identical subunits of M_r 45,000. The lectin not only agglutinates erythrocytes in a Gal-inhibitable manner but also possesses α-galactosidase activity. Upon dissociation of the lectin into monomers the agglutinating activity is lost whereas the α-galactosidase activity remains. Since no sequence information is available for the mung bean lectin it is not clear whether it is related to other lectins or galactosidases.

3.6.6 Gramineae family

Two distinct lectins, called ZMA-I and ZMA-II, have been isolated from maize (*Zea mays*) kernels. ZMA-I and ZMA-II have been described as GalNAc-interactive and mannose-interactive lectins, respectively. Preparations of native ZMA-I contained predominantly proteins of M_r 23,000 whereas ZMA-II consisted mainly of M_r 12,500 proteins. Both lectins apparently consist of M_r 11,000 and M_r 12,000 polypeptides. Since no sequence information is available, the possible relationships between the maize lectins and other plant lectins remain to be established. It is evident, however, that the maize lectins are not related to the chitin-binding Gramineae lectins described above.

3.6.7 Gymnosperm seed lectins

The only gymnosperm lectins isolated and characterized thus far were purified from seeds of *Araucaria brasiliensis* (pinhao). Pinhao seeds contain two distinct lectins. Lectin I (M_r 200,000) consists of ten subunits of M_r 20,000 whereas lectin II (M_r 200,000) is a hexamer composed of six subunits of M_r 34,000. Both lectins are glycosylated and have a similar amino acid composition. Lectin I and lectin II exhibit specificity towards mannose, glucose and their oligomers.

ACKNOWLEDGEMENTS

Supported in part by grants from the Catholic University of Leuven (OT/94/17) and the National Fund for Scientific Research (G.0223.97). W.P. is Research Director and E.V.D. Postdoctoral Fellow of this fund.

REFERENCES

Allen, A.K., Bolwell, G.P., Brown, D.S., Sidebottom, C. and Slabas, A.R. (1996) Potato lectin: a three-domain glycoprotein with novel hydroxyproline-containing sequences and sequence similarities to wheat-germ agglutinin. *International Journal of Biochemistry and Cell Biology*, **28**, 1285-1291.

Barbieri, L., Batelli, G.B. and Stirpe, F. (1993) Ribosome-inactivating proteins from plants. *Biochimica et Biophysica Acta*, **1154**, 237-282.

Barbieri, L., Valbonesi, P., Bonora, E., Gorini, P., Bolognesi, A. and Stirpe, F. (1997) Polynucleotide:adenosine glycosidase activity of ribosome-inactivating proteins: effects on DNA, RNA and poly(A). *Nucleic Acids Research*, **25**, 518-522.

Barondes, S.H. (1988) Bifunctional properties of lectins: lectins redefined. *Trends in Biochemical Sciences*, **13**, 480-482.

Barre, A., Van Damme, E.J.M., Peumans, W.J. and Rougé, P. (1996) Structure-function relationship of monocot mannose-binding lectins. *Plant Physiology*, **112**, 1531-1540.

Broekaert, W.F., Van Parijs, J., Leyns, F., Joos, H. and Peumans, W.J. (1989) A chitin-binding lectin from stinging nettle rhizomes with antifungal properties. *Science*, **245**, 1100-1102.

Broekaert, W.F., Marien, W., Terras, F.R.G., De Bolle, M.F.C., Proost, P., Van Damme, J., Dillen, L., Claeys, M., Rees, S.B., Vanderleyden, J. and Cammue, B.P.A. (1992) Antimicrobial peptides from *Amaranthus caudatus* seeds with sequence homology to the cysteine/glycine-rich domain of chitin-binding proteins. *Biochemistry*, **31**, 4308-4314.

Carrington, D.M., Auffret, A. and Hanke, D.E. (1985) Polypeptide ligation occurs during post-translational modification of concanavalin A. *Nature*, **313**, 64-67.

Collinge, D.B., Kragh, K.M., Mikkelsen, J.D., Nielsen, K.K., Rasmussen, U. and Vad, K. (1993) Plant chitinases. *The Plant Journal*, **3**, 31-40.

Goldstein, I.J., Hughes, R.C., Monsigny, M., Osawa, T. and Sharon, N. (1980) What should be called a lectin. *Nature*, **285**, 66.

Goldstein, I.J. and Poretz, R.D. (1986) Isolation, physicochemical characterization and carbohydrate-binding specificity of lectins, in *The Lectins: Properties, Functions and Applications in Biology and Medicine*, (eds I.E. Liener, N. Sharon and I.J. Goldstein), Academic Press, USA, pp. 33-248.

Kieliszewski, M.J., Showalter, A.M. and Leykam, J.F. (1994) Potato lectin: a modular protein sharing sequence similarities with the extensin family, the hevein lectin family, and snake venom disintegrins (platelet aggregation inhibitors). *The Plant Journal*, **5**, 849-861.

Kocourek, J. and Horejsi, V. (1983) Note on the recent discussion on definition of the term 'lectin', in *Lectins: Biology, Biochemistry, Clinical Biochemistry* (eds T.C. Bog-Hansen and G.A. Spengler), Walter de Gruyter, Berlin and New York, pp. 3-6.

Konami, Y., Uno, T., Fujii, M., Yamamoto, K., Osawa, T. and Irimura, T. (1995) A high degree of sequence homology in the putative carbohydrate recognition domains of pokeweed mitogen and wheat germ agglutinin: poly-N-acetyllactosamine-binding lectins from different species. *Glycobiology*, **5**, 663-670.

Lee, H.-I., Broekaert, W.F. and Raikhel, N.V. (1991) Co- and post-translational processing of the hevein preproprotein of latex of the rubber tree (*Hevea brasiliensis*). *The Journal of Biological Chemistry*, **266**, 15944-15948.

Lerner, D.R. and Raikhel, N.V. (1992) The gene for stinging nettle lectin (*Urtica dioica* agglutinin) encodes both a lectin and a chitinase. *The Journal of Biological Chemistry*, **267**, 11085-11091.

Lord, J.M., Roberts, L.M. and Robertus, J.D. (1994) Ricin: structure, mode of action, and some current applications. *The FASEB Journal*, **8**, 201-208.

Meins F, Jr., Neuhaus J.-M., Sperisen, C. and Ryals, J. (1992) The primary structure of plant pathogenesis-related glucanohydrolases and their genes, in *Genes involved in plant defense*, (ed T. Boller and F. Meins Jr.), Vienna, New York, Springer Verlag, pp. 245-282.

Mirkov, T.E., Wahlstrom, J.M., Hagiwara, K., Finardi-Filho, F., Kjemtrup, S. and Chrispeels, M.J. (1994) Evolutionary relationships among proteins in the phytohemagglutinin-arcelin-α-amylase inhibitor family of the common bean and its relatives. *Plant Molecular Biology*, **26**, 1103-1113.

Peumans, W.J., De Ley, M. and Broekaert, W.F. (1985) Isolation and partial characterization of a new lectin from seeds of the greater celandine (*Chelidonium majus*). *Plant Physiology*, **78**, 379-383.

Peumans, W.J. and Van Damme, E.J.M. (1995) Lectins as plant defense proteins. *Plant Physiology*, **109**, 347-352.

Peumans, W.J., Verhaert, P., Pfüller, U. and Van Damme, E.J.M. (1996) Isolation and partial characterization of a small chitin-binding lectin from mistletoe (*Viscum album*). *FEBS Letters*, **396**, 261-265.

Pusztai, A. and Bardocz, S. (1996) Biological effects of plant lectins on the gastrointestinal tract: metabolic consequences and applications. *Trends in Glycoscience and Glycotechnology*, **8**, 149-165.

Raikhel, N.V., Lee, H.-I. and Broekaert W.F. (1993) Structure and function of chitin-binding proteins. *Annual Reviews in Plant Physiology and Plant Molecular Biology*, **44**, 591-615.

Raina, A. and Datta, A. (1992) Molecular cloning of a gene encoding a seed-specific protein with nutritionally balanced amino acid composition from *Amaranthus. Proceedings of the National Academy of Sciences USA*, **89**, 11774-11778.

Rougé, P., Cambillau, C. and Bourne, Y. (1991) The three-dimensional structure of legume lectins, in *Lectin Reviews* (eds D.C. Kilpatrick, E. Van Driessche and T.C. Bog-Hansen), Sigma Chemical Co., St.Louis, USA, Vol. 1, pp. 143-159.

Sharma, V. and Surolia, A. (1997) Analyses of carbohydrate recognition by legume lectins: size of the combining site loops and their primary activity. *Journal of Molecular Biology*, **267**, 433-445.

Sharon, N. and Lis, H. (1990) Legume lectins - a large family of homologous proteins. *The FASEB Journal*, **4**, 3198-3208.

Stinissen, H.M. and Peumans, W.J. (1985) Recent advances in the biochemistry, cell biology, physiology, biosynthesis and genetics of Gramineae lectins. *Biochimie und Physiologie der Pflanzen*, **180**, 85-106.

Transue, T.R., Smith, A.K., Mo, H., Goldstein, I.J. and Saper, M.A. (1997) Structure of benzyl T-antigen disaccharide bound to *Amaranthus caudatus* lectin. *Nature Structural Biology*, **10**, 779-783.

Van Damme, E.J.M. and Peumans, W.J. (1996) Progress in structural analysis of plant lectins through molecular cloning, in *Lectins: Biology, Biochemistry, Clinical Biochemistry* (eds E. Van Driessche, P. Rougé, S. Beeckmans and T.C. Bog-Hansen), Textop, Denmark, vol. 11, pp. 17-27.

Van Damme, E.J.M., Barre, A., Verhaert, P., Rougé, P. and Peumans, W.J. (1996) Molecular cloning of the mitogenic mannose/maltose-specific rhizome lectin from *Calystegia sepium*. *FEBS Letters*, **397**, 352-356.

Van Damme, E.J.M., Roy, S., Barre, A., Rougé, P., Van Leuven, F. and Peumans, W.J. (1997) The major elderberry (*Sambucus nigra*) fruit protein is a lectin derived from a truncated type 2 ribosome-inactivating protein. *The Plant Journal*, **12**, 1251-1260.

Van Damme, E.J.M., Peumans, W.J., Pusztai, A. and Bardocz, S. (eds) (1998) *Handbook of Plant Lectins. Properties and Biomedical Applications*, John Wiley & Sons, U.K., ISBN 0-471-96445-X in press.

Van Driesche, E. (1988) Structure and function of Leguminosae lectins, in *Advances in Lectin Research* (ed H. Franz), VEB Verlag Volk und Gesundheit, Berlin, Germany, Vol. 1, pp. 73-134.

Van Parijs, J., Joosen, H.M., Peumans, W.J., Geuns, J.M. and Van Laere, A.J. (1992) Effect of the lectin UDA (*Urtica dioica* agglutinin) on germination and cell wall formation of *Phycomyces blakesleeanus* Burgeff. *Archives of Microbiology*, **158**, 19-25.

Wang, M.-B., Boulter, D. and Gatehouse, J. (1994) Characterization and sequencing of cDNA clone encoding the phloem protein PP2 of *Cucurbita pepo*. *Plant Molecular Biology*, **24**, 159-170.

Watkins, W.M. and Morgan, W.T.J. (1952) Neutralization of the anti-H agglutinin in eel serum by simple sugars. *Nature*, **1269**, 825-826.

Young, M.N., Johnston, R.A.Z., Szabo, A.G. and Watson, D.C. (1989) Homology of the D-galactose-specific lectins from *Artocarpus integrifolia* and *Maclura pomifera* and the role of an unusual small polypeptide subunit. *Archives of Biochemistry and Biophysics*, **270**, 596-603.

Young, N.M. and Oomen, R.P. (1992) Analysis of sequence variation among legume lectins. A ring of hypervariable residues forms the perimeter of the carbohydrate-binding site. *Journal of Molecular Biology*, **228**, 924-934.

29. Lipoxygenases

ROD CASEY
John Innes Centre, Norwich Research Park, Colney, Norwich NR4 7UH, UK

1. INTRODUCTION

The capacity of soybean seed to catalyze the oxidation of lipids was recognized 65 years ago (Andre and Hou, 1932) and the activity named “lipid oxidase”. The activity has also been described as “lipoxidase” and “carotene oxidase” (Summer and Summer, 1940; Theorell *et al.*, 1947), but the term lipoxygenase (E.C.1.13.11.12; linoleate: oxygen oxidoreductase) is now generally accepted. Lipoxygenase (LOX) from soybean seeds was among the first enzymes to be crystallized (Theorell *et al.*, 1947), yet a crystal structure for what is now known as isoform I of soybean LOX did not emerge until almost 50 years later (Boyington *et al.*, 1993; Minor *et al.*, 1993). Over that 50 year period the perception of LOX has changed from a biochemical oddity (originally thought to contain neither a metal co-factor nor a prosthetic group of any description) to a key component of plants’ responses to stresses, including pest and pathogen attack (Farmer and Ryan, 1992; Croft *et al.*, 1993), nitrogen status (Grimes *et al.*, 1993), wounding (Saravitz and Siedow, 1996) and possibly drought (Bell and Mullet, 1991).

Axelrod (1974) suggested that LOX is ubiquitous in plants and that inability to demonstrate LOX activity in particular species, or organs, is a consequence of the sensitivity of the detection method, rather than the absence of activity. This is certainly true in many cases, although LOX activity is also a function of the physiological and developmental state of the organ or tissue under examination (see, for example, Eiben and Slusarenko, 1994); activity, and gene expression, often is detectable only after stress treatments or at particular stages of development. It is probably the case that all organs of all plants produce LOX under some conditions.

In vegetative tissues, such as leaves, LOX perform at least three functions. They provide the hydroperoxide substrates for the synthesis of jasmonate, an elicitor of defence gene expression (Vick and Zimmerman, 1983; Farmer and Ryan, 1992) and for the production of volatile aldehydes that act as

P.R. Shewry and R. Casey (eds.), Seed Proteins, 685–708.

antimicrobial compounds (Croft *et al.*, 1993). They also act as so-called vegetative storage proteins, temporary stores of nitrogen that accumulate in response to nitrogen sink limitation (Bunker *et al.*, 1995). The amounts of LOX in non-stressed leaves, stems, roots and other vegetative organs are generally low, but a number of storage organs, particularly grain legume seeds, have sufficiently large amounts under normal growing conditions to enable the LOX polypeptides to be readily detected on an SDS gel of total protein (Figure 1). This article concentrates on seeds; LOX activity has been reported in the seeds of a wide range of plants, including legumes, cereals, brassicas, cucurbits, and solanaceous species (see Table 1).

Table 1 Plant species from which seed LOX have been characterized

Species	Reference
Barley (*Hordeum vulgare*)	van Aarle *et al.*, 1991
Broad bean (*Vicia faba*)	Nicholas and Drapron, 1977
Buckwheat (*Fagopyrum esculentum*)	Henderson *et al.*, 1990
Chickpea (*Cicer arietinum*)	Borthakur and Ramadoss, 1986
Cowpea (*Vigna unguiculata*)	Truong *et al.*, 1979
French bean (*Phaseolus vulgaris*)	Eiben and Slusarenko, 1994
Lupin (*Lupinus albus*)	Andrianarison *et al.*, 1990
Maize (*Zea mays*)	Poca *et al.*, 1990
Pea (*Pisum sativum*)	Eriksson and Svensson, 1970
Peanut (*Arachis hypogea*)	Sanders *et al.*, 1975
Pearl Millet (*Pennisetum glaucum*)	Nagarathna *et al.*, 1992
Rapeseed (*Brassica napus*)	Khalyfa *et al.*, 1990 Meshehdani *et al.*, 1990
Rice (*Oryza sativa*)	Shastry and Rao, 1975
Vetch (*Vicia sativa*)	Andrianarison *et al.*, 1991
Wheat (*Triticum aestivum*)	Shiiba *et al.*, 1991
Winged bean (*Psophocarpus tetragonolobus*)	Truong *et al.*, 1982a

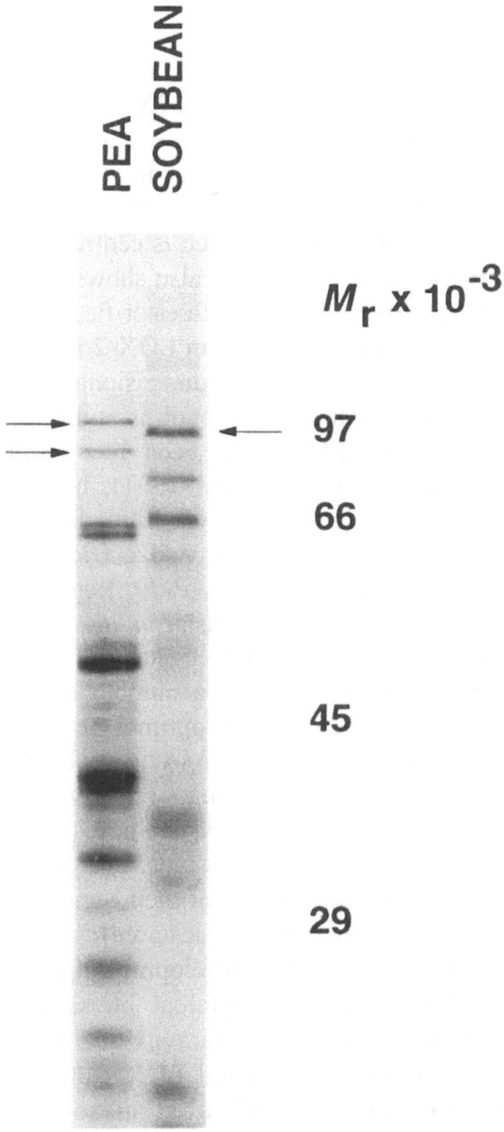

Fig. 1 SDS gel electrophoresis of total protein extracts of soybean (cv. 'Maple Amber') and pea seeds (cv. 'Birte'), stained with Coomassie blue. The arrows denote the positions of lipoxygenase polypeptides; the three soybean LOX are not well-resolved on such a gel. The positions adopted by marker proteins of known M_r are indicated.

2. LOX ISOFORMS

Early studies of seed LOX assumed them to be a single molecular species, but the realization that LOX can exist as a number of isoforms within a given plant organ came from the work of Christopher *et al.* (1970, 1972) on soybean seed LOX, that demonstrated four isoforms - LOX-1, -2, -3a and -3b. Subsequent studies, including analyses of genetic variants that lack LOX-3 (Kitamura *et al.*, 1983), suggested that LOX-3a and -3b are derived from the same parent molecule and that there are only three major isoforms of soybean seed LOX. LOX-1 differs markedly from LOX-2 and -3 from soybean seeds in a number of properties, including stability, but most notably in the pH value for optimum activity, which is centred around 9.0 for LOX-1 and 6.5 to 7.0 for LOX-2 and -3. LOX-1 also shows a marked preference for charged fatty acids and little reactivity with esterified fatty acids. Most plant LOX reported to date are more like soybean LOX-2 and -3 in their properties, and in this sense LOX-1, the three-dimensional structure and action mechanism of which are known in some detail, is atypical of plant LOX.

Multiple forms of seed LOX have been recognized in several other species, including barley (Doderer *et al.*, 1991), broad bean (Nicolas and Drapron, 1977), chickpea (Borthakur and Ramadass, 1986; Sanz *et al.*, 1992), kidney bean (Sanz *et al.*, 1993), maize (Belefant and Fong, 1991; Jensen *et al.*, 1997), pea (Yoon and Klein, 1979; Regdel *et al.*, 1995), peanut (Sanders *et al.*, 1975), rapeseed (Kermasha *et al.*, 1991), rice (Ida *et al.*, 1983), wheat (Shiiba *et al.*, 1991) and winged bean (Truong *et al.*, 1982b). Isoforms are recognized on the basis of a number of parameters, including their amino acid sequence; their pH for optimal activity; differences in net surface charge that permit their separation by ion-exchange chromatography, isoelectric focusing or chromatofocusing; sensitivity to Ca^{2+}; substrate preference; turnover numbers; apparent K_m values; their ability to cooxidize pigments and to generate carbonyl compounds; heat stability; and the nature of their products, which is a consequence of their regiospecificity (the position along the fatty acid chain where oxygen is inserted). Although it is clear from defence responses and from studies of developmental regulation that different vegetative LOX isoforms can play different physiological roles (see, for example, Eiben and Slusarenko, 1994; Saravitz and Siedow, 1996), it is not obvious why seeds should contain several LOX isoforms. Several mutants that lack one or more seed isoforms exist in soybean (Hildebrand and Hymowitz, 1982; Kitamura *et al.*, 1983; Davies and Nielsen, 1986), pea (Wu *et al.*, 1995) and rice (Suzuki *et al.*, 1993) and, at least in the case of soybean, would appear to have unimpaired agronomic performance (Pfeiffer *et al.*, 1992), suggesting that some seed LOX may be dispensable for normal growth and development.

3. THE BIOCHEMISTRY OF LOX

The general reaction catalyzed by all lipoxygenases is the formation of *cis, trans* conjugated hydroperoxides from any compound containing a *cis,cis*-1,4-pentadiene system, such as that found in the commonly occurring polyunsaturated fatty acids [linoleic (18:2); linolenic (18:3) and arachidonic (20:4) acids] (Figure 2). In theory, the oxygen can be added to either end of the pentadiene system, producing 9- and/or 13-hydroperoxides from linoleic or linolenic acids. The newly-formed conjugated bond system in the hydroperoxide absorbs strongly at 234nm (E=25 $mM^{-1}cm^{-1}$) and affords a means of assaying LOX activity. An alternative, but less sensitive, assay is by measurement of oxygen uptake using a Clark-type electrode. Either method is satisfactory with relatively pure enzyme preparations, but both are subject to interference from other compounds or activities in crude extracts. It should be noted that the specific activity of a given enzyme preparation will be a function of the assay method.

Although the substrate for LOX can, in principle, be any compound containing the type of methylene-interrupted double bond system shown in Fig. 2, plant seed LOX show a marked preference for 18-carbon polyunsaturated fatty acids (linoleic and linolenic), which may reflect their relative abundance in plant triacylglycerols. Most seed LOX, however, can also utilise arachidonic acid as substrate (see Boyington *et al*., 1993: Heydeck *et al*., 1991; Nikolaev *et al*., 1990; van Aarle *et al*., 1991; Zhang and Hamberg, 1996), as can potato tuber LOX (Shimizu *et al*., 1984). The products of LOX action on arachidonic acid are significant; potato tuber LOX can further effect the conversion of the 5-hydroperoxide of arachidonic acid into leukotriene A4, a pharmacologically important compound involved in inflammatory and allergic responses in mammals (Shimizu *et al*., 1984).

Most LOX exhibit some degree of regiospecificity and stereospecificity, depending in part on the specific isoform under examination and in part on the reaction conditions. For instance, soybean seed LOX-1 produces almost exclusively 13-hydroperoxy linoleic acid at its optimum pH for activity (pH 9) (Axelrod *et al*., 1981), whereas soybean LOX-2 and -3 generate roughly equal amounts of 9- and 13-hydroperoxide. Virtually all of the 13-hydroperoxylinoleic acid generated by LOX-1 is composed of *Z, E* (*cis, trans*) geometrical isomers, in the S-configuration with respect to optimal isomers (Nikolaev *et al*., 1990). Such specificity is important; it influences the nature of secondary compounds derived from the hydroperoxides (see below) and it underlies the potential for LOX as catalysts in the provision of enantiomerically pure ('chiral') feedstocks for chemical syntheses.

Some LOX, including those from lupin (Andrianarison *et al*., 1990), rice (Zhang and Hamberg, 1996) and wheat (Heydeck *et al*., 1991), can catalyze

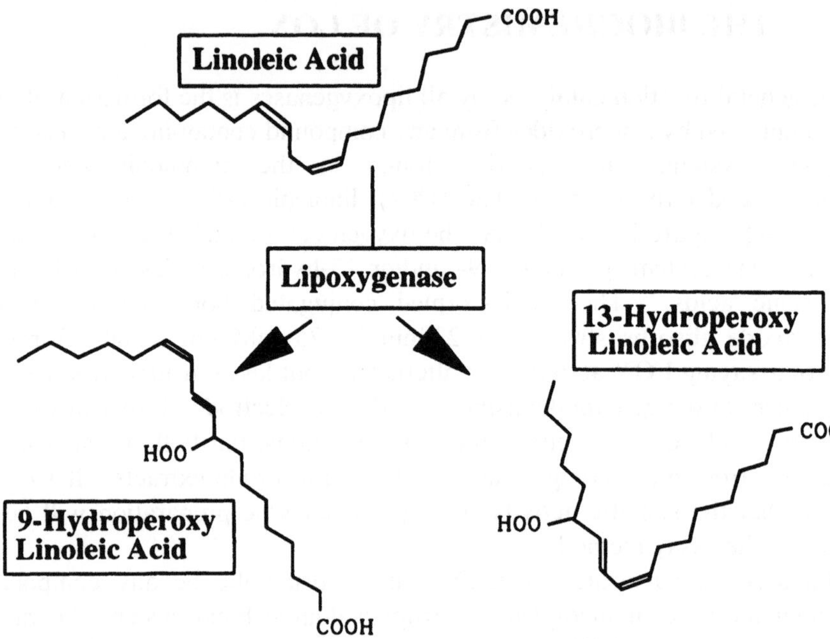

Fig. 2 The formation of 9- and 13-hydroperoxides from linoleic acid by lipoxygenases.

double oxygenation of fatty acid substrates, albeit in low yield and at high enzyme concentrations.

Although it is recognized that different LOX possess different regiospecificities for a given substrate, the phenomenon is not completely understood. Mutagenesis of human 15-LOX has indicated the critical role played by the nature of two amino acids at the base, and two in the wall, of the "fatty acid pocket", presumably by allowing increased entry of the fatty acid backbone to align a different double bond at the active site (Sloane *et al.*, 1991; Prigge *et al.*, 1996) (see Fig. 3). Differences in these important amino acids between pea LOX-2 and -3 (Ealing and Casey, 1988;1989) may also underly their different regiospecificities (Wu *et al.*, 1995). Zhang and Hamberg (1996) and Prigge *et al.* (1996) have suggested that the direction of insertion of the fatty acid - i.e., carboxyl or methyl end in the hydrophobic pocket - can also determine regiospecificity (see Fig. 4). Using a series of C12-C22 trienoic acids as substrates for soybean LOX-1, Hatanaka *et al.* (1990) showed that the hydroperoxy group was introduced at the 6th carbon from the methyl end (the so-called ω6 position, equivalent to the 13-position shown in Fig. 2) irrespective of the chain length. Although the 20:3 fatty acid was a better substrate at pH 9 than the naturally-occurring 18:3 linolenic acid, this was a pH-dependent phenomenon and 18:3 was the best substrate at

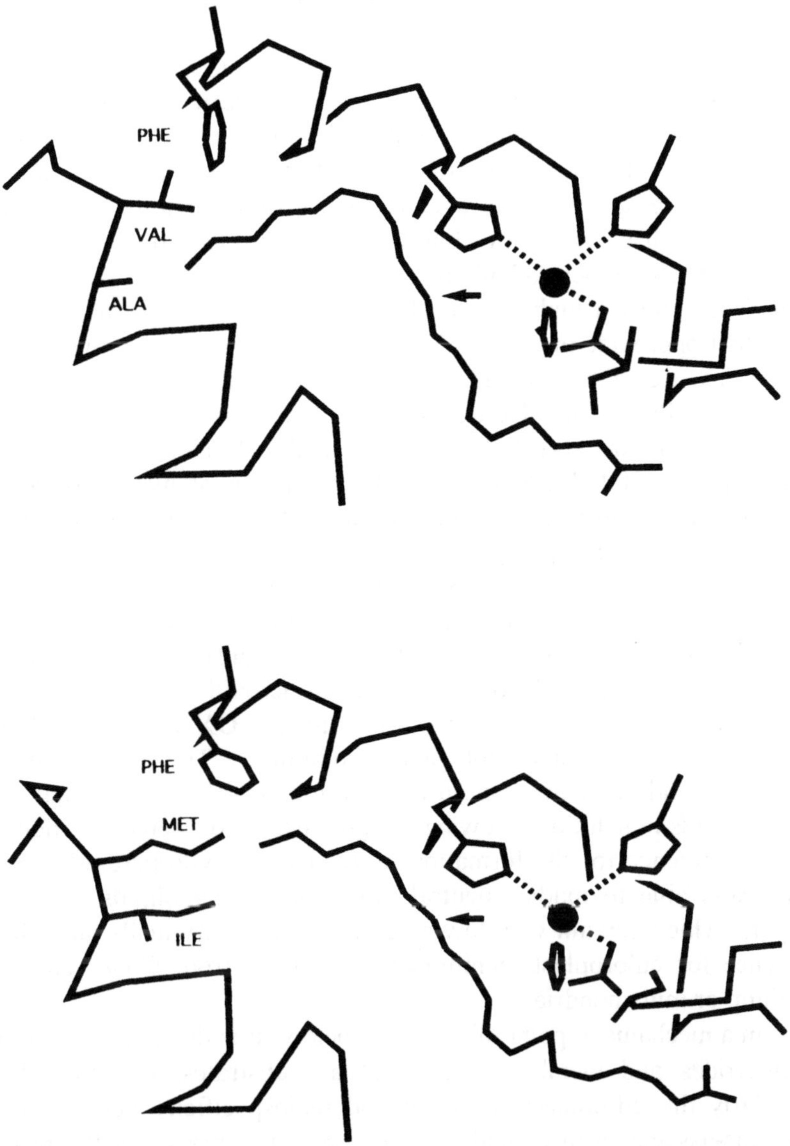

Fig. 3 The importance of specific amino acids at the base of the fatty acid pocket of human lipoxygenases in determining regiospecificity. The arrows denote the site of oxygen insertion. Adapted from Prigge *et al.* (1996), with permission.

pH 7. Hatanaka *et al.* (1990) concluded that the substrate is placed in an appropriate spatial orientation within the active site largely as a consequence of the recognition of the ω-terminal and pentadiene sections of the fatty acid and that the ionization state of the terminal carboxyl group plays a role in accelerating the activity of LOX-1. These are matters that are being clarified through analyses of three-dimensional structures and site-directed mutagenesis experiments and may reflect differences in flexibility within the active sites of different LOX isoforms (Skrzypczak-Jankun *et al.*, 1997). Control of stereospecificity is a consequence of the nature of the reaction mechanism, as discussed below, but can also be influenced by the structure of the active site (Skrzypczak-Jankun *et al.*, 1997).

LOX can show a high degree of activity with free fatty acids; yields of linoleic hydroperoxide in excess of 80% can be achieved and soybean enzyme preparations are stable enough to be immobilized and used as synthetic reagents (see Piazza and Nuñez, 1995). The substrate for LOX *in vivo*, however, may not necessarily be free fatty acid and much interest has centred on the capacity of LOX to catalyze the formation of hydroperoxides from more complex substrates. Piazza and Nuñez (1995) showed that acylglycerols and phosphoglycerides can act as substrates for LOX-1, although the rate of trilinolein oxidation was very slow compared to that of linoleic acid. In contrast, the diacylglycerol, dilinolein, reacted at a comparable rate to linoleic acid, although a significant proportion of the products were compounds other than hydroperoxides. Maccarone *et al.* (1994) demonstrated that soybean LOX-2, unlike LOX-1, can form fatty acid hydroperoxides from plant biological membranes. The activity of LOX-2 with linoleic and linolenic acids in biomembranes differed from that with the free fatty acids in solution, showing a higher degree of stereospecificity and a marked preference for the formation of 13- over 9-hydroperoxide. Soybean LOX-2 was able to oxidize neutral lipids and phospholipids of membrane biolayers, which contain very low amounts of free fatty acids, and showed a preference for chloroplast membranes over those from Golgi, endoplasmic reticulum, or mitochondria.

From a mechanistic point of view (see below) it is difficult to imagine how acylglycerides and membranes can act as substrates, but such findings, particularly the differences in stereo and regiospecificity between free and membrane-associated fatty acids, call for some reappraisal of the features of the reaction catalyzed *in vivo* and of the extent to which *in vitro* analyses can be extrapolated to biological responses.

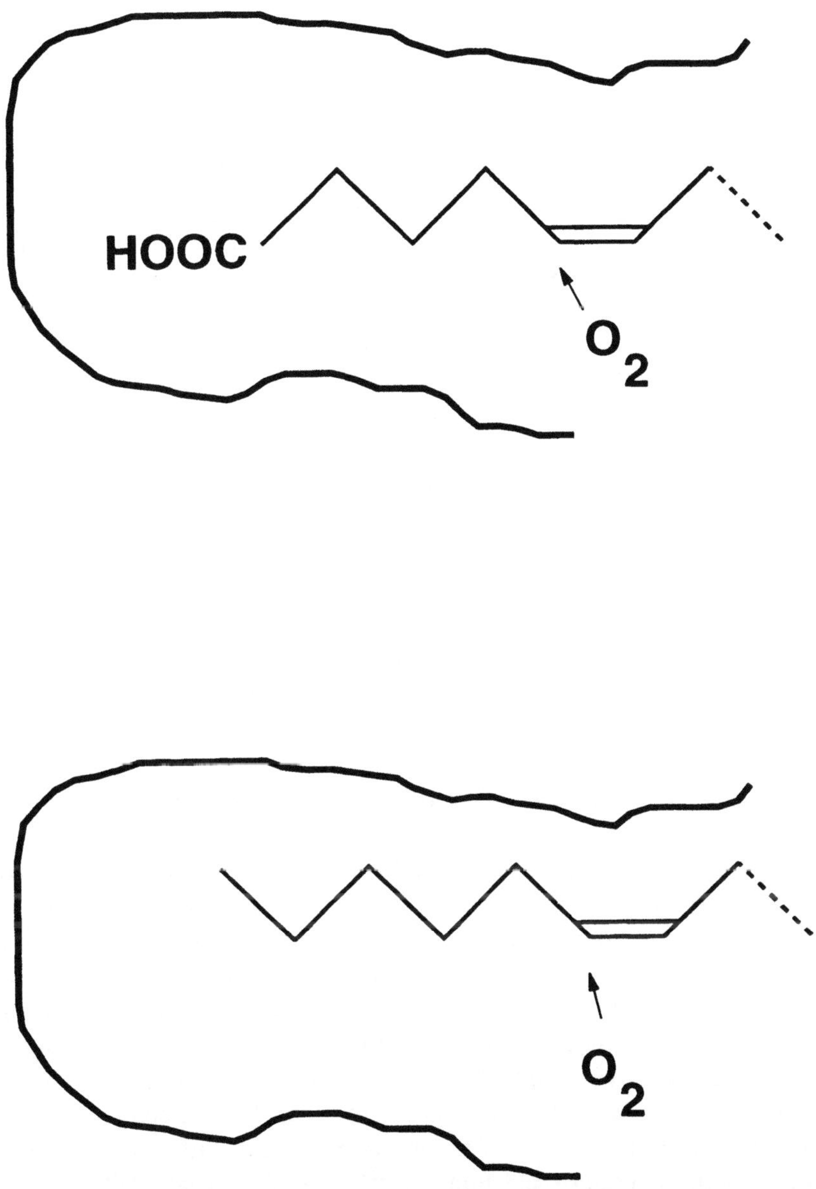

Fig. 4 Diagrammatic representation of how the two possible orientations of the same substrate in the lipoxygenase hydrophobic pocket can give rise to different hydroperoxides. The arrows show the sites of hydroperoxidation. Adapted from Zhang and Hamberg (1996), with permission.

3.1 Secondary reactions

LOX has the ability, in the presence of polyunsaturated fatty acids, to carry out the bleaching of pigments and to catalyze the formation of protein disulphide bonds through its so-called co-oxidation reactions. These important processes that affect the quality of foods (Robinson *et al.*, 1995) and are significant in breadmaking (see later), are believed to be associated with secondary product formation, rather than the direct oxidation of pigments or proteins by fatty acid hydroperoxides. LOX isoforms vary in their ability to carry out co-oxidation reactions, a fact that must reflect some intrinsic, but as yet undefined, property of the enzyme. For instance, one of the pea seed LOX isoforms catalyzes pigment co-oxidation whereas the other does not (Yoon and Klein, 1979); the isoform that catalyzes pigment co-oxidation also generates carbonyl compounds during the course of the formation of hydroperoxides, a phenomenon that has been attributed to the formation of free hydroperoxy radical-containing species, which then decompose to form oxodienoic acids. It seems that such radical intermediates are more likely to form, and dissociate from the enzyme, with some LOX isoforms than others (see Siedow, 1991). In the presence of oxygen, LOX-2 from soybean shows little co-oxidation activity, whereas LOX-3 is a good co-oxidizer; LOX-1 has such activity only at low oxygen tensions. Since soybean flour is routinely added to wheat flour in breadmaking, to take advantage of its LOX-associated co-oxidation properties, it is important to recognize that not all of the soybean LOX isoforms will play a positive role as bread "improvers".

3.2 Reaction mechanism and the active site

In overall terms, the chemistry of hydroperoxide formation by LOX is straightforward; a proton is abstracted from the methylene bridge between the two adjacent double bonds in the *Z, Z*-1,4-pentadiene structure and molecular oxygen, that finally becomes hydroperoxide, is placed two carbon atoms along the chain from the methylene group; this is accompanied by a change in the position of the double bond to form (Fig. 5) the *Z,E*-conjugated hydroperoxide that gives rise to the characteristic absorbance at 234nm. The reaction also shows a degree of absolute (optical, S/R) stereochemistry that varies according to the source of the enzyme, the particular isoform, and the reaction conditions (Andre and Funk, 1986; Gardner, 1989; Nikolaev *et al.*, 1990; Wu *et al.*, 1995, for example). A detailed account of the reaction mechanism is inappropriate to this discussion; despite a considerable amount of work (see, for example, Veldink *et al.*, 1977; Veldink and Vliegenthart, 1985), and a crystal structure for soybean LOX-1 at 1.4A resolution

(Boyington *et al.* 1993; Minor *et al.*, 1996), the exact details of the reaction mechanism are still to be resolved (see Glickman and Klinman, 1996; Minor *et al.*, 1996).

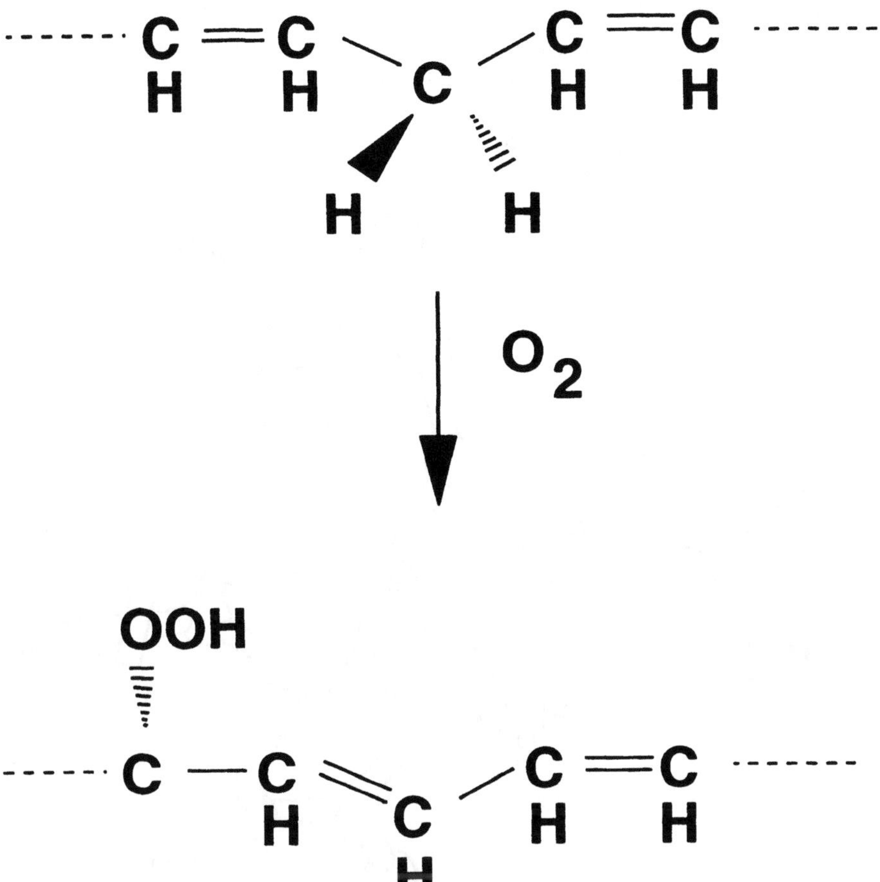

Fig. 5 The stereochemistry of the primary reaction catalyzed by lipoxygenases, showing detail only of the pentadiene structure (cf. Fig. 29.2).

The nature of the active site of the enzyme is clear from the high-resolution crystal structure (Minor *et al.*, 1996). The enzyme contains a single atom of non-haem iron which is essential for activity. Despite a report that LOX contains pyrroloquinoline quinone as a cofactor (van der Meer and Duine, 1988), this has subsequently been dismissed as an artefact (Michaud-Soret *et al.*, 1990; Veldink *et al.*, 1990) and it is now accepted that iron is the sole cofactor in LOX. In soybean LOX-1, the active site has five ligands at distances less than 2.6Å: the side chains of His^{499}, His^{504} and His^{690}; one oxygen atom from the carboxylate of the C-terminal Ile^{839}; and a water molecule. The side chain of Asn[694] is weakly associated with the Fe to complete the co-ordination sphere (see Fig. 6). Because of uncertainties in the binding mode of substrate, it is difficult to evaluate the several chemical mechanisms that have been proposed (Nelson and Sietz, 1994) but the high-resolution structure can impose some limitations on them. It appears, for instance, to make formation of an organo-Fe intermediate improbable, which is consistent with spectroscopic evidence (Funk *et al.*, 1990).

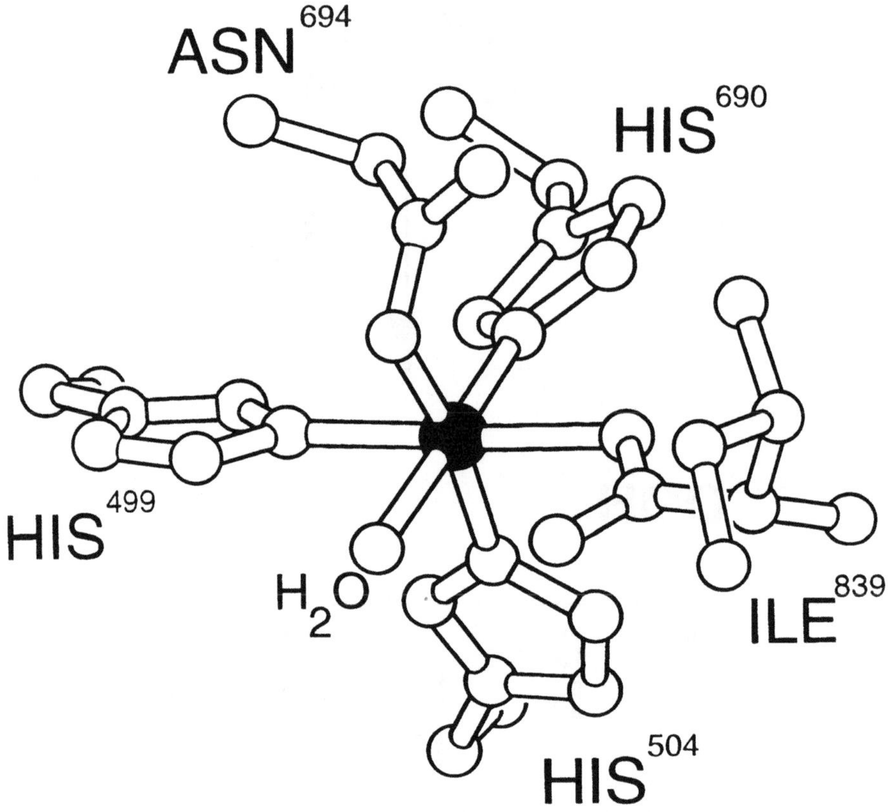

Fig. 6 Representation of the structure of the soybean LOX-1 active site, showing the Fe ligands. The Fe is shaded. Adapted from Minor *et al.* (1996), with permission. The Asn and water are not covalently bound to Fe in the structure for LOX-3 (Skrzypczak-Jankun *et al.*, 1997), but can be brought to a bonding distance without any disturbance to their environment.

The resting, inactive form of LOX is in the Fe(II) (high-spin) state and is activated by conversion to high-spin Fe(III); this is achieved by the hydroperoxide product, and thus the time-course of catalysis always shows an initial lag phase (Schilstra *et al.*, 1994). The high-spin Fe(III) form has a characteristic pale yellow colour; in the presence of an excess of product, a small proportion of the enzyme develops a purple colour, but the purple form is relatively unstable and rapidly decays to the yellow form at room temperature. The catalytic significance of the purple form is unclear. Following the binding of linoleic acid substrate, the hydrogen is abstracted from the C11 methylene group (see Fig. 5) and transferred to an enzymatic group, possibly the water (hydroxide) ligand (Fig. 6; Minor *et al.*, 1996); the LOX Fe(III) is reduced to Fe(II) in the process. The resultant pentadienyl radical, or an isomer thereof, then reacts with oxygen to generate peroxyl radical which is subsequently reduced [by Fe(II)] to peroxide anion and protonated to generate the fatty acid hydroperoxide; this regenerates the active Fe(III) form of the enzyme.

There are two important features of the reaction. The observed stereochemistry (Fig. 5) requires that oxygen insertion occurs on the opposite face of the pentadiene unit to hydrogen abstraction, and this is likely to be achieved through a control of the stereochemistry of oxygen approach, possibly by steric interference (Minor *et al.*, 1996). Secondly, oxygen enters the reaction pathway only after abstraction of hydrogen from substrate; it is proposed (Glickmann and Klinman, 1996) that molecular oxygen is not productively bound by LOX, but reacts directly with the substrate radical (see above). Such a pathway, involving substrate instead of oxygen activation, is a unique mechanism for a metallo-oxygenase.

4. PROTEIN STRUCTURE

All seed LOX have a predicted M_r 90,000-97,000, containing approximately between 840 and 870 amino acids. Although a large number of predicted amino acid sequences are known for plant LOX, many of them represent vegetative LOX. Only eight seed LOX sequences have been published to date, and all of these are predicted from DNA sequences; that for maize LOX is not full-length. They show a moderately high degree of sequence similarity, even between sequences from widely divergent species (Fig. 7), and the critical amino acids that form part of the active site are totally conserved. Alteration of these amino acids, either by site-directed mutagenesis (Steczko *et al.*, 1992) or through natural variation (Wang *et al.*, 1994), completely eliminates activity.

The three-dimensional structure of soybean LOX-1 (Boyington *et al.*, 1993; Minor *et al.*, 1993, 1996) shows it to have two structural domains (Plate 11); an *N*-terminal β-barrel of 146 amino acids and a 693-residue helical bundle that includes the active site. The amino-terminal domain is probably absent from mammalian LOX (see Prigge *et al.*, 1996), and this region shows a relatively high degree of variability among seed LOX

	Pea LOX-2	Pea LOX-3	Soybean LOX-1	Soybean LOX-2	Soybean LOX-3	Rice	Barley
Pea LOX-2							
Pea LOX-3	86						
Soybean LOX-1	79	71					
Soybean LOX-2	84	75	84				
Soybean LOX-3	75	86	71	76			
Rice	54	55	54	55	55		
Barley	56	56	55	57	57	75	
Maize*	58	58	57	59	59	77	83

*Not full-length

Fig. 7 Homology matrix showing percentage identities between plant seed LOX amino acid sequences based on BESTFIT comparisons. References to sequences are: pea LOX-2, Ealing and Casey (1989); pea LOX-3, Ealing and Casey (1988); soybean LOX-1, Shibata *et al.* (1987); soybean LOX-2, Shibata *et al.* (1988); soybean LOX-3, Yenofsky *et al.* (1988); rice, Ohta *et al.* (1992); barley, van Mechelen *et al.* (1995); maize, Jensen *et al.* (1997).

sequences (Fig. 8). The active site Fe is inaccessible to solvent and the three Fe-liganded histidines (see earlier) are located in two π-helices, which form the central sections of long α-helices (π-helices are unusual structures in which residues i and i+5 are H-bonded; see Gaffney, 1996). A number of channels/ "pockets" exist within the three-dimensional structure of the enzyme to permit access of the substrates - oxygen and fatty acid - to the active site iron. It is still not clear which of these represents the main path of entry for the fatty acid and possible that more than one might be used (Minor *et al.*, 1996).

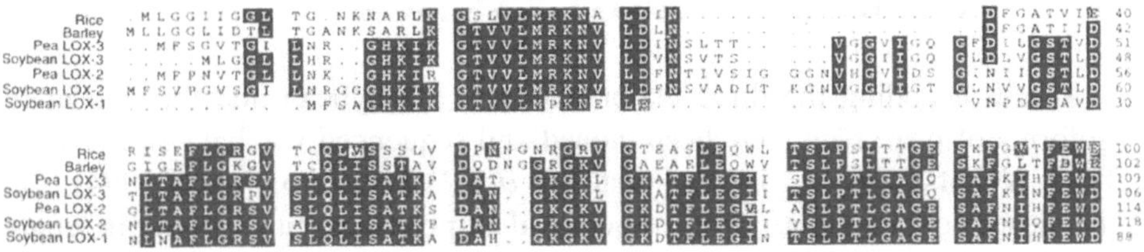

Fig. 8 Alignment of the *N*-terminal regions of plant seed LOX showing variation in sequence within the first 60 amino acids. Dots represent gaps introduced by the PILEUP programme to align the sequences.

Within the plant and animal LOX sequences published to date, there are 56 conserved amino acid residues, including the Fe ligands. Many of these will have been conserved because of the role they play in permitting substrate access to the active site, or in maintaining critical structure. For instance, 26 of these conserved amino acids are to be found in four regions which play important roles in LOX structure by anchoring helices that contain Fe - His ligands or stabilizing the loop that contributes the active-site carboxylate ligand (Prigge *et al.*, 1996). Modelling LOX structures based on that of soybean LOX-1 (Prigge *et al.*, 1996) has given much insight into LOX structure and activity and suggested a number of mutagenesis experiments with which to test theories on the control of positional specificity and enzyme stability. The structure of soybean LOX-3 at 2.6Å resolution (Skrzypczak-Jankun *et al.*, 1997) shows it to be similar in general topology to LOX-1; although there are many sequences differences between LOX-1 and LOX-3 (see Fig. 7), they seem to be scattered over the whole molecule.

5. LOCALIZATION OF LOX

All the available evidence suggests that seed LOX are cytosolic enzymes. Although some vegetative LOX (from rice and *Arabidopsis*) have putative chloroplast targeting sequences (see Shibata and Axelrod, 1995), no such sequences are associated with any of the seed LOX shown in Fig. 7. Immunocytochemical studies (Vernooy-Geritsen *et al.*, 1984) suggested that LOX was localized within the cytosol of soybean cotyledon storage parenchyma cells, although it should be noted that these studies employed germinating seeds and may not necessarily reflect the situation in developing or mature seeds.

6. GENE STRUCTURE, ORGANIZATION, EXPRESSION AND GENETICS

Genes encoding the three major isoforms of soybean and the two major isoforms of pea seeds have been cloned and sequenced (Shibata *et al.*, 1987, 1988; Yenofsky *et al.*, 1988; Forster *et al.*, 1994; Knox *et al.*, 1994). All five genes contain eight introns, at equivalent positions, but varying in length; the disposition of introns bears no relation to the higher order structural elements within the LOX primary sequence (Prigge *et al.*, 1996). The first intron falls within the first 146 amino acids (the *N*-domain), the remainder within the second domain, apparently at random; indeed, intron VII interrupts the coding sequence between the two active site ligands His^{499} and His^{504}. Such a disposition gives little insight into the evolution of LOX gene structure.

In pea, measurements of gene copy number suggest that each of LOX-2 and LOX-3 is encoded by two or three genes (Domoney *et al.*, 1991) and studies of the inheritance of LOX polypeptides show that the LOX-2 and LOX-3 genes are closely linked at a single genetic locus (*Lox*) (North *et al.*, 1989). During the course of pea seed development, both classes of gene are maximally active at mid-late development, producing approximately equal amounts of mRNA (Domoney *et al.*, 1990).

Although direct measurements of the numbers of soybean LOX genes have not been reported, the observation that a single point mutation in a soybean LOX-2 gene eliminates all LOX-2 protein and activity (Wang *et al.*, 1994) suggests that a single gene encodes LOX-2 in soybean, although it does not rule out the possibility that the mutant genotype contains multiple copies of the same mutation.

Studies of soybean lines that lack each of the seed isoforms (Hildebrand and Hymowitz, 1982; Kitamura *et al.*, 1983; Davies and Nielsen, 1986, 1987) have shown that the genes encoding LOX-1 and LOX-2, but not LOX-3, are very tightly linked. The soybean seed LOX genes are expressed during seed mid-maturation (Start *et al.*, 1986). The mutants that lack LOX-1 and LOX-3 have a considerable reduction in, or lack of, the corresponding transcripts (Start *et al.*, 1986). The molecular basis of the LOX-2 mutant, however, is protein instability as a consequence of mutation of an active-site histidine, rather than alterations in gene activity or mRNA accumulation (Wang *et al.*, 1994).

A rice seed LOX mutation has also been shown to behave as a single recessive trait (Suzuki *et al.*, 1996). Such mutants in rice and soybean are of considerable commercial significance because of the role of LOX in generating pleasant and unpleasant volatile aroma compounds.

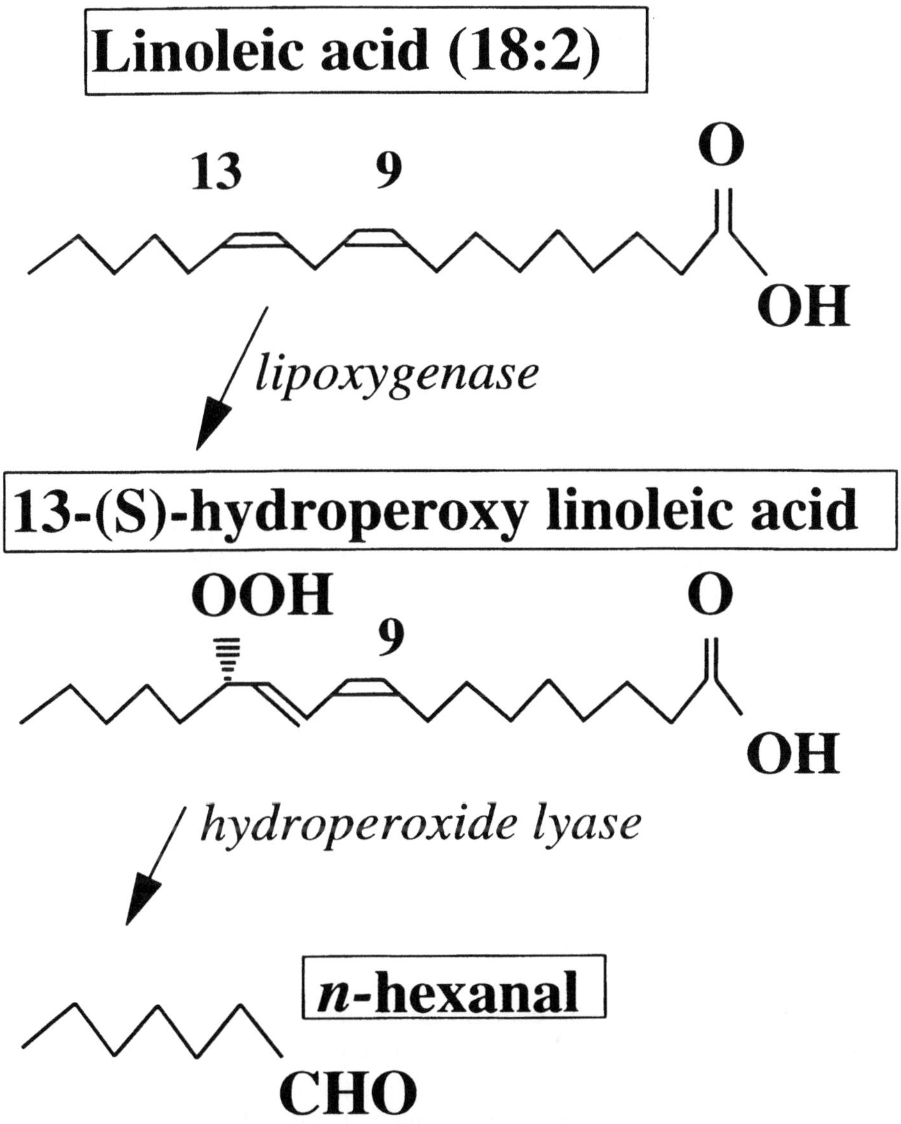

Fig. 9 Production of *n*-hexanal from linoleic acid *via* LOX and hydroperoxide lyase.

7. THE COMMERCIAL IMPORTANCE OF LIPOXYGENASE

There are two major sources of commercial interest in seed LOX; its ability to generate precursors to tastes and aromas (West, 1996), and its capacity to co-oxidize biological molecules. Both of these can contribute to the quality of foods and may be linked; co-oxidative decomposition of pigments, for instance, can itself give rise to volatile aroma compounds. The generation of highly oxidizing hydroperoxide species will also have an impact on the anti-oxidant content of seeds that have high LOX activity. The major interest in terms of aroma generation is in the production of rancid, or "beany" off-flavours, in the form of volatile aldehydes derived from the hydroperoxides; amongst these, *n*-hexanal (see Fig. 9) is the most potent off-flavour.

Volatile aldehyde production reduces the acceptability of rice and soybean products, and considerable effort has been expended on the production of commercial soybean lines that lack LOX-2. Because LOX-2 produces 13-hydroperoxides, it is a major source of hexanal (see Fig. 9) in soybean products; soymilk and soyflour preparations from lines lacking LOX-2 have higher consumer acceptability (Davies *et al*., 1987).

Removal of LOX-1 or LOX-3 does not improve the quality of soybean products and there is evidence to suggest that LOX-3 removal is counterproductive in terms of hexanal production (Hildebrand *et al*., 1990). LOX can carry out co-oxidation of protein thiol groups to form disulphide bonds and it is probably this phenomenon that underlies the use of LOX-active soybean flour as a baking additive. Addition of soybean LOX to wheat flour improves dough rheology by increasing mixing tolerance (Frazier, 1979; Cumbee *et al*., 1997), possibly by increasing the degree of polymerization of the high molecular weight glutenin polypeptides (Casey, 1998) that play an important role in the bread-making process.

8. CONCLUSION

LOX are relatively minor components, comprising at most 1% of the protein in soybean, and considerably less in other seeds. Nevertheless, their activity has significance for the food industry, through the formation of off-flavours in seed (particularly protein) products, the decolourizing of pigments and the potential to compromise anti-oxidant status.

Seed LOX appear to be dispensable in terms of plant physiology and agronomic performance, facilitating the improvement of the quality of legume and cereal seeds by the selective removal of particular LOX isoforms, whilst retaining those vegetative LOX that play a role in plant defence and stress responses.

ACKNOWLEDGEMENTS

I am extremely grateful to Dr. Marcus Durrant for his help with Figures 3 and 6 and to Prof. L. Mario Amzel for providing Plate 11. This work was supported by a grant from the Biotechnology and Biological Sciences Research Council to the John Innes Centre.

REFERENCES

Andre, E. and Hou, K.-W. (1932) The presence of a lipoid oxidase in soybean, *Glycine soya* Lieb. *Comptes Rendues d'Academie des Sciences (Paris)*, **194**, 645-47.

Andre, J.C., and Funk, O. (1986) Determination of stereochemistry in the fatty acid hydroperoxide products of lipoxygenase catalysis. *Analytical Biochemistry*, **158**, 316-21.

Andrianarison, R.H., Rabinovitch-Chable, H. and Beneytout, J.L. (1991) Oxodiene formation during the *Vicia sativa* lipoxygenase-catalyzed reaction; occurrence of dioxygenase and fatty acid lyase activities associated in a single protein. *Biochemical and Biophysical Research Communications,* **180**, 1002-9.

Andrianarison, R.H., Tixier, M. and Beneytout, J.L. (1990) Positional specificity of a *Lupinus albus* lipoxygenase in relation to enzyme concentration and effect of a double dioxygenation product of arachidonic acid. *Biochemical and Biophysical Research Communications*, **172**, 985-92.

Axelrod, B. (1974) Lipoxygenases. American Chemical Society Advanced Chemistry Series, **136**, 324-48.

Axelrod, B., Cheesbrough, T.M. and Lasko, S. (1981) Lipoxygenase from soybeans. *Methods in Enzymology*, **71**, 441-51.

Belefant, H. and Fong, F. (1991) Lipoxygenases in developing *Zea mays* kernels. *Plant Physiology and Biochemistry*, **29**, 99-104.

Bell, E. and Mullet, J.E. (1991) Lipoxygenase gene expression is modulated in plants by water deficit, wounding, and methyl jasmonate. *Molecular and General Genetics*, **230**, 456-62.

Borthakur, A. and Ramadoss, C.S. (1986) Aerobic formation of ketodiene from linoleic acid catalyzed by one of the two forms of lipoxygenase isolated from Bengal Gram (*Cicer arietinum*). *Journal of Agricultural and Food Chemistry*, **34**, 1016-8.

Boyington, J.C., Gaffney, B.J. and Amzel, L.M. (1993) The three-dimensional structure of an arachidonic acid 15-lipoxygenase. *Science*, **260**, 1482-6.

Bunker, T.W., Koetje, D.S., Stephenson, L.C., Creelman, R.A., Mullet, J.E. and Grimes, H.D. (1995) Sink limitation induces the expression of multiple soybean vegetative lipoxygenase mRNAs while the endogenous jasmonic acid level remains low. *The Plant Cell*, **7**, 1319-31.

Casey, R. (1998) Lipoxygenases in breadmaking. Proceedings of the 1st European Symposium on Enzymes in Grain Processing. TNO, The Netherlands, pp.188-194.

Christopher, J.P., Pistorius, E.K. and Axelrod, B. (1970) Isolation of an isozyme of soybean lipoxygenase. *Biochimica et Biophysica Acta*, **198**, 12-19.

Christopher, J.P., Pistorius, E.K. and Axelrod, B. (1972) Isolation of a third isozyme of soybean lipoxygenase. *Biochimica et Biophysica Acta*, **284**, 54-62.

Croft, K.P.C., Jüttner, F. and Slusarenko, A.J. (1993) Volatile products of the lipoxygenase pathway evolved from *Phaseolus vulgaris* (L.) leaves inoculated with *Pseudomonas syringae* pv *phaseolicola*. *Plant Physiology*, **101**, 13-24.

Cumbee, B., Hildebrand, D.F. and Addo, K. (1997) Soybean lipoxygenase isozymes effects on wheat flour dough rheological and breadmaking properties. *Journal of Food Science*, **62**, 281-94.

Davies, C.S. and Nielsen, N.C. (1986) Genetic analysis of a null-allele for lipoxygenase-2 in soybean. *Crop Science*, **26**, 460-3.

Davies, C.S. and Nielsen, N.C. (1987) Registration of soybean germplasm that lacks lipoxygenase isozymes. *Crop Science*, **27**, 370-1.

Davies, C.S., Nielsen, S.S. and Nielsen, N.C. (1987) Flavor improvement of soybean preparations by genetic removal of lipoxygenase-2. Journal of the American Oil Chemists Society, **64**, 1428-33.

Doderer, A., Kokkelink, I., van der Veen, S., Valk, B.E., Schram, A.W. and Douma, A.C. (1992) Purification and characterization of two lipoxygenase isoenzymes from germinating barley. *Biochimica et Biophysica Acta*, **1120**, 97-104.

Domoney, C., Casey, R., Turner, L. and Ellis, N. (1991) *Pisum* lipoxygenase genes. *Theoretical and Applied Genetics*, **81**, 800-805.

Domoney, C., Firmin, J.L., Sidebottom, C., Ealing, P.M., Slabas, A. and Casey, R. (1990) Lipoxygenase heterogeneity in *Pisum sativum*. *Planta*, **181**, 35-43.

Ealing, P.M. and Casey, R. (1988) The complete amino acid sequence of a pea (*Pisum sativum*) seed lipoxygenase predicted from a near full-length cDNA. *Biochemical Journal*, **253**, 915-8.

Ealing, P.M. and Casey, R. (1989) The cDNA cloning of a pea (*Pisum sativum*) seed lipoxygenase. Sequence comparisons of the two major pea seed lipoxygenase isoforms. *Biochemical Journal*, **264**, 929-32.

Eiben, H.G. and Slusarenko, A.J. (1994) Complex spatial and temporal expression of lipoxygenase genes during *Phaseolus vulgaris* (L.) development. *The Plant Journal*, **5**, 123-35.

Eriksson, C.E. and Svensson S.G. (1970) Lipoxygenase from peas, purification and properties of the enzyme. *Biochimica et Biophysica Acta*, **198**, 449-59.

Farmer, E.E. and Ryan, C.A. (1992) Octadecanoid precursors of jasmonic acid activate the synthesis of wound-inducible proteinase inhibitors. *The Plant Cell*, **4**, 129-34.

Forster, C., Knox, M., Domoney, C. and Casey, R. (1994) *lox1:Ps:2*, a *Pisum sativum* seed lipoxygenase gene. *Plant Physiology*, **106**, 1227-8.

Frazier, P.J. (1979) Lipoxygenase action and lipid binding in breadmaking. *Bakers Digest*, **53**, 8-29.

Funk, M.O., Carroll, R.T., Thompson, J.F., Sands, R.H. and Dunham, W.R. (1990) Role of iron in lipoxygenase catalysis. *Journal of the American Chemical Society*, **112**, 5375-76.

Gaffney, B.J. (1996) Lipoxygenases: structural principles and spectroscopy. *Annual Review of Biophysics and Biomolecular Structure*, **25**, 431-59.

Gardner, H.W. (1989) Soybean lipoxygenase-1 enzymically forms both (9S)- and (13S)-hydroperoxides from linoleic acid by a pH-dependent mechanism. *Biochimica et Biophysica Acta*, **1001**, 274-81.

Glickman, M.H. and Klinman, J.P. (1996) Lipoxygenase reaction mechanism: demonstration that hydrogen abstraction from substrate precedes dioxygen binding during catalytic turnover. *Biochemistry*, **35**, 12882-92.

Grimes, H.D., Tranbarger, T.J. and Franceschi, V.R. (1993) Expression and accumulation patterns of nitrogen-responsive lipoxygenases in soybeans. *Plant Physiology*, **103**, 457-66.

Hatanaka, A., Kajiwara, T., Matsui, K. and Ogura, M. (1990) Studies on the substrate specificity of soybean lipoxygenase-1 using an entire series of (ω3Z, ω6Z, ω9Z)-C_{12}~C_{22}-trienoic acids and trienols. *Zeitschriftes für Naturforschung*, **45c**, 1161-4.

Henderson, H.M., Eskin, N.A.M. and Borsa, J. (1990) Buckwheat lipoxygenase: inactivation by gamma-irradiation. *Food Chemistry*, **38**, 97-103.

Heydeck, D., Wiesner, R., Kühn, H. and Schewe, T. (1991) On the reaction of wheat lipoxygenase with arachidonic acid and its oxygenated derivatives. *Biomedica et Biochimica Acta*, **50**, 11-15.

Hildebrand, D.F. and Hymowitz, T. (1982) Inheritance of lipoxygenase-1 activity in soybean seeds. *Crop Science*, **22**, 851-3

Hildebrand, D.F., Hamilton-Kemp, T.R., Longhrin, J.H., Ali, K. and Anderson, R.A. (1990) Lipoxygenase 3 reduces hexanal production from soybean seed homogenates. *Journal of Agricultural and Food Chemistry*, **38**, 1934-6.

Ida, S., Masaki, Y. and Morita, Y. (1983) The isolation of multiple forms and product specificity of rice lipoxygenase. *Agricultural and Biological Chemistry*, **47**, 637-41.

Jensen, A.B., Poca, E., Rigaud, M., Freyssinet, G. and Pagès, M. (1997) Molecular characterization of L2 lipoxygenase from maize embryos. *Plant Molecular Biology*, **33**, 605-14.

Kermasha, S., Khalyfa, A., Alli, I. and Lee, B. (1991) Purification and characterization of lipoxygenase isozymes from canola (*Brassica napus* cv Westar) seed. *Journal of Food Biochemistry*, **15**, 219-38.

Khalyfa, A., Kermasha, S. and Alli, I. (1990) Partial purification and characterization of lipoxygenase of canola seed (*Brassica napus* var. Westar). *Journal of Agricultural and Food Chemistry*, **38**, 2003-8.

Kitamura, K., Davies, C.S., Kaizuma, K. and Nielsen, N.C. (1983) Genetic analysis of a null-allele for lipoxygenase-3 in soybean seeds. *Crop Science*, **23**, 924-7.

Knox, M., Forster, C., Domoney, C. and Casey, R. (1994) Structure of the *Pisum sativum* seed lipoxygenase gene *lox1:Ps:3*. *Biochimica et Biophysica Acta*, **1214**, 341-3.

Maccarrone, M., van Aarle, P.G.M., Veldink, G. and Vliegenthart, J.F.G. (1994) In vitro oxygenation of soybean biomembranes by lipoxygenase-2. *Biochimica et Biophysica Acta*, **1190**, 164-9.

Meshehdani, T., Pokorný, J., Davídek, J. and Pánek, J. (1990) The lipoxygenase activity of rapeseed. *Die Nahrung*, **34**, 727-34.

Michaud-Soret, I., Daniel, R., Chopard, C., Mansuy, D., Cucurou, C., Ullrich, V. and Chottart, J.C. (1990) Soybean lipoxygenases-1, -2a, -2b and -2c do not contain PQQ. *Biochemical and Biophysical Research Communications*, **172**, 1122-8.

Minor, W., Steczko, J., Bolin, J.T., Otwinowski, Z. and Axelrod, B. (1993) Crystallographic determination of the active site iron and its ligands in soybean lipoxygenase-1. *Biochemistry*, **32**, 6320-3.

Minor, W., Steczko, J., Stec, B., Otwinowski, Z., Bolin, J.T., Walter, R. and Axelrod, B. (1996) Crystal structure of soybean lipoxygenase L-1 at 1.4Å resolution. *Biochemistry*, **35**, 10687-701.

Nagarantha, K.C., Shetty, S.A. Prakash, H.S., Bhat, S.G. and Shetty, H.S. (1992) Lipoxygenase activity in fresh, stored, and germinating pearl millet seeds resistant and susceptible to downy mildew disease. *Canadian Journal of Botany*, **70**, 2028-31.

Nelson, M.J. and Seitz, S.P. (1994) The structure and function of lipoxygenase. *Current Opinions in Structural Biology*, **4**, 878-84.

Nicolas, J. and Drapron, R. (1977) Some physicochemical characteristics of horse bean lipoxygenase (*Vicia faba* L.). *Annales des Technologie et Agriculture*, **26**, 119-32.

Nikolaev, V., Reddanna, P., Whelan, J., Hildenbrandt, G. and Reddy, C.C. (1990) Stereochemical nature of the products of linoleic acid oxidation catalyzed by lipoxygenases from potato and soybean. *Biochemical and Biophysical Research Communications*, **170**, 491-6.

North, H., Casey, R. and Domoney, C. (1989) Inheritance and mapping of seed lipoxygenase polypeptides in *Pisum*. *Theoretical and Applied Genetics*, **77**, 805-8.

Ohta, H., Shirano, Y., Tanaka, K., Morita, Y. and Shibata, D. (1992) cDNA cloning of rice lipoxygenase L-2 and characterization using an active enzyme expressed from the cDNA in *Escherichia coli*. *European Journal of Biochemistry*, **206**, 331-6.

Pfeiffer, T.W., Hildebrand, D.F. and TeKrony, D.M. (1992) Agronomic performance of soybean lipoxygenase isolines. *Crop Science*, **32**, 357-62.

Piazza, G.J. and Nuñez, A. (1995) Oxidation of acylglycerols and phosphoglycerides by soybean lipoxygenase. *Journal of the American Oil Chemists Society*, **72**, 463-6.

Poca, E., Rabinovitch-Chable, H., Cook-Moreau, J., Pages, M. and Rigaud, M. (1990) Lipoxygenases from *Zea mays* L.. Purification and physicochemical characteristics. *Biochimica et Biophysica Acta*, **1045**, 107-14.

Prigge, S.T., Boyington, J.C., Gaffney, B.J. and Amzel, L.M. (1996) Structure conservation in lipoxygenases: structural analysis of soybean lipoxygenase-1 and modeling of human lipoxygenases. *Proteins: Structure, Function and Genetics*, **24**, 275-91.

Regdel, D., Schewe, T. and Kuehn, H. (1995) Comparative characteristics of lipoxygenase isoenzymes from green pea seeds. *Biochemistry (Moscow)*, **60**, 715-21.

Robinson, D.S., Wu, Z., Domoney, C. and Casey, R. (1995) Lipoxygenases and the quality of foods. *Food Chemistry*, **54**, 33-43.

Sanders, T.H., Pattee, H.E. and Singleton, J.A. (1975) Lipoxygenase isozymes of peanut. Lipids, **10**, 681-5.

Sanz, L.C., Perez, A.G. and Olias, J.M. (1992) Purification and catalytic properties of chickpea lipoxygenases. *Phytochemistry*, **31**, 2967-72.

Sanz, L.C., Perez, A.G., Rios, J.J. and Olias, J.M. (1993) Positional specificity of ketodienes from linoleic acid aerobically formed by lipoxygenase isoforms from kidney bean and pea. *Journal of Agricultural and Food Chemistry*, **41**, 696-9.

Saravitz, D.M. and Siedow, J.N. (1996) The differential expression of wound-inducible lipoxygenase genes in soybean leaves. *Plant Physiology*, **110**, 287-99.

Schilstra, M.J., Veldink, G.A. and Vliengenthart, J.F.G. (1994) Kinetic analysis of the induction period in lipoxygenase catalysis. *Biochemistry*, **32**, 7686-91.

Shastry, B.S. and Rao, M.R.R. (1975) Studies on lipoxygenase from rice bran. *Cereal Chemistry*, **52**, 597-603.

Shibata, D. and Axelrod, B. (1995) Plant lipoxygenases. *Journal of Lipid Mediators and Cell Signalling*, **12**, 213-28.

Shibata, D., Steczko, J., Dixon, J.E., Hermodson, M, Yazdanparast, R. and Axelrod, B. (1987) Primary structure of soybean lipoxygenase-1. *Journal of Biological Chemistry*, **262**, 10080-5.

Shibata, D., Steczko, J., Dixon, J.E., Andrews, P .C., Hermodson, M. and Axelrod, B. (1988) Primary structure of soybean lipoxygenase-2. *Journal of Biological Chemistry*, **263**, 6816-21.

Shiiba, K., Negishi, Y., Okada, K. and Nagao, S. (1991) Purification and characterization of lipoxygenase isozymes from wheat germ. *Cereal Chemistry*, **68**, 115-22.

Shimizu, T., Rådmark, O. and Samuelsson, B. (1984) Enzyme with dual lipoxygenase activities catalyzes leukotriene A4 synthesis from arachidonic acid. *Proceedings of the National Academy of Sciences, USA*, **81**, 689-93.

Siedow, J.N. (1991) Plant lipoxygenase: structure and function. *Annual Review of Plant Physiology and Plant Molecular Biology*, **42**, 145-88.

Skrzypczak-Jankun, E., Amzel, L.M, Kroa, B.A. and Funk, M.O. Jr., (1997) Structure of soybean lipoxygenase L3 and a comparison with its L1 isoenzyme. *Proteins: Structure, Function and Genetics*, **29**, 15-31.

Sloane, D.L., Leung, R., Craik, C.S. and Sigal, E. (1991) A primary determinant for lipoxygenase positional specificity. *Nature, London*, **354**, 149-52.

Start, W.G., Ma, Y., Polacco, J.C., Hildebrand, D.F., Freyer, G.A and Altschuler, M. (1986) Two soybean seed lipoxygenase nulls accumulate reduced levels of lipoxygenase transcripts. *Plant Molecular Biology* **7**, 11-23.

Steczko, J., Donoho, G.A., Clemens, J.C., Dixon, J.E. and Axelrod, B. (1992) Conserved histidine residues in soybean lipoxygenase: functional consequences of their replacement. *Biochemistry*, **31**, 4053-7.

Sumner, J.B. and Sumner, R.J. (1940) The coupled oxidation of carotene and fat by carotene oxidase. *Journal of Biological Chemistry*, **134**, 531-33.

Suzuki, Y., Nagamine, T., Kobayashi, A. and Ohtsubo, K. (1993) Detection of a new rice variety lacking lipoxygenase-3 by monoclonal antibodies. *Japanese Journal of Breeding*, **43**, 405-9.

Suzuki, Y., Nagamine, T. and Okuno, K. (1996) Genetic analysis of a null-allele for lipoxygenase-3 in rice seeds. *Euphytica*, **91**, 99-101.

Theorell, H., Holman, R.T. and Akeson, A. (1947) Crystalline lipoxidase. *Acta Chemica Scandinavica*, **1**, 571-6.

Truong, Van Den, Raymundo, L.C. and Mendoza, E.M.T. (1979) Cowpea lipoxygenase: Part 1. Physiochemical characterization. *Bulletin of the Philippine Biochemical Society*, **2**, 1-12.

Truong, Van Den, Raymundo, L.C. and Mendoza, E.M.T. (1982a) Winged bean lipoxygenase - Part 1: Isolation and purification. *Food Chemistry*, **8**, 187-201.

Truong Van Den, Raymundo, L.C. and Mendoza, E.M.T. (1982b) Winged bean lipoxygenase - Part 2: physicochemical properties. *Food Chemistry*, **8**, 277-89.

van Aarle, P.G.M., de Barse, M.M.J., Veldink, G. and Vliegenthart, J.F.G. (1991) Purification of a lipoxygenase from ungerminated barley. Characterization and product formation. *FEBS Letters*, **280**, 159-62.

van der Meer, R.A. and Duine, J.A. (1988) Pyrroloquinoline quinone (PQQ) is the organic cofactor in soybean lipoxygenase-1. *FEBS Letters*, **235**. 194-200.

van Mechelen, J.R., Smits, M., Douma, A.C., Rouster, J., Cameron-Mills, V., Heidekamp, F. and Valk B.E. (1995) Primary structure of a lipoxygenase from barley grain as deduced from its cDNA sequence. *Biochimica et Biophysica Acta*, **1254**, 221-5.

Veldink, G.A., Boelens, H., Maccarrone, M., van der Lecq, F., Vliegenthart, J.F.G., Paz, M.A., Flückiger, R. and Gallop, P.M. (1990) Soybean lipoxygenase-1 is not a qunioprotein. *FEBS Letters*, **270**, 135-8.

Veldink, G.A. and Vliegenthart, J.F.G. (1985) Lipoxygenases, nonheme iron-containing enzymes. *Advances in Inorganic Biochemistry*, **6**, 139-61.

Veldink, G.A., Vliegenthart, J.F.G. and Boldingh, J. (1977) Plant lipoxygenases. *Progress in the Chemistry of Fats and Other Lipids*, **15**, 131-66.

Vernooy-Gerritsen, M., Leunissen, J.L.M., Veldink, G.A. and Vliegenthart, J.F.G. (1994) Intracellular localization of lipoxygenases-1 and -2 in germinating soybean seeds by indirect labelling with protein A-colloidal gold complexes. *Plant Physiology*, **76**, 1070-9.

Vick, B.A. and Zimmerman, D.C. (1983) The biosynthesis of jasmonic acid: a physiological role for plant lipoxygenase. *Biochemical and Biophysical Research Communications*, **111**, 470-77.

Wang, W.H., Takano, T., Shibata, D., Kitamura, K. and Takeda, G. (1994) Molecular basis of a null mutation in soybean lipoxygenase 2: substitution of glutamine for an iron-ligand histidine. *Proceedings of the National Academy of Sciences, USA*, **91**, 5828-32.

West, S. (1996) Flavour production with enzymes, in *Industrial Enzymology*, 2nd edn. (eds. T. Godfrey and S. West), MacMillan, London, pp. 211-24.

Wu, Z., Robinson, D.S., Domoney, C. and Casey, R. (1995) High-performance liquid chromatographic analysis of the products of linoleic acid oxidation catalyzed by pea (*Pisum sativum*) lipoxygenases. *Journal of Agricultural and Food Chemistry*, **43**, 337-42.

Yenofsky, R.L., Fine, M. and Liu, C. (1988) Isolation and characterization of a soybean (*Glycine max*) lipoxygenase-3 gene. *Molecular and General Genetics*, **211**, 215-22.

Yoon, S. and Klein, B.P. (1979) Some properties of pea lipoxygenase isoenzymes. *Journal of Agricultural and Food Chemistry*, **27**, 958-62.

Zhang, L.-Y. and Hamberg, M. (1996) Specificity of two lipoxygenases from rice: unusual regiospecificity of a lipoxygenase isoenzyme. *Lipids*, **31**, 803-89.

30. Thionins

FRANCISCO GARCÍA-OLMEDO
Laboratorio de Bioquímica y Biología Molecular, Departamento de Biotecnología - UPM, ETS Ingenieros Agrónomos, E-28040 Madrid, Spain.

1. CHARACTERISTICS

1.1 A single peptide family

The general designation of thionins has been proposed for a family of homologous peptides that includes purothionins, which were first isolated from wheat seeds (Balls *et al.*, 1942) and their homologues from various taxa that have been named viscotoxins and crambins (see García-Olmedo *et al.*, 1989). The crystalline protein material obtained from lipid extracts of wheat endosperm was designated "purothionins" because of its high sulphur content (Balls *et al.*, 1942). This material was found to have bactericidal and fungicidal properties (Stuart and Harris, 1942), to inhibit fermentation of wheat mashes (Balls and Harris, 1944), and to be toxic to laboratory animals (Coulson *et al.*, 1942). The toxic properties of mistletoes were ascribed to the viscumin lectin and to a mixture of small basic proteins, designated "viscotoxins" (Samuelsson, 1974). In a study of the seeds of the Abyssinian cabbage (*Crambe abyssinica*), a high-sulphur crystalline protein was obtained from the aqueous acetone extracts and designated "crambin" (Van Etten *et al.*, 1965).

The thionins have become the subject of intensive structural studies, both in crystalline form and in solution, as they are excellent model systems in the development and refinement of novel methods for the elucidation of the three-dimensional structure of macromolecules. Recombinant DNA and other techniques have been applied to the study of relevant aspects of their molecular biology, such as synthesis and deposition, genetic control and gene structure, *in vitro* activities and possible physiological function.

P.R. Shewry and R. Casey (eds.), Seed Proteins, 709–726.

1.2 Distribution

The original purothionin from hexaploid wheat was later found to be heterogenous, as it was fractionated into two closely-migrating bands by electrophoresis (Nimmo *et al.*, 1968; Fisher *et al.*, 1968; García-Olmedo *et al.*, 1968). In a survey of 22 diploid, tetraploid and hexaploid species of the *Aegilops-Triticum* group, the presence of thionins was demonstrated in the endosperms of all of them, and their electrophoretic mobilities in diploid and tetraploid species suggested that at least one variant per diploid genome complement should be present in hexaploid wheat (Carbonero and García-Olmedo, 1969). Genetic analysis and the determination of amino acid sequences confirmed the existence of three different thionins in the endosperm of hexaploid wheat (reviewed in García-Olmedo *et al.*, 1989). Two thionin components, designated α and β-hordothionin, were found in the endosperm of barley (*Hordeum vulgare*) by Redman and Fisher (1969). The nucleotide sequences of the corresponding cDNA and genomic DNA clones are in agreement with the presence of more than one thionin in this barley tissue (Hernandez-Lucas *et al.*, 1986; Ponz *et al.*, 1986; Rodriguez-Palenzuela *et al.*, 1988). Two thionin variants from oats endosperm (*Avena sativa*) have been also sequenced by Bekes and Lasztity (1981).

The viscotoxin from leaves and stems of European mistletoe (*Viscum album*, Loranthaceae) was found to be a mixture of closely related components. Similar toxins, such as phoratoxins A and B from *Phoradendron tomentosum*, denclatoxin B from *Dendrophtora clavata* and ligatoxin A from *Phoradendron liga*, were also characterized within the Loranthaceae (reviewed in García-Olmedo *et al.*, 1989). Additional leaf thionins have been identified in *Pyrularia pubera*, a parasitic plant from the Santalaceae (Vernon *et al.*, 1985), and from barley (Böhlmann and Apel, 1987; Gausing, 1987).

The crambin reported by Van Etten *et al.* (1965) was also found to be a mixture of two variants, whose primary structures were homologous to the thionins and the viscotoxins (Teeter *et al.*, 1981; Vermeulen *et al.*, 1987).

1.3 Thionin types

The mature thionin peptides are generally 45 to 47 amino acids in length. An unrooted phylogenetic tree allowed the classification of the available amino acid sequences from the thionins (either directly determined or deduced from cDNAs) into at least five types (I-V), one of which (type V) lacks the C-terminal nonapeptide (García-Olmedo *et al.*, 1989; Castagnaro *et al.*, 1992).

The original purothionin type isolated from wheat endosperm (Balls *et al.*, 1942) has four disulphide bridges and is highly basic, with no negatively charged residues. The sequences that are known of this type comprise 45 amino-acid residues, 8 of which are in the central disulphide loop. The second type has been isolated from the leaves of *Pyrularia pubera* (Vernon *et al*., 1985) and barley (Gausing, 1987; Bohlmann and Apel, 1987). This type has four disulphide bridges at the same positions as those of the type I thionin, but the molecules are less basic, with some negatively-charged residues, and their central disulphide loop contains one or two more amino acid residues. The third type includes the viscotoxins and phoratoxins from mistletoes (Loranthaceae), and has the following distinctive features: three disulphide bridges that are conserved with respect to the previous types; fewer basic amino acid residues; a sequence with 46 residues, 9 of which are in the central disulphide loop. The crambins isolated from the Abyssinian cabbage (Cruciferae) represent the fourth type, which has the same sequence length and disulphide-bridge arrangement as type II thionins, but the molecules are neutral, with a low proportion of charged amino acid residues.

The fifth type is quite divergent: the 2nd and 8th cysteines of type I thionins are missing through point mutation and deletion, respectively, thus disrupting the first and second disulphide bridges and potentially allowing the formation of a new bridge between the unpaired cysteines. This new type, which is also neutral, has been identified in a cDNA library derived from developing wheat kernels and is also present in related *Aegilops* species (Castagnaro *et al*. 1992, 1995).

It can be noted that type I and II thionins are closer to each other than to the other types, not only in their gross architecture but also at the level of their amino acid sequences, and that the same is true for types III and IV. Although the different types have been mostly investigated in particular tissues and organs of certain species or groups of related species, the observed divergence between the types cannot be correlated with the evolutionary relationships among the taxa because the types defined here do not represent equivalent subsets of the protein family. Thus, type I, II and III thionins may coexist in a single species, and the thionin isolated from *Pyrularia pubera*, a parasitic plant closely related to the mistletoes, is closer to the barley-leaf thionins than to the viscotoxins and phoratoxins. In *Viscum album* there are thionin genes that are expressed in seeds and in leaves, and in *Crambe abyssinica*, thionin variants distinct from the seed-specific crambin are found that are ubiquitously expressed (Schrader-Fischer and Apel, 1993, 1994). In *Arabidopsis thaliana* there is a thionin gene that is expressed in flowers, siliques and rosette leaves, and a second one that is only expressed at basal levels in seedlings and rosette leaves (Epple *et al*. 1995).

If only the disulphide-bound structure is considered, all known thionins could be classified into three groups: a group with 4 disulphide bonds, which would include types I and II, a group with only 3 of the above disulphide bonds (types III and IV) and a group which presumably has only 2 of the above bonds, plus a novel bond that would involve one cysteine from each of the two above bonds (Figs. 1,2).

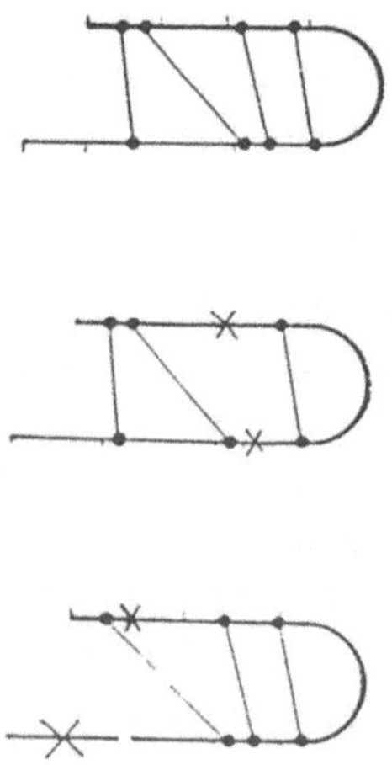

DISULPHIDE STRUCTURE	CHARGES +/-	RESIDUES Total/Loop	TYPE
1	10/0	45/8	I
	9/3	47/10	II
	7/1	46/9	
2	6/0-2	46/9	III
	5/1-2		
	2/2		IV
3	2/2	38/9	V

Fig. 1. Structural features of thionins.

DISULPHIDE BRIDGES

			Type
Avena sativa	TH-β	KSCCKNTLGRNCYNLCRARG--APKLCSTVCRCKLTSGLSCPKDFPK	I
Triticum aestivum	TH-β	KSCCKSTLGRNCYNLCRARG--AQKLCANVCRCKLTSGLSCPKDFPK	I
A. sativa	TH-α	KSCCRNTLGRNCYNLCRSRG--APKLCATVCRCKISSGLSCPKDFPK	I
Hordeum vulgare	TH-β	KSCCRSTLGRNCYNLCRVRG--AQKLCANACRCKLTSGLSCPKGFPK	I
T. aestivum	TH-α1	KSCCRTTLGRNCYNLCRSRG--AQKLCSTVCRCKLTSGLSCPKGFPK	I
	TH-α2	KSCCRSTLGRNCYNLCRARG--AQKLCAGVCRCKISSGLSCPKGFPK	I
H. vulgare	TH-α	KSCCRSTLGRNCYNLCRVRG--AQKLCAGVCRCKLTSSGKCPTGFPK	I
	TH-DB4	KSCCKATLARNCYNTCHFAGG-SRPVCAGACRCKIISGPKCPSDYPK	II
	TH-DC4	KSCCKDTLARNCYNTCRFAGG-SRPVCAGACRCKIISGPKCPSDYPK	II
	TH-DG3	KSCCKNTTGRNCYNACRFAGG-SRPVCATACGCKIISGPTCPRDYPK	II
Pyrularia pubera	TH-PY	KSCCRNTWARNCYNVCRLPGTISREICAKKCRCKIISGTTCPSDYPK	II
Phoradendron tomentosum	P-A	KSCCPTTTARNIYNTCRFGGG-SRPVCAKLSGCKIISGTKCDSGWNH	III
	P-B	KSCCPTTTARNIYNYCRFGGG-SRPICAKLSGCKIISGTKCDSGWDH	III
Ph. liga	L-A	KSCCPSTTARNITNTCRLTHT-SRPTCASLSGCKIISGSTCDSGWNH	III
Dendrophora clavata	D-B	KSCCPTTAARNQYNICRLPGT-PRPVCAALSGCKIISGTGCPPGYRH	III
Viscum album	V-1PS	KSCCPBTTGRBIYBTCRFGGG-SRZVCARISGCKIISASTCPS-YPDK	III
	V-A2	KSCCPNTTGRNIYNTCRFGGG-SREVCASLSGCKIISASTCPS-YPDK	III
	V-B	KSCCPNTTGRNIYNTCRLGGG-SRERCASLSGCKIISASTCPS-YPDK	III
	V-A3	KSCCPNTTGRNIYNACRLTGA-PRPTCAKLSGCKIISGSTCPS-YPDK	III
Arabidopsis thaliana	THi2.2	KICCPTKDDRSVYFVCMLSVS-SQFYCLLKSKCKNTSQTICPPGYTN	III
	THi2.1	KICCPSNQARNGYSVCRIRFS--KGRCMQVSGCQNSSD-TCPRGWVN	III
Crambe abyssinica	CR-1	TTCCPSIVARSNFNVCRLPGT-PEALCATYTGCIIIPGATCPGDYAN	IV
	CR-2	TTCCPSIVARSNFNVCRLPGT-SEAICATYTGCIIIPGATCPGDYAN	IV
T. aestivum	TH-V1	VDCGANPFKVACFNSCLLGPS-TVFQCADFCACRLPAG---------	V

Fig. 2. Amino acid sequence alignment of representative examples of thionin types.

1.4 Solubility

Because of their solubility in petroleum-ether, thionins from wheat endosperm were first thought to be lipoproteins in which the protein moiety was associated with a lecithin-like lipoid (Balls *et al.*, 1942). The protein could be precipitated from the lipid extract as a hydrochloride that was no longer soluble in organic solvents and was soluble in water and aqueous alcohols. Lipid-thionin complexes were converted to a petroleum-ether-insoluble, chloroform-soluble form by precipitation with acetone. Digalactosyl diglyceride (DGDG), a component of the acetone extract, was able to restore petroleum-ether solubility when added back to the chloroform-soluble form (Hernandez-Lucas *et al.*, 1977). The chloroform-soluble thionin preparation obtained by acetone precipitation contained phosphatidylethanolamine, phosphatidylcholine, and other polar lipids. The apoprotein was not soluble in chloroform by itself, so it was assumed that at least some of these lipids were required for chloroform solubility (Hernandez-Lucas *et al.*, 1977). When flour from the tetraploid wheat *T. turgidum* was extracted with petroleum ether supplemented with DGDG, the yield of thionin was increased significantly, suggesting that the differences in yield between tetraploid and hexaploid wheats reported previously were probably due to interspecific differences in the levels of DGDG (García-Olmedo *et al.* 1968; Hernandez-Lucas *et al.*, 1977). Petroleum ether treatment of rye flour had repeatedly failed to extract any thionin, but when the solvent was complemented with acetone-extracted wheat lipids, a good yield of rye thionin was obtained (Hernandez-Lucas *et al.*, 1978). The idea that DGDG was limiting the yield of thionin extracted with petroleum ether was also consistent with the fact that extraction with aqueous solvents, such as 0.05N H_2SO_4 or 1M NaCl, was more efficient than with the organic solvents (Fernandez de Caleya *et al.*, 1976).

1.5 Biosynthesis and subcellular location.

Barley endosperm thionins are synthesized by membrane-bound polysomes as much larger precursors that undergo at least two pocessing steps (Ponz *et al.*, 1983). Using monospecific antibodies raised against the mature protein, two types of precursors were identified: one was detected as an *in vitro* translation product that could not be detected *in vivo*, and the other was detected by *in vivo* labelling (Fig. 3). Pulse-labelling experiments showed conversion of the second precursor into the mature protein (Ponz *et al.* 1983). As predicted from the study of their biosynthesis, the nucleotide sequences of the cDNAs corresponding to α and β thionins from barley endosperm were found to encode precursors that were much larger than the mature protein

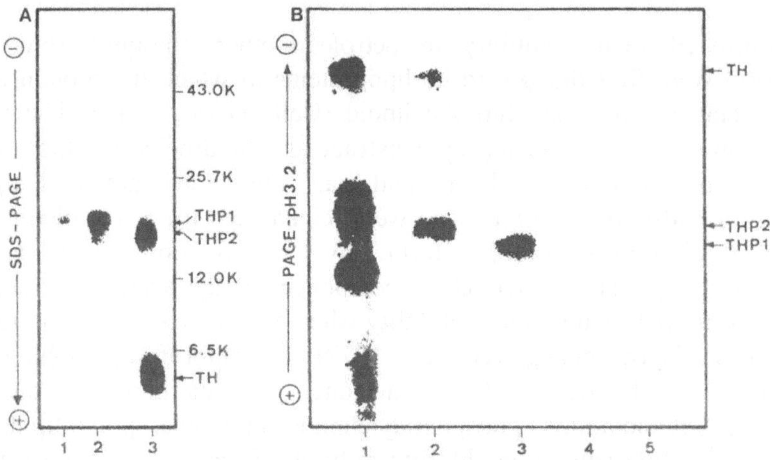

Fig. 3. Thionin biosynthesis. Comparison of *in vivo* and *in vitro* products selected with monospecific antibodies. **A**, Sodium dodecyl sulphate polyacrylamide electrophoresis (SDS-PAGE) of alkylated products: 1, *in vitro* precursor THP1 labelled with [^{35}S]methionine; 2, as in 1 but labelled with [^{35}S]cysteine; 3, *in vivo* precursor (THP2) and mature thionin (TH). **B**, Polyacrylamide gel electrophoresis (PAGE) at pH 3.2 of reduced, non-alkylated products: 1, total *in vivo* extract; 2, *in vivo* products, THP2 and TH; 3, *in vitro* precursor THP1; 4,5 displacement of precursors by non-radioactive thionin (5 μg). (From Ponz *et al.*, 1983, with permission)

(Ponz *et al.*, 1986; Hernandez-Lucas *et al.*, 1986). The deduced structures of these precursors consisted of an *N*-terminal signal peptide, followed by the mature protein and a *C*-terminal acidic protein, as shown in Fig. 4. The same precursor structure was later found for thionins of the other types (Gausing, 1987; Bohlmann and Apel, 1987; Castagnaro *et al.*, 1992; Schrader-Fischer and Apel, 1993, 1994), which strongly suggests that all types of thionins have the same biosynthetic pathway as that of type I thionins.

Cellular fractionation studies of developing barley endosperm by Ponz *et al.* (1983) indicated that type I thionins were intracellular and in a labile association with the particulate fraction. Subsequent studies, using immunogold detection by electron microscopy, have shown that type I thionins are in the periphery of the protein bodies in electron-dense ovoidal structures (Carmona *et al.*, 1993a).

A recent report (Romero *et al.*, 1997) has shown that in barley leaves, mature thionins accumulated in the vacuoles, while the acidic peptide was not detected in any cell fraction. Both purified vacuoles and an acid (pH 5.5) extract from leaves liberated the acidic peptide. An *Mr* 70,000 proteinase that effected this cleavage was purified from the acid extract. Processing by both

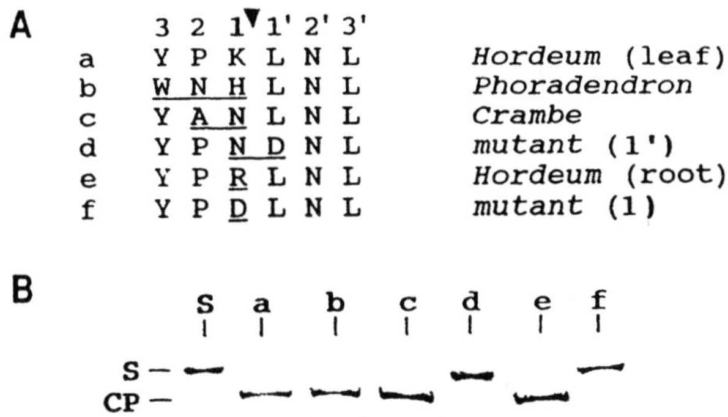

Fig. 4. Structure of the thionin precursor (Ponz *et al.*, 1986) and of the α-thionin gene from barley endosperm. (From Rodriguez-Palenzuela *et al.*, 1988, with permission).

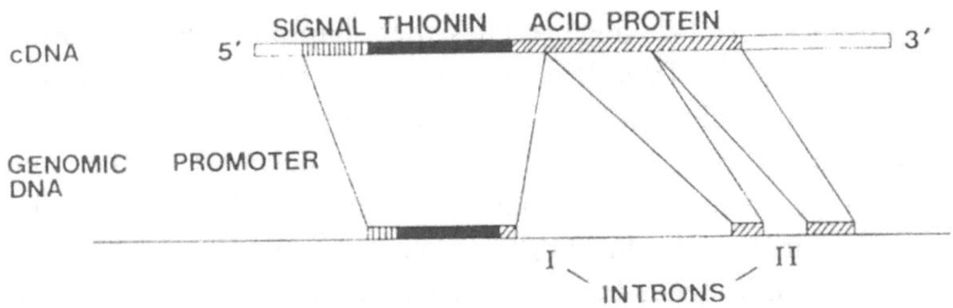

Fig. 5. Processing of cleavage-site sequences from thionin precursors. **A**, amino acid sequences near the cleavage site (arrowhead) of the indicated thionin precursors and mutants. A complete precursor coding sequence was expressed in *Escherichia coli* as a fusion with the MalE protein. **B**, SDS-PAGE of the proteins listed in **A** with the purified processing proteinase from barley. (From Romero *et al.*, 1997; with permission)

lysed vacuoles and by the purified proteinase was inhibited by Zn^{2+} and by Cu^{2+}, but not by inhibitors of previously described vacuolar processing thiol or aspartic proteinases. *In vivo* processing of the thionin precursor in leaf sections was also inhibited by Zn^{2+} and Cu^{2+}. Variants of a fusion protein with altered processing sites that represented those of thionin precursors from different taxa were readily processed by the proteinase, whereas changing the polarity of either the *C*-terminal or *N*-terminal residues of the processing site prevented cleavage by the enzyme (Fig. 5).

2. GENETICS

2.1 Chromosomal locations

Using wheat aneuploids (nulli-tetrasomic and ditelosomic lines), as well as wheat-rye and wheat-barley chromosomal addition lines, genes encoding thionins have been associated with specific chromosomes of these species (García-Olmedo *et al.*, 1982). Three genes (*Pur-A1*, *Pur-B1* and *Pur-D1*), which corresponded to the β, α1, and α2 thionin variants from wheat endosperm (type I), respectively, were identified in the long arms of chromosomes 1A, 1B and 1D, through the electrophoretic analysis of the appropriate aneuploids and the characterization of the isolated proteins (Fernandez de Caleya *et al.*, 1976). In a similar manner, a gene for an endosperm thionin was located in the long arm of chromosome 1R of rye (Sanchez-Monge *et al.*, 1979). Southern-blot analyses of genomic DNAs of wheat and barley, using cDNA probes for type I thionins, were consistent with the presence of 1-2 gene copies per haploid genome (Rodriguez-Palenzuela *et al.*, 1988). Type II genes have been located in chromosome 6H of barley by Southern-blot analysis of DNAs from wheat-barley addition lines (Bohlmann *et al.*, 1988); but there are discrepancies as to the number of copies of this type present: while Bohlmann *et al.* (1988) estimated about 100 genes/haploid genome, Gausing (1987) gave a lower estimate of 9-11 genes. More recently, type V genes have been located within a few kb of the type I genes in wheat, with 1 or 2 copies per haploid genome. Although the amino acid sequence of the mature type V thionin is quite different from those of the other types, the *C*-terminal acidic peptide of the corresponding precursor is less divergent than the mature protein and closer to the type I than to the type II peptide (Castagnaro *et al.*, 1992, 1995).

The gene for α-hordothionin, a type I thionin from barley endosperm, has two introns, 420 and 91 nucleotides long, that interrupt the sequence encoding the *C*-terminal, acidic peptide of the precursor (Rodriguez-Palenzuela *et al.*, 1988)(Fig. 3). Genomic clones of type II thionins have two introns in similar positions as those of type I clones (Bohlmann *et al.*, 1988).

2.2 Gene expression

In barley endosperm, a maximum steady state concentration of thionin mRNA occurred between 13 and 16 days after anthesis. Thus, synthesis of these proteins seemed to take place during the cell-proliferation phase of endosperm development and to cease at the beginning of the cell-enlargement phase. When a fusion of the cauliflower mosaic virus 35S RNA gene promoter with the α-hordothionin gene (comprising coding regions and introns) was expressed in transgenic tobacco, introns from this monocot gene were properly spliced in the dicot species (Carmona *et al.*, 1993b).

The expression of type-II thionin genes has been investigated in barley leaves and a number of interesting responses of these genes to external stimuli have been described. Large amounts of messenger for type II thionins were detected in dark-grown barley seedlings (Gausing, 1987; Bohlmann and Apel, 1987). Steady state messenger levels seemed to be higher in the lower 1/3 of the leaf (younger cells) than in the upper 2/3 (older cells) and to decline sharply upon illumination (Gausing, 1987). The effect of light has been further investigated by Reimann-Philipp *et al.* (1989a), who have postulated the mediation of two photoreceptors, phytochrome and a blue-light-absorbing photoreceptor. Synthesis of thionins concomitantly ceased upon illumination, but the previously accumulated thionin was rather stable (Reimann-Philipp *et al.*, 1989a). The inhibitory effect of light can be overcome by stress- and pathogen-induced signals, as it has been shown that fungal infection induces a transient expression of the thionin genes in the leaves (Bohlmann *et al.*, 1988); Ebrahim-Nesbat *et al.*, 1989) and that the chlorides of divalent cations (Mg^{2+}, Mn^{2+}, Cd^{2+}, Zn^{2+}) elicit a more permanent response (Fisher *et al.*, 1989).

3. BIOLOGICAL PROPERTIES

3.1 Redox properties

Thionin from wheat endosperm can substitute for thioredoxin *f* from spinach chloroplasts in the dithiothreitol-linked activation of chloroplast fructose-1,6-bisphosphatase (Wada and Buchanan, 1981). Under the standard assay conditions, the thionin was only 2% as active as authentic thioredoxin *f*. Nevertheless, activity could be improved by increasing the time of preincubation and the concentration of reductant, suggesting that the thionin could be effectively reduced by thioredoxin *f* (Wada and Buchanan, 1981). This led to experiments which implicate thionins in plant redox metabolism. Johnson *et al.* (1987) have reported a thioredoxin system, consisting of a homogeneous preparation of thioredoxin *h* and partially purified thioredoxin reductase (NADPH), which effectively reduced thionin with NADPH as the hydrogen donor. The reduced thionin, in turn, was capable of activating fructose-1,6-bisphosphatase, suggesting a possible role of thionins as secondary thiol messengers in the redox regulation of enzymes. In the opinion of these authors, the redox properties of thionins could also explain their toxicity.

A property of thionins that might be related to its redox properties is their ability to form selective disulphide bridges with other proteins (Diaz *et al*, 1992; Piñeiro *et al*., 1995). The enzymes β-glucuronidase and neomycin phosphotransferase II were inhibited by thionins through the formation of disulphide-linked adducts and the inhibition was reversed by DTT.

3.2 Interactions with artificial and biological membranes

The ability of thionins to induce leakage of intracellular material was first demonstrated in bacteria and in yeast. The effect could be reversed by certain divalent cations, such as Ca^{2+}, Zn^{2+}, or Fe^{2+} (Reviewed in García-Olmedo *et al*., 1989). The effects of thionins of types I and IV on cultured mammalian cells occurred at the minimum cytotoxic concentration that caused leakage of Rb^{1+} and of uridine. Concentrations of thionins that had no detectable effects on the cultured cells lead to inhibition of translation by antibiotics such as hygromycin B that do not normally cross the plasma membrane (Carrasco *et al*., 1981). As in the case of yeast, Ca^{2+} and Mg^{2+} could revert the action of thionin. The effect of thionins on fungal membranes has been investigated in *Neurospora crassa*, where a similarity in the minimum concentrations required to cause leakage and growth inhibition was observed (Guihard *et al*., 1993; Thevissen *et al*., 1996).

The effects of thionins on smooth-muscle contraction and on insect flight muscle, as well as the sensitivity to thionins of A31 cells infected with the Moloney strain of murine leukemia virus, are all probably related to interactions of thionins with the cell membrane (reviewed by García-Olmedo *et al*., 1989). It has been recently reported that thionins induce leakage and aggregation of artifitial, negatively-charged membranes under conditions in which other plant toxic peptides have no effect (Caaveiro *et al*., 1997).

3.3 Antimicrobial properties and other *in vitro* activities

The toxicity of thionins to different kinds of organisms and to cells in culture has been investigated for several decades. Gram-positive bacteria and, to a lesser extent, Gram-negative bacteria, bakers yeast, and some human pathogenic fungi, were found to be sensitive to a crystalline mixture of type I α-thionins from wheat endosperm, whereas the mycelial fungi tested were found to be insensitive (Stuart and Harris, 1942). After these initial findings, the toxicity to bacteria (Fernandez de Caleya *et al*., 1972; Cammue *et al*., 1992; Florack *et al*., 1993; Molina *et al*., 1993), to yeast (Balls and Harris, 1944; Nose and Ichikawa, 1968; Okada and Yoshizumi, 1970, 1973; Hernandez-Lucas *et al*., 1974), and to fungi (Bohlmann *et al*., 1988; Reimann-Philipp *et al*., 1989b; Terras *et al*., 1992; Molina *et al*. 1993) has been further demonstrated for different thionins. The antifungal activity of thionins is inhibited by Ca^{2+} at concentrations that other divalent and monovalents cations had no effect (Okada and Yoshizumi, 1970; Terras *et al.*, 1992; Cammue *et al.*, 1992).

Type-I thionins were also found to be toxic to mice, guinea-pigs and rabbits when injected intravenously or intra-peritoneally, but not upon oral administration (Coulson *et al.*, 1942). Type-III thionins, isolated from the leaves of the mistletoes and related species, were also found to be toxic on parenteral administration to mice and cats (see Samuelsson, 1974). At sublethal doses they produced hypotension, bradycardia and a negative ionotropic effect on the heart muscle. Intraarterial administration, in higher doses, produced vasoconstriction in arteries of skin and skeletal muscle (Samuelsson, 1974). Cytotoxic effects on cultured mammalian cells have been reported for different thionin types (Carrasco *et al.*, 1981; Vernon *et al.*, 1985).

3.4 Possible role in plant defence

The hypothesis that thionins might play a role in the protection of plants against pathogens was proposed by Fernandez de Caleya *et al.* (1972), who investigated the susceptibility to wheat endosperm thionins of phytopathogenic bacteria in the genera *Pseudomonas, Xanthomonas, Agrobacterium, Erwinia*, and *Corynebacterium*. Purified genetic variants of these thionins differed in activity and showed some degree of specificity. Bohlmann *et al.* (1988) have shown that both endosperm (type I) and leaf (type II) thionins from barley inhibit the fungi *Thielaviopsis paradoxa*, a pathogen of sugar cane, and *Drechslera teres*, a pathogen of barley, at concentrations of 5 x 10^{-4}M. Fungal and bacterial pathogens included in a recent survey were inhibited by thionins at concentrations in the 10^{-6}-10^{-5}M range, which are similar to those found in certain plant tissues (Molina *et al.* 1993).

Direct evidence of a defense role for the thionins is lacking at present. Although thionin mRNA is transiently induced in barley upon infection with *Erysiphe graminis* (Bohlmann *et al.*, 1988) and slight differences in the localization of thionins seems to occcur in the cell walls of susceptible and resistant barley cultivars (Ebrahim-Nesbat *et al.*, 1989), the resistance gene and the thionin genes are located in different chromosomes and the thionin mRNA is induced to similar levels in both susceptible and resistant cultivars. Furthermore, the pre-induction levels of thionins seem to be quite high, due to their low turn-over (Reimann-Philipp *et al.*, 1989a). Transgenic expression in tobacco plants of a barley thionin gene showed reduced lession size when the plants were challenged with two strains of *Pseudomonas syringae* (Carmona *et al.*, 1993b), whereas other strains did not seem to be affected (Florack, 1993). More recently, overexpression of an endogenous thionin has been reported to enhance resistance of *Arabidopsis thaliana* against *Fusarium oxysporum* (Epple *et al.*, 1997) and thionin-sensitive mutants of *P. solanacearum* were found to be avirulent (Titarenko *et al.*, 1997).

4. 3D STRUCTURE

The thionins are among the best characterized proteins with respect to their three-dimensional structure, both in crystals and in solution. Because of their peculiar features they have become model molecules in the development and refinement of novel methods for the elucidation of three-dimensional structures. Thus, an X-ray diffraction method based on the anomalous scattering of sulphur was specifically developed to solve the structure of crambin (Hendrickson and Teeter, 1981). This molecule was also used to test the utility of molecular dynamics with interproton distance restraints for structure determination (Brünger *et al.*, 1987; Clore *et al.* , 1986), and to show that structures obtained for a protein in solution with NMR data can be used to solve the crystal structure of the same protein by molecular replacement (Brünger *et al.*, 1987). This wealth of information indicates that type I, III, and IV thionins, in spite of their extreme divergence, have essentially the same three-dimensional shape, which resembles the Greek capital letter gamma (Γ). The molecules are amphipathic, quite rigid and present very similar three-dimensional structures in solution and in crystal form (Fig. 6). The long arm is formed by two antiparallel α-helices and the short arm by a β-sheet consisting of two short antiparallel β-strands. The hydrophobic residues are clustered at the outer surface of the long arm of theΓ, whereas hydrophilic residues mainly occur at the inner surface of it and at the outer surface of the corner of the Γ (Clore *et al.*, 1986; Hendrickson and Teeter, 1981). It has been proposed that type I thionins have a binding site for phospholipids, which may be implicated in their toxic activity (Rao *et al.*, 1995; Stec *et al.*, 1995).

5. EVOLUTION

The evolution of thionins, particularly that leading to changes in the numbers of disulphide bridges, merits special attention because it represents a general evolutionary problem. Thus, divergence between type I and V thionins has occurred through a putative process of accelerated evolution that has affected the amino acid sequence of the mature thionin but not the precursor

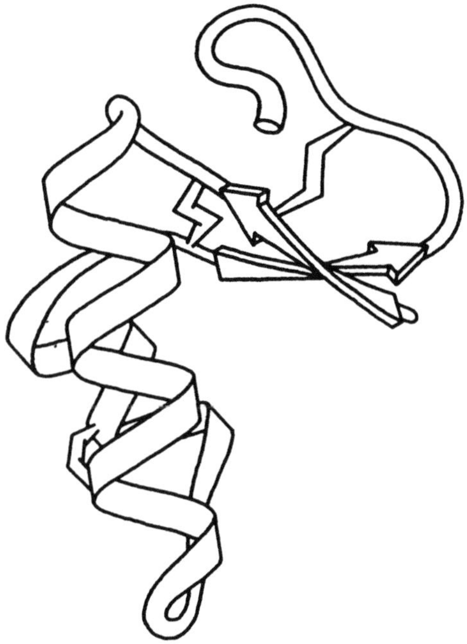

Fig. 6. Structure of crambin (Whitlow and Teeter, 1985)

domains corresponding to the *N*-terminal signal peptide and the long *C*-terminal acidic peptide (Castagnaro *et al*., 1992, 1995). This process involved a deletion and a non-synonymous nucleotide substitution rate equal to the synonymous rate in the thionin sequence.

Percentages of sequence divergence for binary comparisons of equivalent domains within and between types I and V have been calculated (Castagnaro *et al*., 1992, 1995). The coding sequences of two type-V thionins have been absolutely conserved during the approximately 10,000 years that the D genomes of the diploid and the alloploid *Triticum* and *Aegilops* species have been evolving separately, whereas the introns have diverged, especially the larger one, which has suffered one major and several minor deletions. Percentages of divergence between type I and V thionins in the mature-protein domain (59-69%) are about twice as great than those occurring in the other domains, namely signal peptide, *C*-terminal acidic peptide, and introns (21-36%). In contrast, divergence in the mature protein domain of type V thionins is equal to or lower than within this domain of type I thionins, and is certainly much lower than in the corresponding introns.

Type I thionins have four disulphide bridges, whereas type V thionins have only three. It is possible that a temporary loss of function due to mutation of one cysteine (gain or loss) in the duplicated gene might have resulted in a period of accelerated evolution. Mutation of a second cysteine (loss or gain) would have then led to a mature thionin with an even number of cysteines (a common feature of all known thionins) and to a recovery of function that would in turn impose a slower rate of evolution. The extreme divergence in the proportion of charged residues suggests that the functions of the two types, although similar, might be exerted in different contexts or in different cellular environments.

ACKOWLEDGEMENTS

I thank L. Lamoneda for technical assistance and Pilar Carbonero for critically reading the manuscript. Financial support by Comisión Interministerial de Ciencia y Tecnologia PB92-0325 is greatfully acknowledge.

REFERENCES

Balls, A.K., Hale, W.S. and Harris, T.H. (1942) A crystalline protein obtained from a lipoprotein of wheat flour. *Cereal Chemistry* **19**:279-288.

Balls, A.K. and Harris, T.H. (1944) The inhibitory effect of a protamine from wheat flour on the fermentation of wheat mashes. *Cereal Chemistry* **21**:74-79.

Bekes, F. and Lasztity, R. (1981) Isolation and determination of amino acid sequence of avenothionin, a new purothionin analogue from oat. *Cereal Chemistry* **58**: 360-361.

Bohlmann, H. and Apel, K. (1987) Isolation and characterization of cDNAs coding for leaf-specific thionins closely related to the endosperm-specific hordothionin of barley (*Hordeum vulgare* L.). *Molecular and General Genetics* **207**: 446-454.

Bohlmann, H. Clausen, S., Behnke, S., Giese, H., Hiller, C., Reimann-Philipp, U., Schrader, G., Barkholt, V. and Apel, K. (1988) Leaf-specific thionins of barley - a novel class of cell wall proteins toxic to plant-pathogenic fungi and possibly involved in the defense mechanisms of plants. *EMBO Journal* **7**: 1559-1565.

Brünger, A.T., Campbell, R.L., Clore, G.M., Gronenborn, A.G., Karplus, M., Petsko, G.A. and Teeter, M.M. (1987) Solution of a protein crystal structure with a model obtained from NMR interproton distance restraints. *Science* **235**: 1049-1053.

Caaveiro, J.M.M., Molina, A., Gonzalez-Mañas, J.M., Rodriguez-Palenzuela, P., García-Olmedo, F. and Goñi, F.M. (1997) Differential effects of five types of antipathogenic plant peptides on model membranes. *FEBS Letters*. **410**: 338-342.

Cammue, B.P.A., De Bolle, M.F.C., Terras, F.R.G., Proost, P., Van Damme, J., Rees, S.B., Vanderleyden, J. and Broekaert, W.F. (1992) Isolation and characterization of a novel class of plant antimicrobial peptides from *Mirabilis jalapa* L. seeds. *Journal of Biological Chemistry* **267**: 2228-2233.

Carbonero, P. and García-Olmedo, F. (1969) Purothionin in *Aegilops-Triticum* spp. *Experientia* **25**: 1110.

Carmona, M.J., Hernandez-Lucas, C., San Martín, C., Gonzalez, P. and García-Olmedo, F. (1993a) Subcellular localization of type I thionins in the endosperms of wheat and barley. *Protoplasma* **173**: 1-7.

Carmona, M.J., Molina, A., Fernandez, J.A., Lopez-Fando, J.J. and García-Olmedo, F. (1993b) Expression of the α-thionin gene from barley in tobacco confers enhanced resistance to bacterial pathogens. *Plant Journal* **3**: 457-462.

Carrasco, L., Vazquez, D., Hernandez-Lucas, C., Carbonero, P. and García-Olmedo, F. (1981) Thionins: plant peptides that modify membrane permeability in cultured mammalian cells. *European Journal of Biochemistry* **116**: 185-189.

Castagnaro, A., Maraña, C., Carbonero, P. and García-Olmedo, F. (1992) Extreme divergence of a novel wheat thionin generated by a mutational burst specifically affecting the mature protein domain of the precursor. *Journal of Molecular Biology* **224**: 1003-1009.

Castagnaro, A., Segura, A. and García-Olmedo, F. (1995) High conservation among sequences encoding type-V thionins in wheat and *Aegilops*. *Plant Physiology* **107**: 1475-1476.

Clore, G.M., Nilges, M., Sukumaran, D.K., Brünger, A.T., Karplus, M. and Gronenborn, A.M. (1986) The three-dimensional structure of α-1-purothionin in solution: combined use of nuclear magnetic resonance, distance geometry and restrained molecular dynamics. *EMBO Jorunal* **5**: 2729-2735.

Coulson, E.J., Harris, T.H. and Axelrod, B. (1942) Effect on small laboratory animals of the injection of the crystalline hydrochloride of a sulfur protein from wheat flour. *Cereal Chemistry* **19**:301-307.

Diaz, I., Carmona, M.J. and García-Olmedo, F. (1992) Effects of thionins on β-glucuronidase in vitro and in plant protoplasts. *FEBS Letters* **296**: 279-282.

Ebrahim-Nesbat, F., Behnke, S., Kleinhofs, A. and Apel, K. (1989) Cultivar-related differences in the distribution of cell-wall-bound thionins in compatible and incompatible interactions between barley and powdery mildew. *Planta* **179**: 203-210.

Epple, P., Apel, K., Böhlmann, H. (1995) An *Arabidopsis thaliana* thionin gene is inducible via a signal transduction pathway different from that for pathogenesis-related proteins. *Plant Physiology* **109**: 813-820.

Epple, P., Apel, K., Bohlmann, H. (1997) Over expression of an endogenous thionin gives enhanced resistance of *Arabidopsis thaliana* against *Fusarium oxysporium*. *Plant Cell* **9**: 509-520.

Fernandez de Caleya, R., Gonzalez-Pascual, B., García-Olmedo, F. and Carbonero, P. (1972) Susceptibility of phytopathogenic bacteria to wheat purothionins *in vitro*. *Applied Microbiology* **23**: 998-1000.

Fernandez de Caleya, R., Hernandez-Lucas, C., Carbonero, P. and García-Olmedo, F. (1976) Gene expression in alloploid: genetic control of lipopurothionins in wheat. *Genetics* **83**: 687-699.

Fisher, N., Redman, G.G. and Elton, G.A.H. (1968) Fractionation and characterization of purothionin. *Cereal Chemistry* **45**:48-57.

Fisher, R., Behnke, S. and Apel, K. (1989) The effect of chemical stress on the polypeptide composition of the intracellular fluid of barley leaves. *Planta* **178**: 61-68.

Florack, D.E.A., Visser, B. De Vries, P.M., Van Vuurde, J.W.L. and Stiekema, W.J. (1993) Analysis of the toxicity of purothionins and hordothionins for plant pathogenic bacteria. *Netherdands Journal of Plant Pathology* **99**: 259-268.

García-Olmedo. F., Rodriguez-Palenzuela, P., Hernandez-Lucas, C., Ponz, F., Maraña, C., Carmona, M.J., Lopez-Fando, J.J., Fernandez, J.A. and Carbonero P. (1989) The thionins: a protein family that includes purothionins, viscotoxins and crambin. In: Oxford Surveys of Plant Molecular and Cell Biology 6 (B.J. Miflin, ed.) pp. 31-60.

García-Olmedo, F., Sotelo, I. and García-Faure, R. (1968) Identificación de productos de *Triticum aestivum* en las pastas alimenticias. IV. Lipoproteínas solubles en éter de petróleo. *Anales del Instituto Nacional de Investigaciones Agronómicas* **17**: 433-443.

García-Olmedo, F., Carbonero, P. and Jones, B.L. (1982) Chromosomal locations of genes that control wheat endosperm proteins. In: Advances in Cereal Science and Technology 5 (Y. Pomeranz, ed) pp. 1-47.

Gausing, K. (1987) Thionin genes specifically expressed in barley leaves. *Planta* **171**: 241-246.

Guihard, G., Bénédetti, H., Besnard, M. and Letellier, L. (1993) Phosphate effux through the channels formed by colicins and phage T5 in *Escherichia coli* cells is responsible for the fall in cytoplasmic ATP. *Journal of Biological Chemistry* **268**: 17775-17780.

Hendrickson, W.A. and Teeter, A.M. (1981) Structure of the hydrophobic protein crambin determined directly from the anomalous scattering of sulphur. *Nature* **290**: 107-113.

Hernandez-Lucas, C., Carbonero, P. and García-Olmedo, F. (1978) Identification and purification of a purothionin homologue from rye (*Secale cereale* L.). *Journal of Agricultural and Food Chemistry* **26**: 794-796.

Hernandez-Lucas, C., Fernandez de Caleya, R. and Carbonero, P. (1974) Inhibition of brewer's yeasts by wheat purothionins. *Applied Microbiology* **28**: 165-168.

Hernandez-Lucas, C., Fernandez de Caleya, R., Carbonero, P. and García-Olmedo, F. (1977). Reconstitution of petroleum ether soluble wheat lipopurothionin by binding of digalactosyl diglyceride to the chloroform-soluble form. *Journal of Agricultural and Food Chemistry* **25**: 1287-1289.

Hernandez-Lucas, C., Royo, J., Paz-Ares, J., Ponz, F., García-Olmedo, F. and Carbonero, P. (1986) Polyadenylation site heterogeneity in mRNA encoding the precursor of the barley toxin α-hordothionin. *FEBS Letters* **200**: 103-105.

Johnson, T.C., Wada, K., Buchanan, B.B. and Holmgren, A. (1987) Reduction of purothionin by the wheat seed thioredoxin system. *Plant Physiology* **85**: 446-451.

Molina, A., Goy, P.A., Fraile, A., Sanchez-Monge, R. and García-Olmedo, F. (1993) Inhibition of bacterial and fungal plant pathogens by thionins of types I and II. *Plant Science* **92**: 169-177.

Nimmo, C.C., O'Sullivan, M.T. and Bernardin, J.E. (1968) The relation of a "globulin" component of wheat flour to purothionin. *Cereal Chemistry* **45**:28-36

Nose, Y. and Ichikawa, M. (1968) Studies on the effects of flour extract on baker's yeast. *Journal of Fermentation and Technology* **46**: 915-925.

Okada, T. and Yoshizumi, H. (1970) A lethal toxic substance for brewing yeast in wheat and barley. II. Isolation and some properties of toxic principle. *Agricultural and Biological Chemistry* **34**: 1089-1094.

Okada, T. and Yoshizumi, H. (1973) The mode of action of toxic protein in wheat and barley on brewing yeast. *Agricultural and Biological Chemistry* **37**: 2289-2294.

Piñeiro, M., Diaz, I., Rodriguez-Palenzuela, P., Titarenko, E. and García-Olmedo, F. (1995) Selective disulphide linkage of plant thionins with other proteins. *FEBS Letters* **369**: 239-242.

Ponz, F., Paz-Ares, J., Hernandez-Lucas, C., Carbonero, P. and García-Olmedo, F. (1983) Synthesis and processing of thionin precursors in developing endosperm from barley (*Hordeum vulgare* L.). *EMBO Journal* **2**: 1035-1040.

Ponz, F., Paz-Ares, J., Hernandez-Lucas, C., García-Olmedo, F. and Carbonero, P. (1986) Cloning and nucleotide sequence of a cDNA encoding the precursor of the barley toxin α-hordothionin. *European Journal of Biochemistry* **156**: 131-135.

Rao, U., Stec, B. and Teeter, M.M. (1995) Refinedment of purothionins reveals solute particles important for lattice formation and toxicity. I. α1-purothionin revisited. *Acta Crystallographica D.* **51**: 904-913.

Redman, D.G. and Fisher, N. (1969) Purothionin analogues from barley flour. *Journal of the Science of Food and Agriculture* **20**: 427-432.

Reimann-Philipp, U., Behnke, S., Batschauer, A., Schafer, E. and Apel, K. (1989a) The effect of light on the biosynthesis of leaf-specific thionins in barley, *Hordeum vulgare*. *European Journal of Biochemistry* **182**: 283-289.

Reimann-Philipp, U., Schrader, G., Martionoia, E., Barksholt, V. and Apel, K. (1989b) Intracellular thionins of barley. A second group of leaf thionins closely related to but distinct from cell wall-bound thionins. *Journal of Biological Chemistry* **264**: 8978-8984.

Rodriguez-Palenzuela, P., Pintor-Toro, J.A., Carbonero, P. and García-Olmedo, F. (1988) Nucleotide sequence and endosperm-specific expression of the structural gene for the toxin α-hordothionin in barley (*Hordeum vulgare* L.). *Gene* **70**: 271-281.

Romero, A., Alamillo, J.M. and García-Olmedo, F. (1997) Processing of thionin precursors in barley leaves by a vacuolar proteinase. *European Journal of Biochemistry* **243**: 202-208.

Samuelsson, G. (1974) Mistletoe toxins. *Systematic Zoology* **22**: 566-569.

Schrader-Fischer, G. and Apel, K. (1993) cDNA-derived identification of a novel thionin precursor in *Viscum album* that contain highly divergent thionin domains but conserved signal and acidic polypeptide domains. *Plant Molecular Biology* **23**: 1233-1242.

Schrader-Fischer, G. and Apel, K. (1994) Organ-specific expression of highly divergent thionin variants that are distinct from the seed-specific crambin in the crucifer *Crambe abyssinica*. *Molecular and General Genetics* **245**: 380-389.

Sanchez-Monge, R., Delibes, A., Hernandez-Lucas, C., Carbonero, P. and García-Olmedo, F. (1979) Homoeologous chromosomal location of the genes encoding thionins in wheat and rye. *Theoretical and Applied Genetics* **54**: 61-63.

Stec, B., Rao, U. and Teeter, M.M. (1995) Refinement of purothionins reveals solute particles important for lattice formation and toxicity. II. structure of β-purothionin et 1.7 Å resolution. *Acta Crystallographica D.* **51**: 914-924.

Stuart, L.S. and Harris, T.H. (1942) Bactericidal and fungicidal properties of a crystalline protein isolated from unbleached wheat flour. *Cereal Chemistry* **19**:288-300.

Teeter, M.M., Mazer, J.A. and L'Italien, J.J. (1981) Primary structure of the hydrophobic plant protein crambin. *Biochemistry* **20**: 5437-5443.

Terras, F.R.G., Schoofs, H., De Bolle, M.F.C., Van Leuven, F., Rees, S.B., Vanderleyden, J., Cammue, B.P.A. and Broekaert, W.F. (1992) Analysis of two novel classes of plant antifungal proteins from radish (*Raphanus sativus* L.) seeds. *Journal of Biological Chemistry* **267**: 15301-15309.

Thevissen, K., Ghazi, A., De Samblanx, G.W., Brownlee, C., Osborn, R.W. and Broekaert, W.F. (1996) Fungal membrane responses induced by plant defensins and thionins. *Journal of Biological Chemistry* **271**: 15018-15025.

Titarenko, E., Lopea-Solanilla, E., García-Olmedo, F. and Rodriguez-Palenzuela, P. (1997) Mutants of *Ralstonia (Psudomonas) solanacearum* sensitive to antimicrobial peptides are altered in their LPS structure and are avirulent in tobacco. *Journal of Bacteriology* (in press).

Van Etten, C.H., Nielsen, H.C. and Peters, J.E. (1965) A crystalline polypeptide from the seed of *Crambe abyssinica*. *Phytochemistry* **4**:467-473.

Vermeulen, J.A.W.H., Lamerichs, R.M.J.N., Berliner, L.J., De Marco, A., Llinas, M., Boelens, R., Alleman, J. and Kaptein, R. (1987) ^{1}H-NMR characterization of two crambin species. *FEBS Letters* **219**: 426-430.

Vernon, L.P., Evett, G.E., Zeikus, R.D. and Gray, W.R. (1985) A toxic thionin from *Pyrularia pubera*: purification, properties, and amino acid sequence. *Archives of Biochemistry and Biophysics* **238**: 18-29.
Wada, K. and Buchanan, B.B. (1981) Purothionin: a seed protein with thioredoxin activity. *FEBS Letters* **124**: 237-240.
Whitlow, M. and Teeter, M.M. (1985) Energy minimization for tertiary structure prediction of homologous proteins: α-purothionin and viscotoxin A3 models from crambin. *Journal of Biochemical Structural Dynamics* **2**: 831-848.

31. Antifungal Proteins

RUPERT W. OSBORN[1] and WILLEM F. BROEKAERT[2]
[1]*ZENECA Agrochemicals, Jealott's Hill Research Station, Bracknell, Berkshire RG42 6ET, UK.* [2]*F.A. Janssens Laboratory of Genetics, Katholieke Universiteit Leuven, Kardinaal Mercierlaan 92, B-3001 Heverlee-Leuven, Belgium*

1. INTRODUCTION

Plants protect themselves from micro-organisms, including fungi, using a variety of pre-formed and inducible defensive mechanisms. Prevalent amongst the weapons used are antimicrobial (AMPs) or antifungal proteins (AFPs). In the latter case these can be loosely defined as proteins or peptides which directly interfere with the growth and spread of fungi that are present on or inside a plant tissue.

AFPs have been identified in all plant tissues studied but seeds have proven to be a particularly rich source of proteins or peptides with this activity. The most abundant AFPs are likely to act primarily as storage proteins and for some, e.g. 2S albumins and thionins, this role has already been highlighted in other chapters of this book. With other AFPs their localisation profile is not consistent with a storage protein role and it has been proposed that these proteins therefore must play an important role in protecting the developing seed or seedling against soil-borne fungi.

Seed proteins with antifungal activity have been identified primarily through the use of *in vitro* bioassays in which the growth of the fungus is compared in the presence and absence of the particular protein of interest. In some cases this has been undirected and proteins purified for another purpose have been subsequently shown to possess antifungal activity. More recently, the use of antifungal bioassays to direct the purification of proteins with this activity has lead to the identification of many examples of small cysteine-rich peptides which can be categorised into diverse classes based on their amino acid sequences. In addition to these two activity-directed routes, seed antifungal proteins have also been identified based on homology to known classes of pathogenesis-related (PR) proteins. PR-proteins were initially classified based solely upon their structural homology and were identified as proteins that were synthesised at higher rates in pathogen-challenged tissues,

P.R. Shewry and R. Casey (eds.), Seed Proteins, 727–751.

predominantly leaves. Several members of the well characterised PR-protein families have now been demonstrated to exhibit antifungal activity, to be present in seeds and, in many cases, to be induced upon germination.

This chapter aims to review the many different classes of antifungal proteins found in plant seeds and to give an overview of our current knowledge of their biological activities. For simplicity we have categorised the proteins into structural classes .

2. PR-TYPE PROTEINS

Perhaps the best insight into the biological role of antifungal proteins in seed development has been gained by studying physiological changes in AFPs during cereal caryopsis development. These studies have not only identified which members of the various PR-protein families aré present in the seed but have attempted to correlate their biological activity and expression profiles with protection of the seed from fungal attack.

PR-proteins have been classified into at least 5 groups based on their structural homology and several members of each of the well-characterised families have been shown to exert antimicrobial effects (van Loon *et al.*, 1994). Best studied are the hydrolytic chitinase and β-1,3-glucanase enzymes. In a number of studies it has been shown that various PR-type proteins, as well as other proteins with antifungal properties such as ribosome-inactivating proteins (RIPs), accumulate differentially during seed development and germination. In addition, as with the classical PR-proteins, the expression of some of these proteins in germinating seeds is increased following fungal infection.

2.1 Chitinase and β-1,3-glucanase

The first chitinase to be purified to homogeneity from plant seeds was a M_r 30,000 enzyme derived from wheat embryos (Molano *et al.*, 1979) and this particular enzyme was later shown to possess antifungal properties (Broekaert *et al.*, 1988). Following on from this work, other chitinases have now been isolated from a variety of plant seeds and in most cases these enzymes have been shown to be homologous to a family of PR-proteins called PR-3. A PR-3-like chitinase from barley endosperm has been crystallised and its 3-D structure determined by X-ray crystallography (Hart *et al.*, 1993). The structure of this enzyme is characterised by ten helical segments and by a long cleft running through the molecule which is thought to bind the chitin substrate.

β-1,3-glucanases are enzymes with M_r of around 30,000 which hydrolyse β-1,3 bonds in β-1,3 glucans and β-1,3-1,6 glucans. They are homologous to PR-2 proteins and many of these enzymes have been shown to exert antifungal activity, particularly when acting in concert with chitinases. Both chitinases and β-1,3-glucanases have been shown to bring about structural alterations to fungal cell walls and a causal relationship between these alterations and growth inhibitory effects has been suggested. Determination of the 3-D structure of a β-1,3-glucanase from barley seeds by X-ray crystallography has revealed that this enzyme adopts an α/β barrel structure with eight consecutive β-sheet/α-helix folds (Varghese *et al.*, 1994). The enzyme features a long cleft which is able to accommodate a linear stretch of eight β-1,3-linked glycosyl residues.

In developing pea pods Mauch and co-workers identified two chitinases and two β-1,3-glucanases which were differentially regulated (Mauch *et al.*, 1988). During fruit development and maturation the activities of both sets of enzyme increased dramatically in contrast to total protein levels which declined strongly. As a result, the basic protein comprising mainly these two enzymes accounted for about 2-5% of the total seed protein in mature pods. Chitinase and glucanase activities were also induced when immature pea pods were challenged by fungal infection or ethylene. However, the expression profiles of the two isoforms for each class of protein were very different under the three conditions. Chitinase Ch1 and β-1,3-glucanase G2 were strongly induced in immature pods in response to the various stresses, whereas chitinase Ch2 and β-1,3-glucanase G1 accumulated during the course of normal maturation.

A number of other studies have supported this observation that seeds contain different isoforms of enzymes like chitinase and β-1,3-glucanase and that these proteins may have different biological functions. In barley seeds five proteins with chitinase activity were identified and three purified to homogeneity and shown to inhibit fungal growth *in vitro* (Swegle *et al.*, 1992). Of the five proteins, two were found primarily in aleurone and endosperm tissues and the other three were enriched in the embryo. During the early period of barley seed imbibition, two of the embryo-associated chitinases are released into the surrounding aqueous medium. When germinating maize seeds were challenged with fungal pathogens three chitinase and one β-1,3-glucanase isoforms were co-ordinately induced (Cordero *et al.*, 1994). An additional β-1,3-glucanase isoform which is expressed in maize embryo and radicle tissues was also identified but, in contrast, its level of accumulation was not increased upon fungal infection. It may be, therefore, that these proteins play a role in seed germination which is not related to their antifungal activity. More recently, studies using promoter-reporter fusions in tobacco have provided direct evidence that a tobacco I β-1,3-glucanase plays a role in weakening cell walls and facilitating radicle emergence (Vogeli-Lange *et al.*, 1994, Leubner-Metzger *et al.*, 1995). The induction of this class I β-1,3-glucanase gene begins just prior to the onset of endosperm rupture and the induction is localised exclusively in the micropylar region of the endosperm. Addition of ABA to the medium is known to delay germination of tobacco seeds and this also inhibits the accumulation of β-1,3-glucanase (Leubner-Metzger *et al.*, 1995).

2.2 Permatins (PR-5 type)

The first permatin to be identified in plant seed was purified from maize and called zeamatin. This basic protein of M_r 22,000 strongly inhibits the growth of a number of fungi and is stored in high amounts in mature seeds (Roberts and Selitrennikoff, 1990). Subsequently, a number of homologues have been identified in other cereal grains and also shown to possess antifungal activity (Borgmeyer *et al.*, 1992, Huynh *et al.*, 1992, Malehorn *et al.*, 1994).

Amino acid sequence analysis of the seed permatins showed that they are related to a large family of plant proteins, the thaumatin-like proteins, which also includes the PR-5 proteins. Many of these other proteins are induced in vegetative tissue upon pathogen attack or by various stress conditions and have also been shown to also possess antifungal activity (Vigers *et al.*, 1992). The family also includes a bifunctional trypsin/α-amylase inhibitor from maize seed. This latter protein is highly homologous to the maize seed permatin, yet no protease or α-amylase inhibitory activity has been demonstrated with the permatin (Huynh *et al.*, 1992).

Using an *Arabidopsis* thaumatin-like protein gene as a probe the full length cDNA for a zeamatin-like protein (*Zlp*) was identified in a library made from developing maize seeds (Malehorn *et al.*, 1994). The protein is synthesised as a preprotein with a 21-amino acid signal sequence and, unlike some thaumatin-like proteins, including thaumatin itself, it does not contain a *C*-terminal prodomain. Analysis of *Zlp* expression by northern and western blotting showed that it was predominant in non-embryo tissues of developing seeds. Expression was largely limited to the seed although a low basal level was detected in leaves. A similar expression profile has been found in sorghum, where levels of the permatin sormatin increased during caropysis development through to physiological maturity (30 days after anthesis, DAA) and decreased subsequently (Seetharaman *et al.*, 1996). The expression pattern was also mirrored in sorghum by that of chitinase and glucanase enzymes. Upon germination, the level of sormatin increased dramatically together with the level of chitinase.

The 3-D crystal structures of zeamatin and the thaumatin I isoform have been determined and comparisons confirm the overall similarity between the two proteins (Batalia *et al*., 1996, Ogata *et al*., 1992). The proteins consist mainly of β-sheets which are organised to form two β-sandwiches. A third domain comprises four short α-helical domains held together by four disulphide bonds. The most pronounced difference between the two proteins is that the sandwich-cleft is highly acidic in zeamatin and much less so in thaumatin I.

The antifungal spectrum of the seed permatins is broad and includes filamentous and yeast-like fungi. At low concentrations these proteins cause hyphal rupture and it is likely that they act by directly permeabilising the fungal plasma membrane (Roberts and Selitrennikoff, 1990). No enzymatic activity has been found to be associated with these proteins. The potent antifungal activities of these proteins and the concentrations at which they are found in seeds makes it reasonable to suggest that they play a direct role in antifungal defence. For example, zeamatin is present in maize seeds at levels of approximately 50 μg per g dried seed. This is considerably higher than the concentration at which zeamatin exerts antifungal effects *in vitro* (3-10 μg/ml).

2.3 PR-1 and PR-4 type proteins

Examples of other antifungal PR-type proteins identified in plant seeds include a PR-1 homologue in germinating maize seeds (Casacuberta *et al*., 1991) and a barley seed PR-4 homologue (Hejgaard *et al*., 1992). The PR-1 homologue was upregulated when maize seedlings were infected with the fungus *Fusarium moniliforme* but has not been tested directly for antifungal activity. The PR-4 homologue, called CBP N or barwin, was purified from barley seed extracts using chitin affinity chromatography and shown to be a basic protein of M_r 14,000 with homology to tobacco PR-4 and the hevein precursor protein from rubber tree (Hejgaard *et al*., 1992). Two very similar proteins were also identified in barley leaves infected with powdery mildew, *Erysiphe graminis*. The barley seed and leaf PR-4 homologues were all found to inhibit the growth of the fungus *Trichoderma* in microtitre plate assays and to act synergistically with a barley seed chitinase and a barley seed PR-5 protein. In the seed, low concentrations of CBP N were detected in the embryo and starchy endosperm and it seems likely that the aleurone layer is the main site of expression. Other cereals of the tribe Triticeae all contain immunochemically-related proteins of similar size but no cross-reacting proteins could be detected in maize, sorghum or rice grain extracts using CBP N antiserum (Hejgaard *et al*., 1992).

While there is, to date, no information on the 3-D structure of a PR-1-like protein, that of the PR-4-like protein, barwin, from barley seeds has been determined by ^{1}H NMR spectroscopy (Ludvigsen and Poulsen, 1992). Barwin displays an open-shell-like structure formed by two β-sheets, each consisting of four anti-parallel β-strands. Furthermore, a putative binding site for chitin was identified which is consistent with the ability of this protein to bind chitin. However, it is not known whether this property is related to its antifungal activity.

2.4 Ribosome-inactivating proteins

In addition to the PR-related proteins many of these studies also identified type-1 ribosome-inactivating proteins (RIPs) as potentially playing a role in protecting seeds against fungal infection. RIPs are present in many plant tissues and can be classified into two groups based on whether they contain only a catalytic A chain (type-1) or an A chain and a lectin-related B subunit linked by a single disulphide bond (type-2) (Stirpe *et al.*, 1992). These proteins inactivate ribosomes by cleaving a single *N*-glycosidic bond on one adenine base in ribosomal RNA. The B chain of type-2 RIPs specifically recognises galactose or *N*-acetylglucosamine and is able to bind to cells and deliver the catalytic A-chain into the cytoplasm. As a result, the type-2 RIPs are potent cytotoxins. However, to date only one RIP has been shown to have direct antifungal activity; the type-1 RIP from barley seeds (Leah *et al.*, 1991). The intrinsic activity of this protein alone against a number of different pathogenic fungi is low but in combination with either chitinase or β-1,3-glucanase there is a clear synergistic enhancement of inhibition of fungal growth.

The cereal grain type-1 RIPs are proteins of approximately M_r 30,000. Unlike their dicot homologues which are synthesised as preproteins or preproproteins, the monocot type-1 RIPs lack a signal sequence and are presumed to be located in the cytoplasm (Bass *et al.*, 1992, Leah *et al.*, 1991). Using DNA from barley embryos, three genes per haploid genome were identified for the M_r 30,000 type-1 RIP (Leah *et al.*, 1991). In whole developing seeds, mRNA for the barley seed RIP could be detected 23 days after anthesis (DAA) and increased to a maximum about 30 DAA. This expression profile was mirrored by that of a M_r 26,000 chitinase which was also purified and shown to have antifungal activity. In contrast, an M_r 32,000 β-1,3-glucanase was expressed at low levels throughout seed development. Upon germination, mRNAs for the chitinase and β-1,3-glucanase increased whereas no mRNA was detected for the RIP. RIP mRNA was confined to the starchy endosperm and this observation, together with its expression profile, suggests that it may function primarily as an albumin storage protein in barley. The role of RIPs in seed as antifungal proteins may therefore be only a secondary function. The finding that the maize albumin b-32 is a member of the RIP family further strengthens this hypothesis (Bass *et al.*, 1992). The expression of b-32 is controlled by the transcriptional activator *Opaque-2* which has been shown to be a DNA binding protein that affects expression of the major seed storage protein genes.

2.5 Correlation between AFP levels and resistance to grain molds

A few studies have tried unsuccessfully to correlate the presence of PR-related antifungal proteins in cereal grains to resistance to fungal infection during seed development. Considerable variability in the levels of chitinases, glucanases and permatins has been observed among different cultivars of maize, sorghum and wheat (Darnetty *et al.*, 1993). However, while there are quantitative as well as qualitative differences in the isozymes present for each protein class, there are no clear overall relationships between the patterns and fungal resistance. The developmental stage at which the seeds are sampled for their AFP content may be an important factor. Seetharaman *et al.* (1996) have suggested that sampling at or slightly before physical maturity may be much more representative than sampling mature, combine harvested grain. In their study of sorghum caryopses, they were able to demonstrate a good correlation between sormatin levels and mold resistance ratings when cultivars were sampled at 30 DAA but not at two weeks after harvest (>50 DAA). In contrast, chitinase levels did not correlate to mold resistance ratings at any of the time points sampled.

3. CYSTEINE-RICH ANTIFUNGAL PEPTIDES

In addition to the larger PR-type proteins described above, a large number of small cysteine-rich peptides have been identified in seeds, mainly through the screening of extracts for antifungal activity. Some of these peptide families are widespread and have been found in many different plant seeds, in other cases only a few specific examples have been described so far. In some cases their expression appears to be restricted to the seed itself, while other peptide families are expressed in a range of plant organs including seeds.

3.1 Plant defensins

Plant defensins are a family of peptides isolated from a wide range of plants. They are highly basic peptides of M_r about 5000 which contain 8 cysteine residues, all of which are involved in disulphide bridges. Their structural and functional properties resemble those of insect and mammalian defensins and it was on this basis that the term plant defensin was proposed (Broekaert *et al.*, 1995). In general, plant defensins appear to be among the most widespread antifungal peptides in plants and are particularly prevalent in seeds.

The first members of this family were isolated from wheat and barley grains and termed γ-thionins because they showed some homology to α- and β-thionins (Mendez *et al.*,1990; Collila *et al.*, 1990). Proteins homologous to γ-thionins were subsequently isolated from sorghum seeds and shown to inhibit insect α-amylases (Nitti *et al.*, 1995). The first peptides of this protein family that were shown to exert antifungal activity were isolated from radish seeds using this activity to monitor purification (Terras *et al.*, 1992b). The peptides, called Rs-AFPs (*Raphanus sativus* antifungal proteins), are readily released into the surrounding medium during germination and can protect the young seedling from fungal attack (see below)(Terras *et al.*, 1995). Subsequent to this work a large number of other plant defensins exhibiting antifungal activity have been characterised from various plants, mostly from seeds (Moreno *et al.*, 1994; Osborn *et al.*, 1995; Terras *et al.*, 1993b). In addition, homologous peptides with unknown biological activity have also been identified via cDNA cloning (Choi *et al.*, 1995; Epple *et al.*, 1997; Ishibashi *et al.*, 1990; Meyer *et al.*, 1996; Milligan and Gasser, 1995; Sharma and Lönneborg, 1996; Stanchev *et al.*, 1996).

When the amino acid sequences of different plant defensins are aligned there is remarkably little sequence conservation outside the cysteine residues and some glycine residues (Fig. 1). However, despite this low overall homology at the amino acid sequence level, NMR studies on a number of different plant defensins have shown that they share a similar fold which is dominated by an α-helix packed against a triple-stranded β-sheet (Bruix *et al.*, 1993; Fant *et al.*, 1997)(Fig. 2). The particular cysteine-stabilised αβ motif present in plant defensins which determines their overall fold is also found in other peptides, e.g. insect defensin A (Cornet *et al.*, 1995). Indeed, it was this homology which prompted this class of plant proteins to be termed defensins in the first place.

A large number of plant defensins have now been characterised to determine their antimicrobial activity spectrum. When comparing different plant defensins for their relative activity on various fungi, marked differences in activity spectrum can be observed. However, they are generally active against a broad range of fungal species. Based on their effects on fungal growth they can be divided into at least two groups. The "morphogenic" plant defensins cause reduced hyphal elongation with a concomitant increase in hyphal branching, whereas the "nonmorphogenic" plant defensins reduce the rate of hyphal elongation but do not induce marked morphological distortions. In common with many of the peptides described latter in this chapter, the antifungal activity of plant defensins, whether morphogenic or not, is reduced

```
Rs-AFP1              Q K L C - E R P S G T W S G V C G N N N A C K N Q C I N L E K - A R H G S C N Y V F P A H K C I C Y F P C
Rs-AFP2              Q K L C - Q R P S C T W S G V C G N N N A C K N Q C I R L E K - A R H G S C N Y V F P A H K C I C Y F P C
Hs-AFP1          D G V K L C - D V P S G T W S G H C G S S S K C S Q Q C K D R E H F A Y G G A C H Y Q F P S V K C F C K R Q C
Ah-AMP1                  L C H E R P S Q T W S G N C G N T A H C D K Q C Q D W E K - A S H G A C H K R E N H W K C F C Y F N C
Dm-AMP1                E L C - E K A S K T W S G N C G N T G H C D N Q C K S W E G - A A H G A C H V R N G K H M C F C Y F N C
Ct-AMP1                N L C - E R A S L T W T G N C G N T G H C D T Q C R N W E S - A K H G A C H K R - G N W K C F C Y F D C
γ1-H                   R I C - R R R S A G F K G P C V S N K N C A Q V C M - Q E G - W G G G N C D G P L - - R R C K C M R R C
γ1-P                   K I C - R R R S A G F K G P C M S N K N C A Q V C Q - Q E G - W G G G N C D G P F - - R R C K C I R Q C
pSAS10                 K T C - E N L V D T Y R G P C F T T G S C D D H C K N K E H - L L S G R C R D D V - - - R C W C T R N C

Consensus motif        - - C - - - - - - - W - G - C - - - - - C - - - C - - - E - - - - - G - C - - - - - - - - C - C - - - C
                                           F
                                           Y
```

Fig. 1 Amino acid sequences of 9 different plant seed defensins. Complete amino acid sequences were derived from protein sequencing of the following plant defensins: Rs-AFP1 and Rs-AFP2 (Terras *et al.*, 1992b); Hs-AFP1, Ah-AMP1, Dm-AMP1 and Ct-AMP1 (Osborn *et al.*, 1995); γ1-P (Collila *et al.*, 1990); γ1-H (Mendez *et al.*, 1990). The sequence for pSAS10 was derived from cDNA clones with the omission of the putative signal peptide (Ishibashi *et al.*, 1990). Gaps introduced for optimal alignment are indicated by hyphens.

by increasing the ionic strength of the fungal growth assay medium. This antagonism was found to be due to cations, with divalent cations being at least one order of magnitude more potent than monovalent cations (Terras *et al.*, 1992b, 1993b).

In addition to their antifungal activity, some plant defensins also possess antibacterial activity. However, unlike some of the other classes of antifungal peptides described below, this activity is relatively weak with a narrow spectrum. For example, the plant defensin from *Clitoria ternatea* seed, Ct-AMP1, is active against *Bacillus subtilis* but did not affect the growth of three other Gram-positive bacteria that were tested (Osborn *et al.*, 1995).

Exactly how plant defensins inhibit fungal growth is not known at present. However, it has been shown that their primary action is not to cause permeabilisation of fungal membranes through a direct interaction with the lipid bilayer. Using the plant defensins from radish seed (Rs-AFP2) and dahlia seed (Dm-AMP1), permeabilisation of *Neurospora crassa* hyphae to isoaminobutyric acid was only measurable when the plant defensin concentration was raised ten-fold above the minimal concentration required for growth inhibition (Thevissen *et al.*, 1996). In addition, these plant defensins had no effect on electrical current measured in artificial membrane systems (Thevissen *et al.*, 1996) and a plant defensin from potato tubers did not cause permeabilisation of artificial phospholipid vesicles (Caaveiro *et al.*, 1997). Binding studies with the radiolabelled plant defensin from *Heuchera* seed, Hs-AFP1, have recently shown that it binds to specific high affinity sites on the plasma membrane of *Neurospora crassa* hyphae (Thevissen *et al.*, 1997). This interaction is consistent with the presence of protein receptors for plant defensins in fungal plasma membranes and it is presumed that the

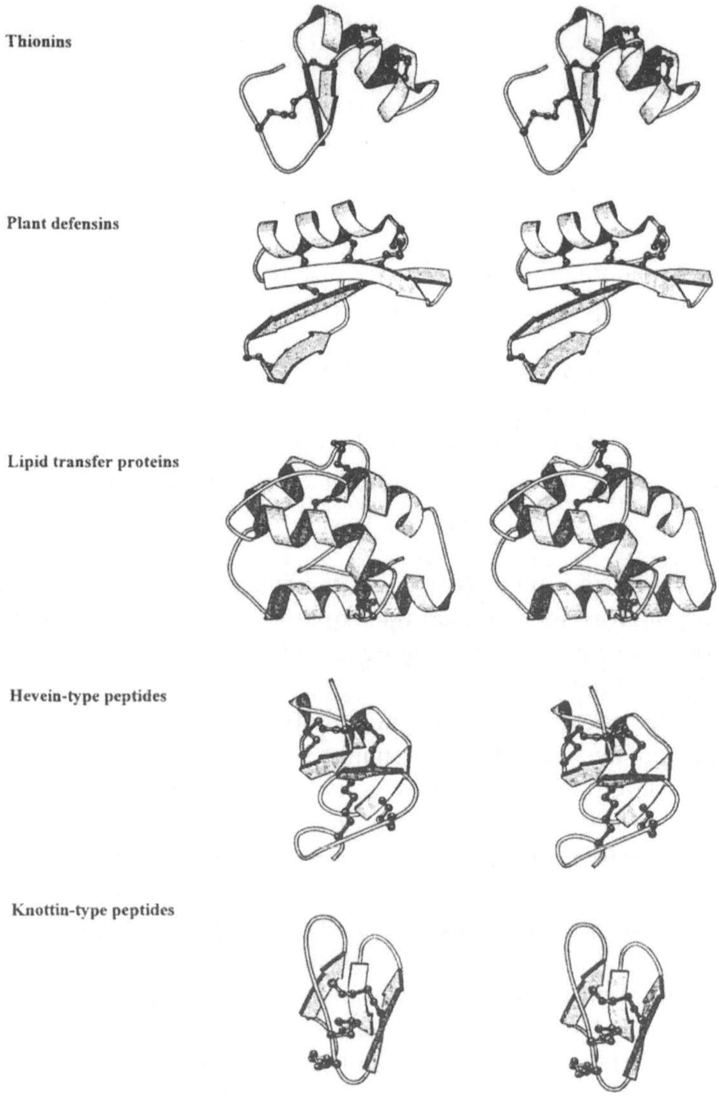

Fig. 2 Three-dimensional structures of different antifungal peptides. Peptide backbones are shown in molscript presentation with helices representing α-helices and arrows representing β-strands. Disulphide bridges are shown as black tubes with balls at the C or S atom positions. References for the structures shown are highlighted in the text. Reproduced with permission from Broekaert *et al.* (1997).

peptides exert their antifungal effects by binding to these receptors. The function of these receptors is presently under investigation.

There have been few studies to identify the number or location of plant defensin genes. In *Arabidopsis*, analysis of ESTs has shown that there is a small multigene family for plant defensins, at least one member of which encodes a seed-expressed isoform (Epple *et al.*, 1995). From their cDNA structures, it is apparent that the majority of plant defensins are synthesized as preproteins. Several also have a *C*-terminal domain which may act as a propeptide but this seems to be unusual. The fact that the seed defensins are synthesized as preproteins would suggest an extracellular or vacuolar localization. In the case of the radish seed defensins (Rs-AFP1 and 2), immunolocalisation studies on semi-thin seed sections showed them to be most abundant in the outer cell wall layer lining the surface of the cotyledons, hypocotyl and endosperm (Terras *et al.*, 1995).

The extracellular localization of plant defensins in seed is not consistent with a role as a storage protein and it has been proposed that their biological role is the protection of seeds or seedlings from soil-borne fungi (Terras *et al.*, 1995). Support for this hypothesis comes from observations that germinating radish seeds cause a halo of growth inhibition when placed on a solid medium inoculated with fungus (Fig. 3). This growth inhibition halo could be mimicked by applying drops containing 1 μg of purified Rs-AFP and was only present if the seed coat was perforated either by the emerging seedling or by a mechanical incision. The growth inhibition halo was not present when a protease, pronase E, was present in the plate medium and it was also absent from around the intact seed when germination was inhibited by abscisic acid in the medium. Analysis of the imbibition solution showed that the Rs-AFPs are preferentially released when the seed coat is perforated. The plant defensins were found to account for about 30% of the released protein although they are only present as 0.5 % of total seed protein. Furthermore, the amount of Rs-AFP released from a single seed was estimated at greater than 1 μg which is sufficient to cause the observed growth inhibition halo. Taken together, these observations show the release of these peptides from a germinating seed could provide protection against fungal infection and suggest that they may contribute to the overall seedling survival rate.

3.2 Lipid transfer proteins

Lipid transfer proteins (LTPs) of plants form a family of homologous proteins containing 90 to 93 amino acid residues with 8 cysteines which form four intra-chain disulphide bonds. They were first identified based on their ability to transfer radiolabelled phospholipids from artificial liposomes to mitochondria (Kader, 1997). Due to the lack of specificity for the phospholipids that they can transport, these proteins are often referred to as nonspecific lipid transfer proteins (nsLTPs). They have been studied in a wide

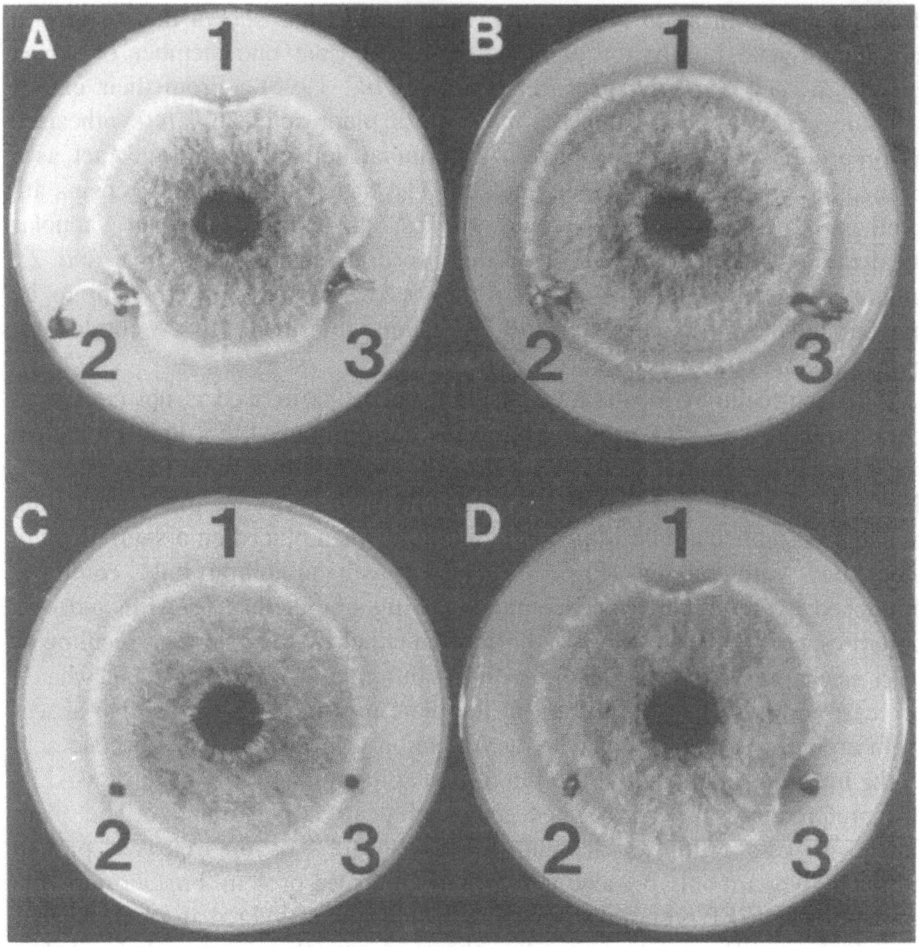

Fig. 3 Release of antifungal compounds by germinating radish seeds. The figure shows the effect of either 1 μg purified Rs-AFP1 (position 1), radish seeds with an intact seed coat (position 2) or radish seeds with an incised seed coat (position 3) on the growth of the fungus *Pyrenophora tritici-repentis* under different conditions. A, assay plates containing five-cereal agar; B, assay plates containing five-cereal agar supplemented with 50 μg/ml Pronase E; C, assay plate as in (A) but the Rs-AFP1 solution and the seeds were autoclaved; D, assay plate contains five-cereal agar supplemented with 100 μM abscisic acid. Figure reproduced with permission from Terras *et al.* (1995).

range of plants and are found to be expressed in a variety of plant organs including the seed.

The first LTP to be reported with antifungal activity was purified from radish seed (Terras *et al.*, 1992a). Subsequently, a number of other antifungal LTPs have been described from different plant sources (Molina *et al.*, 1993b; Nielsen *et al.*, 1996) including a highly active protein, Ace-AMP1, from onion seed (Cammue *et al.*, 1995). When LTPs from different plant species are compared at the amino acid sequence level, about 30% of the residues appear to be conserved. The conserved positions include the 8 cysteines as well as 12 positions which are invariably occupied by hydrophobic or aromatic residues.

Three-dimensional structures have been determined for the wheat, maize and onion seed LTPs using proton NMR (Gincel *et al.*, 1994, Gomer *et al.*, 1996, Tassin *et al.*, 1998) and by X-ray diffraction for the LTP from maize seedlings (Shin et al., 1995)(Fig. 2). This has revealed a common structure consisting of a bundle of four α-helices linked by flexible loops. The α-helices surround a hydrophobic tunnel which, in most cases, can accommodate a fatty acyl chain as shown by crystallography of an LTP-palmitate complex (Shin *et al.*, 1995). The exception is Ace-AMP1 which has a narrow hydrophobic cavity that is obstructed by bulky aromatic residues (Tassin *et al.*, 1998). Interestingly, Ace-AMP1 also lacks the ability to transfer phospholipids in the classical lipid transfer activity assays (Cammue *et al.*, 1995), supporting the view that this hydrophobic tunnel plays a role in the lipid-protein interaction.

All plant nsLTP cDNA/genes characterized so far encode preproteins, with the exception of Ace-AMP1 which is synthesized as a preproprotein with a carboxy-terminal propeptide (Cammue *et al.*, 1995). The finding that LTPs are synthesised as preproteins and therefore are not cytosolic is inconsistent with the original role proposed for these proteins, namely in the intracellular trafficking of phospholipids from the endoplasmic reticulum, the site of phospholipid biosynthesis, to other cellular organelles. This hypothesis has now, therefore, been largely abandoned and it is considered the main role of LTPs is as a defense mechanism against microbial infection.

The antifungal activities of the different LTPs can differ significantly. For example, the onion seed LTP is highly active against a broad range of fungi (with IC_{50} values ranging from 1 to 6 μg/ml), whereas the radish seed nsLTP is only moderately active (IC_{50} values ranging from 7 to 100 μg/ml) and maize and wheat seed nsLTPs are virtually inactive against most fungi (Cammue *et al.*, 1995). Also, the sensitivity of their antifungal activity to the presence of cations varies with the nsLTP type. The onion seed nsLTP was almost as potent in a low ionic strength medium as in the same medium supplemented with 1 mM Ca^{2+} and 50 mM K^{+}, whereas the radish nsLTP showed drastically reduced antifungal activity in the presence of cations (Cammue *et al.*, 1995). The onion seed LTP, which exhibits strong antifungal activity, was also shown to be inhibitory to Gram-positive bacteria but not to a range of Gram-negative bacteria (Cammue *et al.*, 1995). In contrast, none of the LTPs from seeds of onion, radish, maize or wheat caused lysis of human erythrocytes, nor did they affect the viability of cultured human fibroblasts (Cammue *et al.*, 1995).

Plant seeds contain a number of proteins which are structurally related to the LTPs and which also possess some antifungal activity, albeit weak in comparison to proteins like Ace-AMP1. These include the dicot seed storage proteins 2S albumins, cereal trypsin/α-amylase inhibitors (discussed in Chapters 24 and 26) and puroindolines from wheat endosperm (Dubreil *et al.*, 1998; Terras *et al.*, 1992b, 1993a, 1993b). Although there is no significant sequence homology between the different classes, there is a similarity when the cysteine pattern is considered. All these proteins also share the ability to strongly potentiate the antimicrobial activity of thionins (Molina *et al.*, 1993b; Terras *et al.*, 1993a; Dubreil *et al.*, 1998). Even the wheat seed LTP, which displays little or no antifungal activity, was able to increase the potency of thionins (Dubreil *et al.*, 1998). The only exception is the onion seed LTP homologue, Ace-AMP1, which is a very potent inhibitor but lacks the lipid transfer activity and the ability to act synergistically with thionins. Indeed, Ace-AMP1 appears to be a unique member of the LTP class and may have a different mode of action. Whereas the standard LTPs and proteins like the 2S albumins appear to directly permeabilise fungal membranes, Ace-AMP1 may act through a membrane-located receptor. The ability of Ace-AMP1 to permeabilise phospholipid bilayers in artificial vesicles was significantly less than that of puroindoline, although it is far superior in its antimicrobial activity (Tassin *et al.*, 1997).

3.3 Thionins

Thionins represent a family of peptides, most of which comprise 45 to 47 amino acids. They are covered in depth in Chapter 30, but because of their well documented antimicrobial activity, it is necessary to also mention them in the context of other seed antifungal proteins.

It has been known for over 50 years that thionins inhibit bacterial and fungal growth *in vitro*. In subsequent studies, the antimicrobial activity spectrum of thionins has been analyzed more extensively and shown to include several Gram-positive and Gram-negative plant pathogenic bacteria (Cammue *et al.*, 1992; Molina *et al.*, 1993a), as well several pathogenic fungi, with IC_{50} values generally ranging from 1 to 15 μg/ml (Cammue *et al.*, 1992; Molina *et al.*, 1993a). The antifungal activity of thionins is inhibited by Ca^{2+} at concentrations above 5 mM, but not by Mg^{2+} or Ba^{2+} at concentrations up to 10 mM, nor by monovalent cations at concentrations up to 50 mM (Terras *et al.*, 1992b; Cammue *et al.*, 1995). The majority of antimicrobial activity testing has been carried out with the eight-cysteine type thionins. More recently, a six-cysteine type thionin from radish was compared with an eight-cysteine type thionin from barley seed. In these tests on the fungus *Fusarium culmorum*, the radish thionin was much less potent (Terras *et al.*, 1996). It is, however, not known whether this is also the case with other fungi nor whether it can be generalised for other six-cysteine type thionins.

As mentioned above, thionins act synergistically with other antifungal seed proteins which also act directly on fungal membranes. The potentiation of the activity of thionins by proteins such as 2S albumins is shown by measuring changes in ion-fluxes. A thionin from barley seed has been shown to mediate a transient influx of Ca^{2+} into *Neurospora crassa* hyphae, with a steady efflux of K^+ and a steady alkalinization of the culture medium, the latter possibly due to increased H^+ influx. All these ion fluxes could be observed within 1 min of addition of the protein (Thevissen *et al.*, 1996). By measuring K^+ efflux from hyphae of *Fusarium culmorum* it could be demonstrated that a combination of thionin and 2S albumin led to a synergistic increase in the efflux (Terras *et al.*, 1993a). Interestingly, the radish six-cysteine type thionin does not act synergistically with 2S albumins.

Studies on the subcellular localisation of thionins in barley and wheat seed have shown that they are localised in the vacuolar compartments (protein bodies) (Bohlmann, 1994) where it is likely that they function both as storage proteins for the seed and also as a defensive mechanism.

3.4 Hevein-like AFPs

Many plant proteins are able to bind to chitin and many of these contain a common cysteine-rich domain of about 40 residues that has been shown to be responsible for the binding (Iseli *et al.*, 1993). In recent years a number of plant proteins have been isolated and shown to consist of essentially just this chitin-binding domain. The best studied example is hevein which is a 43 amino acid chitin-binding peptide and which is one of the most abundant proteins in rubber tree latex (Van Parijs *et al.*, 1991). Hevein has weak antifungal activity but the other examples, which have been isolated from various plant seeds, have been found to be much more potent inhibitors of fungal growth *in vitro*.

The first examples were purified from seeds of *Amaranthus caudatus* and designated Ac-AMP1 and Ac-AMP2 (Broekaert *et al.*, 1992). The two peptides are 29 and 30 amino acids in length, respectively, and are identical except for the addition of a single amino acid at the carboxy-terminus. The peptides show a high degree of homology to the chitin-binding domains of proteins like chitinases and also to hevein but lack the domain which includes last two cysteines (Fig. 4a). Two further examples were subsequently identified from seeds of sweet pepper (Broekaert *et al.*, 1994) and Japanese morning glory (Koo et al., unpublished; Genbank accession number U40076). In both these cases the peptides have eight cysteines and therefore resemble more closely hevein (Fig. 4a).

```
(A)

Ac-AMP1          VGECVR---GR-CPSGMCCSQFGYCGKGPKYCG
Ac-AMP2          VGECVR---GR-CPSGMCCSQFGYCGKGPKYCGR
Pn-AFP2           QQCGRQASGRLCGNGLCCSQWGYCGSTAAYCGAG--CQSQCKS
Ca-AMP1          QEQCGNQAGGRACANRLCCSQYGYCGSTRAYCGVG--CQSNCGR
hevein            EQCGRQAGGKLCPNNLCCSQWGWCGSTDEYCSPDHNCQSNCKD
To chit           EQCGSQAGGARCASGLCCSKFGWCGNTNDYCGPG-NCQSQCPGG

Consensus motif --C-----G--C----CCS--G-CG----YC

(B)

Mj-AMP1                QCIGNGGRCNENVGPPYCCSGF-CLRQPGQGYGYCKNR
Mj-AMP2                 CIGNGGRCNENVGPPYCCSGF-CLRQPNQGYGVCRNR
PAFP-s                AGCIKNGGRCNASAGPPYCCSSY-CFQIAGQS-GVCPNRDQEGPYPGPVYKCQCFGFR
AAI                     CIPKWNRCGPKMDGVPCCEPYTCTSD---YYGNCS
ω-conotoxin GVIA        CKSPGSSCSPTSYN--CCRS--CNP----YTKRCYG

Consensus motif         C------C--------CC----C----------C
```

Fig. 4 Amino acid sequences of hevein-like AFPs, Mj-AMPs and related proteins. (A) Amino acid sequences of hevein-like AFPs (Ac-AMP1 and Ac-AMP2 from amaranth seeds, Ca-AMP1 from sweet pepper seeds, Pn-AFP2 from Japanese morning glory seeds) aligned with hevein and the N-terminal amino acid sequences from tobacco chitinase. (B) Amino acid sequences of Mj-AMP1 and Mj-AMP2 aligned with PAFP-s from *Phytolacca* seed and representatives from the 'knottin' family. See text for protein sequence references. Gaps introduced for optimal alignment are indicated by hyphens.

The cDNAs encoding the amaranth peptide Ac-AMP2 and Pn-AFP2 from Japanese morning glory show that both proteins are synthesised as preproproteins with an amino-terminal signal sequence and a carboxy-terminal propeptide (De Bolle *et al.*, 1993). Northern blot analysis showed that transcripts for the Ac-AMPs were only present in developing seeds and therefore synthesis of these peptides is likely to be confined to the seed. Larger transcripts which cross-hybridised with the Ac-AMP2 RNA probe were identified in leaves and roots but these probably encode proteins like chitinases which contain the chitin-binding domain homologous to the Ac-AMPs. It was initially proposed that the carboxy-terminal prodomain may have a targeting function and direct the Ac-AMPs into protein bodies in the seed. However, expression of Ac-AMP2 with and without the prodomain in transgenic plants showed that in both cases the expressed peptide was located in the apoplast (De Bolle *et al.*, 1996). It is, therefore, unlikely that the prodomain plays a role in cellular trafficking and consequently the Ac-AMPs are thought to be located extracellularly in the seed.

The antimicrobial spectra of the Ac-AMPs and the homologue from sweet pepper seeds are relatively broad and include a range of filamentous fungi and

Gram-positive bacteria but not Gram-negative bacteria. Their potency is, however, dependent on the ionic strength of the media in which they are tested and in particular on the concentration of divalent cations (Broekaert *et al.*, 1992). In media containing 1 mM $CaCl_2$ and 50 mM KCl, these peptides do not inhibit fungal growth at concentrations below 100 μg/ml, whereas in low ionic strength media they are generally inhibitory at concentrations of 10 μg/ml and below.

The antimicrobial peptides have also been shown to bind chitin which is, perhaps, not surprising given their amino acid sequences. However, it is not known whether this is linked to their antifungal activity. Indeed, their wide antimicrobial spectrum perhaps argues against this and it is perhaps more likely that they act by interfering directly with the plasma membrane or some component of the plasma membrane.

3.5 *Mirabilis jalapa* antimicrobial peptides (Mj-AMPs)

Two peptides which share some common features with the hevein-type peptides described above have been isolated from seeds of the four o'clock plant, *Mirabilis jalapa* (Cammue *et al.*, 1992). These peptides, termed Mj-AMP1 and Mj-AMP2, are 37 and 36 amino acids in length, respectively, and contain 6 disulphide-linked cysteines. The two peptides are 90% identical to one another and share homology with an antifungal peptide, PAFP-s, from *Phytolacca americana* (pokeweed) seed (Lin *et al.*, 1992) and a group of peptides collectively termed 'knottins' (Chagolla-Lopez *et al.*, 1994)(Fig. 4b). The knottins include an α-amylase inhibitor from amaranth seeds, nerve-acting peptides from spider and *Conus* snails, and the race-specific elicitor AVR9 from the fungus *Cladosporium fulvum*. The fold of these proteins is characterised by a compact triple-stranded β-sheet with a long loop connecting the first and second β-sheets. In addition, the cysteine motif of the knottins is similar to the first 6 cysteines of the hevein-like AFPs, and these two groups are structurally related (Fig. 2).

The Mj-AMPs are expressed as preproteins and are encoded by a low copy number gene family (De Bolle *et al.*, 1995). In developing seeds, mRNA was only detected at the onset of desiccation (40 DAA) and reached a maximum level in dry seeds. Furthermore, the expression of Mj-AMPs seems to be restricted to the seed itself as no cross-hybridising mRNA could be detected in healthy or stressed tissues of *M. jalapa* other than seeds at a late developmental stage. This expression profile is very similar to that described above for the amaranth peptides.

The Mj-AMPs have been shown to inhibit a broad range of fungi and Gram-positive bacteria but not the two Gram-negative bacteria included in the tests (Cammue *et al.*, 1992). As with the Ac-AMPs, their antifungal activity is also strongly antagonised by inorganic cations.

3.6 *Impatiens balsamina* antimicrobial peptides (Ib-AMPs)

The smallest plant antimicrobial peptides identified to date have been recently purified from *Impatiens balsamina* seeds (Tailor *et al.*, 1997). Four isoforms, called Ib-AMP1 to 4, were identified using antifungal bioassays to follow the fractionation of crude seed extracts. The peptides are highly basic and contain four disulphide-linked cysteines. No significant homology to other peptides has been found but their cysteine motif does resemble that of the amino-terminal portion of eight-cysteine type thionins.

In media of low ionic strength all four of the Ib-AMPs exert broad spectrum activity against a range of fungi and Gram-positive bacteria (Tailor *et al.*, 1997). However, in medium containing 1 mM $CaCl_2$ and 50 mM KCl only Ib-AMP4, which is the most basic isoform, still inhibits fungal growth below 100 μg/ml.

Most intriguing of all is the finding that all four Ib-AMP isoforms are encoded by a single preproprotein (Tailor *et al.*, 1997). Thus, the predicted Ib-AMP precursor protein consists of a presequence followed by 6 mature peptide domains, each of which is flanked by acidic prodomains ranging from 16 to 35 amino acids in length. This is the only example so far described in plants of a single precursor giving rise to multiple antimicrobial peptide isoforms. However, a very similar multipeptide structure has been found for apidaecins, which are pathogen-inducible antimicrobial peptides from honey bee (Casteels-Josson *et al.*, 1993). The concatenation of small peptide domains thus appears to be a very effective way of enhancing their simultaneous expression.

The Ib-AMPs are expressed late in seed development and accumulate in the dry seed. In northern blot experiments no cross-hybridising mRNA was detected in germinating seed but the peptides themselves were detected by immunoblotting up to 48 hours after germination (Tailor *et al.*, 1997). The Ib-AMPs are estimated to account for approximately 0.5% of total protein in mature *I. balsamina* seed and are presumed to play a role in protecting the germinating seed. At present it is not known whether these peptides are also expressed in other parts of the plant or whether, like the hevein-like and knottin-type peptides, their expression is limited to the seed itself.

3.7 *Zea mays* basic peptide (MBP-1)

A four cysteine disulphide-linked peptide has also been isolated from maize seed and shown to inhibit the spore germination and hyphal extension of a number of important maize seed pathogens (Duvick *et al.*, 1992). The 33 amino acid peptide shares no significant homology with other known peptides and also has a distinct cysteine-motif. Its antifungal activity has also been shown to be sensitive to cations but, unlike most of the other small seed-derived AFPs, MBP-1 has also been found to inhibit at least one Gram-negative bacterium. Based on its relatively low amount in the seed (0.014% of total protein) MBP-1 is unlikely to serve as a storage protein. It may, therefore, also play a role in protecting the maize seed from microbial infection.

3.8 *Macadamia integrifolia* antimicrobial peptide (MiAMP1)

An example of another distinct antimicrobial peptide class has been isolated from *Macadamia* nut kernels (Marcus *et al.*, 1997). This peptide is 76 amino acids long and was shown to inhibit the growth of several fungi and Gram-positive bacteria, but not Gram-negative bacteria. The mature peptide is derived from a preprotein containing a signal peptide and no significant homology has been found to any protein sequence present in databases.

4. CONCLUSIONS

Plant seeds have provided a rich source of antifungal proteins and peptides. These include homologues of most of the PR-protein families as well as representatives from at least eight distinct families of cysteine-rich peptides. In many cases their expression patterns and localisation strongly suggest a direct role in protecting both the developing seed and the germinating seedling from fungal infection. However, to date, direct evidence to substantiate this link has been difficult to obtain. Only in a few studies have the authors tried to demonstrate a correlation between levels of seed AFPs and resistance to seed-borne fungal diseases and, where measurements have been made, the results have largely been inconclusive. With other seed AFPs any protective role is probably secondary and their primary role is either in the germination process or to act simply as seed storage proteins. The engineering of plants for disease resistance is gaining momentum and a number of seed AFPs have been successfully shown to provide protection when expressed in transgenic plants (see Terras *et al.*, 1995). It may well be, therefore, that the best insights into the *in vivo* function of these proteins and their contribution to seed survival will come through the manipulation of their expression in transgenic plants.

ACKNOWLEDGEMENTS

The authors acknowledge the financial support of research in their laboratories by the Commission of the European Community (AGRE-0005, AIR2-CT94-1356).

REFERENCES

Bass, H.B., Webster, C., O'Brian, C.R., Roberts, J.K.M. and Boston, R.S. (1992) A maize ribosome-inactivating protein is controlled by the transcriptional activator Opaque-2. *Plant Cell*, **4**, 225-34

Batalia, M.A., Monzingo, A.F., Ernst, S., Roberts, W. and Robertus, J.D. (1996) The crystal structure of the antifungal protein zeamatin, a member of the thaumatin-like, PR-5 protein family. *Nature Structural Biology*, **3**, 19-23

Bohlmann, H. (1994) The role of thionins in plant protection. *Critical Reviews in Plant Sciences*, **13**, 1-16

Borgmeyer, J.R., Smith, C.E. and Huynh, Q.K. (1992) Isolation and characterization of a 25 kDa antifungal protein from flax seeds. *Biochemical Biophysics Research Communication*, **187**, 480-87

Broekaert, W.F., Van Parijs, J., Allen, A.K. and Peumans, W.J. (1988) Comparison of some molecular, enzymatic and antifungal properties of chitinases from thorn-apple, tobacco and wheat. *Physiology and Molecular Plant Pathology*, **33**, 319-31

Broekaert, W.F., Mariën, W., Terras, F.R.G., De Bolle, M.F.C., Proost, P., Van Damme, J., Dillen, L., Claeys, M., Rees, S.B., Vanderleyden, J. and Cammue, B.P.A. (1992) Antimicrobial peptides from *Amaranthus caudatus* seeds with sequence homology to the cysteine/glycine-rich domain of chitin-binding proteins. *Biochemistry*, **31**, 4308-14

Broekaert, W.F., Cammue, B.P.A., Osborn, R.W. and Rees, S.B. (1994) Biocidal chitin binding proteins. International patent application WO94/11511.

Broekaert, W.F., Terras, F.R.G., Cammue, B.P.A. and Osborn, R.W. (1995) Plant defensins: Novel antimicrobial peptides as components of the host defense system. *Plant Physiology*,**108**, 1353-58

Broekaert, W.F., Cammue, B.P.A., De Bolle, M.F.C., Thevissen, K., De Samblanx, G.W. and Osborn, R.W. (1997) Antimicrobial Peptides in Plants. *Critical Reviews in Plant Sciences*, **16**, 297-323; published by CRC Press.

Bruix, M,, Jiménez, M.A., Santoro, J., González, C., Colilla, F.J., Méndez, E. and Rico, M. (1993) Solution structure of γ1-H and γ1-P thionins from barley and wheat endosperm determined by ^{1}H-NMR: a structural motif common to toxic arthropod proteins. *Biochemistry*, **32**, 715-24

Caaveiro, J.M.M., Molina, A., Gonzales-Manas, J.M., Rodriguez-Palenzuela, P., Garcia-Olmedo, F., Goni, F.M. (1997) Differential effects of five types of antipathogenic plant peptides on model membranes. *FEBS Letters*, **410**, 338-42

Cammue, B.P.A., De Bolle, M.F.C., Terras, F.R.G., Proost, P., Van Damme, J., Rees, S.B., Vanderleyden, J. and Broekaert, W.F. (1992) Isolation and characterization of a novel class of plant antimicrobial peptides from *Mirabilis jalapa* L. seeds. *Journal of Biological Chemistry*, **267**, 2228-33

Cammue, B.P.A., Thevissen, K., Hendriks, M., Eggermont, K., Goderis, I.J., Proost, P., Van Damme, J., Osborn, RW., Guerbette, F., Kader, J-C. and Broekaert, W.F. (1995) A potent antimicrobial protein from onion seeds showing sequence homology to plant lipid transfer proteins. *Plant Physiology*, **109**, 445-55

Casacuberta, J.M., Puigdoménech, P. and San-Segundo, B. (1991) A gene coding for a basic pathogenesis-related (PR-like) protein from *Zea mays*. Molecular cloning and induction by a fungus (*Fusarium moniliforme*) in germinating maize seeds. *Plant Molecular Biology*, **16**, 527-36

Casteels-Josson, K., Capaci, T., Casteels, P. and Tempst, P. (1993) Apidaecin multipeptide precursor structure: a putative mechanism for amplification of the insect antibacterial response. *EMBO Journal*, **12**, 1569-78

Chagolla-Lopez, A., Blanco-Labra, A., Patthy, A., Sanchez, R. and Pongar, S. (1994) A novel α-amylase inhibitor from Amaranth (*Amaranthus hypocondriacus*) seeds. *Journal of Biological Chemistry*, **269**, 23675-680.

Choi, Y., Ahn, J.H., Choi, Y.D. and Lee, J.S. (1995) Tissue-specific and developmental regulation of a gene encoding a low molecular weight sulphur-rich protein in soybean seeds. *Molecular and General Genetics*, **246**, 266-68

Colilla, F.J., Rocher, A. and Mendez, E. (1990) γ-Purothionins: amino acid sequences of two polypeptides of a new family of thionins from wheat endosperm. *FEBS Letters*, **270**, 191-94

Cordero, M.J., Raventos, D. and San Segundo, B. (1994) Differential expression and induction of chitinases and β-1,3-glucanases in response to fungal infection during germination of maize seeds. *Molecular Plant-Microbe Interactactions*, **7**, 23-31

Cornet, B., Bonmatin, J-M., Hétru, C., Hoffmann, J.A., Ptak, M. and Vovelle, F. (1995) Refined three-dimensional solution structure of insect defensin A. *Structure*, **3**, 435-48

Darnetty, Leslie, J.F., Muthukrishnan, S., Swegle, M., Vigers, A.J. and Selitrennikoff, C.P. (1993) Variability of antifungal proteins in the grains of maize, sorghum and wheat. *Physiologia Plantarium*, **88**, 339-49

De Bolle, M.F.C., David, K.M.M., Rees, S.B., Vanderleyden, J., Cammue, B.P.A. and Broekaert, W.F. (1993) Cloning and characterization of a cDNA encoding an antimicrobial chitin-binding protein from amaranth, *Amaranthus caudatus*. *Plant Molecular Biology*, **22**, 1187-90

De Bolle, M.F.C., Eggermont, K., Duncan, R.E., Osborn, R.W., Terras, F.R.G. and Broekaert, W.F. (1995) Cloning and characterization of two cDNA clones encoding seed-specific antimicrobial peptides from *Mirabilis jalapa* L. *Plant Molecular Biology*, **28**, 713-21

De Bolle, M.F.C., Osborn, R.W., Goderis, I.J., Noe, L., Acland, D., Hart, C.A., Torrekens, S., Van Leuven, F. and Broekaert, W.F. (1996) Antimicrobial peptides from *Mirabilis jalapa* and *Amaranthus caudatus*: expression, processing, localization and biological activity in transgenic tobacco. *Plant Molecular Biology*, **31**, 993-1008

Dubreil, L., Gaborit, T., Bouchet, B., Gallant, D., Broekaert, W.F., Quillien, L. and Marion, D. (1998) Spatial and temporal distribution of puroindolines and non specific lipid transfer proteins in *Triticum aestivum* seeds. Relationships with their *in vitro* antifungal properties. (in preparation)

Duvick, J.P., Rood, T., Rao, A.G. and Marshak, D.R. (1992) Purification and characterization of a novel antimicrobial peptide from maize (*Zea mays* L.) kernels. *Journal of Biological Chemistry*, **267**, 18814-20

Epple, P., Apel, K. and Bohlmann, H. (1995) An *Arabidopsis thaliana* thionin gene is inducible via a signal transduction pathway different from that for pathogenesis-related proteins. *Plant Physiology*, **109**, 813-20

Epple, P., Apel, K. and Bohlmann, H. (1997) ESTs reveal a multigene family for plant defensins in *Arabidopsis thaliana*. *FEBS Letters*, **400**, 168-72

Fant, F., Vranken, W.F., Martins, J.C. and Borremans, F.A.M. (1997) Solution conformation of *Raphanus sativus* antifungal protein 1 (Rs-AFP1) by ^{1}H nuclear magnetic resonance. Resonance assignments, secondary structure and global fold. *Bulletin de la Societe de Chimie Belge*, **106**, 51-57

Gincel, E., Simorre, J-P., Caille, A., Marion, D., Ptak, M. and Vovelle, F. (1994) Three-dimensional structure in solution of a wheat lipid-transfer protein from multidimensional ^{1}H-NMR data. A new folding for lipid carriers. *European Journal of Biochemistry*, **226**, 413-22

Gomar, J., Petit, M-C., Sodano, P., Sy, D., Marion, D., Kader, J-C., Vovelle, F. and Ptak, M. (1996) Solution structure and lipid binding of a nonspecific lipid transfer protein extracted from maize seeds. *Protein Science*, **5**, 565-77

Hart, P.J., Monzingo, A.F., Ready, M.P., Ernest, S.R. and Robertus, J.D. (1993) Crystal structure of an endochitinase from *Hordeum vulgare* L. seeds. *Journal of Molecular Biology*, **229**, 189-93

Hejgaard, J., Jacobsen, S., Bjørn, S.E. and Kragh, K.M. (1992) Antifungal activity of chitin-binding PR-4 type proteins from barley grain and stressed leaf. *FEBS Letters*, **307**, 389-92

Huynh, Q.K., Borgmeyer, J.R. and Zobel, J.F. (1992) Isolation and characterization of a 22 kDa protein with antifungal properties from maize seeds. *Biochemical and Biophysical Research Communications*, **182**, 1-5

Iseli, B., Boller, T. and Neuhaus, J-M. (1993) The *N*-terminal cysteine-rich domain of tobacco class I chitinase is essential for chitin binding but not for catalytic or antifungal activity. *Plant Physiology*, **103**, 221-26

Ishibashi, N., Yamauchi, D. and Minamikawa, T. (1990) Stored mRNA in cotyledons of *Vigna unguiculata* seeds: nucleotide sequence of cloned cDNA for a stored mRNA and induction of its synthesis by precocious germination. *Plant Molecular Biology*, **15**, 59-64

Kader, J-C. (1997) Lipid-transfer proteins: a puzzling family of plant proteins. *Trends Plant Sciences*, **2**, 66-70

Leah, R., Tommerup, H., Svendsen, I. and Mundy, J. (1991) Biochemical and molecular characterization of three barley seed proteins with antifungal properties. *Journal of Biological Chemistry*, **266**, 1564-73

Leubner-Metzger, G., Fründt, C., Vögeli-Lange, F. and Meins, F. Jr (1995) Class I β-1,3-glucanases in the endosperm of tobacco during germination. *Plant Physiology*, **109**, 751-59

Lin, Z-P., Hu, Z. and Ye, H-C (1992) Screening valuable genes from wild species of plants, in *Conservation of Plant Genes*, Academic Press, pp. 241-46

Ludvigsen, S. and Poulsen, F.M. (1992) Three-dimensional structure in solution of barwin, a protein from barley seed. *Biochemistry*, **31**, 8783-89

Malehorn, D.E., Borgmeyer, J.R., Smith, C.E. and Shah, D.M. (1994) Characterization and expression of an antifungal zeamatin-like protein (*Zlp*) gene from *Zea mays*. *Plant Physiology*, **106**, 1471-81

Marcus, J.P., Goulter, K.C., Green, J.L., Harrison, S.J. and Manners, J.M. (1997) Purification, characterisation and cDNA cloning of an antimicrobial peptide from *Macadamia integrifolia*. *European Journal of Biochemistry*, **244**, 743-49

Mauch, F., Hadwiger, L.A. and Boller, T. (1988) Antifungal hydrolases in pea tissue. I. Purification and characterization of two chitinases and two β-1,3-glucanases differentially regulated during development and in response to fungal infection. *Plant Physiology*, **87**, 325-33

Méndez, E., Moreno, A., Colilla, F., Pelaez, F., Lìmas, G.G., Mendez, R., Soriano, F., Salinas, M. and de Haro, C. (1990) Primary structure and inhibition of protein synthesis in eukaryotic cell-free system of a novel thionin, γ-hordothionin, from barley endosperm. *European Journal of Biochemistry*, **194**, 533-39

Meyer, B., Houlné, G., Pozueta-Romero, J., Schantz, M-L. and Schantz, R. (1996) Fruit-specific expression of a defensin-type gene family in bell pepper. Upregulation during ripening and upon wounding. *Plant Physiology*, **112**, 615-22

Milligan, S.B .and Gasser, C.S. (1995) Nature and regulation of pistil-expressed genes in tomato. *Plant Molecular Biology*, **28**, 691-711

Molano, J., Polacheck, I., Duran, A. and Cabib, E. (1979) An endochitinase from wheat germ. *Journal of Biological Chemistry*, **254**, 4901-7

Molina, A., Ahl-Goy, P., Fraile, A., Sánchez-Monge, R. and García-Olmedo, F. (1993a) Inhibition of bacterial and fungal plant pathogens by thionins of types I and II. *Plant Science*, **92**, 169-77

Molina, A., Segura, A. and García-Olmedo, F. (1993b) Lipid transfer proteins (nsLTPs) from barley and maize leaves are potent inhibitors of bacterial and fungal plant pathogens. *FEBS Letters*, **316**, 119-22

Moreno, M., Segura, A. and García-Olmedo, F. (1994) Pseudothionin-St1, a potato peptide active against potato pathogens. *European Journal of Biochemistry*, **223**, 135-39

Nielsen, K.K., Nielsen, J.E., Madrid, S.M. and Mikkelsen, J.D. (1996) New antifungal proteins from sugar beet (*Beta vulgaris* L.) showing homology to non-specific lipid transfer proteins. *Plant Molecular Biology*, **31**, 539-52

Nitti, G., Orrù, S., Bloch, C. Jr, Morhy, L,. Marino, G. and Pucci, P. (1995) Amino acid sequence and disulphide-bridge pattern of three γ-thionins from *Sorghum bicolor*. European *Journal of Biochemistry*, **228**, 250-56

Ogata, C.M., Gordon, P.F., de Vos, A.M. and Kim, S-H. (1992) Crystal structure of a sweet tasting protein thaumatin I, at 1.65 Å resolution. *Journal of Molecular Biology*, **228**, 893-908

Osborn, R.W., De Samblanx, G.W., Thevissen, K., Goderis, I., Torrekens, S., Van Leuven, F., Attenborough, S., Rees, S.B. and Broekaert, W.F. (1995) Isolation and characterisation of plant defensins from seeds of Asteraceae, Fabaceae, Hippocastanaceae and Saxifragaceae. *FEBS Letters*, **368**, 257-62

Roberts, W.K. and Selitrennikoff, C.P. (1990) Zeamatin, an antifungal protein from maize with membrane-permeabilizing activity. *Journal of General Microbiology*, **136**, 1771-78

Seetharaman, K., Waniska, R.D. and Rooney, L.W. (1996) Physiological changes in sorghum antifungal proteins. *Journal of Agricultural Food Chemistry*, **44**, 2435-41

Sharma, P. and Lönneborg, A. (1996) Isolation and characterization of a cDNA encoding a plant defensin-like protein from roots of Norway spruce. *Plant Molecular Biology*, **31**, 707-12

Shin, D.H., Lee, J.Y., Hwang, K.Y., Kim, K.K. and Suh, S.W. (1995) High-resolution crystal structure of the non-specific lipid-transfer protein from maize seedlings. *Structure*, **3**, 189-99

Stanchev, B.S., Doughty, J., Scutt, C.P., Dickinson, H. and Croy, R.R.D. (1996) Cloning of *PCP1*, a member of a family of pollen coat protein (*PCP*) genes from *Brassica oleracea* encoding novel cysteine-rich proteins involved in pollen-stigma interactions. *Plant Journal*, **10**, 303-13

Stirpe, F., Barbieri, L., Battelli, M.G., Soria, M. and Douglas, A.L. (1992) Ribosome-inactivating proteins from plants: present status and future prospects. *Biotechnology*, **10**, 405-12

Swegle, M., Kramer, K.J. and Muthukrishnan, S. (1992) Properties of barley seed chitinases and release of embryo-associated isoforms during early stages of imbibition. *Plant Physiology*, **99**, 1009-14

Tailor, R., Acland, D.P., Attenborough, S., Cammue, B.P.A., Evans, I.J., Osborn, R.W., Ray, J., Rees, S.B. and Broekaert, W.F. (1997) A novel family of small cysteine-rich antimicrobial peptides from seed of *Impatiens balsamina* is derived from a single precursor protein. *Journal of Biological Chemistry*, **272**, 24480-87

Tassin, S., Broekaert, W.F., Marion, D., Acland, D.P., Ptak, M., Vovelle, F. and Sodano, P. (1998) Solution structure of Ace-AMP1, a potent antimicrobial protein extracted form onion seeds. Structural analogies with non-specific lipid transfer proteins. *Biochemistry* (in press)

Terras, F.R.G., Goderis, I.J., Van Leuven, F., Vanderleyden, J., Cammue, B.P.A. and Broekaert, W.F. (1992a) *In vitro* antifungal activity of a radish (*Raphanus sativus* L.) seed protein homologous to nonspecific lipid transfer proteins. *Plant Physiology*, **100**, 1055-58

Terras, F.R.G., Schoofs, H.M.E., De Bolle, M.F.C., Van Leuven, F., Rees, S.B.,Vanderleyden, J., Cammue, B.P.A. and Broekaert, W.F. (1992b) Analysis of two novel classes of antifungal proteins from radish (*Raphanus sativus* L.) seeds. *Journal of Biological Chemistry*, **267**, 15301-09

Terras, F.R.G., Schoofs, H.M.E., Thevissen, K., Osborn, R.W., Vanderleyden, J., Cammue, B.P.A. and Broekaert, W.F (1993a) Synergistic enhancement of the antifungal activity of wheat and barley thionins by radish and oilseed rape 2S albumins and by barley trypsin inhibitors. *Plant Physiology*, **103**, 1311-19

Terras, F.R.G., Torrekens, S., Van Leuven, F., Osborn, R.W., Vanderleyden, J., Cammue, B.P.A. and Broekaert, W.F. (1993b) A new family of basic cysteine-rich plant antifungal proteins from Brassicaceae-species. *FEBS Letters*, **316**, 233-40

Terras, F.R.G., Eggermont, K., Kovaleva, V., Raikhel, N.V., Osborn, R.W., Kester, A., Rees, S.B., Torrekens, S., Van Leuven, F., Vanderleyden, J., Cammue, B.P.A. and Broekaert, W.F. (1995) Small cysteine-rich antifungal proteins from radish: Their role in host defense. *The Plant Cell*, **7**, 573-88

Terras, F.R.G., Torrekens, S., Van Leuven, F. and Broekaert, W.F. (1996) A six-cyteine type thionin from the radish storage organ displays weak *in vitro* antifungal activity against *Fusarium culmorum*. *Plant Physiology and Biochemistry*, **34**, 599-603

Thevissen, K., Ghazi, A., De Samblanx, G.W., Brownlee, C., Osborn, R.W. and Broekaert, W.F. (1996) Fungal membrane responses induced by plant defensins and thionins. *Journal of Biological Chemistry*, **271**, 15018-25

Thevissen, K., Osborn, R.W., Acland, D.P. and Broekaert, W.F. (1997) Specific, high affinity binding sites for an antifungal plant defensin on *Neurospora crassa* hyphae and microsomal membranes. *Journal of Biological Chemistry*, **272**, 32176-81

van Loon, L.C., Pierpoint, W.S., Boller, T. and Conejero, V. (1994) Recommendations for naming plant pathogenesis-related proteins. *Plant Molecular Biology Reporter*, **12**, 245-64

Van Parijs, J., Broekaert, W.F., Goldstein, I.J. and Peumans, W.J. (1991) Hevein: an antifungal protein from rubber-tree (*Hevea brasiliensis*) latex. *Planta*, **183**, 258-64

Varghese, J.N., Garrett, T.P.J., Colman, P.M., Chen, L., Høj, P.B. and Fincher, G.B. (1994) Three-dimensional structures of two plant β-glucan endohydrolases with distinct substrate specificities. *Proceedings of the National Academy Science USA*, **91**, 2785-89

Vigers, A.J., Wiedemann, S., Roberts, W.K., Legrand, M., Selitrennikoff, C.P. and Fritig, B. (1992) Thaumatin-like pathogenesis-related proteins are antifungal. *Plant Science*, **83**, 155-61

Vögeli-Lange, R., Fründt, C., Hart, C.M., Beffa, R., Nagy, F. and Meins, F. Jr (1994) Evidence for a role of β-1,3-glucanase in dicot seed germination. *Plant Journal*, **5**, 273-78

32. LEA Proteins

ANDREW C. CUMING
Department of Biology, Leeds University, Leeds, UK

1. INTRODUCTION

1.1 Embryonic development

Embryogenesis in plants can be thought of as a series of stages leading to the formation of a mature, differentiated, miniature plant within the seed. Embryogenesis is initiated by fertilisation, in which the fusion of a sperm nucleus and an egg nucleus gives rise to the diploid zygote. The fusion of a second sperm nucleus with the two polar nuclei within the embryo sac produces the lineage that ultimately becomes the triploid endosperm.

The development of the zygote passes through an initial stage of cellular proliferation, in which the early differentiation of the characteristic tissues of the embryo occurs. The differentiation stage of development culminates in the formation of a viable embryo that is normally held in a state of primary dormancy within the developing seed, thus preventing inappropriately premature germination.

The dormant embryo remains metabolically active, passing through the maturation stage of development. This is characterised by an increase in the dry mass of the embryo, as it accumulates storage compounds, and in the majority of cases maturation is accompanied by a progressive dehydration of the seed. The final product of the plant's reproductive process is thus a metabolically inert, dry and highly resilient propagule that can be dispersed in both space and time. The dehydrated state allows the seed to be stored dry, without significant loss of viability, at least until the next growing season. Frequently storage is possible for many years, until germination is triggered by the uptake of water, in combination with environmental dormancy-breaking cues.

P.R. Shewry and R. Casey (eds.), Seed Proteins, 753–780.

1.2 Embryonic gene expression

The successive stages of embryonic development are underpinned by changes in the pattern of gene expression, as gene products required for, and specific to each phase of development are synthesised. Over the last 20 years, the identification of stage-specific genes, the analysis of their products and of their modes of regulation has been a major area of research in seed biology. Much of this research has focused on the period of embryonic maturation. During this period, the principal storage compounds of seeds are laid down, and understanding and manipulating this process is of clear and direct agricultural relevance. Additionally, the study of seed maturation allows insights into the control of seed dormancy and its breakage, and the acquisition of desiccation tolerance by plant tissues. The class of embryonic proteins known as LEA proteins is thought to be intimately associated with this last property.

The term "LEA protein" is an acronym, coined by Galau, Hughes and Dure (1986). It describes a subset of gene products, seen to accumulate to high levels during the maturation phase of cotton seed development. The proteins and their cognate messenger RNAs were thus *L*ate *E*mbryogenesis *A*bundant. The LEA gene products were distinct from storage proteins, in that their mRNAs began to accumulate relatively later during embryonic development than did those encoding storage proteins. Moreover, their mRNAs persisted at a high level throughout the period of desiccation, remaining abundantly present within the dry embryo: typically transcripts encoding storage proteins are degraded during the last stages of embryogenesis. In developing a nomenclature for this group of proteins, Galau *et al.* proposed that LEA proteins were encoded by *Lea* mRNAs, which were in turn the products of *Lea* genes. I shall adhere to this nomenclature throughout this chapter.

1.3 The subclasses of LEA proteins

The identification of *Lea* mRNAs as a distinct class of the cellular mRNA population followed from the cloning of a number of *Lea* cDNA sequences, which were used as hybridisation probes in order to define their temporal pattern of accumulation during embryogenesis (Galau *et al.*, 1986). The subsequent sequence analysis of these clones enabled the amino acid sequences of the LEA proteins to be deduced (Baker *et al.*, 1988), and to be compared with sequences emerging from studies of seed development in other species. It rapidly became clear from these comparisons that the sequences obtained from cotton were representative of a broad class of highly conserved genes expressed in a wide range of plants with the same developmental specificity.

Comparisons of the deduced polypeptide sequences of the various *Lea* cDNA and genomic clones have led to the establishment of three major subclasses of LEA protein. These have been designated the Group 1, Group 2 and Group 3 LEA proteins (Dure *et al*., 1989). Additionally, two further, less frequently represented classes have been identified as Group 4 and Group 5 LEA proteins. The properties of these classes are summarised in Table 1, and described in detail in section 2, below.

Table 1 Characteristics of LEA protein classes.

Group	Consensus	Structural features	Other features	Example
1	GGQTRREQLGEEGYSQMGRK	Apparently random-coil	No cysteine	Wheat "Em" protein
2	EKKGIMDKIKEKLPG $^{V}/_{T}$DEYGNP	Amphipathic α-helices	Serine-rich stretches	"Dehydrins"
3	TAQAAKEKAXE	Amphipathic α-helices		Cotton LEA D-7, D-29
4	Conserved N-termini	*N*-terminal α-helices *C*-terminal random coil		Cotton LEA D-113
5		Heat insoluble, probably globular		Cotton LEA D-34 Maize Rab-28

It is now clear that the synthesis and accumulation of LEA proteins is not an exclusive feature of seed maturation. *Lea* mRNA and LEA protein accumulation can be observed to occur in vegetative tissues of a very wide range of plants, in response to a number of environmental and hormonal cues. The common feature linking these cues is that they are all associated with the imposition of water deficit stress. Thus, in addition to the onset of dehydration by developing seeds, LEA protein synthesis can be induced by drought stress, cold-acclimation and freezing-stress of vegetative tissues. The expression of the *Lea* genes can also be induced by treatment of tissues with the plant growth regulator abscisic acid (ABA) - centrally implicated in water stress responses in plants - and jasmonic acid (JA), a component in a number of responses to biotic and abiotic stresses. Consequently, there is a general consensus among researchers (supported almost entirely by correlative and circumstantial evidence, rather than by direct experimental demonstration) that the LEA proteins are responsible for ameliorating the consequences of water deficit stress within cells. This is a consequence of the removal of water by evaporative or osmotic dehydration or by the formation of intra- or extracellular ice. These proteins are therefore supposed to mediate the acquisition of desiccation tolerance in anhydrobiotic structures (principally seeds, and the vegetative tissues of so-called "resurrection plants"), which survive extremes of desiccation.

2. PROTEIN SEQUENCE AND STRUCTURE OF LEA PROTEINS

2.1 General characteristics

The primary structures of a large number of LEA proteins have been established, largely by deduction from cloned DNA sequences. From these studies it is possible to highlight some features common to the Group 1, 2, 3 and 4 LEA proteins. Typically, these proteins lack cysteine residues and are predominantly composed of charged and uncharged, polar amino acid residues. Analysis using the Kyte-Doolittle hydropathy algorithm shows that the LEA proteins are overwhelmingly hydrophilic (Figure 1). This is borne out experimentally by an examination of their physical properties, particularly that of their ability to remain in solution when extracts of embryonic proteins are boiled for 10-20 minutes. Under such conditions, proteins are denatured and the majority of cellular proteins aggregate through illegitimate hydrophobic interactions and can be precipitated by brief centrifugation. Lacking extensive hydrophobic tracts, most LEA proteins are not precipitated by this treatment - a characteristic that facilitates their purification substantially. However, very little is known of the higher orders of structure within these groups of protein. It is clear, from sequence comparisons, that a high degree of sequence homology exists between members of these protein families throughout the plant kingdom, particularly within sequence domains that have been observed to be repeated, to different extents, within individual members of each family. A high degree of sequence conservation is strongly suggestive of evolutionary parsimony, where a strongly conserved sequence motif has specific and important structural and/or functional consequences.

Unfortunately there is, as yet, little direct evidence to support the structural models for LEA proteins that have been developed both from a consideration of their amino acid sequences and from a supposition of their biological function. Although there is persuasive circumstantial evidence indicating a function in the acquisition of desiccation tolerance by the developing embryo, direct evidence that unambiguously supports this role is conspicuous by its scarcity. Nevertheless, the nature of the scientific process is that it relies on hypothetical models being advanced, that are amenable to experimental test, and in the subsequent sections, I shall outline the current hypotheses concerning the structure-function relationships of the LEA proteins.

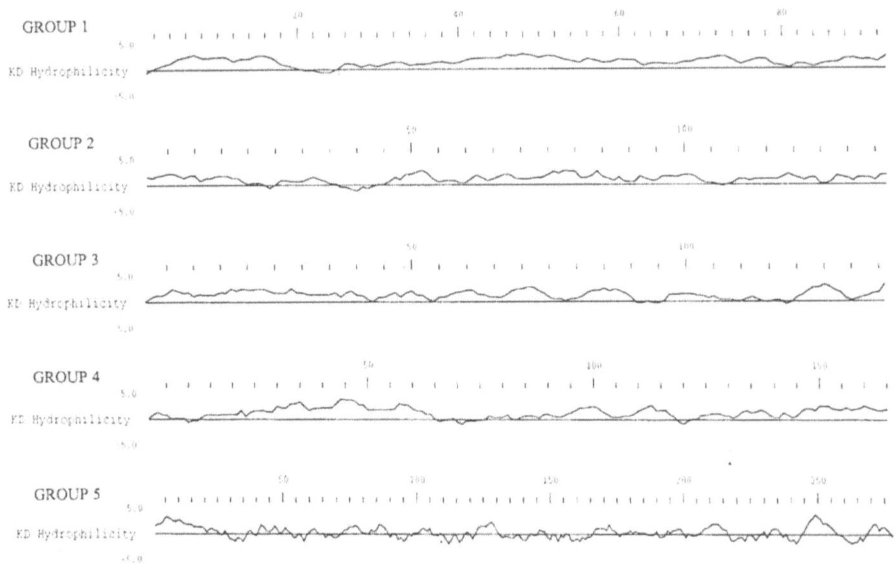

Fig. 1 Hydropathy of LEA proteins. The hydropathic properties of the residues within representatives of the five groups of LEA proteins are illustrated, as a plot derived from the Kyte-Doolittle hydropathy algorithm. Hydrophilic residues fall above the line, hydrophobic residues fall below the line. Group 1 LEA proteins are represented by the wheat Em protein; Group 2 by the barley dehydrin, dhn-7; Group 3 by the cotton LEA D-7 protein; Group 4 by the cotton LEA D-113 protein; Group 5 by the maize rab-28 protein.

2.2 Group 1 LEA proteins.

The Group 1 LEA proteins are typified by the "Em" protein of wheat embryos (Grzelczak *et al.*, 1982). Sequence analysis of the original cotton *Lea* cDNA clones revealed that one of these, designated *Lea* D-19, exhibited high sequence homology with that previously obtained for the wheat protein (Litts *et al.*, 1987). Em homologues have now been identified in a wide range of plants. The highest degree of sequence conservation within this group of proteins is seen in a 20-amino acid domain, GGQTRREQLGEEGYSQMGRK. In most of the Group 1 LEA proteins characterised to date this domain is present only once. However, in the barley Em gene family, of which four members have been cloned, this sequence has undergone duplication, being present once in HVB19.1a and HVB19.1b, three times in HVB19.3 and in four copies in HVB19.4 (Rap *et al.*, 1995), and in soybean a similar duplication of this motif has been observed (Calvo *et al.*, 1997).

Until recently, it could be argued that that there was more experimentally based structural information available for the Group 1 proteins than for any other class. However, this was only true insofar as all the available evidence indicates that these proteins have no defined structure! McCubbin *et al.* (1985) undertook a detailed study of the physicochemical properties of the wheat Em protein, in which circular dichroism (CD) spectroscopy indicated the polypeptide to exist largely as a random coil. The primary sequence is distinguished by a relatively large number of glycine residues (16 out of 93) which, by virtue of allowing free rotation about the peptide bond, would afford the polypeptide chain a high degree of flexibility. Russouw *et al.* (1997) have carried out similar studies with an homologous protein isolated from pea embryos, and found no significant changes to occur in the CD spectra before and after heating to 80 C. This indicated that the heat solubility of the protein stems from its inability to be irreversibly heat-denatured (although it could be argued that a protein existing as a random coil represents a permanently denatured structure). Interestingly, however, a spectral transition was observed as the salt concentration was varied between 0.1M and 0.5M NaCl, suggesting that the protein could undergo an ionically mediated conformational change. The nature of this change remains enigmatic.

2.3 Group 2 LEA proteins

The Group 2 LEA proteins are exemplified, in cotton, by the *Lea* D-11 gene product. This protein was found to be highly homologous with a sequence isolated from rice, encoding a mRNA for a *Mr* 21,000 polypeptide accumulating in response to the plant growth regulator abscisic acid, and hence designated *Rab21* (for *R*esponsive to *AB*A: Mundy and Chua, 1988; this has since been redesignated *Rab 16*). Other homologues included a number of sequences isolated as cDNA clones corresponding to mRNAs induced in maize and barley seedlings upon the onset of dehydration stress, and consequently designated "dehydrins" (Close *et al.*, 1989). This nomenclature has achieved considerable popularity, and has been used by the majority of authors in preference to "Group 2 LEA".

Over 100 dehydrin homologues, ranging in relative molecular mass between 9,000 and 200,000, have been identified as cloned sequences from a range of plants extending through widely diverged taxa. These include the green alga *Chlorella*, gymnosperms and both dicotyledonous and monocotyledonous angiosperms. Additionally, immunologically-related proteins have been observed in pteridophytes, bryophytes and even in cyanobacteria, although in these taxa the putative dehydrins remain to be definitively identified by sequence analysis.

The properties of dehydrins have been recently and comprehensively reviewed by Close (1996, 1997). Three characteristic sequence "modules" define this class of protein. All dehydrin proteins contain a lysine-rich 15-mer with the consensus sequence EKKGIMDKIKEKLPG. This domain, designated the "K-domain", may be present in multiple copies, the number varying between one and 11. Typically, the K-domain located nearest the carboxy-terminus of the polypeptide retains the closest correspondence with the consensus sequence. A second motif, located near the *N*-terminus, is the so-called "Y-domain", with the consensus sequence $^{V}/_{T}$DEYGNP. This sequence is similar to the nucleotide binding domain found in the *GroEL* and *GroES* chaperonins, although no such function has yet been identified among the dehydrin proteins. Finally, the third feature commonly observed in dehydrins is a stretch of contiguous serine residues - the "S-segment" - which can be sites of protein phosphorylation (Vilardell *et al.*, 1990). The phosphorylation of the S-segment has been suggested to be associated with the provision of a signal for the nuclear localisation of some dehydrin species. (Goday *et al.*, 1994). These motifs are separated within the dehydrin sequence by stretches of largely polar amino acids for which no secondary structure can be predicted, and which are suggested to provide extended, flexible domains - the "Φ-domains" (Close, 1997).

Not all of the dehydrin-like proteins contain all of these motifs, and Close (1996) has classified this family of proteins into five subgroups based on the presence, order and number of the Y, S and K-domains. Of these, the most prevalent class comprises the "Y_nSK_2" proteins, (where n = 1, 2 or 3). The synthesis of these proteins is strongly induced by dehydration of plant tissue. Synthesis of a second class, which contain only K-domains (the "K_n" subgroup) is induced both by dehydration and by low-temperature stress, whilst that of a third class (the "K_nS" group) is induced by both stresses. A small number of proteins are identified as falling into two additional classes ("SK_n" and "Y_2K_n") and Close (1996) identifies two further proteins which appear to be hybrid in nature: an "SK_3S" protein from *Stellaria*, and a "YK_n" protein from spinach.

Like the Group 1 LEA proteins, the dehydrins appear to be largely unstructured proteins. However, physical analysis of a purified maize dehydrin by CD spectroscopy reportedly revealed an α-helical content of approximately 10-15%, with 75% of the polypeptide appearing disordered (Ceccardi *et al.*, 1994; Close, 1996). Secondary structure prediction models indicate the principal component likely to contribute the helical content of the protein to be the two K-domains. This sequence has the potential to form an amphipathic α-helix in which the hydrophilic and hydrophobic residues are disposed on opposite faces of the helix: a similar structural motif (derived from an unrelated sequence) can be observed in the Group 3 LEA proteins (Dure *et al.*, 1989; Dure, 1993), and the properties of such helices in both

Group 2 and Group 3 LEA proteins are discussed at greater length in section 2.4.

2.4 Group 3 LEA proteins

In common with the Em-like and dehydrin classes of LEA proteins, the group 3 LEA proteins are also represented among the late embryogenesis-abundant proteins of other plant groups. These proteins also share a characteristic repetitive amino acid sequence motif. In the Group 3 LEA proteins this comprises an 11-mer that appears to have undergone an original duplication, and subsequent limited sequence divergence, in an ancestral gene to produce a contiguous pair of 11 amino acid sequences. Subsequent duplication and divergence of the 22-mer have given rise to additional copies of the 11-residue sequence (Dure *et al.*, 1989). Although these individual sequence elements have not retained identical amino acids in each position, there have clearly been constraints on the type of substitution allowed in the evolution of the gene family. This provides a strong indication that the motif has important structural and functional consequences for this class of LEA proteins.

Following a comparison of the 11-mer repeat found in a range of Group 3 LEA proteins, Leon Dure and his colleagues have developed a model structure that might be adopted by proteins of this class. The overall consensus sequence derived for the motif shows that there is a clear preference for particular classes of residue in specific positions in the sequence (Dure *et al.*, 1989; Table 2). Thus in positions 1,2,4,5 and 9 residues whose side-chain is terminated by a methyl group are clearly preferred. Residues 3, 7 and 11 have polar side-chains, and residues 6 and 8 tend to be positively charged. Only in position 10 does there appear to be no constraint on the chemical property of the amino acid. Dure (1993) has used computational structure prediction programs to derive a structural model of a

Table 2 Conservation of properties within the Group 3 LEA repeat motif.
A consensus sequence for the LEA Group 3 tandemly repeated 11-mer sequence was derived from Group 3 LEAs of five plant species (Reproduced from Dure, 1993 by kind permission of the author and publishers)

Position	1	2	3	4	5	6	7	8	9	10	11
	23 Thr 12 Ala 6 Lys 2 Met	28 Ala 12 Thr 2 Gly 1 Met	20 Gln 19 Glu 4 Gly	29 Ala 5 Thr 3 Ser 3 Tyr 1 Glu 1 Lys 1 Cys	26 Ala 13 Thr 2 Val 1 Ser 1 Gly	34 Lys 2 Arg 5 Gln 1 Asn 1 Asp	22 Gln 11 Glu 9 Asp 1 Gly	36 Lys 3 Arg 3 Thr 1 Ser	28 Ala 15 Thr	12 Ser 8 Ala 5 Gln 4 Gly 3 Val 2 Met 2 Arg 2 Tyr 2 Glu 1 His 1 Lys 1 Cys	25 Glu 3 Asp 9 Gln 1 Asn 3 Ala 2 Gly
Consensus	Thr/Ala	Ala/Thr	Gln/Glu	Ala/Thr	Ala/Thr	Lys/Arg	Gln /Glu-Asp	Lys/Arg	Ala/Thr	Any	Glu-Asp/ Gln
	35/43	40/43	39/43	34/43	39/43	36/43	42/43	39/43	43/43		38/43
Sequence used for modelling	Thr	Ala	Gln	Ala	Ala	Lys	Glu	Lys	Ala	Gly	Glu

protein domain composed of a contiguous series of such motifs. Using the consensus peptide sequence $(TAQAAKEKAGE)_3$ the resulting structure comprised an amphipathic α-helix, in which the methylated residues at positions 1,2,5 and 9 formed a "hydrophobic stripe" along one face of the helix, whilst the polar and charged residues in positions 3.6,7,8 and 11 were distributed around the wider, hydrophilic face of the helix (Plate 12 panels a and b). It was predicted that additional stability could be conferred on the helical structure by the close juxtaposition of the charged residues K_8 and E_{11}, which could be linked by ionic bridges, and by a charge-dipole interaction between Q_3 and E_7.

Amphipathic helices such as this provide a means whereby intra- and inter-molecular interactions can occur. Such structures exist in mammalian lipoproteins, in which amphipathic helices associate to form bundles: in apolipoprotein E, which also contains an 11-mer repeating motif, a 4-helix bundle is formed by the hydrophobic interactions occurring between the side-chains of non-polar residues lying along one face of the helix, whilst the highly hydrophilic faces of the helices provide a highly polar globular surface to the external environment. Such a conformation has been predicted for the cotton Group 3 LEA D-29 protein (Dure *et al*., 1989), and for a Group 3 LEA protein from the resurrection plant, *Craterostigma plantagineum*, both of which have the potential to form four amphipathic helices. In these potential structures, a significant number of the methylated residues are relatively bulky, allowing the helices to associate at sufficient distance to form a 4-helix bundle, with the non-polar side chains packed into the centre.

However, these two proteins appear somewhat anomalous, by comparison with the majority of Group 3 LEA sequences, in which the methylated residues are principally alanine, threonine and valine - all with small side-chains and consequently possessing a relatively "flat" hydrophobic surface (Dure, 1993). Amphipathic helices with this composition could potentially form dimers, through the close association of the flat hydrophobic surfaces (Plate 12 panel c), but a higher-order complex would be sterically unfavourable: without bulkier side-chains able to interact at (relatively) long range, there would be excessive charge repulsion between the polar surfaces brought into close proximity. Dure has reported that the cotton LEA D-7 protein exhibits dimer-like behaviour in the course of its purification, but this has yet to be formally demonstrated. The dimeric structure formed by two amphipathic Group 3 α-helices would, uniquely, produce a right-handed coiled-coil structure, rather than the classical left-handed coiled-coil normally formed by the association of right-handed α-helices. Clearly, structural studies will be necessary to determine the nature of higher-order structures within the Group 3 LEA proteins.

Similar amphipathic helices have been predicted to occur in the Group 2 LEA proteins (dehydrins). Interestingly, under conditions of increased ionic strength, a purified maize dehydrin (section 2.2) appeared to undergo a conformational change, interpreted as a reversible dimerization, mediated by hydrophobic association of two monomers, *in vitro* (Close, 1996). Since this Group 2 dehydrin lacks any contiguous stretch of hydrophobic residues within its primary structure, it must be presumed that any such hydrophobic interaction must have been through hydrophobic interfaces created in the secondary structures adopted by the polypeptide. Further evidence for the formation of higher-order structures derives from observations made during the purification of a Group 2 LEA protein accumulated in cold-acclimated spinach leaves. The "COR85" polypeptide (*Mr* = 85,000) - was shown to migrate as an oligomer of *Mr ca.* 350,000 oligomer in non-denaturing polyacrylamide gel electrophoresis, and in size exclusion (gel-filtration) chromatography (Kazuoka and Oeda, 1994).

In contrast, structural analysis of a recombinant dehydrin synthesised in *E. coli*, the *Mr* 16,000 "Dsp16" polypeptide of the "resurrection plant", *Craterostigma plantagineum*, did not fully accord with the structural predictions (Lisse *et al.*, 1996). In dilute aqueous solution, neither CD spectroscopy nor NMR analysis revealed a significant α-helical component within the protein. Indeed, no defined three-dimensional structure could be discerned, the data suggesting that the protein adopted an unfolded, although not wholly random-coil, conformation. Conformational change suggestive of helix formation could be observed in the protein following the addition of the cosolvent, trifluoroethanol. This is widely used in protein structural studies as a solvent of highly hydrophobic molecules and is known to promote the formation of helical structures by altering the interactions that occur between the amino acids and the solvent molecules. Whether the structure adopted by a recombinant (and thus highly pure) Dsp16 protein in aqueous buffers is a true representation of its state in dehydrating plant tissues is not clear. Certainly, the cytosol will provide a far more complex environment than the test-tube, in which concentrations of this and other LEA proteins, and of ionic and non-ionic solutes will be increasing during desiccation. Roberts *et al.* (1993) have reported that the Group 3 LEA D-7 protein and the Group 4 LEA D-113 protein accumulate to concentrations as high as 0.23 and 0.27 mM, respectively, in the cytosol of mature cotton embryos. Interestingly, Walters *et al* (1997) have reported that the proteins of dry wheat embryos that remain soluble, after boiling (principally a mixture of Group 1, 2 and 3 LEA proteins), are tenaciously associated with non-covalently bound sugars, probably sucrose and raffinose, which are abundant in dry embryos. The sugar-protein complex would necessarily involve extensive hydrogen bonding between the sugar moiety and the polar side-chains of the LEA proteins. This might reasonably be expected to modify the nature of the interactions in which the potential structure-forming elements of the polypeptide chain could participate.

The extensive sequence conservation of the potentially helix-forming elements of the Group 2 and Group 3 LEA proteins does suggest a structural or functional role. Of particular interest in this regard is the identification of non-plant homologues of these motifs. Recently, a sequence was identified in a protein from a quail heart cell line containing a repeated motif homologous with the consensus from the Group 3 LEA proteins (Niu *et al.*, 1996). The remainder of this avian sequence contained no other similarity with the plant sequences. Additionally, Close (1997) has reported a yeast protein with immunological cross-reactivity to an antibody directed against the dehydrin "K-domain", although searches of the recently published complete yeast genome DNA sequence revealed no open reading frames with dehydrin homology.

A consideration of the evolution of protein structures indicates that most proteins have evolved by the assembly, and continual shuffling, of modular sequence elements, contributing specific structural motifs to build up the individual three-dimensional forms of the existing population of enzymes and structural proteins. It is not unreasonable to postulate that the amphipathic helix represents such a structural module (exemplified by the presence of such structures in proteins as diverse as LEA proteins and apolipoproteins). Such sequences might therefore be found as a conserved structural element in other, unrelated, proteins where such a specific 3-dimensional motif is required for protein function.

2.5 Group 4 LEA proteins

This group is represented in cotton by the LEA D113 protein (Baker *et al.*, 1988). Related gene products have been identified in other species, accumulating in non-seed tissues subjected to dehydration stress, or undergoing acclimation to low temperatures. These include the sunflower "ds11" gene product and the tomato "LE25" protein. The Group 4 LEA proteins appear bipartite in nature: the *N*-terminal domains of the group 4 family members are relatively conserved in their amino acid sequences, and structure prediction suggests that this part of the molecule is capable of forming long α-helices (Ingram and Bartels, 1996). When such predictive analysis was applied to the tomato LE25 polypeptide sequence, two such helices were predicted. Helical wheel displays of the amino acid residues indicated that the side-chains were clustered in domains around the circumference of the helix (Imai *et al*, 1996). In the first helix, four distinct sectors were predicted, moving clockwise from a hydrophobic domain, through a sector of positive charge, followed by a predominantly negatively charged domain opposite the hydrophobic region. Another positively charged sector completed the circle. This structure is similar to the amphipathic helices bearing a hydrophobic stripe, seen in the Group 2 and Group 3 LEA proteins. The second helix also exhibited a non-random disposition of different amino-acid side chains, in this instance two distinct domains - one positively charged and the other negative - lying on opposite faces of the helix.

The *C*-terminal domains of the different members of this class of protein do not exhibit substantial amino acid sequence conservation, and they differ in length, relative to one another. However, in each case, structure prediction suggests that the *C*-terminal domains comprise a highly flexible and hydrophilic random coil, distinguished by a relatively high content of glycine residues.

2.6 Group 5 LEA proteins

Of the six *Lea* mRNA sequences published by Baker *et al.* (1988), the LEA D-34 polypeptide sequence is a representative of the Group 5 LEA proteins. A second cotton protein of this class (LEA D-95) has since been recognised. Other members of this class include the ABA and dehydration-inducible pcC27-45 sequence of *Craterostigma plantagineum* (Ingram and Bartels, 1996), a carrot embryogenesis-related protein (ECP31) and its *Arabidopsis* homologue (AtECP31: Yang *et al.*, 1996) and the maize *rab28* gene product (Pla *et al.*, 1991). These proteins are distinctively different from the other four groups of LEA protein, in that they contain a significantly higher proportion of hydrophobic residues (Figure 1) and do not share the property of boiling-solubility. This suggests that they adopt a globular conformation, but the nature of this structure remains unknown.

3. LEA PROTEINS: A ROLE IN ANHYDROBIOSIS?

3.1 The nature and distribution of anhydrobiosis

Anhydrobiosis is a property exhibited by living organisms that survive prolonged periods of extreme desiccation. This ability is usually restricted to specialised structures, such as cysts, spores or seeds. However, within the plant kingdom the capacity for anhydrobiosis is widespread in the "adult" or vegetative forms of lower plants, notably algae and bryophytes. These plants colonise habitats subject to sporadic availability of water, and they lack specialised organ systems, such as roots, for the retrieval of moisture deep in the soil. Among higher plants, anhydrobiosis is limited largely to seeds and some forms of pollen.

The progressive dehydration of seeds is a natural process, representing an important adaptation of land plants - the ability to produce resilient propagules, to ensure the survival of the species during times of environmental stress. Anhydrobiosis is also occasionally exhibited by vegetative tissues among a group of plants designated "resurrection plants", on the basis of their ability to recover from a completely dehydrated state without apparent deleterious consequences.

3.2 The cellular consequences of dehydration

Cellular dehydration can be caused by evaporative water loss, by exposure to medium of high salt concentration (ionic stress) or by low temperature, in which water is removed through the formation of ice crystals. The intracellular consequences of water loss, through whatever means, will be identical. In the short term, plants respond to osmotic stress by osmoregulation - the accumulation of osmotically active, but non-ionic solutes ("compatible solutes") within the cell. These may include sugars, or compounds such as proline or glycine-betaine, all of which have been demonstrated to accumulate in moderately stressed plant tissue (McCue and Hanson, 1990).

Further dehydration can have severe consequences for the macromolecular components of the cell. The ionic concentration of the cytosol will increase, with contingent perturbation of concentration-dependent biochemical processes (enzyme reactions) and of macromolecular interactions stabilised by ionic bridges. Additionally, following the withdrawal of bulk water which acts as a solvent for the free diffusion of molecules within the cell, and the consequent biochemical activities within it, the removal of "bound water" from proteins can cause irreversible denaturation (see 3.5).

Dehydration can also cause serious disruption of biological membranes. Phospholipid molecules in lipid-water systems are typically arranged in bimolecular sheets stabilised by the hydrophobic interactions between the fatty acid chains, and the hydrophilic association of the polar head groups with the aqueous phase. When the water content is reduced to about 20%, a conformational change occurs. The only water molecules remaining are those adsorbed to the polar head groups, and the bilayer assumes an hexagonal array of water-filled "channels", as the hydrophobic interactions become the dominant intermolecular forces. This results in the breakdown of cellular compartmentation and consequent leakage of the cellular contents occurring upon rehydration. In order to survive the extremes of desiccation typically achieved in anhydrobiotic structures (as low as 5% original moisture content in many seeds) it is necessary to ameliorate the deleterious effects of water loss in some way.

3.3 Correlative associations between LEA proteins and desiccation

Since the earliest identification of LEA proteins, hypotheses regarding their functions have revolved around a potential function in the acquisition of desiccation tolerance. Embryonic maturation - the time of LEA protein accumulation - is a period when the seed is undergoing a progressive and programmed desiccation..

Alternative functions for the majority of LEA proteins have been largely discounted. No enzymatic function has been assigned to any of the LEA proteins described above, and although novel enzymatic activities have been identified for some proteins which accumulate with the same developmental specificities (for example, an aldose reductase and a glucose dehydrogenase in barley embryos: Bartels *et al.*, 1991; Alexander *et al.*, 1994), the LEA proteins accumulate to a much higher cellular concentration than is typically the case for enzymes. Additionally, the predicted structural flexibility inherent in the Group 1 and Group 2 LEA proteins, at least, would appear to rule out an enzymatic role.

A storage function for LEA proteins also seems unlikely, although not impossible. The LEA proteins appear to be principally located within the cytosol, in some instances in the nucleus (Goday *et al*. 1994) and (in the case of a cold-regulated LEA protein, expressed in leaves) targeted to chloroplasts (Lin and Tomashow, 1992). In contrast, the storage proteins typical of the seeds of higher plants are characteristically found packaged within membrane-bound protein bodies (a feature that facilitates their mobilization during germination by proteolytic enzymes that might otherwise degrade cytosolic proteins). However, the amino acid compositions of many LEA proteins are not inconsistent with a storage function, being rich in glutamate and glutamine. Additionally, these proteins disappear rapidly during seed germination, thus providing their constituent amino acids for immediate use - probably more rapidly than those released from embryonic storage globulins. Thus, even though these proteins may not comprise "classical" storage proteins, it would appear that they could provide at least an accessory storage function, providing "ready-use" amino acids for new protein synthesis early in imbibition.

The strongest support for a primary desiccation-protectant role derives from the observed accumulation of these proteins at a time when the developing embryo is acquiring desiccation tolerance. Bartels *et al.* (1988) demonstrated that immature barley embryos acquired desiccation tolerance between 12 and 18 days after pollination, and that this could be precociously induced by the application of ABA to isolated embryos. Coincidentally, ABA induced the accumulation of LEA proteins, which were not present in desiccation-intolerant embryos.

The significance of this correlation becomes less clear, however, when so-called "recalcitrant" seeds are examined. Such seeds typically do not acquire desiccation tolerance, rapidly losing viability when dehydrated. It has been reported that notwithstanding their dehydration sensitivity, embryos of recalcitrant and semi-recalcitrant seeds nevertheless accumulate Group 2 dehydrins during seed development and in response to dehydration or to ABA treatment (Bradford and Chandler, 1992; Finch-Savage *et al.*, 1994).

Further correlative evidence indicating that LEA protein accumulation comprises at least a component of dehydration stress survival by plants has been derived from studies of dehydrating vegetative tissues. This is most spectacularly demonstrated by the properties of the "resurrection plant" *Craterostigma plantagineum.* The vegetative tissues of this plant are capable of surviving complete desiccation, a process that is accompanied by the accumulation of a collection of proteins homologous with the LEA proteins identified in developing seeds (Ingram and Bartels, 1996). A similar accumulation of LEA proteins is observed following salt-stress. As is the case during seed development, the accumulation of these proteins in vegetative tissues is ABA-induced.

Anhydrobiotic lower plants also exhibit responses in common with the desiccating seeds of angiosperms. A Group 3 LEA protein has been shown to accumulate in tissues of the liverwort *Riccia fluitans* as it undergoes a transition from a submerged to a terrestrial existence (Hellwege *et al.*, 1996) and immunological studies have shown that desiccation tolerant mosses constitutively express dehydrin proteins (Bewley *et al*, 1993; A. Cuming, unpublished data). These observations, together with the identification of a cold-induced Group 3 LEA protein homologue in the green alga *Chlorella* (Honjoh *et al.*, 1995; Joh *et al.*, 1995), imply that the molecular responses to water deficit stress represent a very ancient evolutionary solution to the colonisation of terrestrial habitats by plants.

A number of LEA proteins have been identified as components of the low-temperature stress response of higher plants. Many species undergo acclimation to low temperatures, a process involving the expression of novel genes associated with frost tolerance (Hughes and Dunn, 1996). Whilst many of these genes encode proteins predicted to effect alterations in membrane lipid fluidity (increases in fluidity being necessary to avert low-temperature-induced phase changes in cellular membranes, leading to leakiness and loss of compartmentation), a subset of cold-induced proteins is homologous with Group 2 and Group 3 LEA proteins. A chilling-tolerant genetic line of the black-eyed bean, which shows reduced electrolyte leakage (a parameter of membrane integrity) at low temperatures, accumulates an *Mr* 35,000 dehydrin that is not seen in chilling-sensitive lines (Ismail *et al.*, 1997). Freezing, like drought stress and ionic stress, causes cellular dehydration. The accumulation of LEA proteins in response to this stress, and the ability of ABA simultaneously to induce cold-hardening and LEA protein accumulation, points to a common function in protecting the plant from the consequences of dehydration.

3.4 Direct evidence for a desiccation-protective function.

Attempts to determine experimentally whether LEA proteins stabilise macromolecular structures and exert a protective influence in anhydrobiosis have utilised two approaches: *in vitro* studies with purified components and *in vivo* studies with transgenic plants. Neither approach has produced compelling evidence to support the current hypotheses. Studies carried out *in vitro* have sought to determine whether the addition of a purified LEA protein to the freezing-susceptible enzyme lactate dehydrogenase (LDH) renders it more tolerant of freeze-thaw cycles. Thus, two proteins of the Group 2 (dehydrin) class, associated with low-temperature stress - COR-85 from spinach (Kazuoka and Oeda, 1994) and WSC-120 from wheat (Houde *et al.*, 1995) - have been reported to provide a measure of cryoprotection to LDH in freeze-thaw activity assays. The native COR-85 protein, purified from plants, was reported to be twice as effective as lactoglobulin and three times more effective than bovine serum albumin (BSA) (a commonly used cryoprotectant) in preventing loss of LDH activity. The WSC-120 protein was produced as a recombinant protein in *E. coli*, and found to be as effective as BSA, and considerably better than sucrose in the preservation of LDH. Additionally, Close (1996) has reported that a maize dehydrin has also been very effective in the cryoprotection of LDH, in the added presence of compatible solutes. However, these data have yet to be published.

The results of transgenic experiments have been contradictory. These have typically involved the overexpression of LEA protein genes in transgenic plants, accompanied by physiological assays predictive of enhanced tolerance to osmotic stress. These include both growth tests, and indicators of cellular integrity, such as electrolyte leakage, which results from the breakdown of the phospholipid bilamellar structure of cell membranes. Using three different LEA protein genes, expressed in transgenic tobacco, Iturriaga *et al.* (1992) were unable to observe any difference between transgenic and control plants. However, Artus *et al.* (1996) reported that constitutive expression of the *Cor15a* gene in transgenic *Arabidopsis* enhanced the freezing tolerance of non-cold-acclimated chloroplasts by 2 °C, and also modified the freezing tolerance of protoplasts isolated from non-acclimated plants. These results were far from unambiguous, however. At lower temperatures (between -5 °C and -8 °C) the protoplasts from transgenic plants performed better than those from control plants, whilst at slightly higher temperatures (-2 °C to -4 °C) the reverse was the case.

Xu *et al.* (1996) overexpressed the barley Group 3 *Lea* gene *HVA-1* in transgenic rice, and reported that the transgenic seedlings grew faster than controls when recovering from osmotic stress induced by imbibition with 0.1M NaCl or 0.2M mannitol. Additionally, transgenic seedlings showed delayed wilting and faster growth rates under salt-stress conditions. Finally, in an heterologous test of LEA protein function, the expression of the tomato Group 4 *Lea* gene *Le-25* in yeast produced cells showing improved growth rates in media containing high salt concentrations (1.2M NaCl), as a result of the transformed cells showing a shorter "lag phase" before the resumption of growth after ionic challenge. These cells also showed a greater tolerance to freezing (Imai *et al.*, 1996). The authors suggested that the LE-25 protein might act as an ion scavenger, on the basis of the charge distribution around its predicted *N*-terminal α-helices.

If, overall, few firm conclusions can be drawn from attempts to enhance dehydration- and salt-tolerance by the expression of single *Lea* genes, this is hardly surprising. Dehydration stress has multiple consequences for plants, and the acquisition of desiccation tolerance is likely to be a complex, multifactorial process. Under these circumstances, the effect of overexpressing a gene for a single component of the stress-response is likely to prove, at best, marginal.

3.5 Hypothetical functions of LEA proteins

How could a class of relatively uncharacterized (and in some cases apparently unstructured) proteins exert such a protective function? The hypotheses that have been advanced relate to the stabilisation of macromolecular structure in environments of reduced water content. The biological activities of macromolecules (most notably of enzymes) critically require that these molecules retain their characteristic three-dimensional structures. The tertiary structures of proteins in dilute solution depend on intramolecular interactions between the amino acid side chains, and the interaction of surface residues with the molecules of the surrounding solvent (water). The solvent also interacts with other, non-protein solutes, which are typically excluded from the interior structure of protein molecules and form an ordered hydration state with which the exterior surface of the protein can interact through hydrophilic interactions (both hydrogen-bonding and ionic). Within the interior of proteins, hydrophobic interactions dominate. Under such conditions, protein unfolding is not thermodynamically favoured, without an increased energetic input (as occurs in thermal denaturation), or unless the hydration state of the mixture is seriously perturbed, as occurs in the dehydration of biological systems.

Under circumstances of extreme desiccation, therefore, it is critical for proteins to retain an hydration shell around their surfaces, in order to retain their stable three-dimensional structure. One possible function of LEA proteins may be the provision of such a minimal shell of hydration around cellular macromolecules. For the Group 1 LEA proteins, it has been calculated that these polypeptides have a remarkably high potential for hydration: several times greater than that for "normal" cellular proteins (McCubbin *et al.*, 1985). This is a function of the remarkably high number of charged and uncharged polar residues within their structure. The *C*-terminal domains of the Group 4 LEA proteins and the "Φ-domains" of the dehydrins should have similar properties. Additionally these proteins, existing as random coils, are highly flexible and consequently can adopt almost any three-dimensional form. Such proteins would be capable of "coating" other macromolecules, infiltrating between them by virtue of their flexibility and conforming to their surface characteristics. In this way, the Group 1 and Group 4 LEA proteins would, through their retention of bound water and their "sharing" of this hydration shell with other proteins, provide the minimum of hydration necessary to maintain the structures of other macromolecules within an increasingly desiccated state. Even under extremes of desiccation, the loss of their tightly bound water component should not prevent the Group 1 LEA proteins from undertaking a continued role in the solvation of other macromolecules: since their own stability does not reside in a defined, water-dependent three-dimensional conformation, these proteins could still provide a layer of their own hydroxylated residues to interact with the surface groups other proteins, acting as a non-evaporable "replacement water".

A similar function can be envisaged for membrane stabilisation. Desiccation of lipid bilayer systems leads to loss of integrity as a result of bilayer-hexagonal phase transitions. Critical to bilayer formation is the maintenance of the polar association of phospholipid head groups with the aqueous phase. The coating of the polar surface with a hydrogen-bonding replacement would promote the stability of these critical structures. Such a role in stabilising both membranes and proteins is generally believed to be undertaken by sugars, such as trehalose in yeast, and sucrose and raffinose in cereal embryos (Crowe *et al*, 1996; Hoekstra *et al.*, 1997). The action of these sugars in the preservation of anhydrobiotic structures is believed to be through both direct interactions with other macromolecules and by contributing to the formation of a vitrified state (Leopold *et al.*, 1992) in which immobilisation of the cellular contents is achieved. Unlike sugars, the Group 1 LEA proteins would not be subject to crystallisation, and the observation that LEA proteins are tightly, but adventitiously, associated with sugars in dry wheat embryos (Walters *et al.*, 1997) suggests that these two classes of compound may act synergistically in providing alternative hydrogen bonding opportunities in highly desiccated tissues such as dry seeds.

For the Group 2 and Group 3 LEA proteins, hypotheses concerning their biological function have centred upon their potential for amphipathic helix formation. Two principal interactions have been proposed: the sequestration of ions, whose intracellular concentrations might otherwise become unacceptably high within the dehydrating cell, and charge-mediated interactions with other macromolecular components. Dure *et al.* (1989) and Dure (1993) have suggested that the amphipathic helices of the Group 3 LEA proteins could form bundles, through the interaction of their "hydrophobic stripes", thereby presenting a highly charged - and therefore ion-binding - surface to the exterior. Based on a calculated intracellular concentration of *ca.* 0.3 mM for the cotton LEA D-7 protein (Roberts *et al*, 1993), Dure (1993) suggests that this protein alone could sequester 1.5 mM phosphate salts.

The amphiphilic helices of the Group 2 (dehydrin) proteins might act similarly. Alternatively, the hydrophobic surfaces of the Group 2 helices might interact with other hydrophobic surfaces, such as lipids, in the manner of the apolipoproteins which they resemble, or with hydrophobic domains of proteins that become unfolded as desiccation proceeds (Close, 1997). In this latter action, the dehydrins might behave in a manner similar to the chaperonins, typical of heat shock responses, in preventing illegitimate intermolecular hydrophobic associations. A substantial proportion (*ca.* 75%) of Group 2 LEA protein sequence remains, like the Group 1 LEA proteins, apparently structureless. Thus in addition to a helix-mediated hydrophobic interaction, there is ample opportunity for the hydrophilic, unstructured (Φ) domains to engage in interactions with the soluble phase of the cell, or with other polar surfaces. This potential action has been described as a "detergent-like" means of maintaining the solvation of macromolecular surfaces (Close, 1997).

4. *LEA* GENES AND THEIR EXPRESSION

If the study of the *Lea* proteins has raised more questions than answers concerning the nature of the acquisition of desiccation tolerance in seeds or vegetative tissues, the same cannot be said for the study of their cognate genes. In contrast, the study of the regulation of *Lea* gene expression has contributed signnificantly to our understanding of the way in which the plant growth regulator ABA elicits the expression of specific genes during embryogenesis. This bears on the important processes of the imposition and maintenance of dormancy, and the control of pre-harvest sprouting.

4.1 4.1 *Lea* genes are members of multigene families

As is the case for many classes of protein in higher eukaryotes, the LEA proteins of plants are encoded by genes that exist as members of multigene families. Such families characteristically originate by duplication of an ancestral DNA sequence, with subsequent divergence both of sequence and of specificity of expression. Inspection of the coding sequences of the genes encoding the LEA proteins of Groups 1, 2 and 3 provides ample evidence of this process, with different numbers of repeated amino acid motifs being widespread among them. The individual *Lea* genes exhibit variation within families, not only through their sequence rearrangements, but also in their patterns of expression. Different members of each family have characteristic patterns of expression, both in terms of the developmental stage at which the genes are activated, and in terms of the hormonal or environmental stimuli that elicit their expression.

4.2 Induction of *Lea* gene expression during seed development

A common feature of the LEA protein genes that are expressed during seed development is that their expression is modulated by the plant growth regulator ABA. This is also the case for those *Lea* genes that are expressed in vegetative tissues, in response to environmental stress, the synthesis of ABA being one of the earliest responses to water stress by plants (Chandler and Robertson, 1994). In seeds, premature desiccation or the imposition of an osmotic stress also causes the induction of *Lea* genes (Morris, *et al*., 1990). However, while the molecular mechanisms which operate to bring about osmotic-stress-mediated *Lea* gene expression appear to be identical with those triggered by ABA, an increase in intracellular ABA concentration is not necessarily a prerequisite for the stress-induction of seed-specific *Lea* gene expression (Morris *et al*., 1990; Espelund *et al*., 1995).

Hormone-regulation of *Lea* genes has been extensively studied at the molecular level, and a number of regulatory DNA sequences have been identified in the promoters of these genes that interact with sequence-specific transcription factors. The first such genes to be characterized in this way were the wheat Em gene (encoding a Group 1 LEA protein; Marcotte *et al*., 1989) and the rice *rab16* gene (encoding a Group 2 dehydrin; Mundy *et al*., 1990). In each case, functional analysis of the promoters of these genes by transient expression in an ABA-responsive, embryo-derived cell culture demonstrated the key role of an ABA-responsive sequence element ("ABRE") containing the conserved core sequence ACGT. In the majority of ABREs analysed to date, this occurs in the hexameric sequence $^{T}/_{C}$ACGTG. The ACGT motif is the core of the palindromic "G-box" - CCACGTGG - and such elements have been found to occur in a large number of plant promoters. These act to promote transcription through interactions with DNA-binding proteins of the so-called "bZip" class (Guiltinan *et al.*, 1990). These transcription factors are dimeric proteins in which the constituent polypeptides each contain two distinctive domains: a basic, DNA binding domain (the "b-domain") and an α-helical domain characterised by leucine residues that lie along one face of the helix and are thus able to promote dimerisation with a similar helix through hydrogen-bonding interaction (the "leucine-zipper" domain).

The presence of a G-box motif alone is apparently able to direct ABA-responsive gene expression: Skriver *et al.* (1991) demonstrated the presence of multiple G-box elements upstream of a "minimal" promoter was sufficient to confer ABA inducibility on a reporter gene in transient expression experiments. Additionally, mutations in the ACGT core that abolish binding of bZip proteins also abolish the ABA-responsiveness of the promoter (Guiltinan *et al.*, 1990). However, although all ABA-regulated genes contain G-boxes, it is clear that not all G-box elements necessarily confer ABA-inducibility, *in vivo*. These sequence motifs are widespread among plant genes, many of which do not exhibit ABA-responsive gene expression. The G-box, therefore, should be viewed as a part of the transcriptional complex that forms when a *Lea* gene is expressed in an ABA-responsive manner. Several factors will determine whether a G-box element is a component of an ABA-responsive element, and one of the most important of these is the sequence context in which the ACGT core sequence is situated. The bases on either side of the core sequence dramatically affect the affinity with which different bZip factors will bind to the element (Williams *et al.*, 1992). Thus, in order for a gene to be active in a particular cell type, the cell must contain the correct collection of G-box-binding factors required for recognition of the ABRE. Additionally, because the bZip proteins bind G-boxes as dimers, which may be either homodimeric or heterodimeric in nature, it is clear that considerable specificity may be conferred on a gene simply through the interaction of the appropriate G-box sequence with the correct bZip protein.

There may be other *cis*-acting sequences that combine with G-box elements to modulate ABA-mediated gene expression. Shen *et al.* (1996) recently described the presence of two novel sequence elements in the barley Group 3 *Lea* genes *HvA1* and *HvA22*. Described as "coupling elements", these were required to be present in combination with a G-box motif to drive ABA-mediated expression of a recombinant reporter. Such elements have since been reported to occur in a number of other *Lea* genes, in a wide range of plants (Quatrano *et al.*, 1997).

Other factors also regulate the ABA-responsiveness of *Lea* gene expression. In seeds, one of these additional requirements is the presence of a second transcription factor, encoded in maize by the *VP-1* gene (McCarty *et al*., 1991) and in *Arabidopsis* by the *ABI-3* gene (Giraudat *et al*., 1992). This gene is a highly conserved regulator of seed-specific gene expression. The expression of most ABA-responsive genes in maturing embryos requires the presence of the VP-1/ABI-3 protein. However, this is not universal: for example, the Group 5 *Lea* gene, *rab28*, and the Group 3 *Lea* gene *mlg3* are expressed in an ABA-responsive manner in embryos of maize *vp-1* mutants (Pla *et al*., 1991; Thomann *et al*., 1992).

The *VP-1* and *ABI-3* genes have been shown to encode a transcriptional activator protein required for the transactivation of ABA-responsive genes in seeds (McCarty *et al*., 1991). Additionally, the activity of these genes is required for developing seeds to enter into primary dormancy: the phenotype of the maize *vp-1* mutation is that of vivipary, or precocious germination of the developing kernel prior to dehydration of the cob. Some of the *Lea* genes are clearly dependent on the presence of the VP-1/ABI-3 protein for their expression, since they are seed-specific in their pattern of expression (Parcy *et al*., 1994; Hattori *et al*., 1995). The Group 1 Em class of genes displays this characteristic, being expressed in embryos and aleurone layers of developing grains and early during germination, but not during normal vegetative development. In transgenic tobacco, the highly active wheat Em promoter directs transgene expression in seeds (Marcotte *et al*., 1989), but not in the highly ABA-responsive guard cells of mature leaves (Taylor *et al*., 1995).

The means by which the VP-1/ABI-3 protein activates ABA-responsive gene expression in seeds is not known. No sequence-specific DNA-binding activity has been associated with this protein, although it appears to activate both bZip-ABRE mediated gene expression and the ABA-responsiveness of some genes that lack a G-box motif, such as the *C-1* gene of maize (Hattori *et al*., 1992). In this latter case the *cis*-acting sequence required for ABA-mediated expression resides in an element known as the "Sph" or "Ry" element, (CATGCATG). This is found in the promoters of a number of other seed genes, including *Lea* genes and those encoding storage globulins (Baumlein *et al*., 1992). The expression of embryonic storage proteins is also ABA-inducible (Dure *et al*., 1981) and in maize and *Arabidopsis* has been shown to be *VP-1/ABI-3*-dependent; Thomann *et al*., 1992; Parcy *et al*., 1994).

For genes regulated by a bZip-ABRE interaction, the VP-1 protein appears to enhance bZip binding in some way. Its ability to do so resides in a conserved, basic domain "BR-2" (for "Basic region 2": an 18 amino acid stretch highly conserved among the *VP-1* homologues isolated from other plant species; Hill *et al*., 1996). It has recently been suggested that the VP-1 protein may interact indirectly with the other components of the ABA-response complex. A rice protein "GF-14" has been identified as a potentially capable of mediating an association between VP-1 and bZip-ABRE complexes (Quatrano *et al*., 1997). This protein is a member of the class of proteins generically known as "14-3-3" proteins, which are believed to undertake a wide range of protein-protein associations in cellular signal transduction pathways (Ferl, 1996).

Given the modular nature of plant promoters in general, the plethora of sequence-specific DNA binding proteins within nuclei, and the complexity of inter- and intracellular signal transduction pathways and the interactions between them, it is to be expected that a still more complex description of the ABA-regulation of *Lea* gene expression in seed tissues will emerge in the next few years.

4.3 *Lea* gene expression in vegetative tissues.

The expression of *Lea* genes in vegetative tissues is a consequence of environmental stress stimuli. This frequently has an ABA-component, the accumulation of ABA being presumed to provide the cellular signal required for activation of specific gene expression. In many cases, the *Lea* genes that are expressed in vegetative tissues in response to application of ABA or imposition of environmental stress are encoded by DNA sequences distinct from those which are expressed during seed maturation. In this respect, the *Lea* gene families provide model examples of the evolutionary process, in which duplication and divergence of a DNA sequence allow the duplicated sequence to acquire new biological properties and thereby confer upon its host organism a selective advantage.

The stress-induced *Lea* genes whose regulation has been studied in detail all appear to contain G-box-like ABREs. These interact with bZip proteins in a manner essentially similar to that seen in seed development, with the specific difference that they do not require transactivation by a VP-1-like factor. Interestingly, a number of these "vegetative *Lea* genes" have been found to respond differentially to stress- and hormone-induction, with individual gene family members clearly having undergone divergence in their means of regulation. Thus, the *Arabidopsis* cold-induced Group 2 *Lea* genes, *Cor15a* and *Cor15b*, are a pair of tandemly linked homologues (separated by only 1kb in the genome) whose products are targeted to chloroplasts (Wilhelm and Tomashow, 1995). The *Cor15a* gene is strongly up-regulated by low temperature, drought stress and by the application of ABA, whereas the *Cor15b* gene is drought-insensitive, whilst remaining cold- and ABA-inducible. Presumably such differences in expression result from sequence differences in the two promoters which participate in different associations with transcription factors. Both genes contain G-box motifs within their promoters, and the *Cor15a* gene is reported to have an additional element (a "C-repeat": TGGCCGAC) which is seen in other cold-regulated genes (Baker *et al.*, 1994). This element has some similarity to the "coupling elements" described by Shen *et a*l. (1996), and it may be that it is differences in elements of this type, outside the core ACGT motif but interacting with it, which specify the subtle differences in the regulation of the two genes.

A notable exception to the seed-specificity of VP-1/ABI-3-activated transcription of *Lea* genes has been demonstrated in a lower plant. The promoter of the seed-specific wheat Em gene is able to drive transgene expression in the vegetative tissue of the moss *Physcomitrella patens* (Knight *et al.*, 1995). ABA-induction of the gene in moss utilises the same responsive sequences as are used in wheat, implying that the transcriptional machinery - including, presumably, homologues of the VP-1 and bZip Em binding protein EmBP-1 - has been conserved through the *ca.* 200 million years of evolution between the bryophytes and cereals. It is interesting to speculate whether the presence of a VP-1/ABI-3-class of protein in vegetative tissues of mosses underlies the constitutive expression of LEA-type genes in these organisms, which has been taken to contribute to their desiccation tolerance (Bewley *et al.*, 1993).

5. SUMMARY

The Late Embryogenesis Abundant proteins that accumulate during the maturation stage of seed development are a group of typically hydrophilic proteins encoded by small multigene families. The expression of these genes is elicited during embryogenesis by the plant growth regulator abscisic acid, and by the imposition of osmotic stress. The *Lea* genes appear to be highly conserved in evolution, both in terms of the conservation of the amino acid sequences they encode, and in their specificity of expression. A subset of LEA proteins accumulate in vegetative tissues in an ABA-dependent manner, and in response to a variety of dehydrative stimuli. Although there is little direct experimental evidence available, there is strong correlative support for the hypothesis that these proteins are intimately associated with the acquisition of desiccation tolerance in seed maturation, and the amelioration of water stress in vegetative tissues.

REFERENCES

Alexander, R., Alamillo, J.M. Salamini, F. and Bartels, D. (1994) A novel embryo-specific barley cDNA clone encodes a protein with homologies to bacterial glucose and ribitol dehydrogenase. *Planta* **192,** 519-525

Artus, N.N, Uemura, M., Steponkus, P.L., Gilmour, S.J., Lin, C.T. and Tomashow, M.F. (1996) Constitutive expression of the cold-regulated *Arabidopsis thaliana* COR15A gene affects both chloroplast and protoplast freezing tolerance. *Proceedings of the National Academy of Science, USA* **93,** 13404-13409

Baker, J., Steele, C. and Dure, L.S. III (1988) Sequence and characterization of 6 LEA proteins and their genes from cotton. *Plant Molecular Biology* **11,** 277-291

Baker, S.S., Wilhelm, K.S. and Tomashow, M.F. (1994) The 5'-region of *Arabidopsis thaliana* COR15A has cis-acting elements that confer cold-regulated and ABA-regulated gene expression. *Plant Molecular Biology* **24,** 701-713

Bradford, K. and Chandler, P.M. (1992) Expression of "dehydrin" proteins in embryos and seedlings of *Zizania palustris* and *Oryza sativa* during dehydration. *Plant Physiology* **99,** 488-494

Baumlein, H., Nagy, I., Villaroel, R. Inzé, D. and Wobus, U. (1992) *Cis*-analysis of a seed protein gene promoter: the conservative RY repeat CATGCATG within the legumin box is essential for tissue-specific expression of a legumin gene. *The Plant Journal* **2,** 233-239

Bartels, D., Engelhardt, K., Roncarati, R., Schneider, K., Rotter, M. and Salamini, F. (1991) An ABA and GA modulated gene expressed in the barley embryo encodes an aldose reductase related protein. *EMBO JOURNAL* **10,** 1037-1043

Bartels, D., Singh, M. and Salamini, F (1988) Onset of desiccation tolerance during development of the barley embryo. *Planta* **178,** 485-492

Bewley, J.D., Reynolds, T.L. and Oliver, M.J. (1993) Evolving strategies in the adaptation to desiccation. **In:** *Response of Plants to Cellular Dehydration During Environmental Stress* (T.J. Close and E.A. Bray, eds.) pp. 193-201. American Society of Plant Physiologists, Rockville, MD.

Calvo, E.S., Wurtele, E.S. and Shoemaker, R.C. (1997) Cloning, mapping and analyses of the Em-like gene family in soybean [*Glycine max* (L) Merr.]. *Theoretical and Applied Genetics* **94,** 957-967

Ceccardi, T.L., Meyer, N.C. and Close, T.J. (1994) Purification of a maize dehydrin. *Protein Expression and Purification* **5,** 266-269

Chandler, P.M. and Robertson, M. (1994) Gene expression regulated by abscisic acid and its relation to stress tolerance. *Ann. Rev. Plant Physiology Plant Molecular Biology* **45,** 113-141

Close, T.J. (1996) Dehydrins - emergence of a biochemical role of a family of plant dehydration proteins. *Physiol. Plant* **97,** 795-803

Close, T.J. (1997) A commonality in the responses of plants to dehydration and low temperature. *Physiol. Plant.* **100,** 291-296

Close, T.J., Kortt, A.A. and Chandler, P.M. (1989) A cDNA-based comparison of dehydration-induced proteins (dehydrins) in barley and corn. *Plant Molecular Biology* **13,** 95-108

Crowe, L.M., Reid, D.S. and Crowe, J.H. ('1996) Is trehalose special for preserving dry material? *Biophysical. Journal* **71,** 2087-2093

Dure, L.S. III (1993) A repeating 11-mer amino acid motif and plant desiccation. *The Plant Journal* **3,** 363-369

Dure, L.S. III, Crouch, M., Harada, J., Ho, T-H.D. Mundy, J., Quatrano, R.S., Thomas, T. and Sung, Z.R. (1989) Common amino acid sequence domains among the LEA proteins of higher plants. *Plant Molecular Biology* **12,** 475-486

Dure, L.S. III and Galau, G.W. (1981) Developmental biochemistry of cottonseed embryogenesis and germination. 13. Regulation of biosynthesis of principal storage proteins. *Plant Physiology* **68,** 187-194

Espelund, M., Debedout, J.A., Outlaw, W.H. and Jakobsen, K.S. (1995) Environmental and hormonal regulation of barley late embryogenesis abundant LEA messenger RNAs is via different signal transduction pathways. *Plant Cell and Environment.* **18,** 943-949

Ferl, R.J. (1996) 14-3-3 proteins and signal transduction. *Annual Review of Plant Physiology and Plant Molecular Biology* **77,** 49-73

Finch-Savage, W.E., Pramanik, S.K. and Bewley, J.D. (1994) The expression of dehydrin proteins in desiccation-sensitive (recalcitrant) seeds of temperate trees. *Planta* **193,** 478-485

Galau, G.W., Hughes, D.W. and Dure, L.S. III (1986) Abscisic acid induction of cloned cotton late embryogenesis abundant (LEA) messenger RNAs. *Plant Molecular Biology* **7,** 155-170

Giraudat, J. , Hauge, B.M., Valon, C., Smalle, J., Parcy, F. and Goodman, H.M. (1992) Isolation of the *Arabidopsis ABI3* gene by positional cloning. *The Plant Cell* **4,** 1251-1261

Goday, A. Jensen, A.B., Culianez-Macia, F.A. Alba, M.M., Figueras, M. Serratosa, J., Torrent, M. and Pages, M. (1994) The maize abscisic acid-responsive protein RAB-17 is located in the nucleus and interacts with nuclear localization signals. *The Plant Cell* **6,** 351-360

Grzelczak, Z.F., Sattolo, M.H., Hanley-Bowdoin, L.K., Kennedy, T.D. and Lane, B.G. (1982) Synthesis and turnover of proteins and messenger RNA in germinating wheat embryos. *Canadian Journal of Biochemistry.* **60,** 389-397

Guiltinan, M.J., Marcotte, W.R. and Quatrano, R.S. (1990) A plant leucine zipper protein that recognizes an abscisic acid response element. *Science* **250,** 267-271

Hattori, T., Terada, T. and Hamasuma, S. (1995) Regulation of the *Osem* gene by abscisic acid and the transcriptional activator VP1: analysis of the *cis*-acting promoter elements required for regulation by abscisic acid and VP1. *The Plant Journal* **7,** 913-925

Hattori, T., Vasil, V., Rosenkrans, L. Hannah, L.C., McCarty, D.R. and Vasil, I.K. (1992) The *viviparous-1* gene and abscisic acid activate the *C1* regulatory gene for anthocyanin biosynthesis during seed maturation in maize. *Genes and Development* **6,** 609-618

Hellwege, E.M., Dietz, K.J. and Hartung, W (1996) Abscisic acid causes changes in gene expression involved in the induction of the landform of the liverwort *Riccia fluitans* L. *Planta* **198,** 423-432

Hill, A., Nantel, A., Rock, C.D. and Quatrano, R.S. (1996) A conserved domain of the *viviparous-1* gene product enhances the DNA binding activity of the bZip protein EmBP-1 and other transcription factors. *Journal of Biological Chemistry* **271,** 3366-3374

Hoekstra, F.A., Wolkers, F., Buitnik, J., Golovina, E.A., Crowe, J.H. and Crowe, L.M. (1997) Membrane stabilization in the dry state. *Comparative Biochemistry and Physiology* **117,** 335-341

Honjoh, K., Yoshimoto, M., Joh, T., Kajiwara, T., Miyamoto, T. and Hatano, S. (1995) Isolation and characterization of hardening-induced proteins in *Chlorella vulgatis* C-27 - Identification of late embryogenesis abundant proteins. *Plant Cell Physiology* **36,** 1421-1430

Houde, M., Daniel, C., Lachapelle, M., Allard, F., Laliberté, S. and Sarhan, F. (1995) Immunolocalization of freezing-tolerance-associated proteins in the cytoplasm and nucleoplasm of wheat crown tissues. *The Plant Journal* **8,** 583-593

Hughes, M.A. and Dunn, M.A. (1996) The molecular biology of plant acclimation to low temperature. *Journal of Experimental Botany* **47,** 291-306

Imai, R., Chang, L., Ohta, A., Bray, E.A. and Takagi, M. (1996) A *Lea* class gene of tomato confers salt and freezing tolerance when expressed in *Saccharomyces cerevisiae. Gene* **170,** 243-248

Ingram, J., and Bartels, D. (1996) The molecular basis of desiccation tolerance in plants. *Annual Review of Plant Physiology and Plant Molecular Biology* **47,** 377-403

Ismail, A.M., Hall, A.E. and Close, T.J. (1997) Chilling tolerance during emergence of cowpea associated with a dehydrin and slow electrolyte leakage. *Crop Science* **37,** 1270-1277

Iturriaga, G., Schneider, K., Salamini, F. and Bartels, D. (1992) Expression of desiccation-related proteins from the resurrection plant *Craterostigma plantagineum* in transgenic tobacco. *Plant Molecular Biology* **20,** 555-558

Joh, T., Honjoh, K., Yoshimoto, M., Funabashi, J., Miyamoto, T. and Hatano, S (1995) Molecular cloning and expression of hardening-induces genes in *Chlorella vulgaris*C-27: the most abundant clone encodes a late embryogenesis abundant protein. *Plant Cell Physiology* **36,** 85-95

Kazuoka, T. and Oeda, K. (1994) Purification and characterization of COR-85 oligomeric complex from cold-acclimated spinach. *Plant Cell Physiology* **35,** 601-611

Knight, C.D., Seghal, A., Atwal, K., Wallace, J.C., Cove, D.J., Coates, D., Quatrano, R.S., Bahadur, S., Stockley, P.G. and Cuming A.C. (1995) Molecular responses to abscisic acid and stress are conserved between moss and cereals. *The Plant Cell* **7,** 499-506

Leopold, A.C., Bruni, F. and Williams, R.J. (1992) Water in dry organisms. **In:** *Water and Life*. (G. Somero, C.B. Osmond and R.J. Williams, eds.) pp. 161-169. Springer-Verlag, Berlin.

Lin, C.T. and Tomashow, M.F. (1992) DNA sequence of a complementary DNA for cold-regulated *Arabidopsis* gene *COR15* and characterization of the COR-15 polypeptide *Plant Physiology* **99,** 519-525

Lisse, T., Bartels, D., Kallitzer, H.R. and Jaenicke, R. (1996) The recombinant dehydrin-like desiccation stress protein from the resurrection plant *Craterostigma plantagineum* displays no defined 3-dimensional structure in its native state. *Biological Chemistry* **377,** 555-561

Litts, J.C., Colwell, G.W., Chakerian, R.L. and Quatrano, R.S. (1987) The nucleotide sequence of a cDNA clone encoding the wheat Em protein. *Nucleic Acids Research* **15,** 3607-3618

Marcotte, W.R., Russell, S.H. and Quatrano, R.S. (1989) Abscisic acid responsive sequences from the *Em* gene of wheat. *The Plant Cell* **1,** 969-976

Morris, P.C., Kumar, A., Bowles, D.J. and Cuming. A.C. (1990) Osmotic stress and abscisic acid regulate expression of the wheat *Em* genes. *European Journal of Biochemistry* **190,** 625-630

Mundy, J. and Chua, N-H (1988) Abscisic acid and water stress induce the expression of a novel rice gene. *EMBO JOURNAL* **7,** 2279-2286

Mundy, J., Yamaguchi-Shinozaki, K. and Chua, N-H (1990) Nuclear proteins bind conserved elements in the abscisic acid responsive promoter of a rice *rab* gene. *Proceedings of the National Academy of Science, USA* **87,** 1406-1410

McCarty, D.R., Hattori, T., Carson, C.B., Vasil, V., Lazar, M. and Vasil, I.K. (1991) The Viviparous-1 developmental gene of maize encodes a novel transcriptional activator. *Cell* **66,** 895-905

McCubbin, W.D., Kay, C.M. and Lane, B.G. (1985) Hydrodynamic and optical properties of the wheat-germ Em protein. *Canadian Journal of Biochemistry and Cell Biology* **63,** 3607-3618

McCue, K.F. and Hanson, A.D. (1990) Drought and salt tolerance - towards understanding and application. *Trends in Biotechnology* **8,** 358-362

Niu, S., Antin, P.B. and Morkin, E. (1996) Cloning and sequencing of a developmentally regulated avian messenger RNA containing the LEA motif found in plant seed proteins. *Gene* **175,** 187-191

Parcy, F., Valon, C., Raynal, M., Gaubier-Comella, P., Delseny, M. and Giraudat, J. (1994) Regulation of gene expression programs during *Arabidopsis* seed development: Roles of the *ABI3* locus and of endogenous abscisic acid. *The Plant Cell* **6,** 1567-1582

Pla, M., Gomez, J., Goday, A. and Pages, M (1991) Regulation of the abscisic acid responsive gene *rab28* in maize viviparous mutants. *Molecular and General Genetics* **230,** 394-400

Quatrano, R.S., Bartels, D., Ho, T-H.D. and Pages, M. (1997) New insights into ABA-mediated processes. *The Plant Cell* **9,** 470-475

Rap, S., Espelund, M., Saabøe-Larssen, S., Hollung, K., Helliesen, E. and Jakobsen K.S. (1995) Evolution of the Group 1 late embryogenesis abundant (*Lea*) genes - Analysis of the *LeaB19* gene family in barley. *Plant Molecular Biology* **28,** 1039-1054

Roberts, J.K., DeSimone, N.A., Lingle, W.L. and Dure, L.S. III (1993) Cellular concentrations and uniformity of cell-type accumulation of 2 LEA proteins in cotton embryos. *The Plant Cell* **5,** 769-780

Russouw, P.S., Farrant, J., Brandt, W. and Lindsey, G.G. (1997) The most prevalent protein in a heat-treated extract of pea (*Pisum sativum*) embryos is a LEA Group 1 protein; its conformation is not affected by exposure to high temperature. *Seed Science Research* **7,** 117-123

Shen, Q., Zhang, P. and Ho, T-H.D. (1996) Modular nature of abscisic acid (ABA) response complexes: Composite promoter units that are necessary and sufficient for ABA induction of gene expression in barley. *The Plant Cell* **8,** 1107-1119

Skriver, K., Olsen, F.L., Rogers, J.C. and Mundy, J.C. (1991) *Cis*-acting DNA elements responsive to gibberellin and its antagonist abscisic acid. *Proceedings of the National Academy of Science, USA* **88,** 7266-7270

Taylor, J.E., Renwick, K.F., Webb, A.A.R., McAinsh, M.R., Furini, A., Bartels, D., Quatrano, R.S., Marcotte, W.R. and Hetherington, A.M. (1995) ABA-regulated promoter activity in stomatal guard cells. *The Plant Journal* **7,** 129-134

Thomann, E.B., Sollinger, J., White, C. and Rivin, C.J. (1992) Accumulation of Group 3 late embryogenesis abundant proteins in *Zea mays* embryos - roles of abscisic acid and the *viviparous-1* gene product. *Plant Physiology* **99,** 607-614

Vilardell, J., Goday, A., Freire, M.A., Torrent, M., Martinez, M.C., Torne, J.M. and Pages, M. (1990) Gene sequence, developmental expression and protein phosphorylation of *rab17* in maize. *Plant Molecular Biology* **14,** 423-432

Walters, C., Ried, J.L. and Walker-Simmons, M.K. (1997) Heat soluble proteins extracted from wheat embryos have tightly bound sugars and unusual hydration properties. *Seed Science Research* **7,** 125-134

Wilhelm, K.S. and Tomashow, M.F. (1993) *Arabidopsis thaliana* COR15b, an apparent homolog of COR 15A, is strongly responsive to cold and ABA, but not drought. *Plant Molecular Biology* **23,** 1073-1077

Williams, M.E., Foster, R. and Chua, N-H (1992) Sequences flanking the hexameric G-box core CACGTG affect the specificity of protein binding. *The Plant Cell* **4,** 485-496

Xu, D., Duan, X., Wang, B., Hong, B., Ho, T-H. D. and Wu, R. (1996) Expression of a late embryogenesis abundant protein gene, HVA1, from barley confers tolerance to water deficit and salt stress in transgenic rice. *Plant Physiology* **110,** 249-257

Yang, H., Saitou, T., Komeda, Y., Harada, H, and Kamada, H. (1996) Late embryogenesis abundant protein in *Arabidopsis thaliana* homologous to carrot ECP31. *Physiologia Plantarum* **98,** 661-666.

33. Seed Oleosins

MAURICE M. MOLONEY
University of Calgary, Department of Biological Sciences, 2500 University Drive N.W., Calgary, AB, T2N 1N4, Canada

1. INTRODUCTION

Seeds are well adapted for storage of key nutritional components necessary for the establishment of a new sporophyte. While seeds store a wide variety of organic and inorganic molecules, above all they devote energy towards the deposition of protein, carbohydrate and lipid. These molecules may be deposited in different proportions and in different tissues and organs of the seed. Thus, for example, storage lipid in the form of triacylglyceride may accumulate to levels of approximately 40% of the seed dry weight in a number of oilseeds or as little as 2% of seed dry weight in some cereal grains.

In spite of the diversity of storage products deposited in seeds, it appears that oil deposition in all seeds follows similar steps and results in the production of oil-storage organelles known as oil-bodies, oleosomes or occasionally spherosomes. Most seeds store neutral lipid in the form of triacylglyceride. A single known exception to this is the jojoba seed which stores wax esters formed by condensation of a fatty acid and fatty alcohol (Huang *et al.*, 1992).

Oil-bodies as simple organelles have been described in some detail since the mid-sixties (Frey-Wyssling *et al.*, 1963; Trelease, 1969). In electron microscopy, the intensely lipophilic core of the organelle distinguishes it from all other subcellular structures (Trelease, 1969; Wanner and Theimer, 1978). Further characterization of these unique organelles has revealed that they are surrounded by a membrane (Yatsu and Jacks, 1972; Jacks *et al.*, 1967). This membrane appeared to be a half-unit phospholipid membrane with the lipophilic aliphatic chains oriented into the lumen of the oleosome. This information was obtained initially from quantitative estimation of oleosome membrane thickness compared to that of adjacent structures such as glyoxysomes with typical unit-membrane profiles (Trelease, 1969; Yatsu and Jacks, 1972) and complemented by direct chemical measurements of phospholipid content (Slack *et al.*, 1980; Jacks *et al.*, 1990).

P.R. Shewry and R. Casey (eds.), Seed Proteins, 781–806.

While neutral and phospholipids account for the major proportion of matter found in oil-bodies, it became clear that a further component was present. In most seed oil-bodies studied there was a significant protein component varying from around 1% of the oil-body in corn and peanut (Tzen and Huang, 1992; Jacks *et al.*, 1990) to at least 4% of oil-body weight in *Brassica* species (Huang, 1992).

The protein component of oil-bodies is remarkably homogeneous, comprising polypeptides derived from one gene family. All of these can be classified together on the basis of evolutionary conservation, lipophilicity profiles and secondary structure. They comprise a family of proteins referred to as *oleosins*.

Oleosins of different species are relatively low molecular weight proteins which mainly cluster around M_r 20,000 (Bergfeld *et al.*, 1978; Moreau *et al.*, 1980; Tzen *et al.*, 1993; Yatsu and Jacks, 1972). It is not unusual on polyacrylamide gels to find bands of M_r about 40,000 which cross-react with oleosin antibodies. It is generally believed that these are dimers which have not properly dissociated during sample preparation. To date no seed-specific genes have been isolated that encode oleosin-like proteins of such high molecular weight.

Some reports have suggested that other proteins may be "associated" with oleosomes but this appears to occur in an artefactual manner. Apparently, the association takes place *in vitro*, but not *in vivo* (Kalinski *et al.*, 1992) or may be the result of incomplete washing of prepared oil-bodies from seed extracts. Clearly, the kind of association that oleosins have with oil-bodies is qualitatively different than other proteins studied to date. Oleosins cannot be removed from oil-bodies by simple washing even with high salt or mild detergent treatment whereas any other protein present in significant amounts may be stripped from oil-body preparation by such treatments (Stobart and Stymne, 1990). Even washing oil-bodies with 9M urea does not remove oleosins, but does remove all other contaminating proteins (Millichip *et al.*, 1996). The reason for this avid binding of oleosins to oleosomes is evident once the structure and subcellular trafficking of the protein are considered (see below).

2. SEED OIL-BODIES

2.1 Oil-body function

The role of oil-bodies in plant seeds is primarily one of storage of oil in readiness for key germination signals such as imbibition and hormonal lifting of dormancy. When the switch to germination occurs there is a need to control availability of these precious energetic resources. A physical solution to this need is for the seed cell to ensure that the surface-to-volume ratio of oil-bodies remains high. This would facilitate mass-transfer of triacylglycerides to waiting lipases as germination takes place. This hydrolysis is the first step in providing carbon skeletons for β-oxidation which the seed requires to release energy from the stored oils. It is noteworthy that in most seeds, oil-bodies exist as a large number of small, discrete organelles and that they resist coalescence during seed desiccation when the protoplast shrinks and results in cytoplasmic compression. Direct evidence for this property was reported by Cummins *et al.* (1993). In their experiments, these workers showed that oil-bodies from young embryos that were low in oleosin content undergo coalescence when subjected to controlled desiccation in a dialysis bag. Oil-bodies from mature embryos with a high oleosin titre underwent desiccation but did not coalesce. On re-hydration the oil-bodies re-emulsified to resemble their original hydrated state. Why such coalescence does not occur is an interesting question. The presence and topology of oleosins on oil-bodies may provide part of the answer to this question.

2.2 Distribution of oil-bodies in seeds

In exalbuminous seeds such as those of *Brassica spp* and *Glycine* spp, oil-bodies are generally found in the cotyledons and throughout the embryonic axis. In cereals, oil-bodies are more typically a feature of the scutellum and embryo, but not a constituent of the endosperm. This situation is typified by corn and rice. Oil-bodies are, however, found in the aleurone layer (Aalen *et al.*, 1994). In albuminous dicotyledons such as castor bean, the majority of the stored lipid is in the endosperm. Once again, the oil is stored in discrete oleosomes with oleosins associated with their surface. A gel electrophoretic analysis of oil-body proteins from several common oilseed species is shown in Figure 1.

Interestingly, in seeds which are surrounded by oleaginous fruits such as avocado and olive, the seed cells contain small oleosomes replete with a complement of oleosin isoforms whereas the oil-bodies of the fruit appear to have no oleosin associated with them (Ross *et al.*, 1993). In this context it is noteworthy that the role of the oil deposited in seeds and fruit is entirely different. The former is to nourish the sporophyte, the latter is to attract/nourish animals capable of dispersing the seed. Similarly, in both these examples the seeds are destined to undergo desiccation and cytoplasmic compression. This is not true of the fruit which has no direct role as an energy source (via lipid degradation) for the seed. It is interesting to note that whereas seed oil-bodies are typically of the order of <1μM in diameter (Murphy, 1990; Huang, 1992), fruit oil-bodies are frequently as large as 20μM in diameter (Platt-Aloia and Thomson, 1981). These differences in size correlate with the presence or absence of oleosins suggesting that oleosins may be a primary determinant in these variations.

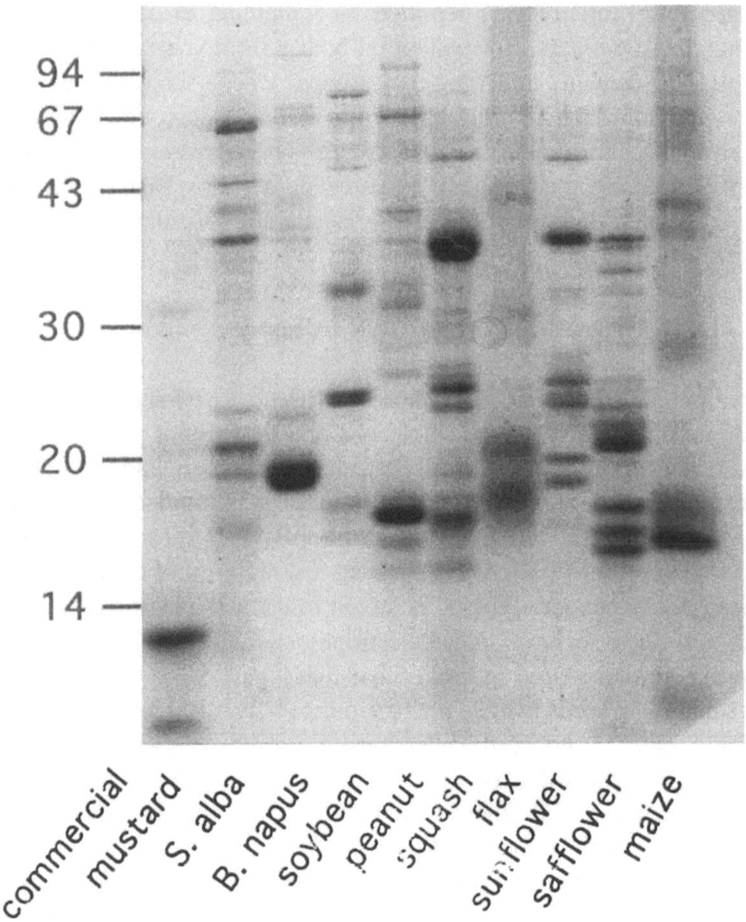

Fig. 1 Oil-body proteins (oleosins) in different seeds and in commercial mustard preparations. The lanes correspond to washed oil-body preparations from the species shown. Note the preponderance of oleosins of M_r about 20,000. Generally multiple isoforms are found. The bands around M_r 40,000 in maize, sunflower, safflower, squash and *Sinapis alba* are non-covalently bound dimers which are not dissociated in hot Laemmli buffer. The lane marked "commercial mustard" is provided to illustrate one circumstance in which oleosins are consumed as food. Although the mustard sample comprises mainly storage protein (2S albumin) a M_r 20,000 oleosin is also evident visualized using a *B. napus* anti-oleosin antibody (figure courtesy of Dr. J. Boothe and Ms. J. Saponja, University of Calgary).

2.3 Nature and timing of deposition of oil-bodies

Oil-bodies in oilseeds can account for over 40% of the total weight of the dry seed. In consequence, the deposition of oil-bodies is a major activity in such seeds. Studies of a number of dicotyledonous oilseeds suggest that the deposition of triacylglyceride begins fairly early in embryo development, probably at the late heart or early torpedo stage. From this developmental stage on neutral lipid synthesis and its manifestation as oil-bodies continues throughout embryo and cotyledonary development. In many species, triacylglyceride synthesis continues well into seed maturation during the period where the seed is undergoing lowered water potential (Murphy *et al.*, 1989b). There has been significant controversy over whether oil-bodies initially form as naked oil globules which then assemble their complement of phospholipid and oleosin (Bergfeld *et al.*, 1978; Stobart *et al.*, 1986; Cummins and Murphy, 1990). This controversy has not yet been resolved through rigorous analysis of oil-body biogenesis *in vivo* due to many technical problems. However, the "naked droplet" hypothesis has been rendered unlikely by several pieces of circumstantial evidence all of which support the idea that oleosomes with their entire complement of phospholipid and oleosin are produced and released from the endoplasmic reticulum. The evidence that, taken together, weighs heavily in favour of "mature oleosome" biogenesis on ER is as follows:

1) Triacylglycerol assembly takes place on the ER. The last enzyme in the pathway is ER-associated (Cao and Huang, 1986; Weselake *et al.*, 1993; Little *et al.*, 1994).

2) Phospholipids are synthesized on the ER and indeed the ER membrane, if delaminated, could provide the required half-unit membrane.

3) Oleosins are synthesized by co-translation on the ER (Loer and Herman, 1993; Hills *et al.*, 1993).

4) Oleosins which are re-routed from normal ER trafficking become unstable *in vivo* (van Rooijen and Moloney, 1995b; Abell et *al.*, 1997).

5) Oleosins synthesized in cells which are unable to make triacylglycerides and oil-bodies become unstable and are rapidly degraded (Abell *et al.*, 1997; Galanis and Moloney, unpublished).

Perhaps the most disquieting phenomenon is this regard is that such a mechanism implies a frequent delamination of ER membrane and blebbing off of assembled oleosomes. Despite the fact that this process must be occurring constantly in maturing seed cells, a convincing electron micrograph capturing the phenomenon has rarely been obtained, and to the author's knowledge never published.

2.4 Fate of oil-bodies during germination

The role of seed oil-bodies is the storage of lipid for germination. In consequence, the activation of germination mobilizes organelles and enzymes required to access and process the available carbon skeletons. The mechanism by which this occurs has not yet been rigorously explained. Clearly the first step in the mobilization of energy from triacylglycerol is hydrolysis catalysed by lipases. Which lipase performs this function is not yet known although strong circumstantial evidence exists for a corn lipase which associates tightly with corn oil-bodies and which appears and persists in concert with the hydrolysis of seed lipid (Fernandez *et al.*, 1988). During the metabolism of oil it is interesting to note that although oleosins are also degraded, they persist somewhat longer than storage proteins (Murphy *et al.*, 1989a). After the hydrolysis of lipid it is possible to observe under the EM the remnants of oil-bodies which exist as "ghosts" in which the half-unit membrane can be stained and compared to existing unit membranes in the cells (Trelease, 1969).

3. PHYSICAL BIOCHEMISTRY OF OLEOSINS

Oleosins are a unique class of protein which are identifiable in different species because of a large number of common features. These features include, but are not limited to, sequence homology.

Sequence comparisons of large numbers of oleosins have been performed (Huang, 1992; Lee *et al.*, 1994). These comparisons confirm that oleosins comprise three distinct domains. The first, at the *N*-terminus of the protein can be 40-60 amino acids in length. In some, but not all cases, this region can be amphipathic. The second region is a highly lipophilic domain of 65-75 amino acids. In normal hydrophobic plots (e.g. Chou-Fasman analysis) this appears to be the longest consistently hydrophobic sequence in the current sequence databases. This stretch is replete with leucines, isoleucines, valines and alanines and a smaller number of phenylalanines. This hydrophobic core of the protein is responsible for most of the unique properties of oleosins. Thirdly, oleosins typically have a *C*-terminal domain of quite variable length, from about 50-90 amino acids. This region frequently contains an amphipathic α-helix domain of about 30-40 amino acids, but this is not universally predicted when such computer-based structural analyses are performed. It is mainly variation in this *C*-terminal domain which accounts for the divergent molecular weights of oleosins in different species.

The conserved hydrophobic core of all oleosins imparts unique and important properties to this protein. Besides the long stretch of lipophilic amino acids, oleosins have a conserved motif which is referred to as the "proline knot". The proline knot is an array of 12 amino acids with three prolines within the sequence. The proline knot effectively bisects the hydrophobic core into two parts of equal length. The preponderance of proline residues is likely to generate a kink in the polypeptide chain and this may permit the two flanking strands to form an anti-parallel arrangement (Huang, 1992; Huang, 1996). It is possible that such an antiparallel arrangement may be stabilized by the pairing of amino acids sharing similar hydrophobicity or hydrophilicity. There is also the possibility of hydrogen bonding between carboxyl and amino groups aligned opposite one another (Huang, 1992). It is, however, difficult to predict the secondary structure adopted by a polypeptide which exists in a non-aqueous environment *in vivo*. Proposals that the lipophilic core might organize as an α-helix (Qu *et al.*, 1990) or a β-strand (Murphy *et al.*, 1991), have been based on theoretical models rather than empirical results. More direct measurements using lipophilic solvents for circular dichroism and FT infrared spectroscopy in solid state have provided evidence that the central core can adopt a β-sheet conformation although these measurements do not prove the anti-parallel conformation (Jacks *et al.*, 1990; Li *et al.*, 1992).

More recently, Lacey *et al.* (1998) have shown using FTIR spectroscopy that the hydrophobic core of oleosins *in situ* appears to adopt an almost completely α-helical structure. This work was performed using sunflower (*Helianthus annuus* L) and safflower (*Carthamus tinctorius*) oil-bodies, although there is no reason to believe that this observation would not be more generally applicable. The study by Lacey *et al.* (1998) is the most direct measurement made so far on oleosin secondary structure in its native environment.

In addition to the questions of secondary and tertiary structure of the oleosin molecule, significant work has been devoted to the question of whether oleosins are functional individually or in some kind of structure associated with other oleosins. This is hypothesized because:

1) Seed oleosins normally comprise at least two (often more) isoforms distinguishable by molecular weight and immunologically (Tzen *et al.*, 1990).

2) In some species oleosin isoforms appear to exist in stoichiometric ratios e.g. corn (Lee *et al.*, 1995).

3) Proteins of higher molecular weight than a simple oleosin are frequently extracted and even retain their structure in denaturing conditions (Li *et al.*, 1992). These are thought to be non-covalently bound homo- or heterodimers (see Figure 1).

It has been proposed that oleosins exist naturally as heterodimers between two immunologically distinct groups of oleosin polypeptides of higher (H form) or lower (L form) molecular weights (Huang, 1996). Perhaps the best experimental evidence to date in support of this is the work of Lee *et al.* (1995) which demonstrated a 1:1 stoichiometric ratio of L and H forms of oleosin in corn, which appeared to be decoupled from relative transcription rates for the two types. Mutants impaired in transcription for one of the H form oleosins still gave rise to approximate 1:1 stoichiometry of the H-form and the L-form. This suggests that the overall amounts of the oleosin isoforms accumulated were interdependent and regulated post-transcriptionally, possibly at the level of protein stability. Whether such protein-protein interactions are needed for oleosin function is still unknown. However, it is clear that such interactions are not absolutely necessary for targeting of oleosin nor its stability on oil-bodies. This was demonstrated by Ting *et al.* (1997) who expressed a corn oleosin gene in yeast. When the yeast was induced nutritionally to produce storage lipid, the plant oleosin accumulated concomitantly and associated specifically with yeast lipid bodies. As only one oleosin isoform was expressed (M_r 19,000; H-form) in the yeast cells it is clear that all the necessary information for targeting and organelle-specific accumulation resides in the individual isoform. This result does not exclude the possibility that association of oleosins could involve homodimers or heterodimers and that either may impart stability.

4. TARGETING AND TOPOLOGY OF OLEOSIN ON MEMBRANES

4.1 Spatial organization of oleosins on mature oil-bodies

It has long been hypothesized that oleosins reside in an amphipathic environment and adopt a tertiary structure which spans two phases, lipid and water (Tzen and Huang, 1992; Murphy *et al.*, 1990). Clearly, the expected topology would require the entire hydrophobic region to be embedded in the neutral lipid phase while the *N*- and *C*-terminal domains, which are much more hydrophilic or amphipathic, would reside on the outside of the organelle, in contact with the cytoplasm. There is now a great deal of experimental evidence to support this idea.

First, it was shown by Tzen and Huang (1992) that on treating oil-bodies with trypsin a protein fragment of about M_r 8,000 is protected. This size is sufficient to account for the central hydrophobic core which typically has 65-75 amino acids. The critical role of oleosin in such an amphipathic environment was also demonstrated by experiments performed by Tzen and Huang (1992). These workers showed that oleosins are determinants of oil-body stability and also that they effectively form a “shell” around individual oil-bodies. In the above trypsinization, the oil-bodies changed their characteristics. After trypsin treatment they easily coalesced and floated rapidly. Average oil-body diameter increased significantly due to coalescence. Interestingly, treatment of intact oil-bodies with phospholipase (PL) had no effect on oil-bodies or their propensity to coalesce. However, if trypsin treatment precedes the PL treatment, the PL results in hydrolysis of the phospholipids destabilizing the oil-bodies. This result clearly indicates that oleosins are present at a density such that they can sterically hinder the access of phospholipase to the oil-bodies.

This "shell" of oleosins around oil-bodies underlines the probable biological role of oil-bodies. In addition to steric hindrance, oleosins (combined with phospholipids) present a substantial surface negative charge. Although oleosins possess an overall positive charge, this is oriented inwards towards the negative charge of the phospholipid. This results in an aggregate negative charge being displayed on the surface. This charge probably limits coalescence and even aggregation. This was demonstrated experimentally by Kühnel *et al.* (1996) who measured the time taken for oil-bodies to float from a homogeneous suspension at different pH. At pH values between 4.5-6.0 there was rapid floatation and clearing of the 'unternatant', presumably due to aggregation. This aggregation was not accompanied by coalescence and proved to be at least partially reversible. These experiments illustrate the role of oleosins in providing a mechanism which limits coalescence of oil-bodies, a property which is valuable to a seed undergoing germination.

In seed cells, it is interesting to note that oleosins are not found in substantial amounts in any other organelle or subcellular fraction other than oil-bodies. This suggests very specific partitioning or (more likely) specific targeting. This targeting is very much integrated with the question of oil-body biogenesis.

4.2 Oil-body biogenesis and oleosin targeting

It was argued earlier that the preponderance of evidence available on oil-body biogenesis in seeds supports the idea that "mature" oil-bodies equipped with TAG, PL and oleosins are assembled and released from the ER. This has been a controversial area, although most of the controversy is based on hypothesis rather than direct empirical observations.

Initial experimental evidence that oleosins are co-translationally associated with the ER came from Hills *et al.* (1993) and Loer and Herman (1993). Both of these papers showed that using a canine pancreatic microsome preparation, oleosin mRNA could be translated *in vitro* and the translation product became ER-associated. Treatment with proteinase K showed that an M_r 15,000 fragment of the M_r 20,000 *in vitro* translation product was protected from proteinase K (PK) digestion. This suggested that a substantial amount of the nascent polypeptide was inserted into ER and thus protected from proteolysis. However, these experiments were never able to determine which domains were protected against proteolysis. In particular, it was not clear if part of the *N*- and part of the *C*-terminal domains were protected or the whole of one domain at the expense of the other. A further question was whether at any time during ER-association the *N*- or *C*-termini were exposed to the ER lumen. In order to determine the precise topology of the oleosin on the ER, Abell *et al.* (1997) performed a series of *in vitro* translations using truncated forms of oleosin mRNA. Based on the sizes of the PK protected fragments it was possible to show that both the *N*- and most of the *C*-terminus of the oleosin nascent polypeptide are exposed on the cytoplasmic side of the ER. This idea is explored in more detail below. The entire hydrophobic core is embedded inside the phospholipid bilayer of the ER and does not appear to protrude into the lumen. The truncation experiments also demonstrate that the signal sequence for ER incorporation resides at the beginning of the hydrophobic core. Taking these results (Abell *et al.*, 1997) with the earlier work of van Rooijen and Moloney (1995b), it is probable that the signal sequence is located close to the interface between the *N*-terminus and central hydrophobic domain. In the *in vitro* translation experiments the loss of 44 amino acids from the *N*-terminus did not dramatically inhibit ER association. However, the loss of 47 amino acids from the *N*-terminus resulted in significant protein instability (van Rooijen and Moloney, 1995b). Despite the uncertainty as to the precise motif which serves as a signal sequence, it is clear that the signal sequence is not cleaved during co-translational association with ER. That this co-translation is similar to classical co-translational models with or without a cleavable signal peptide was demonstrated by Thoyts *et al.* (1995). In this work, it was shown that *in vitro* translation could be arrested in the presence of added (20nM) canine signal recognition particle (SRP) in the absence of recipient microsomes. Once microsomes were present, this arrest was relieved showing that the SRP recognising the nascent oleosin signal sequence was capable of docking on ER SRP receptor and facilitating insertion into ER. A model depicting the key features of oil-body biogenesis and oleosin targeting is given in Figure 2.

While there is evidence for the involvement of an uncleaved signal sequence in oleosin ER insertion, this is not the only signal necessary to ensure that oleosins insert correctly onto oil-bodies. Recent studies by Abell *et al.* (1997) investigated the role of the conserved "proline knot" in regulating trafficking from the ER to oil-bodies. The remarkable conservation of the proline knot in all oleosins studied to date suggests that it must have a pivotal role. But for what step of oleosin accumulation is it essential? Abell *et al.* (1997) showed that by substituting the prolines in the knot with leucines, the co-translational properties of the oleosin were unimpaired. Oleosin polypeptide associated with ER had a slightly altered conformation as judged by proteolysis of exposed sequence, but the entire central core was still inserted into the ER and protected against PK treatment. However, further experiments with this proline knot variant *in vivo* showed that loss of the motif resulted in little or no targeting to oil-bodies, indeed the major outcome was destabilization of the oleosin polypeptide, either during or after targeting to the oil-body. The overall effect was to eliminate oleosin accumulation. Given earlier work by van Rooijen and Moloney (1995b) on targeting of oleosins *in vivo*, it is clear that any aberration in oleosin targeting results in destabilization of the protein. It is tempting to use this evidence as partial confirmation of the hypothetical antiparallel model presented by Huang (1992) but, until it is clear at

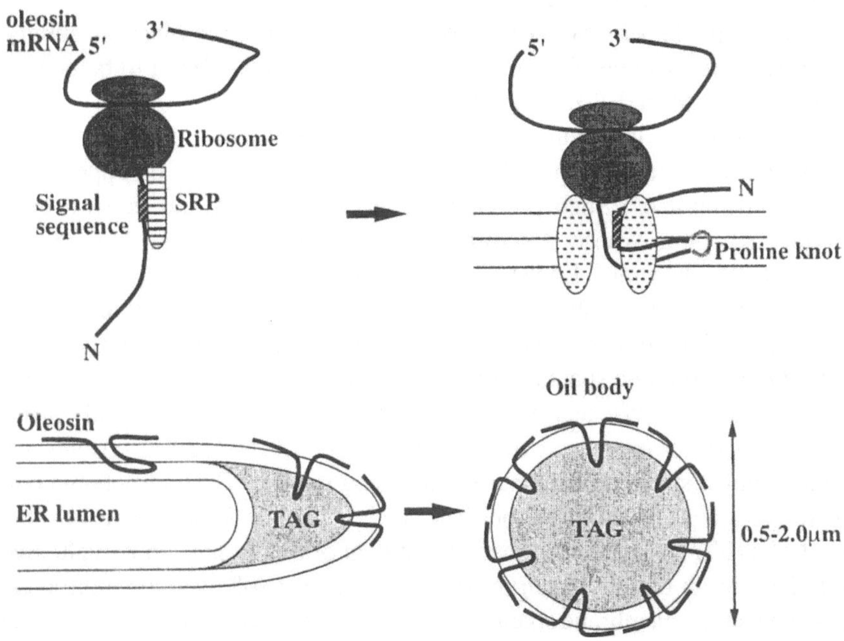

Fig. 2 Model illustrating co-translation and targeting of oleosins to oil-bodies in developing seeds. It is proposed that the emerging signal sequence binds a signal recognition particle (SRP) which attaches the ribosome to the ER. The nascent oleosin polypeptide adopts an unusual conformation on the ER with the *N*- and *C*-termini on the cytoplasmic side and the lipophilic core of the oleosin in the ER lipid-bilayer. The oil-bodies are released from the ER with their complement of oleosins already attached (figure courtesy of Mr. B. Abell, University of Calgary).

what stage the instability occurs, this can only be considered circumstantial evidence. Clearly, however, the conformation of oleosins determined by the proline knot is a key determinant in oleosin targeting and or stability.

4.3 Topology of oleosins during targeting to oleosomes and its implications

The topology of oleosins at their destination (discussed above) raises questions about their topology during trafficking and the relationship between that topology and oleosome biogenesis. This question has been investigated recently by the use of *in vitro* translation experiments on microsomes (Abell *et al.*, 1997). These authors demonstrated unequivocally that the topology of oleosins on the ER during its trafficking is such that the *N*- and *C*-termini are both displayed on the cytoplasmic side of the ER bilayer. It appears that neither the *N*- nor *C*-termini are exposed to the lumen of the ER. The central core does not function simply as a transmembrane spanning domain, allowing the *N*- and *C*-termini to be displayed on opposite sides of the ER membrane.

This topology is unusual as the majority of proteins with a *single* transmembrane spanning domain display their two termini on opposite sides of the ER membrane. Such proteins include glycophorin and the low-density lipoprotein receptor, whose *N*-termini are lumenal and whose *C*-termini are cytoplasmic (Wickner and Lodish, 1985). An alternative topology with the *C*-terminus displayed lumenally and *N*-terminus displayed cytoplasmically occurs with the asialoglycoprotein receptor (Speiss and Lodish, 1986). Some proteins with *multiple* membrane spanning domains appear to retain *N*- and *C*-termini on the cytoplasmic side (Wickner and Lodish, 1985). Oleosins are the only proteins so far reported to retain a cytoplasmic display of *N*- and *C*-termini while containing only a single transmembrane spanning region. The hydrophobic region is, however, unique in its properties due to the "proline knot" discussed above. Abell *et al.* (1997) hypothesized that the removal of this proline knot might switch the topology of an oleosin during trafficking so that it followed the more general models with single membrane spanning domains. This reasoning was based on the idea that the proline knot might provide the structural constraints needed to form the more rigid antiparallel structure proposed by Huang (1992). However, elimination of the conformational constraint of the proline knot did *not* result in a change of overall topology of oleosins on ER. Both *N*- and *C*-termini remained exposed on the cytoplasmic side. The only detectable effect of this substitution of the proline knot on topology was a small change in overall conformation so that less of the *N*-terminus was exposed at the ER surface. In contrast, although the overall topology was unchanged, the targeting and/or stability of the oleosin was drastically impaired, indicating a critical role for the proline knot in targeting. This latter result was demonstrated by expressing the proline knot-minus mutant oleosin in transgenic seeds.

It is intriguing to ask whether the topology of oleosins on the ER has a role in determining the orientation of release of oil-bodies from the ER. It is noteworthy that mammalian apolipoproteins have the opposite topology (i.e. both *N*- and *C*-termini exposed to the lumen) and the release of "oil-bodies" (lipoprotein vesicles) occurs into the secretion pathway (ER lumen) rather than on the cytoplasmic side. This connection has never been investigated experimentally and it clearly requires further work, given its relevance to oleosome biogenesis.

5. MOLECULAR BIOLOGY OF SEED OLEOSINS

5.1 Seed oleosin gene families

Oleosin genes from at least 25 different species have been isolated and sequenced to date. Most (but not all) are available in GenBank. The availability of these sequences enables a number of general ideas about oleosins to be surmised.

1) Oleosins frequently occur in plants as discrete gene families. Even in simple diploid species with a low chromosome number and DNA content such as *Arabidopsis* there are at least four family members comprising seed oleosins.

2) In agreement with the observations by Tzen *et al.* (1990) made at the protein level, coding sequences can frequently be compared horizontally by homology to high and low molecular weight isoforms, with most of the variations in sequence evident at the *N*- and *C*-termini.

3) Oleosins are often major constituents of seeds. By inference, their genes are extremely active at the transcriptional level. Examples of this are the promoters and upstream elements of an *Arabidopsis* oleosin (Plant *et al.*, 1994) and a soybean oleosin (Sarmiento *et al.*, 1997).

4) Regulation of oleosin gene expression is clearly tissue- and organ-specific (Plant *et al.*, 1994) and dependent on hormones (specifically abscisic acid) (Wilen *et al.*, 1993; Plant *et al.*, 1994).

The existence of oleosins as small gene families raises a number of questions relating to their evolution, expression patterns and differential roles (if any).

5.2 Seed oleosin gene expression

There has been significant discussion over several years about the precise timing of oleosin gene expression. Murphy *et al.* (Murphy and Cummins, 1989; Murphy *et al.* 1989b) suggested that in *Brassica* species oleosin accumulation occurs quite late in seed development, in particular, somewhat later than the most active phase of TAG biosynthesis. This idea was reinforced by further work by Cummins *et al.* (1993). Qu and Huang (1990), however, demonstrated that in maize scutellum the pattern of oleosin gene expression was much earlier and coincided quite closely with TAG accumulation. Furthermore, Tzen *et al.* (1993) investigated the accumulation of oleosins in *Brassica napus* and found a concerted accumulation of oleosin and TAG. The discrepancies between these two sets of experiments appear hard to reconcile although significant differences occur in experimental design, most notably an analysis to 36 days post anthesis in Tzen *et al.* (1993) and to 77 days post anthesis in Cummins *et al.* (1993). While it is difficult to explain these differences, it is certainly clear that a seed oleosin promoter from *Arabidopsis* (which is also >90% identical to a *B. napus* promoter;

Lee and Huang, 1991) can furnish expression of a recombinant protein in transgenic *Brassica* zygotic embryos at the heart stage of embryogenesis. This is clearly earlier than storage protein deposition and suggests that, at least at the *transcriptional* level, oleosin gene expression commences early enough to coincide with lipid accumulation and earlier than the deposition of storage proteins (Murphy *et al.*, 1989b). An additional interesting factor in this debate is the possibility that different oleosin gene family members might show overlapping patterns, some expressing earlier than others. This possibility was reported by Thoyts *et al.* (1995) in sunflower (*Helianthus annus*) seeds where two oleosin family members designated SO5 and SOM showed temporally distinct expression patterns with SO5 transcripts accumulating earlier than those for SOM. Such asynchronous expression of oleosin isoforms has also been detected in *Arabidopsis* by Kirik *et al.* (1996). In the latter work an oleosin isoform (*Atol3*) showed expression at a much earlier developmental stage than the two other isoforms studied.

The idea that all oilseeds might require at least two oleosin isoforms which permit heterodimers to form (Tzen *et al.*, 1990) has been proposed and was discussed above. The evidence for this is only circumstantial at this time and based almost entirely on observations on corn scutellum (Lee *et al.*, 1995). Experiments with oleosin expression in yeast (Ting *et al.*, 1997) and with recombinant oleosin expression in plants (van Rooijen and Moloney, 1995a,b) suggest that such heterodimerization is not essential for oleosin targeting.

Despite this, it is intriguing that there is horizontal sequence homology between species for the different isoform types (L-type and H-type) (Tzen *et al.*, 1990). Such conservation would normally imply an essential function which could not be lost throughout evolution without detriment to the life-cycle of the plant. Clearly, such conservation suggests that the different isoforms do not constitute functionally redundant entities. It is possible that the interaction of some portion of the two isoforms might give rise to as yet unidentified functionality such as the creation of a docking site for a lipase during germination (Huang, 1996). This idea has been proposed for some time, but no direct evidence is yet available.

5.3 Seed oleosin gene regulatory sequences

After storage proteins, oleosins are frequently the most abundant proteins in oilseeds, such as *Brassica*, sunflower, safflower and soybean. In corn, oleosins are among the most abundant proteins of the embryo and scutellum. This implies that oleosins promoters may be responsible for high-level transcription during certain phases of seed development. This prediction is verified in studies on oleosin transcriptional regulatory regions ("promoters") from *Arabidopsis*, *B. napus* and soybean. In the case of the *Arabidopsis* promoter, Plant *et al.* (1994) showed that a 2.5 kbp upstream sequence from one *Arabidopsis* oleosin gene sustained expression of β-glucuronidase (GUS) as a reporter at levels in excess of 20nmol methylumbellferone mg $prot^{-1}min^{-1}$. This level is comparable to that which is possible with strong plant promoters such as the cauliflower mosaic virus 35S promoter. Such a comparison was provided by Keddie *et al.* (1994) using an 870bp promoter from a *Brassica* oleosin in transgenic tobacco seeds. In this case, using GUS as a reporter, the specific activity of GUS observed in seeds was very similar for the CaMV 35S promoter and the oleosin promoter. In a similar manner, expression of a soybean oleosin in transgenic rapeseed under the control of a soybean oleosin promoter resulted in the recombinant oleosin comprising about a quarter of the total oleosin complement in the seed (Sarmiento *et al.*, 1997). This further emphasizes the high transcriptional activity supported by such promoters. Interestingly, it has also been demonstrated that an *Arabidopsis* oleosin promoter may be used to drive high level heterologous gene expression in a wide range of oilseeds including flax, soybean, safflower, sunflower, castor bean and cotton (Abenes *et al.*, 1997). Surprisingly, this promoter also functioned at easily detectable levels in the scutellum of corn indicating that this dicot promoter may function in a monocot. This was unexpected given the report of Lee *et al.* (1990) which indicated that the corn oleosin promoter was not functional in transgenic *Brassica* seeds.

It is noteworthy that the oleosin promoter from *Arabidopsis* retains strong tissue specificity. In seeds with significant endosperm such as castor bean, the promoter drives expression only in embryo tissue and not in endosperm. This is in spite of the fact that in castor bean the majority of oil storage is in endosperm. The cellular specificity of oleosin genes expressed in seeds in different species is further emphasized by work from Aalen *et al.* (1994). These authors studied a barley oleosin transcript, B23D. This transcript, which encodes a homologue of the corn M_r 18,000 oleosin, was found to be expressed not *only* in embryos, but also in the aleurone layer of the developing seed. The expression did not extend to the endosperm.

5.4 Hormonal sensitivity of seed-oleosin gene expression

Oleosin genes exhibiting seed-specificity also show strong sensitivity to the influence of hormones and water potential. It was shown by Qu *et al.* (1990) in corn and by Taylor *et al.* (1990) in *Brassica napus* that oleosin gene expression and accumulation is dramatically affected by ABA treatments at levels that appear physiologically meaningful (10^{-5}-10^{-6}M). This sensitivity was studied in more detail by van Rooijen *et al.* (1992b) who demonstrated that both ABA and water stress (imposed using sorbitol) resulted in rapid and sustained oleosin gene expression. The promoters of an *Arabidopsis* oleosin (van Rooijen *et al.*, 1992a; Plant *et al.*, 1994) and one from *B. napus* (Keddie *et al.*, 1994) were further dissected in order to identify key motifs that might be responsible for their inducibility. Both groups identified a number of structural features which suggested the presence of *Ab*scisic acid *r*esponse *e*lements (ABREs) as elements upstream of the transcriptional start. Keddie *et al.* (1994), using a gel retardation assay, were able to establish that a region from the *Brassica* oleosin promoter bound a nuclear protein in embryos that showed (in competition experiments) specificity for ABRE-like sequences. Plant *et al.* (1994) showed that although the overall amplitude of gene expression was affected by promoter length, truncations down to 400 bases upstream of the start codon still allowed for significant ABA-sensitive induction. All the putative ABREs in these promoters are found within this stretch of upstream sequence. The nature of ABA sensitivity of oleosin gene expression was also demonstrated genetically by several authors. Koorneef *et al.* (1989) showed that *Arabidopsis* mutants in ABA synthesis and ABA sensitivity accumulate less storage proteins, but also less oleosins. Parcy *et al.* (1994) further showed that the expression of oleosin transcripts were retarded significantly in an ABA insensitivity mutant (*abi3*) of *Arabidopsis*, only appearing quite late in embryogenesis and with lower amplitude than in wild type embryos.

The relationship between ABA sensitivity and osmotic sensitivity has not yet been fully investigated. It has been shown that treatment of *B. napus* microspore-derived embryos with a high osmoticum (12.5% sorbitol) results not only in greatly increased oleosin gene expression, but also increases in endogenous ABA (Wilen *et al.*, 1990). It is, however, also clear from Plant *et al.* (1994) that the combination of 20μM ABA and 12% sorbitol results in higher oleosin gene expression than with either treatment alone, suggesting that the osmotic effects are not exclusively mediated through endogenous ABA concentrations. Studies by Kirik *et al.* (1996) noted that, in certain *fusca* mutants of *Arabidopsis*, oleosin transcription is reduced in amplitude, but not altered temporally. In the *fus3* mutants, the ABA level was not reduced indicating that the down-regulation in this mutant is independent of ABA levels. Furthermore, *fus3/abi3* double mutants revealed no epistatic relationships.

Overall, oleosin promoters, while showing some similarities in expression patterns to storage protein promoters, are a very distinct group of promoters which could be exploited in unique ways for the genetic engineering of oilseed traits, in particular in controlling the early expression of genes in lipid biosynthesis or protein accumulation. A significant volume of work remains if we are to understand better their regulation and cellular specificity.

6. OLEOSIN EVOLUTION

The role of oleosins can be in part surmised by considering their appearance and evolution. First, although this review focuses on seed oleosins, seeds are not the only structures in which oleosins are found. In the spermatophytes, oleosins are found in both seeds and in some gametophytic tissues. In *Pinus*, oleosins accumulate not only in the sporophytic tissues of the seed, but also in the megagametophytes. In floral tissues there is still significant controversy about whether oleosins found in anthers are primarily in the tapetum (sporophytic tissue), in pollen (gametophyte), or both (Wang *et al.*, 1997; Ross and Murphy, 1996).

Clearly, from *in situ* hybridization experiments (Ross and Murphy, 1996) a substantial proportion of anther-localized oleosin gene expression is in the tapetum. This has also been recently confirmed by Wu *et al.* (1997), who found that the major oleosin-like proteins (M_r 45,000 and M_r 48,000) in the anthers of *Brassica napus* were located in the tapetum rather than in pollen. This is significant in that it places these two major oleosin-like proteins in sporophytic and not gametophytic tissue. This leaves the only clear case of an oleosin expressed in gametophytic tissues to be that of the pine megagametophyte (Lee *et al.*, 1994). Interestingly, when analysed in an unrooted phytogenetic tree, the pine oleosin is found uniquely in a side-branch distinct from all other seed or floral oleosins (Huang, 1996).

If oleosins are associated with both gametophytic and sporophytic tissues, it is not unlikely that they predate the seed system. Although an area of conjecture, it seems likely that oleosin ancestors arose in early terrestrial plants, probably homosporous bryophytes or seedless vascular plants which store neutral lipid as an energy source. The advent of heterospory probably provided a path for the divergent evolution of oleosin into the male and female gametophytic structures. By the arrival of the early spermatophytes of the Gymnospermae, the diverging oleosins were manifest in the non-persistent male gametophyte (pollen) or the persistent megagametophyte which is retained in seeds. The emergence of angiosperms allowed the recruitment of the "seed" oleosins for expression and accumulation in sporophytic or even endospermic tissues. The common denominator for this entire sequence is the need for discrete oil-bodies which survive desiccation without undergoing coalescence. This need is equally manifest in bryophyte homospores and desiccating angiosperm seeds. This proposed evolutionary sequence is essentially without evidence at this time, but should be readily amenable to investigation.

7. BIOTECHNOLOGICAL APPLICATIONS OF SEED OLEOSINS

The properties of oleosins, particularly their intense lipophilicity and our increasing understanding of their targeting and topology, have led to some interesting uses of oleosins in biotechnological applications. These are based upon a number of key properties of oleosins which have been noted earlier. These include:

1) The efficiency of partitioning of oleosin onto oleosomes, but their absence from other cellular organelles.
2) The avidity of oleosins for oleosomes even after they are released from cells.
3) The absence of cleavage of the signal sequence.
4) The variability of the *N*- and *C*-termini of oleosins in comparison with the highly conserved lipophilic core.
5) The high levels of oleosin accumulated in many oilseed species.

Genetic manipulation of oleosin coding sequences has permitted the expression of recombinant forms of oleosins in seeds, which may permit not only the expression, but also the simple isolation of a desired recombinant protein.

7.1 Oleosin-GUS fusions target to oil-bodies

The first evidence that such an approach might work was provided by van Rooijen and Moloney (1995a). In their experiments, these authors created a recombinant oleosin gene comprising the entire genomic coding sequence of an *Arabidopsis* oleosin (van Rooijen *et al.*, 1992a) fused (in the absence of an intervening stop-codon) to a coding sequence for β-glucuronidase from *E. coli*. In this experiment, the two domains (oleosin and β-glucuronidase (GUS)) were separated by a sequence specifying a four amino-acid motif which, in the fusion protein, provides a site for cleavage by the protease, thrombin. In these experiments, it is noteworthy that the GUS domain has a M_r of approximately 67,000. This is much larger than the *Arabidopsis* oleosin used which is M_r about 18,000. When the fusion gene was introduced into *Brassica napus* using *Agrobacterium*-based methods (Moloney *et al.*, 1989), the expression occurred in a seed-specific manner. This resulted in the accumulation of the oleosin-GUS fusion protein ($M_r \sim 85,000$) in seeds. When seed extracts were fractionated it was found that the majority (78%) of the accumulated oleosin-GUS was associated specifically with oil-bodies (van Rooijen and Moloney, 1995a). This demonstrated that an extension of the *C*-terminal end of the protein by as much as M_r 67,000 had no effect on subcellular trafficking of the protein. A parallel experiment in which GUS was fused in-frame to oleosin at the *N*-terminal end resulted in equally unimpaired targeting of the protein to oil-bodies. This latter experiment also verified the lack of a *cleavable* signal sequence as such a cleavage in the *N*-terminal region of the oleosin would have resulted in detachment of the GUS. Figure 3 illustrates a typical configuration of a *C*-terminal recombinant protein fusion to oleosins.

In the *C*-terminal configuration, it was found that the GUS enzyme remained active. This was not entirely unexpected given previous reports of fusion proteins containing GUS (Jefferson, 1987; Datla *et al.*, 1991), but it is nonetheless

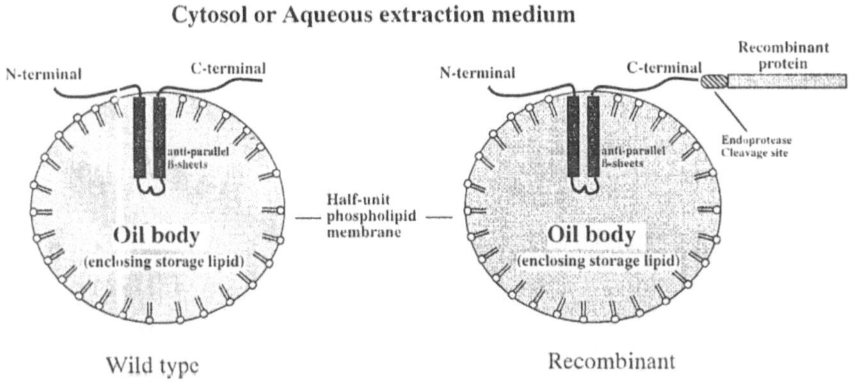

Fig. 3 Configuration of oleosins on wild-type and recombinant oil-bodies based on models of Huang (1992) and van Rooijen and Moloney (1995).

interesting that even when attached avidly to a particle such as an oil-body the enzymatic activity remains. It may even imply that enzyme assembly may occur on the oil-body surface as GUS is believed to function as a homotetramer (Jefferson, 1987). In effect, this association of GUS with an oil-body is a form of biological enzyme immobilization. The enzyme may be recycled by continuous rounds of use and centrifugation (van Rooijen and Moloney, 1995a). Subsequent studies also showed that the recombinant oil-bodies are remarkably robust, resisting mechanical stress, pH change and even elevated temperature (Khhnel *et al.*, 1996). This principle of immobilization of enzymes on oil-bodies is not confined to model enzymes such as β-glucuronidase. Recently, Liu *et al.* (1997) performed similar experiments on a xylanase derived from a rumen fungus, *Neocallimastix patriciarum*. The xylanase (M_r 40,000) was expressed as a functional oleosin fusion which was amenable to recycling. It retained essentially all of the catalytic properties of the native enzyme.

Finally, it was also shown by van Rooijen and Moloney (1995a) that the oleosin-GUS fusion separated by a thrombin cleavage site could also be cleaved when the oil-bodies were suspended in presence of thrombin. This released the GUS into the aqueous fraction where it could be recovered and concentrated. A schematic diagram illustrating this process is given in Figure 4.

7.2 Production of the blood anticoagulant hirudin as an oleosin fusion

The principle demonstrated in van Rooijen and Moloney (1995a) was further tested using a pharmacologically active model, hirudin. Hirudin is a blood anticoagulant produced naturally in the salivary glands of the medicinal leech,

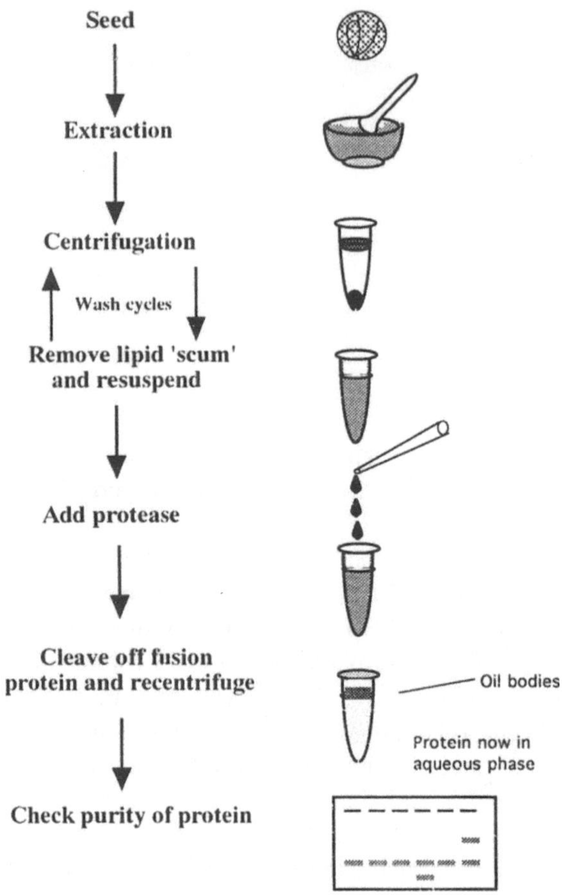

Fig. 4 Schematic diagram of a protein production and recovery system using oleosin fusions as carriers.

Hirudo medicinalis (Markwardt and Walsmann, 1967). The polypeptide is 65 amino acids long and is the most potent inhibitor of thrombin known (Stone and Hofsteenge, 1986). It is also extremely low in allergenicity for mammals (Klocking, 1991). As a result it is widely considered to be an excellent candidate as a therapeutic anticoagulant. Hirudin requires the establishment of three disulphide bridges in order to show biological activity (Rydel *et al.*, 1990). Thus, expression of this protein in plants also tested the ability of the oleosin system to sustain certain aspects of protein folding. Parmenter *et al.* (1995) created an oleosin-hirudin fusion gene which incorporated a cleavage site for Factor Xa between the two domains. Expression of this construct in a seed-specific manner in *B. napus* and *B. carinata* (Chaudhary *et al.*, 1998) resulted in accumulation of the oleosin-hirudin fusion protein. Isolated oil-bodies were shown to accumulate the protein preferentially over other cellular fractions with about 80% of the hirudin (as detected immunologically) being attached to oil-bodies. Immunofluorescence studies also showed that the hirudin moiety was displayed on the oil-body surface. This was further verified by the lability of the Factor Xa cleavage site which was accessible to Factor Xa on re-suspended oil-bodies. Overnight treatment with Factor Xa permitted release of a polypeptide of the correct size into the medium. On centrifugation, this 'unternatant' was tested in a thrombin inhibition assay. The released hirudin proved to be a highly effective inhibitor of thrombin activity indicating that the polypeptide had indeed folded correctly and correctly processed at its *N*-terminus. The latter can be concluded because thrombin inactivation of hirudin depends upon interaction between the *N*-terminus and the active cleft of thrombin (Stone and Maraganore, 1993).

The released hirudin was resolved into two chromatographic peaks which, by mass spectral analysis, were shown to represent a full-length hirudin molecule and a truncated version of hirudin lacking the last two amino-acids from the *C*-terminus (Parmenter *et al.*, 1996). The truncation was the same as had been noted when hirudin is expressed in yeast. This truncation is probably due to the activity of a carboxypeptidase. In yeast, the carboxypeptidases ysc Y and ysc A have been implicated (Heim *et al.*, 1994).

The GUS and hirudin models both show that oleosin fusions can give rise to correctly folded biologically active proteins in seeds. The oleosin-oil body association provides a major advantage in protein purification. In the case of hirudin, two simple chromatographic steps yielded the desired protein at purity of at least 95%. This approach has been attempted for several other proteins of industrial, agricultural and medical value and appears to be widely applicable to different protein classes.

Perhaps the major limitation to this system is that it is not clear whether it might be applicable to production of glycosylated proteins. The work of Abell *et al.* (1997) demonstrates unequivocally that native oleosins are not exposed to the ER lumen. It would, therefore, generally be expected that recombinant proteins attached to oleosins would have similar topology. This would appear to limit the potential of this system for glycosylated proteins. On the other hand, given the specialized glycosylation patterns and constituents found in plants, in particular xylose and fucose residues (Faye, 1988), it may be advantageous to express heterologous proteins on oil-bodies rather than secreting them, thus avoiding glycosylation. Many proteins which are glycosylated in nature retain their activity even in the absence of glycosylation (Krantz, 1991). The usefulness of this seed-oleosin expression technology will ultimately be determined by answers to such questions as well as our ability to scale up extraction and purification processes that permit commercial quantities of desired proteins to be produced from oilseeds.

8. CONCLUDING REMARKS

Oleosins are unique proteins both in composition and function. They have a number of properties which are now well-documented and understood, in particular, their unusual lipophilicity "fingerprint" (the very long stretch of lipophilic amino-acids), their conservation and their relatedness between diverse species. More recently, we have made significant headway in our understanding of the trafficking of these proteins following co-translation and their topology during targeting. The biological roles of oleosins may still contain some surprises, but it is nevertheless clear that their major role is in preventing coalescence of oil-bodies during seed desiccation. There is still significant work in this area to be done. In particular, it must be asked whether oleosins also have important properties in relation to oil mobilization during germination. The nature of the interaction of oil-bodies with glyoxysomes is worthy of serious investigation. The existence of oil-bodies in floral and gametophytic tissues as well as in embryos and endosperms raises some interesting evolutionary questions which require further investigation. Remarkably, although it seems intuitively probable, there is not yet any direct experimental evidence that oleosins predate the spermatophytes.

The properties of seed oleosins are providing a major opportunity for exploitation in the plant biotechnology sector. It they prove to be a useful addition to the tools available to those engaged in molecular farming, this should also necessitate further research to answer many of the outstanding questions on the structure and function of these ubiquitous seed proteins.

REFERENCES

Aalen, R.B., Opsahl-Ferstad, H.G., Linnestad, C. and Olsen, O.A. (1994) Transcripts encoding an oleosin and a dormancy-related protein are present in both the aleurone layer and the embryo of developing barley (*Hordeum vulgare* L.) seeds. *Plant Journal* **5**: 385-396

Abell, B.M., Holbrook, L.A., Abenes, M., Murphy, D.J., Hills, M.J. and Moloney, M.M. (1997) Role of the proline knot motif in oleosin endoplasmic reticulum topology and oil body targeting *Plant Cell* **9**: 1481-1493

Abenes, M., Holbrook, L. and Moloney, M.M. (1997) Transient expression and oil body targeting of an *Arabidopsis* oleosin-GUS reporter fusion protein in a range of oilseed embryos *Plant Cell Reports* **17**: 1-7

Bergfeld, R., Hong, Y.-N., Kuhnl, T. and Schopfer, P. (1978) Formation of oleosomes (storage lipid bodies) during embryogenesis and their breakdown during seedling development in cotyledons of *Synapis alba* L *Planta* **143**: 297-307

Cao, Y.Z. and Huang, A.H.C. (1986) Diacylglycerol acyltransferase in maturing oil seeds of maize and other species *Plant Physiology* **82**: p.813-p.820

Chaudhary, S., Parmenter, D.L. and Moloney, M.M. (1998) Transgenic *Brassica carinata* as a vehicle for the production of recombinant proteins in seeds *Plant Cell Reports* **17**: 195-200

Cummins, I. and Murphy, D.J. (1990) Mechanism of oil body synthesis and maturation in developing seeds. In: *Plant Lipid Biochemistry, Structure and Function*, edited by Harwood, J.L. and Quin, P.J. London: Portland Press, 1990, p. 231-233.

Cummins, I., Hills, M.J., Ross, J.H.E., Hobbs, D.H., Watson, M.D. and Murphy, D.J. (1993) Differential, temporal and spatial expression of genes involved in storage oil and oleosin accumulation in developing rapeseed embryos: implications for the role of oleosins and the mechanisms of oil-body formation *Plant Molecular Biology* **23**: 1015-1027

Datla, R.S.S., Hammerlindl, J.K., Pelcher, L.E., Crosby, W.L. and Selvaraj, G. (1991) A bifunctional fusion between β-glucuronidase and neomycin phosphotransferase: a broad-spectrum marker enzyme for plants *Gene* **101**: 239-246

Faye, L. (1988) Structure, biosynthesis, and function of plant glycoprotein glycans *Current Topics in Plant Biochemistry and Physiology* **7**: 62-82

Fernandez, D.E., Qu, R., Huang, A.H.C. and Staehelin, L.A. (1988) Immunogold localization of the L3 protein of maize lipid bodies during germination and seedling growth *Plant Physiology* **86**: 270-274

Frey Wyssling, A., Grieshaber, E. and Muhlethaler, K. (1963) Origin of spherosomes in plant cells *Journal of Ultrastructure and Molecular Structure Research* **8**: 506-516

Heim, J., Takabayashi, K., Meyhack, B., Märki, W. and Pohlig, G. (1994) C-terminal proteolytic degradation of recombinant desulfato-hirudin and its mutants in the yeast *Saccharomyces cerevisiae European J ournal Biochemistry* **226**: 341-353

Hills, M.J., Watson, M.D. and Murphy, D.J. (1993) Targeting of oleosins to the oil bodies of oilseed rape (*Brassica napus L.*) *Planta* **189**: 24-29

Huang, A.H.C. (1992) Oilbodies and oleosins in seeds *Annual Reviews Plant Physiology Plant Molecular Biology* **43**: 177-200

Huang, A.H.C. (1996) Oleosins and oil bodies in seeds and other organs *Plant Physiology* **110**: 1055-1061

Jacks, T.J., Yatsu, L.Y. and Altschul, A.M. (1967) Isolation and characterization of peanut spherosomes *Plant Physiology* **42**: 585-597

Jacks, T.J., Hensarling, T.P., Neucere, J.N., Yatsu, L.Y. and Barker, R.H. (1990) Isolation and physicochemical characterization of the half-unit membranes of oilseed lipid bodies *Journal of the American Oil Chemists Society* **67**: 353-361

Jefferson, R.A. (1987) Assaying chimeric genes in plants: the GUS gene fusion system *Plant Molecular Biology Reporter* **5**: 387-405

Kalinski, A., Melroy, D.L., Dwivedi, R.S. and Herman, E.M. (1992) A soybean vacuolar protein (P34) related to thiol proteases is synthesized as a glycoprotein precursor during seed maturation *Journal Biological Chemistry* **267**: 12068-12076

Keddie, J.S., Tsiantis, M., Piffanelli, P., Cella, R., Hatzopoulos, P. and Murphy, D.J. (1994) A seed-specific *Brassica napus* oleosin promoter interacts with a *Plant Molecular Biology* **24**: 327-340

Kirik, V., Kölle, K., Balzer, H-J. and Bäumlein, H. (1996) Two new oleosin isoforms with altered expression patterns in seeds of the *Arabidopsis* mutant *fus3 Plant Molecular Biology* **31**: 413-4170

Klocking, H-P. (1991) Toxicology of hirudin *Semin. Tromb. Hemost.* **17**: 126-129

Koorneef, M., Hanhart, C.J., Hilhorst, H.W.M., Karssen, C.M. (1989) *In vivo* inhibition of seed development and reserve protein accumulation in recombinants of abscisic acid biosynthesis and responsiveness mutants in *Arabidopsis thaliana Plant Physiology* **90**: 463-469

Krantz, S.B. (1991) Erythropoietin *Blood* **77**: 419-434

Kühnel, B., Holbrook, L.A., Moloney, M.M. and an Rooijen, G.J.H. (1996) Oil bodies of transgenic *Brassica napus* as a source of immobilized β-glucuronidase *Journal of the American Oil Chemists Society* **73**: 1533-1538

Lacey, K.J., Wellner, N., Beaudoin, F., Napier, J.A. and Shewry, P.R. (1998) Secondary structure of oleosins in oil bodies isolated from seeds of safflower (*Carthamus tinctorius* L.) and sunflower (*Helianthus annuus* L.) *Biochemistry Journal* **334**: 469-477.

Lee, K., Bih, F.Y., Learn, G.H., Ting, J.T.L., Sellers, C. and Huang, A.H.C. (1994) Oleosins in the gametophytes of *Pinus* and *Brassica* and their phytogenetic relationship with those in the sporophytes of various species *Planta* **193**: 461-469

Lee, K., Ratnayake, C. and Huang, A.H.C. (1995) Genetic dissection of the co-expression of genes encoding the two isoforms of oleosins in the oil-bodies of maize kernel *Plant Journal* **7**: 603-611

Li, M., Smith, L.J., Clark, D.C., Wilson, R. and Murphy, D.J. (1992) Secondary structures of a new class of lipid body proteins from oilseeds *Journal Biological Chemistry* **267**: 8245-8253

Little, D., Weselake, R., Pomeroy, K., Furukawa-Stoffer, T. and Bagu, J. (1994) Solubilization and characterization of diacylglycerol acyltransferase from microspore-derived cultures of oilseed rape *Biochemistry Journal* **304**: 951-958

Loer, D. and Herman, E.M. (1993) Cotranslational integration of soybean (Glycine max) oil body membrane protein oleosin into microsomal membranes *Plant Physiology* **101**: 993-998

Lui, J-H., Selinger, B., Cheng, K-J., Beauchemin, K.A., and Moloney, M.M. (1997) Plant seed oil-bodies as an immobilization matrix for a recombinant xylanase from the rumen fungus *Neocallimastix patriciarum* *Molecular Breeding* **3**: 463-470

Markwardt, F. and Walsmann, P. (1967) Reindarstellung und analyse des thrombin-inhibitors hirudin *Hoppe Seylers Zeitschr. Physiol Chem.* **348**: 1381-1386

Millichip, M., Tatham, A.S., Jackson, F., Griffiths, G., Shewry, P.R. and Stobart, K. (1996) Purification and characterization of oil bodies (oleosomes) and oil-body boundary proteins (oleosins) from the developing cotyledons of sunflower (*Helianthus annuus* L.) *Biochemistry Journal* **314**: 333-337

Moloney, M.M., Walker, J.M. and Sharma, K.K. (1989) High efficiency transformation of *Brassica napus* using *Agrobacterium* vectors *Plant Cell Reports* **8**: 238-242

Moreau, R.A., Liu, K.D.F. and Huang, A.H.C. (1980) Spherosomes of castor bean endosperm. Membrane components, formation, and degradation *Plant Physiology* **65**: 1176-1180

Murphy, D.J. and Cummins, I. (1989) Biosynthesis of seed storage products during embryogenesis in rapeseed, *Brassica napus* *Journal of Plant Physiology* **135**: 63-69

Murphy, D.J., Cummins, I. and Kang, A.S. (1989a) Immunocytochemical and biochemical studies of the mobilisation of storage oil-bodies and proteins in germinating cotyledons of oil seed rape *Brassica napus* *Journal of Food Agriculture* **48**: 209-223

Murphy, D.J., Keen, J.N., O'Sullivan, J.N., Au, D.M.Y. and Edwards, E.W. (1989b) Synthesis of the major oil-body membrane protein in the developing rapeseed (*Brassica napus*) embryos *Biochemistry Journal* **258**: 285-293

Murphy, D.J. (1990) Storage lipid bodies in plants and other organisms *Progress in Lipid Research* **29**: 299-324

Murphy, D.J., Keen, J.N., O'Sullivan, J.N., Au, D.M.Y., Edwards, E.W., Jackson, P.J., Cummins, I. (1991) A class of amphipathic proteins associated with lipid storage bodies in plants. Possible similarities with animal serum apolipoproteins *Biochimica et Biophysica Acta* **1088**: 86-94

Parcy, F., Valon, C., Raynal, M., Gaubier-Comella, P., Delseny, M. and Giraudat, J. (1994) Regulation of gene expression programs during *Arabidopsis* seed development: roles of the ABI3 locus and of endogenous abscisic acid *Plant Cell* **6**: 1567-1582

Parmenter, D., Boothe, J., van Rooijen, G., Yeung, E. and Moloney, M. (1995) Production of biologically active hirudin in plant seeds using oleosin partitioning *Plant Molecular Biology* **29**: 1167-1180

Parmenter, D., Boothe, J.G. and Moloney, M.M. (1996) Production and purification of recombinant hirudin from plant seeds In: *Transgenic Plants: A Production System for Industrial and Pharmaceutical Proteins*, edited by Owen, M.R.L. and Pen, J. John Wiley & Sons Ltd. p. 261-280

Plant, A.L., Van Rooijen, G.J.H., Anderson, C.P. and Moloney, M.M. (1994) Regulation of an *Arabidopsis* oleosin gene promoter in transgenic *Brassica napus* *Plant Molecular Biology* **25**: 193-205

Platt-Aloia, K.A. and Thomson, W.W. (1981) Ultrastructure of the mesocarp of mature avocado fruit and changes associated with ripening *Annals of Botany* **48**: 451-465

Qu, R., Vance, V.B. and Huang, A.H.C. (1990) Expression of genes encoding oleosin isoforms in the embryos of maturing maize kernels *Plant Science* **72**: 223-232

Qu, R.D. and Huang, A.H.C. (1990) Oleosin KD 18 on the surface of oil bodies in maize. Genomic and cDNA sequences and the deduced protein structure *Journal of Biological Chemistry* **265**: 2238-2243

Ross, J.H.E., Sanchez, J., Millan, F. and Murphy, D.J. (1993) Differential presence of oleosins in oleogenic seed and mesocarp tissues in *Plant Science* **93**: 203-210

Ross, J.H.E. and Murphy, D.J. (1996) Charactisation of anther-expressed genes encoding a major class of extracellular proteins in the pollen coat of *Brassicaceae* *Plant Journal* **9**: 625-637

Rydel, T.J., Ravichandran, K.G., Tulinsky, A., Bode, W., Huber, R., Roitsch, C. and J.W. Fenton II (1990) The structure of a complex of recombinant hirudin and human α-thrombin *Science* **249**: 277-280

Sarmiento, C., Ross, J.H.E., Herman, E. and Murphy, D.J. (1997) Expression and subcellular targeting of a soybean oleosin in transgenic rapeseed. Implications for the mechanism of oil-body formation in seeds *Plant Journal* **11**: 783-796

Slack, C.R., Bertaud, W.S., Shaw, B.D., Holland, R., Browse, J. and Wright, H. (1980) Some studies on the composition and surface properties of oil bodies from the seed cotyledons of safflower (*Carthamus tinctorius*) and linseed (*Linum ustatissimum*) *Biochemistry Journal* **190**: 551-561

Spiess, M. and Lodish, H.F. (1986) An internal signal sequence: The asialoglycoprotein receptor membrane anchor *Cell* **44**: 177-185

Stobart, A.K., Stymne, S. and Hoglund, S. (1986) Safflower microsomes catalyse oil accumulation *in vitro*: a model system *Planta* **169**: 33-37

Stobart, A.K. and Stymne, S. (1990) Triacylglycerol biosynthesis In: *Methods in Plant Biochemistry*, edited by Harwood, J.L. and Bowyer, J. p. 19-46.

Stone, S.R. and Hofsteenge, J. (1986) Kinetics of the inhibition of thrombin by hirudin *Biochemistry* **25**: 4622-4628

Stone, S.R. and Maraganore, J.M. (1993) Hirudin and hirudin-based peptides *Methods in Enzymology* **223**: 312-336

Taylor, D.C., Weber, N., Underhill, E.W., Pomeroy, M.K., Keller, W.A., Scowcroft, W.R., Wilen, R.W. (1990) Storage protein regulation and lipid accumulation in microspore embryos of *Brassica napus L* *Planta* **181**: 18-26

Thoyts, P.J.E., Millichip, M.I., Stobart, A.K., Griffiths, W.T., Shewry, P.R. and Napier, J.A. (1995) Expression and *in vitro* targeting of a sunflower oleosin *Plant Molecular Biology* **29**: 403-410

Ting, J.T.L., Balsamo, R.A., Raynayake, C. and Huang, A.H.C. (1997) Oleosin of plant seed oil-bodies is correctly targeted to the lipid bodies in transformed yeast *Journal of Biological Chemistry* **272**: 3699-3706

Trelease, R.N. (1969) Changes and characteristics of lipid bodies during development. Ph.D thesis, Univ of Texas, Austin, TX

Tzen, J.T.C., Lai, Y.K., Chan, K.L. and Huang, A.H.C. (1990) Oleosin isoforms of high and low molecular weights are present in the oil bodies of diverse seed species *Plant Physiology* **94**: 1282-1289

Tzen, J.T.C. and Huang, A.H.C. (1992) Surface structure and properties of plant seed oil bodies *Journal of Cell Biology* **117**: 327-335

Tzen, J.T.C., Cao, Y.Z., Laurent, P., Ratnayake, C. and Huang, A.H.C. (1993) Lipids, proteins, and structure of seed oil bodies from diverse species *Plant Physiology* **101**: 267-276

van Rooijen, G.H.J., Terning, L.I. and Moloney, M.M. (1992a) Nucleotide sequence of an *Arabidopsis thaliana* oleosin gene *Plant Molecular Biology* **18**: 1177-1179

van Rooijen, G.J.H., Wilen, R.W., Holbrook, L.A. and Moloney, M.M. (1992b) Regulation of accumulation of mRNAs encoding a 20kDa oil-body protein in microspore derived embryos of *Brassica napus* *Canadian Journal of Botany* **70**: 503-508

van Rooijen, G.J.H. and Moloney, M.M. (1995a) Plant seed oil-bodies as carriers for foreign proteins *Bio/Technology* **13**: 72-77

van Rooijen, G.J.H. and Moloney, M.M. (1995b) Structural requirements of oleosin domains for subcellular targeting to the oil body *Plant Physiology* **109**: 1353-1361

Wang, T.W., Balsamo, R.A., Ratnayake, C., Platt, K.A., Ting, J.T.J. and Huang, A.H.C. (1997) Identification, subcellular localization, and developmental studies of oleosins in the anther of *Brassica napus* *Plant Journal* **11**: 475-487

Wanner, G. and Theimer, R.R. (1978) Membranous appendices of spherosomes (oleosomes). Possible role in fat utilisation in germinating oil seeds *Planta* **140**: 163-169

Weselake, R.J., Pomeroy, M.K., Furukawa, T.L., Golden, J.L., Little, D.B. and Laroche, A. (1993) Developmental profile of diacylglycerol acyltransferase in maturing seeds of oilseed rape and safflower and microspore-derived cultures of oilseed rape *Plant Physiology* **102**: 565-571

Wickner, W.T. and Lodish, H.F. (1985) Multiple mechanism of protein insertion into and across membranes *Science* **230**: 400-407

Wilen, R.W., Mandel, R.M., Pharis, R.P., Holbrook, L.A. and Moloney, M.M. (1990) Effects of abscisic acid and high osmoticum on storage protein gene expression in microspore embryos of *Brassica napus* *Plant Physiology* **94**: 875-881

Wilen, R.W., Hays, D.B., Mandel, R.M., Abrams, S.R. and Moloney, M.M. (1993) Competitive inhibition of abscisic acid-regulated gene expression by stereoisomeric acetylenic analogs of abscisic acid *Plant Physiology* **101**: 469-476

Wu, S.S.H., Platt, K.A., Ratnayake, C., Wang, T-W., Ting, J.T.L. and Huang, A.H.C. (1997) Isolation and characterization of neutral-lipid-containing organelles and globuli-filled plastids from *Brassica napus* tapetum *Proceedings Of The National Academy Of Sciences Of The United States* **94**: 12711-12716

Yatsu, L.Y. and Jacks, T.J. (1972) Spherosome membranes: Half-unit membranes *Plant Physiology* **49**: 937-943

34. Synthesis, processing and deposition of seed proteins: The pathway of protein synthesis and deposition in the cell

ALLISON R. KERMODE[1] and J. DEREK BEWLEY[2]
[1]Simon Fraser University, Department of Biological Sciences, Burnaby, B.C., Canada V5A 1S6. [2]University of Guelph, Department of Botany, Guelph, Ontario, Canada N1G 2W1

1. INTRODUCTION

The synthesis and deposition of seed storage proteins is subject to strict temporal and spatial regulation; it only occurs in specific tissues/organs and only at specific times during seed development. Cell expansion and deposition of reserves (including storage proteins) follow the histodifferentiation stage of seed development, and occur largely in the absence of further cell divisions. It is during this time that young storage parenchyma cells (particularly within the major reserve organs of seeds - the endosperm, cotyledons or megagametophyte) accumulate a variety of proteins, including storage proteins, acid hydrolases, plant defense proteins, and other reserve materials or metabolites in storage compartments (organelles), some of which are called protein bodies.

Within many seeds, the protein body is a storage vacuole. Thus, vacuoles may perform several functions in seeds - e.g. as temporary storage depots (during seed development) and as sites of macromolecular hydrolysis during the post-germinative phase of the plant lifecycle, when seedling growth must be supported. It is during this latter phase of reserve mobilization that the vacuole accumulates hydrolytic enzymes, many of which are synthesized *de novo*, and the plant cell vacuole is thought of as equivalent to the animal lysosome. Not all hydrolytic enzymes are synthesized exclusively during the post-germinative phase; some are accumulated during seed development. It has become apparent that seed proteins may be compartmentalized into different vacuoles on the basis of their function. Notably, storage proteins that accumulate in protein storage vacuoles are excluded from a second type of vacuole, the lytic vacuole, that is characterized by an acidic pH and the presence of proteolytic enzymes (Rogers, 1998).

P.R. Shewry and R. Casey (eds.), Seed Proteins, 807–841.

There are distinct pathways of protein body formation in seeds; however, many details of protein body biogenesis are lacking. A pathway thought to be characteristic of most dicotyledonous seeds involves the subdivision or fragmentation of a large central vacuole. In this mechanism of protein body formation (also thought to be characteristic of gymnosperm seeds), storage proteins are deposited into subdividing vacuoles via transport through the Golgi complex (e.g. rice globulins [glutelins], some wheat prolamins and legume globulins). However, the protein storage vacuole is not the equivalent of a transformed lytic vacuole (e.g. in pea cotyledonary storage parenchyma cells); rather it arises *de novo* during development and two distinct vacuolar compartments co-exist during a significant period of seed development. For a review of potential mechanisms of protein body biogenesis in dicotyledous seeds, see Robinson and Hinz (1997).

In some cereals there is direct formation of protein bodies from the rough ER. Here, storage proteins are synthesized on the rough ER and are deposited directly within the lumen of this organelle where they remain and accumulate (e.g. prolamins of maize, rice and sorghum). Evidence exists for yet another route (e.g. for some of the wheat prolamins), which begins by assembly of the proteins into protein bodies within the ER. However, these ER-derived protein bodies undergo subsequent internalization within vacuoles by a specific process analogous to autophagy (reviewed in Galili *et al.*, 1993). Thus the wheat prolamin storage proteins exhibit two pathways or routes of transport to protein bodies.

With few exceptions, seed storage proteins undergo several modifications during or after their synthesis including glycosylation, folding and assembly, disulphide bond formation and proteolytic processing. This chapter focuses on the synthesis, processing and deposition of seed proteins and will emphasize the major pathways of protein transport in seeds as well as some of the approaches (ultrastructural, biochemical, and genetic) that have been important for elucidating these pathways.

2. GENERAL PATHWAYS OF PROTEIN TRANSPORT IN PLANT CELLS

Major targets of protein transport in plant cells include the plastid, the mitochondrion, the nucleus and the peroxisome/glyoxysome (Figure 1). Proteins destined for these organelles are translated on "free" polyribosomes of the cytosol; concurrently, or shortly thereafter, they are imported into the appropriate organelle by direct recognition of specific targeting signals. In contrast, transport to the vacuole and cell surface occurs along the endomembrane system (the ER, Golgi apparatus and transport vesicles);

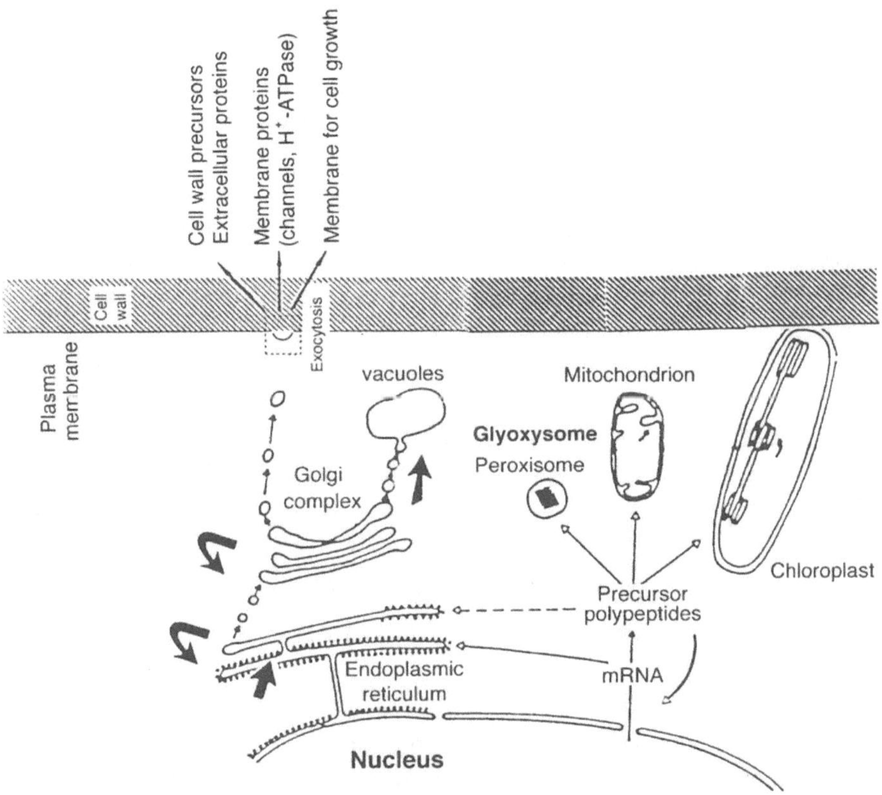

Fig. 1. The major pathways of protein transport within the plant cell. Also shown is the current model of protein transport through the plant secretory pathway. Thin arrows depict signal-independent transport and represent bulk-flow (the "default" pathway). Thick arrows denote transport steps which are signal-mediated, including translocation across the ER membrane, retention of resident proteins in the ER and Golgi complex, as well as transport to protein storage vacuoles and lytic vacuoles. Based on Verner and Shatz (1988). With permission from the American Association for the Advancement of Science.

proteins destined for this so-called "secretory" pathway are synthesized on polyribosomes associated with the ER and enter a common "gateway" by first translocating across the membrane of the endoplasmic reticulum (ER) into the lumen (Figure 1). As well as seed storage proteins, other seed proteins are transported along the "secretory" pathway during seed development including those destined for the tonoplast (the vacuolar membrane), the plasma membrane and the extracellular matrix (cell wall). Protein transport along the secretory pathway is also intensive in tissues of the cereal caryopsis (i.e. the aleurone layer and scutellum) that have a glandular function during post-germinative seedling growth and secrete a large number of hydrolytic enzymes (e.g. α-amylase, β-glucanase and β-xylanase) that mediate endosperm storage reserve mobilization.

The organelles of the secretory pathway are functionally connected by vesicular traffic; proteins move along this intercommunicating system in the lumen or in the membrane of vesicles which form from one compartment and fuse to the next compartment. The major route of transport is from the ER to the Golgi complex and from there to the vacuole or to the cell surface (plasma membrane, cell wall or extracellular space). Thus the protein composition of the ER and Golgi complex includes both permanent residents as well as proteins *en route* to their final destinations.

3. MOST SEED PROTEINS ARE TRANSPORTED ALONG THE SECRETORY TRANSPORT PATHWAY

3.1 Signal-dependent and signal-independent steps along the secretory pathway: Secretion is a default pathway in seeds

All seed proteins destined for transport along the secretory pathway are synthesized on ribosomes associated with the ER. The first step of entry into the pathway is mediated by a *N*-terminal signal peptide on the nascent protein which directs a transmembrane translocation from the cytosol to the lumen of the ER, a step generally accompanied by signal peptide cleavage. The ER contains a mixture of proteins with multiple destinations. Some proteins will become permanent residents of the ER (e.g. molecular chaperones that help storage proteins fold and assemble). In cereals where there is direct formation of protein bodies from the rough ER, there are specific mechanisms for retaining seed proteins after their translocation into the ER lumen. Other seed proteins are exported to the Golgi apparatus, often for subsequent distribution to the cell surface or vacuole. How this complex traffic pattern is organized, and which steps along the pathway occur by default (i.e. are signal-independent), or alternatively, require specific targeting signals, have been subjects for both speculation and debate (reviewed in Chrispeels, 1991; Kermode, 1996). The available evidence from eukaryotic systems (including plant seeds) indicates that each of the steps along the intracellular route leading to constitutive secretion from the cell (i.e. ER to Golgi complex, movement through the Golgi stacks, and movement from the Golgi complex to the cell surface) occurs by default: secretory and plasma membrane proteins are carried along by a non-selective, "bulk-flow" process. Proteins destined for targets other than the cell surface must contain additional positive topogenic information. Thus, additional information is required for selective retention in the ER (or the Golgi complex) or for diversion of proteins away from the bulk-flow pathway, to one of the branch pathways within the *trans*-Golgi network (e.g. to the seed storage vacuole or to lytic vacuoles) (Figure 1). Secretion is the outcome when a seed protein contains no additional targeting information over that of the signal peptide. Disruption of signal-dependent sorting in the *trans*-Golgi network by chemical treatments which affect the pH of the transport pathway (and presumably receptor-ligand interactions, e.g. chloroquine and monensin) also cause mistargeting and secretion (Craig and Goodchild, 1984).

3.2 Signal-dependent translocation of seed proteins across the ER membrane

3.2.1 Role of the signal peptide and identification of other components of the translocation machinery

In vitro protein import systems have established that the signal peptides of seed proteins generally function with a high degree of autonomy, being both necessary and sufficient to direct passenger proteins into the lumen of the ER. Analyses of the *N*-terminal amino acid sequences of seed vacuolar proteins shows that all of them have features that characterize other eukaryotic signal sequences *viz.*, a charged *N*-terminal region, a core of 10 to 15 hydrophobic amino acids and a more polar *C*-terminal sequence (Chrispeels, 1991).

The ER-translocation process in plants is similar to that in animals and fungi (Simon and Jones, 1988). Translocation across the ER membrane is mediated by receptor-like systems (termed the translocon) that involve proteins in the cytosol and on the target membrane; this complex cellular machinery is able to recognize signal peptides, despite their variation in amino acid sequence (reviewed in Rapoport, 1992). There are several distinct components of such a receptor-like system, which act sequentially and hence greatly enhance the specificity of the system. One component is the signal recognition particle (SRP), a ribonucleoprotein complex, consisting of a molecule of RNA and six polypeptide chains including an M_r 54,000 protein with a GTP-binding domain. The M_r 54,000 polypeptide subunit (SRP-54) selectively binds to the signal peptide sequence on the precursor as it emerges from the ribosome, temporarily halting translation. The ternary complex (consisting of the ribosome, nascent precursor and SRP) is targeted to the rough ER membrane due to the affinity of the SRP for an M_r 72,000 integral ER protein (part of the SRP-receptor or docking protein). Contact between

SRP and its receptor results in the dissociation of SRP from both the ribosome and the signal sequence. The ribosome becomes membrane bound (possibly by associating with several ER proteins), translation resumes and the signal sequence is inserted into the translocation site to begin the translocation event. Subsequently, the SRP is released from its receptor (by hydrolysis of a bound GTP molecule) and the SRP is free to enter a new targeting cycle (Figure 2). At least two steps of the translocation event appear to be controlled by the hydrolysis of GTP, again contributing to the specificity of the system. The GTPase cycle of SRP-54 may control signal sequence insertion into the translocation channel.

Solutions to the enigma of how diverse signal peptide sequences/shapes can be recognized by, and accommodated within, the M_r 54,000 polypeptide component of the SRP have been sought (Rothman, 1989). Sequestration of signal peptide sequences may occur in a hydrophobic methionine-rich pocket formed by secondary structural domains (amphipathic α-helices) of the M_r 54,000 SRP protein (the M-domain). More specifically, the unbranched methionine side chains lining this pocket provide projecting flexible "bristles" which recognize and accommodate only the common hydrophobic core of signal peptides, albeit diverse in primary sequence. Less is known about the mechanics of protein translocation; transport through a proteinaceous pore (rather than movement directly through the lipid bilayer) is suggested by studies in yeast and mammals.

Protein synthesis (i.e. translation) and translocation of the protein generally occur simultaneously. Strong evidence argues against a *mechanistic* coupling between these two events; however, their temporal association may nonetheless play an important role: that of ensuring that nascent polypeptides maintain (or acquire) a translocation-competent conformation. Binding of the SRP to the signal peptide, and the arrest of translation until functional insertion into the ER occurs, would largely perform this important function, although accessory proteins may also play a role.

3.2.2 Role of signal peptide cleavage

In most cases, the signal peptide is removed (in a co-translational manner) from the nascent polypeptide chain as it emerges into the ER lumen by a signal peptidase located on the inner face of the membrane. Cleavage generally occurs at the more polar *C*-terminal region of the signal sequence and is thought to allow subsequent processes like protein folding and assembly to proceed normally (Vitale *et al.*, 1993). For example, when the soybean storage protein glycinin is synthesized in an *in vitro* translation system (in the absence of ER membranes) from a synthetic mRNA corresponding to the entire glycinin coding region, the presence of the signal

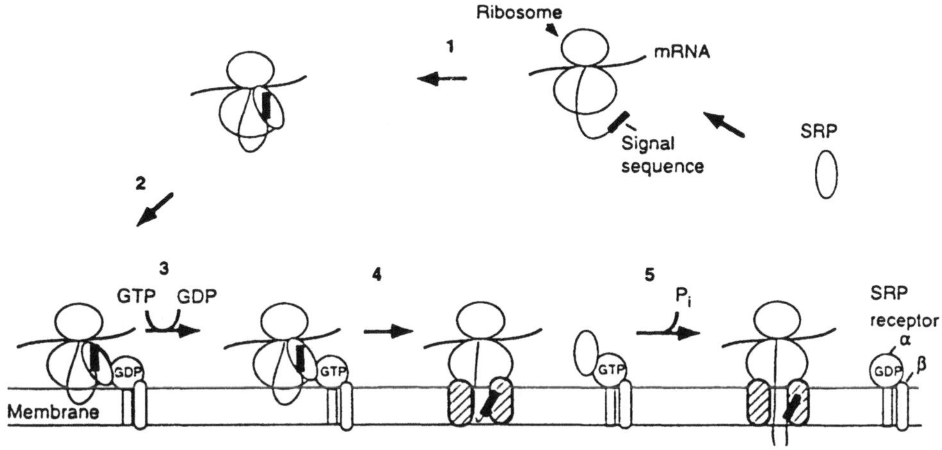

Fig. 2. The ER targeting cycle. The scheme shows the first steps in protein translocation across the ER membrane. When the signal sequence of a growing polypeptide chain has emerged from the ribosome, it is recognized and bound by the SRP (step 1). In step 2, the complex containing the ribosome, nascent chain, and SRP binds to the ER membrane through an interaction of the SRP with its membrane receptor, which consists of two subunits. In step 3, guanosine diphosphate (GDP) is exchanged for GTP at the SRP receptor. In step 4, the SRP is released both from the ribosome and from the signal sequence. The ribosome becomes membrane-bound, probably by its interaction with Sec61p or associated proteins, and translocation begins. In step 5, the SRP is released from its receptor by GTP hydrolysis and can then enter a new targeting cycle. Whether or not the two subunits of the SRP receptor dissociate during the SRP cycle is not known. From Rapoport (1992). With permission from the American Association for the Advancement of Science.

peptide inhibits assembly (trimerization) and results in misfolding of monomers (Dickinson *et al.*, 1987). The *floury-2* mutation of maize results in reduced storage protein synthesis in the endosperm, altered protein body morphology, and the synthesis of M_r 24,000 unprocessed α-zein proteins (Gillikin *et al.*, 1997). The higher molecular mass α-zeins are produced as a consequence of the mutation causing a Ala→Val change in the protein, which prevents cleavage of the signal peptide. Unprocessed mutant protein becomes inserted into the ER membrane (instead of translocating fully to the ER lumen) resulting in perturbation of storage protein packing and protein body morphology in mutant endosperm cells (Gillikin *et al.*, 1997).

3.2.3 Integral membrane proteins are partially translocated across the ER membrane

The membranes surrounding the organelles of the secretory pathway in seeds represent highly specialized compartments and membrane functions are typically carried out by the select group of proteins embedded within, or otherwise associated with, the lipid bilayer. After translocating partly across the ER membrane, these proteins are transported to further compartments of the secretory pathway (along with luminal proteins) in the membranes of the transport vesicles. Transmembrane proteins are either single-pass (i.e. they span the phospholipid bilayer only once and are classed into groups I, II or III) or multi-pass (i.e. they span the bilayer multiple times). Topology is established when the protein first assembles into the membrane. For tonoplast seed proteins, as well as plasma membrane proteins, topology is adopted at the ER membrane; subsequent transport is mediated by vesicles in which protein topology remains unchanged. Thus a domain of a membrane protein that remains cytoplasmic at the ER membrane, is still cytoplasmic after transport of the protein to the tonoplast or to the plasma membrane (reviewed in Austen and Westwood, 1991).

The tonoplast intrinsic proteins, α-TIP and γ-TIP (abundant in the tonoplasts of protein storage vacuoles and lytic vacuoles, respectively) contain several membrane-spanning domains (reviewed in Maurel, 1997). α-TIP of bean seeds is synthesized on the rough ER and its transport to the tonoplast is mediated by the secretory system; its C-terminal transmembrane domain may contain sufficient information for transport to the tonoplast. However, the possibility that a signal-independent default pathway for membrane proteins leads to the tonoplast (rather than to the plasma membrane) is being investigated (Hofte and Chrispeels, 1992).

3.3 Co- and post-translational modifications along the secretory pathway

Many seed proteins undergo extensive modifications during their transit including glycosylation, folding, oligomer assembly and proteolytic processing. Some of these processes have been studied in relation to their role in transport efficiency, protein stability and protein targeting *per se*.

3.3.1 Glycosylation

Glycosylation is a modification which many seed vacuolar proteins undergo *en route* to their target organelle. Many undergo *N*-linked glycosylation, in which the glycan is linked to the amide nitrogen of an asparagine residue. In their mature state, vacuolar glycoproteins often have both high mannose and complex *N*-linked glycans (Figure 3) (reviewed in Chrispeels, 1991; Faye *et al.*, 1992; Vitale *et al.*, 1993; Kermode, 1996). Glycans of the high mannose type (e.g. $Glc_3Man_9GlcNAc_2$) are assembled on dolichol pyrophosphate lipid carriers in the ER lumen. These are transferred (in a co-translational manner) to specific Asn residues (of the acceptor sequence Asn-X-Ser/Thr) on nascent polypeptides by oligosaccharyl transferase. Two ER-resident enzymes (glucosidases I and II) subsequently remove the three terminal glucose residues from the high mannose glycans on most proteins (a notable exception is α-mannosidase of jackbean cotyledons which escapes this modification).

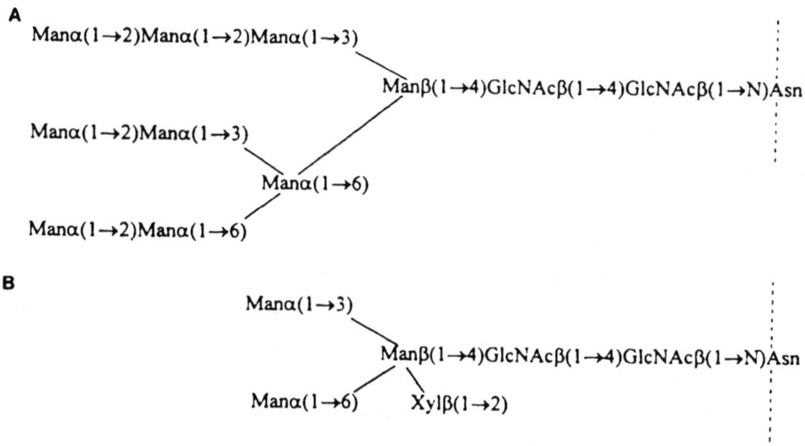

Fig. 3. The glycan side chains of glycoproteins. Mature phaseolin has one complex (lower) and two high-mannose (upper, middle) glycan side chains that arise by differential post-translational processing of the precursor oligosaccharide. Redrawn from Sturm *et al.* (1987). From Bewley and Black (1994). With permission from Plenum Press.

The dolichol pyrophosphate lipid carrier is embedded in the lipid bilayer and the high mannose glycan is built up by the stepwise addition of individual monosaccharides using sugar nucleotides as the donors. Stepwise addition of sugars to yield the structure $Man_{3\text{-}5}GlcNAc_2$-PP-dolichol is thought to occur on the cytosolic face of the ER membrane; following translocation of the entire intermediate to the luminal face, the subsequent addition of sugars is completed. The sugar donors for this latter process are dolichol-P-mannose. The product, dolichol PP-$GlcNAc_2Man_9$, is formed by the addition of four mannose residues on the luminal side; this intermediate must be glucosylated before it can serve as an oligosaccharide substrate for nascent polypeptide chains; the donor for this reaction is dolichol-P-glucose which is itself synthesized from UDP-Glc and dolichol-P. The catalytic subunit of the plant enzyme Glc-P-dolichol synthase (a polypeptide of M_r 39,000) has been purified from mung bean microsomes (Drake *et al*., 1991).

The enzyme oligosaccharyl transferase, which is responsible for transfer of the high mannose chain to the Asn residue of the growing polypeptide chain, is thought to be part of an integral membrane protein complex, possibly comprised of ribophorins I and II and a polypeptide of M_r 48,000 (Kelleher *et al*., 1992). One complex occurs at each protein translocation site (translocon) in the ER membrane; this may play a role in maximizing the efficiency (and accuracy) of glycosylation, allowing rapid continuous monitoring of the nascent and growing polypeptide chain for the presence of glycosylation sites (Vitale *et al*., 1993). As mentioned previously, elaboration of the glycan moiety to the protein normally occurs co-translationally. The coupling of glycosylation with protein synthesis and translocation ensures an efficient transfer of the oligosaccharide as a result of the close proximity of the transferase complex in the ER membrane. Glycosylation influences protein conformation. Thus, it may be important that glycosylation occurs early in the pathway of translocation to ensure that the protein will subsequently fold and assemble properly, rendering it competent for transport out of the ER lumen.

Presence of the consensus tripeptide sequence Asn-X-Ser/Thr within a polypeptide is not sufficient to guarantee glycosylation; in many cases, the site is not glycosylated or is glycosylated with low efficiency (reviewed in Vitale *et al*., 1993). As an example, the bean storage protein phaseolin has two glycosylation sites; the one closest to the *N*-terminus is always glycosylated, but utilization of the second site is not efficient (Sturm *et al*., 1987; Bollini *et al*., 1983). The pattern of glycosylation must be characteristic of the protein since it generally does not depend on the cell type (or host organism) in which it is synthesized. An oligosaccharide chain on the portion of the polypeptide already synthesized can affect co-translational glycosylation, either by directly masking a second subsequent glycosylation site or, indirectly, through an effect on folding (Vitale *et al*., 1993). There is likely to be a complex interplay between folding, assembly, and glycosylation to ensure that a newly-synthesized protein acquires a functional conformation after translocation into the ER lumen (Vitale *et al*., 1993).

Following co-translational transfer of the glycan onto the nascent protein, removal of the three Glc residues in the ER results in the formation of a typical high-mannose glycan [Man_9 $(GlcNAc)_2$]. This may be further modified later in the secretory pathway (in the Golgi complex) i.e., converted from a high mannose glycan to a complex glycan as a result of the sequential action of several glycosidases and glycosyl-transferases (reviewed in Faye *et al*., 1992). The complex glycans of plant glycoproteins are characterized by fucose and xylose sugar residues attached to the proximal GlcNAc and core mannose residues, respectively. This modification renders the glycoprotein resistant to the *Streptomyces* enzyme endoglycosidase H; hence this characteristic has been used as evidence for passage through the Golgi complex (Chrispeels, 1991).

Post-translational processing of *N*-linked glycans of plant proteins occurs in a *cis*-to-*trans* direction (reviewed in Faye *et al*., 1992; Vitale *et al*., 1993). Apparently, glycans must be readily accessible to the appropriate enzymes in the Golgi apparatus; when inaccessible, they remain in a high mannose form as the protein moves through the Golgi complex. A large variety of complex structures may be formed on mature glycoproteins and high mannose chains and different complex chains may be found on the same polypeptide. In general, the complex glycans of seed vacuolar proteins tend to be very similar and are closely related to ([Xyl]Man_3[Fuc]$GlcNAc_2$); following their exit from the Golgi complex, vacuolar glycoproteins contain one or two terminal *N*-acetylglucosamine residues that are removed soon after arrival at their final destination, giving rise to homogeneous complex glycan structures.

What are the general roles of *N*-linked glycosylation and glycan processing? The available evidence indicates that glycans do not contain vacuolar targeting information (reviewed in Chrispeels, 1991; see references therein). Not all vacuolar proteins are glycoproteins (e.g. 11S globulins, legumin, glycinin and others); moreover, some glycosylated precursors lose their glycans during post-translational processing *en route* to the vacuole, e.g. concanavalin A, wheat germ agglutinin and β-glucanase. Tunicamycin inhibits *N*-linked glycosylation in the ER by interfering with the elaboration of glycans (GlcNAc-1-P) onto the dolichol lipid intermediate. However, this antibiotic generally does not cause missorting of (normally glycosylated) vacuolar proteins, including pea vicilin and bean phytohemagglutinin. Likewise, unglycosylated seed vacuolar proteins generated by site-directed mutagenesis (e.g. phytohemagglutinin and barley lectin) are not mistargeted in a heterologous host plant system.

While *N*-linked glycans do not play a role in targeting *per se*, they may play a fundamental role in promoting correct protein folding and, as a consequence, enhance protein stability (e.g. via protection against proteolysis) (Bustos *et al*., 1991). The position of glycan chains and their accessibility to the glucosidases that carry out glucose trimming in the ER has a fundamental effect in controlling the assembly of storage proteins such as phaseolin (Lupattelli *et al*., 1997). For example, when glucose trimming is prevented in bean seeds by the use of the glucosidase inhibitor, castanospermine, the assembly of phaseolin into trimers is accelerated. Conversely, phaseolin polypeptides bearing only partially trimmed glycans are unable to form trimers (Lupattelli *et al*., 1997). A general effect of glycosylation on protein stability is indicated by similar studies of other seed vacuolar proteins (e.g. phytohemagglutinin and barley lectin). Both folding and assembly of denatured subunits of soybean lectin are completely prevented in the presence of the oligosaccharide $Man_9GlcNAc_2Asn$ (M9-Asn). In contrast, the oligosaccharide GM9-Asn ($Glc_{1\text{-}3}Man_9GlcNAc_2Asn$) interferes with polypeptide assembly. Thus, the branch Man α1-2 Man α1-2 Man linked to the 3 position of the β-mannosyl residue of the high mannose chains (Figure 3) may function in the folding of the subunit polypeptides, while assembly of folded subunits is associated with other branches (Nagai *et al.*, 1993).

3.3.2 Folding and oligomer assembly

Among the numerous modifications that seed proteins undergo, one of the most important is the acquisition of a functional, three-dimensional structure and the assembly of polypeptide subunits into oligomers. With few exceptions, these processes take place in the lumen of the ER and in large part will determine a protein's stability and subsequent efficient transport (reviewed in Pelham, 1989; Vitale *et al.*, 1993; Chrispeels, 1991; Kermode, 1996). Many plant vacuolar proteins (e.g. storage proteins) are oligomers (commonly dimers, trimers, tetramers or hexamers), most of which undergo complete oligomerization in the ER. Most lectins and several enzymes destined for the vacuole or the cell surface are active in the ER, which is indicative that they have attained a functional three-dimensional conformation and are properly assembled.

Transport of proteins out of the ER is subject to various rules of ER quality control and is often the rate-limiting step in transport along the secretory pathway. Correct folding and assembly of proteins is necessary for their efficient exit from the ER and subsequent transport (Pelham, 1989). Thus, "sorting" of proteins at this step appears to be conformation-specific and failure to acquire the correct conformation can result in ER retention and/or degradation. The efficiency of protein folding *in vivo* is due, in part, to several resident proteins of the ER which play a role in assisting in both the folding and oligomer assembly of nascent proteins. Folding of most proteins probably begins before synthesis/translocation of the entire protein has been completed, as soon as a protein domain emerges into the ER lumen (reviewed in Vitale *et al.*, 1993). In fact, it appears that a nascent polypeptide chain is accosted on the lumenal side of the ER membrane by a "welcoming" committee of enzymes and factors that facilitate and accelerate folding in several ways (Pelham, 1989). Perhaps the best characterized of these ER - "helper" proteins (termed "molecular chaperones") is BiP (Binding protein, in mammalian cells also known as the glucose-regulated protein, GRP78) (Munro and Pelham, 1986). BiP is related to the cytosolic heat shock proteins (of the Hsp70 family), but is a soluble protein which is targeted to the ER by a cleavable signal peptide; a carboxyterminal (H/K)DEL sequence mediates its retention. ER-localized BiP has been detected in tomato, tobacco, bean

(*Phaseolus vulgaris*), spinach and in the endosperm of maize and aleurone layer of barley (reviewed in Kermode, 1996).

In addition to BiP, several other ER resident proteins potentially involved in facilitating folding and assembly of nascent proteins have recently been identified in plants. One such protein is protein disulphide isomerase (PDI). Others are calnexin, calreticulin (Nelson *et al.*, 1997) and endoplasmin; the latter is an abundant ER-resident also referred to as GRP94 (Vitale *et al.*, 1993; Kermode, 1996). Calnexin interacts with many nascent membrane and soluble proteins of the secretory pathway. It is unrelated to BiP and endoplasmin and is further distinguished in that it is an integral membrane protein.

The main function of BiP (and some of the other molecular chaperones) is probably to promote protein assembly; it may also serve to prevent export of misfolded proteins from the ER (Pelham, 1989). Misfolded and misassembled proteins are produced continuously by the cell; the reactions that result in protein folding, covalent modification, and oligomeric assembly have a finite margin of error. Synthesis of BiP is induced when abnormal proteins accumulate in the ER, and it preferentially associates with such proteins until they either fold correctly or are degraded. Thus, by specifically retaining abnormally-folded and incompletely-assembled proteins, BiP contributes to the selectivity of transport (Pelham, 1989). Abnormal products that are retained in the ER by BiP include mutant and misfolded forms of proteins, chimaeric proteins and some unglycosylated forms of proteins (Kermode, 1996 and references therein). By helping to prevent the aggregation of folding intermediates, BiP may promote protein folding *per se*. More specifically, proteins such as BiP may act as reversible detergents, binding to hydrophobic surfaces, but intermittently utilizing ATP to change their conformation to a non-binding state. In this manner, unfolded proteins would be maintained in solution without sequestering them permanently in a nonfunctional complex; the nascent proteins could thus avoid aggregation or precipitation and be able to achieve their final tertiary and quaternary structures (Pelham, 1989).

BiP is elevated in developing seeds during the period of storage protein synthesis and deposition (e.g. in yellow cedar) (Xia and Kermode, 1998) and in germinated seedlings - in which the constituent cells are likely to have a large flux of protein through the secretory pathway (Denecke *et al.*, 1991). Maize BiP is over-produced during endosperm development in seeds of plants carrying *floury-2*, *defective endosperm-B30*, or *mucronate* mutations which are associated with a reduction in zein accumulation and altered protein body morphology (Fontes *et al.*, 1991; Boston *et al.*, 1991).

The role of BiP and other molecular chaperones depends on their ability to recognize a wide variety of nascent polypeptides that share no obvious sequence similarity while accurately discriminating between properly folded and unfolded structures (Blond-Elguindi *et al.*, 1993 and references therein).

Only some synthetic peptides stimulate the ATPase activity of BiP and alter its oligomeric state. Moreover, BiP preferentially binds synthetic peptides containing aliphatic residues (i.e. those amino acids most likely to be exposed only in unfolded or unassembled proteins) accommodating them in an environment energetically equivalent to the interior of a folded protein. The detailed configuration of *N*-linked glycans on nascent glycoproteins may also provide some ER-molecular chaperones (e.g. calnexin) with a mechanism for monitoring their folding and maturation status.

Most of the proteins specific to the ER, including BiP, endoplasmin, calsequestrin, calreticulum and calnexin have a demonstrated role in the calcium binding and storage function of the ER (Kermode, 1996). Moreover, intraluminal Ca^{2+} binding may be important in the regulation of some of the associations of molecular chaperones with incompletely folded proteins *in vivo*.

Restrictions in the transport of proteins out of the ER can be viewed as a late stage in a series of mechanisms utilized by the cell to ensure faithful expression of genetic information at the level of functional and correctly localized proteins (Hurtley and Helinus, 1989). In plants, the ER may be a major site of degradation of some of the proteins which fail to assume their functional conformation (whether or not they are associated with BiP or a BiP-like protein); other sites of degradation may be the vacuole and the cytosol (reviewed in Vitale, 1993; Kermode, 1996; Taylor, 1997). Tunicamycin treatment leads to reduced transport of certain plant proteins and either their retention in the ER (e.g. unglycosylated pro-ConA) or degradation at that site (e.g. β-fructosidase) (Faye and Chrispeels, 1989). When transiently expressed in tobacco protoplasts, mutant polypeptides of phaseolin (generated by deleting a domain required for trimer formation) occur in ATP-dependent association with ER-localized BiP and it is within the ER lumen that the mutant polypeptides are rapidly degraded (Vitale *et al*., 1993 and references therein). Disruption of protein folding or oligomer assembly may or may not prevent transport of seed proteins beyond the ER to the vacuole. A genetically engineered phaseolin that was modified to contain a helix-breaking epitope ("hi-met" phaseolin) (Hoffman *et al*., 1988) accumulates in transgenic tobacco seeds to a much lower amount than its unmodified counterpart. However, the "hi-met" protein is transport competent and appears to be degraded in the vacuoles (protein bodies) and not in the ER (Herman *et al*., 1994).

3.3.3 Formation of disulphide bonds

Many seed proteins transported along the secretory pathway (e.g. seed storage proteins and extracellular proteins) have disulphide bonds that stabilize their tertiary and quaternary structures. The formation of these bonds is not random, but is rapidly catalyzed in nascent proteins by protein disulphide isomerase (PDI) (reviewed in Gething and Sambrook, 1992). This enzyme interacts with unfolded proteins in the ER and catalyzes thiol oxidation and disulphide exchange reactions. Depending on the nature of the polypeptide substrate and the imposed redox potential, PDI promotes disulphide formation, isomerization or reduction.

PDI of wheat, alfalfa, soybean and tobacco has been characterized (reviewed in Kermode, 1996). The enzymes (e.g. from alfalfa and wheat endosperm and aleurone layer cells) are dimers and contain the two thioredoxin-like active sites characteristic of vertebrate and yeast PDIs, and a C-terminal ER retention signal, KDEL. All forms of PDI are soluble, or only loosely associated with the ER membrane, and are released by incubation at high pH. PDI-depleted microsomes can still import γ-gliadin (a non-glycosylated wheat storage protein), but the protein cannot achieve its correct disulphide-bonded state; reconstitution of microsomes in the presence of purified PDI restores this function (Bulleid and Freedman, 1988).

3.3.4 Proteolytic processing

Seed proteins transported along the secretory pathway often undergo post-translational proteolytic processing, including storage proteins, seed defense proteins and various vacuolar enzymes (e.g. α-mannosidase and thiol proteases). For seed storage proteins, processing begins during their transit to the vacuole, in a pre-vacuolar compartment (see later discussion), but is completed in the protein storage vacuole (reviewed by Muntz, 1996). In some cases, the polypeptide is simply cleaved into two or three smaller polypeptides, all of which remain together in the oligomer; in others, a domain is lost (e.g. from the amino-terminus, carboxy-terminus or from within the polypeptide). The degree of processing is also variable; it may be complete or only partial, with some polypeptides remaining unprocessed (Table 1) (reviewed in Chrispeels and Tague, 1990; Muntz, 1996). In the protein storage vacuole of the developing soybean seed, glycinin (a legumin-type storage protein) occurs as a 12S hexamer in which each of the six subunits is composed of an acidic and a basic polypeptide joined by a single disulphide bond. Proglycinin subunits are assembled into 9S trimers in the ER and, following their intracellular transport through the secretory system, arrive at the vacuole. Within the protein body/vacuole, a specific cleavage occurs, yielding the acidic and basic polypeptides and the trimers assemble into 12S hexamers (Dickinson *et al.*, 1989; reviewed in Bewley and Black, 1994) (Figure 4). Interestingly, the assembly of glycinin trimers into hexamers *in vitro* requires the proteolytic cleavage of proglycinin into its basic and acidic subunits (Dickinson *et al.*, 1990). For this reason cleavage is considered to be a potential regulatory step in the pathway leading to glycinin deposition.

However, in developing pea seeds *in vivo*, subsequent to import into the protein storage vacuole, trimeric 9S prolegumin assembles into 12S hexamers without prior processing (Hinz *et al*., 1997).

Table 1. Types of proteolytic processing of vacuolar proteins[a]

No proteolytic processing[b]	***N*-Terminal domain lost**
Phytohemagglutin	Napin
Soybean agglutinin	Sulphur-rich protein of Brazil nut
β Subunit of soybean β-conglycinin	Tomato proteinase inhibitor I
Tomato proteinase inhibitor II	Potato proteinase inhibitor I
Pea vicilin	Sweet potato sporamin
Phaseolin	α and α ' subunits of soybean β-conglycinin
Patatin	
***C*-Terminal domain lost**	**Internal cleavage with or without loss of a domain**
Soybean glycinin	12S globulins
Wheat germ agglutinin	Sulphur-rich protein of Brazil nut
Barley lectin	Napin
	Castor bean ricin
Cleavage and religation	Castor bean *Ricinus communis* agglutinin
Concanavalin A	Pea lectin
	Pea vicilin

[a] From Chrispeels and Tague (1990). With permission from Plenum Publishing Corp.
[b] except signal peptide.

Processing of glycinin is effected by a one-point cleavage that occurs at the carboxyl side of the asparaginyl residue located at the junction of the subunits, the modification being carried out by a cysteine endoprotease present in the protein storage vacuoles of soybean cotyledons (Shimada *et al*., 1994). Proteolysis is largely attributable to the conformational accessibility of the enzyme to the asparaginyl residue. Putative vacuolar processing enzymes are also involved in the cleavage of other classes of storage proteins, including the 2S albumins of *Arabidopisis*, *Brassica* and other seeds (D'Hondt *et al*., 1993); the thiol protease of barley aleurone layer cells (aleurain) is also processed by a specific endoprotease located within the vacuole and pre-vacuolar compartment (Holwerda *et al*., 1990; Rogers, 1998).

Of all seed storage proteins studied, the 2S albumins undergo some of the most extensive post-translational processing. Maturation is a multi-step process that follows cleavage of the signal peptide, and involves the removal of three peptides: an amino-terminal, an internal and a carboxy-terminal fragment (Crouch *et al*., 1983; Sun *et al*., 1987). In the mature protein,

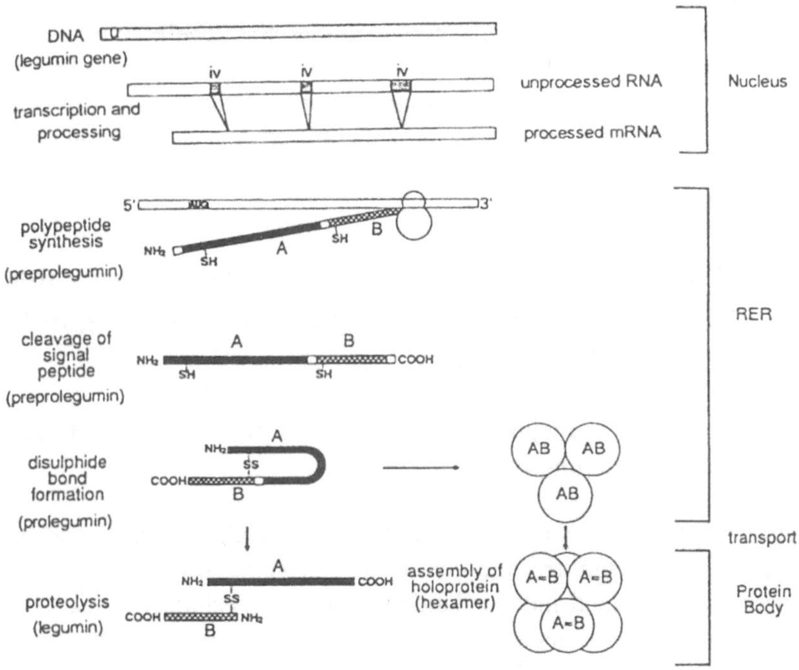

Fig. 4. A representation of the synthesis and processing of the legumin (glycinin) in soybean cotyledons. Following transcription, the mRNA encoding glycinin is transported from the nucleus to the rough endoplasmic reticulum (RER) where it is translated. The mRNAs for the acidic (A) and basic (B) subunits are joined by codons for a 4-amino-acid linker sequence on the primary translation product (preprolegumin). The mRNA also encodes the signal peptide at the amino-terminal end and a pentapeptide at the carboxy-terminal end. The primary product (prolegumin) is processed to yield the A and B subunits joined by the linker sequence and disulphide bonds (S-S); the signal peptide and the pentapeptide are cleaved off. In the final step of processing the mature protein is formed as the linker sequence is removed and the acidic and basic subunits are joined only by the disulphide bonding. The mature acidic subunit contains 278 amino acids (*Mr* ~ 40,000) and the basic subunit 180 amino acids (*Mr* ~ 20,000). Assembly of the subunits within the protein bodies yields the mature hexameric legumin holoprotein. Based on Krochko and Bewley (1988). Reprinted with permission of Longman Group UK (Gale Research Inc.). See also Dickinson *et al.* (1989) for details and Bednarek and Raikhel (1992) for a general review. From Bewley and Black (1994). With permission from Plenum Press.

subunits of M_r 9,000 and 4,000 remain which are linked by disulphide bonds. Post-translational cleavages occur on the *C*-terminal sides of asparagine residues 35 and 74, which are conserved among the 2S albumin precursors in seeds of different plant species (Hara-Nishimura *et al.*, 1993). The two Asn residues are located in the hydrophilic regions of the precursor, suggesting that the processing enzyme can recognize exposed Asn residues on the surface of the protein and effect cleavage on the *C*-terminal side of these residues (Hara-Nishimura *et al.*, 1993).

3.4 Targeting of cereal and dicot seed storage proteins to the protein body/storage vacuole via sorting in the *trans*-Golgi network

3.4.1 Vesicular transport of vacuolar seed proteins through the ER and Golgi complex

All of the transport steps from the ER to the Golgi complex and transport through the Golgi itself are common to most proteins of the secretory pathway and occur via a signal-independent (bulk-flow) mechanism. Seed proteins destined for locations beyond the ER (e.g. rice globulins [glutelins], some wheat prolamins and globulins of leguminous seeds that are destined for the protein storage vacuole) are transported in small membranous vesicles which bud from the ER membrane. These vesicles are directed to the Golgi apparatus which is composed of flattened, membrane-enclosed saccules. Movement through the Golgi compartments is also mediated by the budding and fusion of transport vesicles; the mechanisms involved in protein movement through the Golgi stacks, and how proteins are sorted according to destination and then packaged for delivery, are reviewed elsewhere (Robinson and Hinz, 1997).

Protein-coated vesicles mediate the transport of proteins from the ER to the *cis*-Golgi network (the first compartment of the Golgi complex), a process regulated by GTP-binding proteins. The vesicle-transporting proteins fall largely into two families: the ARF/Sar1 family, which provides the essential components of the machinery involved in early steps of vesicle budding from the ER, and the Ypt/Rab family, which is implicated in the targeting and/or fusion of the ER-transport vesicles with the Golgi complex (reviewed in Kermode, 1996; Kim *et al.*, 1997; Bar-Peled and Raikhel, 1997). A number of peripheral membrane proteins (SAR 1, SEC 23/24 and SEC 13/31) are likely to be involved in the budding of vesicles from the ER and form a complex collectively termed the COP (*co*at *p*rotein) II coat, together with at least one integral ER-membrane protein, SEC 12. Homologs of these genes have been isolated from plants (e.g. *Sar1*, *Sec12*, *Sec13* and *Sec 23*) and studies to characterize the proteins (particularly their subcellular localization) are underway.

3.4.2 Sorting of vacuolar seed proteins at the *trans*-Golgi network: dense vesicles and clathrin-coated vesicles

Signal-dependent sorting of seed proteins away from bulk-flow (e.g. to the storage or lytic vacuole) occurs in the *trans*-Golgi network, a complex tubular reticulum in which proteins are segregated into different transport vesicles to be dispatched to their final destinations. Thus, it is this compartment in which seed protein traffic signals are recognized (in some cases, by specific receptors), allowing them to be correctly sorted to the vacuole.

In mammalian cells, the vesicles that transport proteins from the *trans*-Golgi network to other compartments (via sorting-dependent mechanisms) are covered with a protein coat consisting of clathrin and a subset of characteristic polypeptides (reviewed in Robinson and Hinz, 1997). Coated pits concentrate selected vesicle cargo, usually transmembrane receptors bound to a soluble ligand, while excluding resident membrane proteins. The cytoplasmic face of the coated pit is composed of a layer of clathrin-associated proteins (adaptins) bound to the membrane, surrounded by an attached lattice of clathrin subunits, thought to be assembled by the recruitment of soluble forms of clathrin and adaptins to the membrane. Coated pits progress to form vesicles; once released from their membrane attachment, mature coated vesicles are rapidly uncoated, releasing clathrin and adaptins for another round of coated pit and vesicle formation (Figure 5; Robinson and Hinz, 1997). The uncoated vesicle is now a fusion-competent vesicular intermediate.

The major structural component of the clathrin-coated vesicle is the clathrin triskelion, a three-legged structure comprised of three clathrin heavy chains (M_r 180,000) and three clathrin light chains (M_r 30-40,000), each bound to a heavy chain near the vertex of the triskelion. Clathrin heavy chain may play a structural role in vesicle formation, effecting a change in membrane shape required to produce a curved vesicle. The clathrin light chains, on the other hand, probably play a regulatory role in vesicle consumption by regulating the equilibrium between clathrin coat assembly and disassembly (reviewed in Pryer *et al.*, 1992). For example, disassembly of the clathrin lattice and release from the vesicle surface requires clathrin light chain and is stimulated by the enzymatic action of Hsc70 (uncoating ATPase), a member of the heat shock protein family.

The adaptins of coated pits and vesicles are thought to mediate clathrin binding to membranes and play a role in the selection of vesicle cargo (Figure 5); adaptin complexes are able to bind *in vitro* to the cytoplasmic domains of

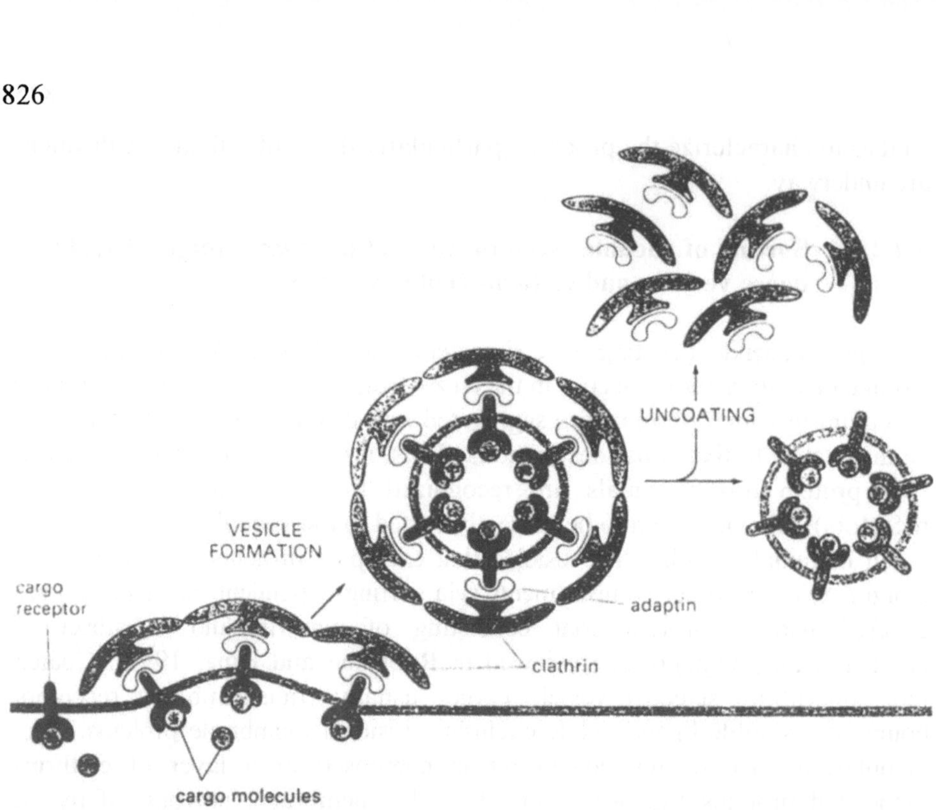

Fig. 5. Selective transport mediated by clathrin-coated vesicles. The adaptins bind both clathrin triskelions and cargo receptors. From Alberts *et al.* (1994). With permission from Garland Publishing Inc.

transmembrane receptors that are packaged into clathrin-coated vesicles (Pryer *et al.*, 1992; Robinson and Hinz, 1997).

In the cells of higher plants, the structural relationships between the ER and Golgi apparatus and between intra-organellar transport pathways are much less clear; however, there is evidence that some features of Golgi structure and vesicular transport are similar to those of animal cells. Both adaptins (e.g. β-adaptin) and the vesicle uncoating ATPase from plants have been isolated and characterized (Holstein *et al.*, 1994; Kirsch and Beevers, 1993). Clathrin-coated vesicles are involved in endocytosis in plant cells; recent evidence suggests that their participation in the sorting of proteins to seed vacuoles is dependent upon whether the proteins are destined for lytic vacuoles or for protein storage vacuoles (Robinson and Hinz, 1997).

Cells of developing seed storage tissues contain two vacuole types that differ in their luminal and membrane protein composition. For example, cotyledonary cells of leguminous seeds (e.g. pea) contain protein storage vacuoles that sequester storage proteins and contain α-TiP in their tonoplast, and also lytic vacuoles which do not contain these proteins, but do have γ-TiP in their tonoplast (Hoh *et al.*, 1995). Likewise, in other plant cells (e.g. barley

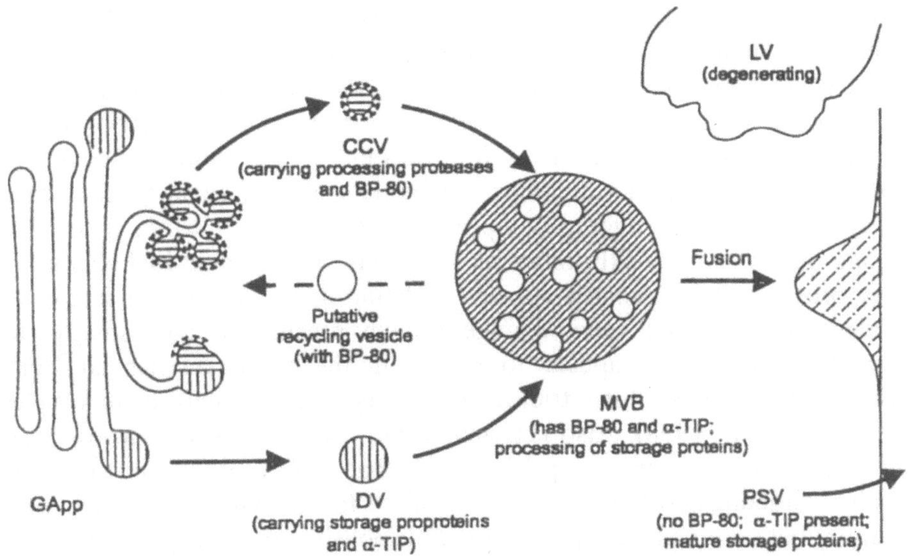

Fig. 6. Current model for sorting of proteins targeted from the *trans*-Golgi network to protein storage vacuoles and to lytic vacuoles, based on studies of protein transport in pea cotyledons. The storage proteins vicilin and legumin and the tonoplast protein α-TIP are transported via dense vesicles; other seed proteins such as acid hydrolases are transported in clathrin coated vesicles by being bound to specific receptors (e.g. BP-80). In pea seeds, both vesicle types are shown to deliver their contents to a pre-vacuolar compartment (multivesiclular bodies) which in turn fuses with the protein storage vacuole. Because the lytic and protein storage vacuoles co-exist in cotyledonary parenchyma cells only during pea seed development, the former are shown in the Figure to degenerate as development proceeds. In other seed types and cells (e.g. aleurone layer cells) where lytic vacuoles persist, or arise post-germinatively, proteolytic enzymes (e.g. aleurain) are initially delivered to the lytic prevacuolar compartment (Rogers, 1998). From Robinson *et al.* (1998). With permission.

root tip cells), there are two populations of vacuoles : one population playing a storage role (and containing barley lectin and α-TIP) and the other population functioning as lytic compartments and containing the thiol protease aleurain and γ-TIP (Rogers, 1998). Sorting of proteins for targeting to these two types of vacuoles is proposed to occur at the *trans*-Golgi network, but to involve different mechanisms and different vesicle types; the current model is presented in Figure 6 (Robinson *et al.*, 1998). In the storage parenchyma cells of pea cotyledons, vicilin and legumin assemble into electron-dense aggregates at the periphery of the Golgi cisternae, forming the nuclei of so-called dense vesicles; these dense vesicles eventually reach the *trans*-Golgi network, possibly by cisternal progression through the Golgi stack. α-TIP is also sequestered within these vesicles, residing within the vesicle membrane. Physical aggregration of storage polypeptides (involving protein-protein interactions) is important for their segregation from other secretory proteins that are destined for secretion or for transport to the lytic vacuole; both acidification of transport vesicles as well as the generally acidic pI of storage proteins may aid in the aggregation process (Hohl *et al.*, 1996). In contrast to storage proteins and α-TIP, other seed proteins such as acid hydrolases are transported in clathrin-coated vesicles by being bound to specific receptors. Although they do not carry storage protein precursors, interestingly the clathrin-coated vesicles appear to bud from the dense vesicles and may represent a mechanism for retrieval of proteins missorted into dense vesicles (Robinson *et al.*, 1998). Also detected within clathrin coated vesicles is the putative vacuolar sorting receptor, BP-80, that binds specifically to the *N*-terminal targeting domains of proteins destined for the lytic vacuole (e.g. pro-aleurain) but not to mutated domains that demolish correct targeting (Kirsh *et al.*, 1994; 1996).

Dense vesicles do not transport their contents directly to the protein storage vacuole, but rather to an intermediate compartment, in pea seeds referred to as the multivesicular body (Robinson *et al.*, 1998); such bodies are thought to fuse with the protein storage vacuole leading to a stratification of storage protein deposits at the tonoplast. While lytic and protein storage vacuoles co-exist in cotyledonary parenchyma cells during pea seed development, the former are absent at the end of development and therefore may degenerate as development proceeds. Because the putative receptor BP-80 is not only enriched in clathrin-coated vesicles, but also in multivesicular bodies, clathrin-coated vesicles are proposed to transport acid hydrolases (particularly those involved in the proteolytic processing of storage proteins) to the protein storage vacuole via the pre-vacuolar compartment.

Acidification of pre-vacuolar compartments by vacuolar H^+-ATPases (V-ATPases) may be essential for protein sorting in plants, similar to their role in animal and yeast cells. A V-ATPase isolated from a Golgi-enriched fraction (and distinguishable from the tonoplast-associated V-ATPase) is required for the transport of vacuolar protein precursors in tobacco cells (Matsuoka *et al.*, 1997).

3.4.3 Identifying vacuolar protein targeting signals on seed proteins

Some seed vacuolar proteins are synthesized as precursors (preproproteins) and undergo a specific post-translational processing event (in which a pro-domain is lost) upon their arrival at the vacuole. The cleaved pro-domain usually contains information which is both necessary and sufficient for vacuolar targeting (reviewed in Kermode, 1996; Chrispeels and

Raikhel, 1992; Bednarek and Raikhel, 1992; Nakamura and Matsuoka, 1993). The pro-domains on these vacuolar proteins may be located at the *N*-terminus (e.g. barley aleurain) or at the *C*-terminus (e.g. barley lectin). Graminaeceous lectins other than those of barley, such as wheat germ agglutinin and rice lectin, contain similar proteolytically processed *C*-terminal domains and these probably serve the same function. Sporamin (a vacuolar storage protein in tuberous roots of sweet potato) undergoes a pattern of proteolytic processing which is similar to that of the cereal lectins. This protein is synthesized as a preproprotein with a short amino-terminal propeptide (NTPP) domain which undergoes cleavage in a post-translational manner, and has a vacuolar targeting function.

Propeptides of seed vacuolar proteins do not necessarily contain targeting information. A short *C*-terminal segment of the Brazil nut 2S albumin is sufficient for targeting to the vacuole (G. Saalbach, pers. commun.); conversely, none of the three propeptide segments of the *Arabidopsis* 2S albumin storage protein (which undergo proteolytic removal) is involved in vacuolar localization (D'Hondt *et al.*, 1993).

There is no obvious homology between the different plant vacuolar targeting peptides; *C*-terminal signals contain short hydrophobic peptides, and in the *N*-terminal signals, the sequence NPIR occurs. Identification of the critical structural features of these peptides important for correct targeting has led to a more detailed view of the sorting process in plants (Nakamura and Matsuoka 1993; Neuhaus *et al.*, 1994; Dombrowski *et al.*, 1993). In sporamin, conserved amino acids in the *N*-terminal part of the propeptide ($N^{26}PIRL^{30}$) constitute the core of the vacuolar-sorting determinant; both the side chain structures of the conserved residues and the higher-order structure of the polypeptide backbone in this region are likely to be recognized by the vacuolar sorting machinery. Analysis of the functional elements of the barley lectin *C*-terminal pro-domain suggest that no specific amino acids are involved; many different alterations and deletions are tolerated by the targeting machinery, although short hydrophobic amino acid stretches seem important.

Much less is known about the targeting mechanisms of vacuolar proteins that do not undergo a specific pattern of processing and propeptide removal. The progress made towards characterizing seed vacuolar proteins, such as phytohemagglutinin and legumin, indicates that different and more complex sorting mechanisms exist for these latter proteins. In legumin from broadbean, both a 281-amino acid *N*-terminal segment and a 76-amino acid *C*-terminal region are sufficient for targeting to plant vacuoles (Saalbach *et al.*, 1991). In phytohemagglutinin of the common bean, an internal segment (amino acids 83 to 133) is sufficient to direct the reporter protein invertase to the plant cell vacuole (Von Schaewen and Chrispeels, 1993).

Not only is there no consensus sorting sequence for seed proteins destined for the plant vacuole, but it is not possible to categorize sorting signals for proteins destined for storage or lytic compartments (reviewed in Robinson and Hinz, 1997). For example, both aleurain (a vacuolar thiol protease in barley aleurone layer cells which resides in lytic vacuoles; Rogers, 1998) and sporamin (which resides in protein storage vacuoles, although not of seed tissues) (Hattori *et al.* 1987), possess the critical motif NPIR. An aspartic proteinase (which participates in the digestion of storage proteins following germination as well as in the processing of barley lectin) contains an NPIR motif at its *N*-teminus, but is present in both storage and lytic vacuoles of barley roots (Paris *et al.*, 1996).

There are at least two different vacuolar sorting mechanisms in tobacco. Use of 0.33 μM wortmannin, an inhibitor of phospholipid biosynthesis, inhibits delivery of proteins to the vacuole mediated by a *C*-terminal signal, but not an *N*-terminal signal (Matsuoka *et al.*, 1995). In addition, soluble and membrane proteins of the seed vacuole may not travel together by the same route; the latter (i.e. tonoplast proteins such as V-ATPase, pyrophosphatase, and γ-TIP) may arrive at their final destination via the plasma membrane (Robinson and Hinz, 1997; also see Gomez and Chrispeels, 1993).

As noted above, a current model for vacuolar transport postulates that storage proteins (e.g. pea legumin and vicilin and possibly barley lectin) use an aggregation mechanism for sorting and transit in dense vesicles to the next (pre-vacuolar) compartment, whereas other proteins, e.g. acid hydrolases, are bound by specific receptors. Receptor-mediated recognition would depend upon the sorting determinant of the vacuolar seed protein having a highly conserved amino acid sequence (e.g. barley proaleurain); a mechanism involving protein aggregation would not require sequence conservation, but would rely on the conservation of the general physical characteristics of the protein. The wide variability in the structure and physical properties of vacuolar sorting motifs on seed proteins is indicative of more than one sorting mechanism. Protein-protein interactions, rather than a *C*-terminal KDEL sequence, mediate retention of prolamins in the ER of the maize endosperm (Geli *et al.*, 1994). Robinson and Hinz (1997) suggest that two factors contribute to self aggregation of vacuolar proteins : one is the presence of hydrophobic residues in sorting signals (Dombrowski *et al.*, 1993; Neuhaus *et al.*, 1994) and the other is a low pH in the sorting compartment which serves to reduce the solubility of storage proteins with a low pI (Casey, 1979).

3.4.4 Other components of the plant vacuolar targeting machinery: A search for receptors

The identification of specific sequences required for vacuolar targeting has initiated a search for receptors mediating the recognition of those signals and an analysis of whether different classes of targeting signals compete for the same receptor (reviewed in Okita and Rogers, 1996; Gal and Raikhel, 1993). An integral membrane protein of M_r 80,000 (BP-80) has been identified in clathrin coated vesicles of developing pea cotyledons (Kirsch *et al.*, 1994; Paris *et al.*, 1996, 1997). The protein binds specifically (at neutral pH) to the *N*-terminal targeting signal of proaleurain (bound to a column) and dissociates from its ligand when the pH is lowered to 4. The vacuolar targeting determinant of sporamin (NPIR) competes (weakly) for binding of the putative receptor protein; a mutant form of this sequence does not. Endopeptidase-B (EP-B), a secreted thiol protease homologous to aleurain also, predictably, does not exhibit an interaction with the M_r 80,000 protein. The receptor recognizes the *C*-terminal sorting determinant of Brazil nut 2S albumin, which may not be entirely compatible with the selectivity expected of this sorting receptor, particularly when considered in the context of a two-vacuole status for plant cells (Robinson and Hinz, 1997).

Sequence analysis of homologs of pea BP-80 from *Arabidopsis*, rice and maize indicate that the receptor is highly conserved in monocotyledonous and dicotyledonous plants. The transmembrane orientation of the protein is indicative of a receptor function; it is a type I integral membrane protein with its carboxy-terminus exposed to the cytoplasm (and containing the Tyr-X-X-hydrophobic residue motif characteristic of transmembrane proteins that interact with clathrin-associated adaptor proteins during protein sorting). The cysteine-rich binding domain is located in the *N*-terminal luminal portion of the protein (Paris *et al.*, 1997; Ahmed *et al.*, 1997). BP-80 localizes to the dilated ends of Golgi cisternae and to pre-vacuolar compartments, capable of fusing with lytic vacuoles. When expressed in a transient manner in tobacco suspension-cultured protoplasts, a truncated form of the receptor lacking cytoplasmic and transmembrane domains is secreted (Paris *et al.*, 1997). The relative importance of the transmembrane and cytoplasmic domains of BP-80 for its targeting to the lytic prevacuole have been defined (L. Jiang and J. Rogers, pers. commun.). The transmembrane domain alone is sufficient to specify correct targeting; substitution of the BP-80 domains (*C*-terminal or transmembrane) with those of α-TIP abolishes targeting to the lytic pre-vacuolar compartment. Thus, the length and/or physical structure of both domains are important for correct targeting of this integral membrane protein.

A different type of putative receptor (AtPEP12) has been cloned from *Arabidopis*. The gene can functionally complement the *pep12* yeast syntaxin mutant and encodes a protein homologous to mammalian syntaxins (Conceicao *et al.*, 1997) which are thought to be involved in the docking of vesicles with their target membrane. According to the current model, a specific v-SNARE in the vesicle membrane interacts with a specific t-SNARE (e.g. syntaxin/PEP12 and SNAP-25) in the target membrane along with the soluble factors α-SNAP and NSF, common components of transport found at all levels of the secretory pathway. The t-SNAREs and v-SNAREs enable the vesicle to dock at the target membrane and permit fusion to occur (reviewed in Kermode, 1996). The specificity lies in the different isoforms of these integral membrane proteins residing in various cell membranes and vesicles (Conceicao *et al*., 1997).

3.5 Biogenesis of ER-derived protein bodies in cereal seeds and mechanisms of storage protein retention

In seeds of some cereals (e.g. maize, rice and sorghum) there is direct formation of protein bodies from the rough ER; the prolamins are retained within the lumen of the rough ER to form protein deposits that remain surrounded by ER membrane (Figure 7) (Larkins and Hurkman, 1978; Krishman *et al*., 1986; Taylor *et al*., 1985; also see reviews by Gal and Raikhel, 1993; Galili *et al*., 1993). In contrast, prolamins of other cereals, such as wheat, barley and oats accumulate (along with globulins) in vacuole-derived protein bodies, and in most cases transit through the Golgi complex to arrive there. Rice seeds accumulate two types of storage proteins : the globulin-like glutelins and the prolamins. The former are transported to vacuoles via the Golgi complex, while the latter aggregate within the ER lumen (Krishman *et al*., 1986; Li *et al*., 1993b). These different pathways exhibited by the two storage protein types appear to be facilitated by a differential targeting of their respective mRNAs onto distinct rough ER membranes (Li *et al*., 1993a). During the period of storage protein deposition and protein body biogenesis, the endosperm cells exhibit two morphologically distinct populations of ER membranes: the cisternal ER and the ER that delimits protein bodies (protein body ER). The specific mRNAs in subcellular fractions enriched in these two ER types are non-randomly distributed. Notably, prolamin transcripts are enriched on the rough ER that delimits the prolamin protein bodies, whereas glutelin mRNAs predominate on the cisternal ER (Li *et al*., 1993a). The molecular basis for segregation of prolamin transcripts to the surface of the protein body ER is unknown, although specialized signal peptides and/or components involved in their recognition (i.e. signal recognition particles or ER membrane proteins) have been suggested to play a role (Li *et al*., 1993a). Another intriguing possibility is that elements of the cytoskeleton are involved. There is a close association of actin filaments with the prolamin-type protein bodies of maize (Abe *et al*., 1991) and a role for the cytoskeletal framework in protein synthesis is firmly established (Singer *et al*., 1989). F-actin may facilitate prolamin-protein body biogenesis by facilitating interaction of prolamin mRNAs with the protein body-ER. Non-random deposition of prolamins may also occur within the endosperm of developing maize where specific prolamin types are inserted

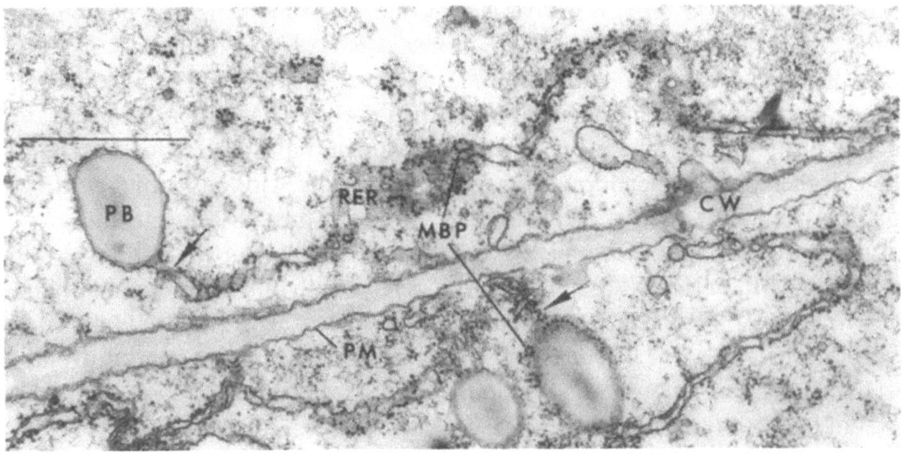

Fig. 7. Protein bodies originating from the rough ER. Electron micrograph of a developing cell of the starchy endosperm of maize to illustrate the formation of protein bodies (PB) forming from rough endoplasmic reticulum (RER) and the proximity of membrane-bound polyribosomes (MBP). Note the continuity between the protein body membranes and extended RER cisternae (arrow). CW, cell wall. From Larkins and Hurkman (1978). With permission from the American Society of Plant Physiologists. Courtesy of B. Larkins.

sequentially into protein bodies; correct assembly of different types of prolamins may be required for the formation of normal protein bodies (Lending and Larkins, 1989; Bagga *et al.*, 1997).

What, then, is the mechanism of retention of storage proteins within the ER? Is it due to their intrinsic structure or physical properties, or do specific ER-factors also play a role? The conventional mechanism for retention of soluble proteins within the ER is the carboxy-terminal tetrapeptide sequence KDEL or HDEL. Yet, it is unlikely that the mechanism involves this specific retention signal since it is not present on any of the storage proteins that are retained in the ER, and it is more likely to partly be a function of the mature protein (Larkins *et al.*, 1979).

In rice seeds, the ER molecular chaperone, BiP, retains prolamins within the ER lumen by promoting the folding and assembly of prolamins into protein bodies (Li *et al.*, 1993a; Muench *et al.*, 1997). Within developing endosperm cells, BiP is abundant at the periphery of the protein inclusion body, but not in the central portion of the protein body or in the cisternal ER membranes connecting the protein bodies. Dissociation of BiP from these bodies is an ATP-dependent process and the biogenesis of protein bodies is proposed to be a sequential process mediated by BiP. According to this model, BiP binds to the nascent prolamin as it emerges through the ER membrane, an interaction that serves to maintain the polypeptide in a competent state for subsequent assembly into protein bodies. Following the completion of protein synthesis, the prolamin-BiP complexes are released into the ER lumen and BiP utilizes ATP to assemble the prolamins onto the protein body surface; dissociated BiP is then recycled. A critical concentration of BiP-prolamin complexes within the ER lumen may trigger the initiation of the BiP-mediated aggregation process. Interestingly, BiP has a higher affinity for rice prolamins than it does for rice glutelins, possibly due to the greater proportion of aliphatic amino acids in the former (Kim and Okita, 1988). Although BiP probably associates transiently with glutelins within the ER lumen, it is not localized to the surface of the cisternal ER where glutelins are synthesized (Li *et al.*, 1993a, 1993b). The involvement of BiP in prolamin protein body formation is also supported by its elevated concentration in the protein bodies of maize mutants defective in zein accumulation.

The structures of cereal prolamins have been characterized in detail (reviewed in Galili *et al.*, 1993; Shewry and Tatham, 1990, and in chapters 2-7 in this volume). Most of the prolamins contain several short blocks of repeated amino acids that are rich in glutamine and proline. In the sulphur-poor prolamins of wheat, barley and rye, and the α-zeins of maize, these repeats comprise almost the entire length of the polypeptide. In contrast, the sulphur-rich prolamins and the high molecular weight (HMW) prolamins of wheat, barley and rye contain a repetitive region, but, in addition, contain 'unique-sequence' regions either at their *N*-terminus (sulphur-rich prolamins) or at their *N*- and *C*-terminal regions (HMW prolamins). These domains are enriched in α-helices and contain cysteine residues which form inter- and/or intra-molecular disulphide bonds. The secondary structure of the *N*-terminal repetitive region may be responsible for ER retention and assembly into protein bodies within the ER, possibly via hydrophobic and hydrogen-bond interactions (Galili *et al.*, 1993). Different retention mechanisms may operate for the maize prolamins, specifically β-, γ- and δ-zeins, as compared to maize α-zeins. For γ-zein, a proline-rich domain composed of repeats of a hexapeptide (PPPVHL) in the *N*-terminal domain is responsible for ER-retention, whereas both the proline-rich repeat and the *C*-terminal cysteine-rich domains are needed for protein body biogenesis (Geli *et al.*, 1994).

In summary, retention in ER-derived protein bodies in cereals may involve multiple mechanisms and these mechanisms may interact (e.g. BiP-mediated assembly/aggregation and physical characteristics of the protein, preventing transport out of the ER). There is evidence for yet another route (e.g. for some of the wheat prolamins), which begins by assembly of the proteins into protein bodies within the ER. However, these ER-derived protein bodies undergo subsequent internalization within vacuoles by a specific process analogous to autophagy (reviewed by Galili *et al.*, 1993). Thus the wheat prolamin storage proteins may be unique in exhibiting two pathways or routes of transport to protein bodies.

4. SUMMARY

Seed proteins undergo several modifications during or after their synthesis including signal peptide cleavage, glycosylation, folding and assembly, disulphide bond formation and proteolytic processing. These co- and post-translational modifications are of fundamental importance to the functions of seed proteins (e.g. the activity of hydrolytic enzymes) and often have a profound influence on the stable accumulation of seed proteins in storage parenchyma cells (e.g. storage proteins). Efficiency of transport along the secretory pathway can depend upon the glycosylation status of the protein as well as its progress in acquiring the correct three-dimensional conformation. In fact, one of the major factors contributing to a protein's half-life in the cell is its ability to undergo normal folding and assembly processes and to assume a functional three-dimensional structure; failure to achieve this results in the activation of the cell's disposal mechanisms to remove non-functional proteins or components. A consideration of these factors also has far-reaching implications for studies aimed at the engineering of seed proteins for greater nutritional value, where success is dependent upon achieving high levels of accumulation of the altered protein in transgenic plant hosts. The ultimate challenge here is to design or manipulate polypeptide sequences to enhance the nutritive value of seed proteins, but maintain a functional and stable three-dimensional conformation; this is impeded by our limited knowledge of the rules that govern post-translational processing of proteins and how these processes relate to a protein's ultimate stablity in the cell. Thus, the structural features of storage proteins that allow for their stable accumulation in cells need to be eludidated. Clearly more information is also required about the different mechanisms of protein body biogenesis in seeds and the diverse mechanisms for targeting of seed proteins to ER-derived protein bodies, protein storage vacuoles and lytic vacuoles.

ACKNOWLEDGEMENTS

We are grateful to David G. Robinson, Liwen Jiang and John Rogers for providing preprints of their most recent data. This chapter is dedicated to the memory of Donald J. Kermode who revealed his true colours late in life and a heart of gold.

REFERENCES

Abe, S., You, W. and Davies, E. (1991) Protein bodies in corn endosperm are enclosed by and enmeshed in F-actin. *Protoplasma*, **165**, 139-149.

Ahmed, S.U., Bar-Peled, M. and Raikhel, N.V. (1997) Cloning and subcellular location of an *Arabidopsis* receptor-like protein that shares common features with protein-sorting receptors of eukaryotic cells. *Plant Physiology*, **114**, 325-336.

Alberts, B., Bray, D., Lewis, J., Raff, M., Roberts, K. and Watson, J.D. (1994) *Molecular Biology of the Cell*, third ed., Garland Publishing Inc., New York.

Austen, B.M. and Westwood, O.M.R. (1991) Protein Targeting and Secretion, Oxford University Press, New York.

Bagga, S., Adams, H.P., Rodriguez, F.D., Kemp, J.D. and Sengupta-Goplan, C. (1997) Coexpression of the maize δ-zein and β-zein genes results in stable accumulation of δ-zein in endoplasmic reticulum-derived protein bodies formed by β-zein. Plant Cell, **9**, 1683-1696.

Bar-Peled, M. and Raikhel, N.V. (1997) Characterization of AtSEC12 and AtSAR1, proteins likely involved in endoplasmic reticulum and Golgi transport. Plant Physiology, **114**, 315-324.

Bednarek, S.Y. and Raikhel, N.V. (1992) Intracellular trafficking of secretory proteins. *Plant Molecular Biology*, **20**, 133-150.

Bewley, J.D. and Black, M. (1994) *Seeds. Physiology of Development and Germination*, second ed., Plenum Press, New York.

Blond-Elguindi, S., Cwirla, S.E., Dower, W.J., Lipshutz, R.J., Sprang, S.R., Sambrook, J.F. and Gething, M.-J.H. (1993). Affinity panning of a library of peptides displayed on bacteriophages reveals the binding specificity of BiP. *Cell*, **75**, 717-728.

Bollini, R., Vitale, A. and Chrispeels, M.J. (1983) *In vivo* and *in vitro* processing of seed reserve protein in the endoplasmic reticulum: evidence for two glycosylation steps. *Journal of Cell Biology*, **96**, 999-1007.

Boston, R.S., Fontes, E.B.P., Shank, B.B. and Wrobel, R.L. (1991) Increased expression of the maize immunoglobulin binding protein homologue b-70 in three zein regulatory mutants. *Plant Cell*, **3**, 497-505.

Bulleid, N.J. and Freedman, R.B. (1988) Defective co-translational formation of disulphide bonds in protein disulphide-isomerase-deficient microsomes. *Nature*, **335**, 649-652.

Bustos, M.M., Kalkan, F.A., Van den Bosch, K.A. and Hall, T.C. (1991) Differential accumulation of four phaseolin glycoforms in transgenic tobacco. *Plant Molecular Biology*, **16**, 381-395.

Casey, R.D. (1979) Immunoaffinity chromatography as a means of purifying legumin from *Pisum* (pea) seeds. *Biochemical Journal*, **177**, 509-520.

Chrispeels, M.J. (1991) Sorting of proteins in the secretory system. *Annual Review of Plant Physiology and Plant Molecular Biology*, **42**, 21-53.

Chrispeels, M.J. and Raikhel, N.V. (1992) Short peptide domains target proteins to plant vacuoles. *Cell*, **68**, 613-616.

Chrispeels, M.J. and Tague, B.W. (1990) Transport and targeting of proteins to protein storage vacuoles (protein bodies) in developing seeds, in *Recent Advances in Development and Germination of Seeds*, (ed. R.B. Taylorson), Plenum Publ. Corp., New York.

Conceicao, A.D.S., Marty-Mazars, Bassham, D.C., Sanderfoot, A.A., Marty, F. and Raikhel, N.V. (1997) The syntaxin homolog AtPEP12p resides on a late post-Golgi compartment in plants. *Plant Cell*, **9**, 571-582.

Craig, S. and Goodchild, D.J. (1984) Golgi-mediated vicilin accumulation in pea cotyledon cells is re-directed by monensin and nigericin. *Protoplasma*, **122**, 91-97.

Crouch, M.L., Tenbarge, K.M., Simon, A.E. and Ferl, R. (1983) cDNA clones for *Brassica napus* seed proteins: evidence from nucleotide sequence analysis that both subunits of napin are cleaved from a precursor polypeptide. *Journal of Molecular and Applied Genetics*, **2**, 273-283.

Denecke, J., Goldman, M.H.S., Demolder, J., Seurinck, J and Botterman, J. (1991) The tobacco luminal binding protein is encoded by a multigene family. *Plant Cell*, **3**, 1025-1035.

D'Hondt, K., Bosch, D., Van Damme, J., Goethals, M., Vandekerckhove, J. and Krebbers, E. (1993) An aspartic proteinase present in seeds cleaves *Arabidopsis* 2S albumin precursors *in vitro*. *Journal of Biological Chemistry*, **268**, 20884-20891.

D'Hondt, K., Van Damme, J., Van Den Bossche, C., Leejeerajumneun, S., De Rycke, K., Derksen, J., Vandekerckhove, J. and Krebbers, E. (1993) Studies of the role of the propeptides of the *Arabidopsis thaliana* 2S albumin. *Plant Physiology*, **102**, 425-433.

Dickinson, C.D., Floener, L.A., Lilley, G.G. and Nielsen, N.C. (1987) Self-assembly of proglycinin and hybrid proglycinin synthesized *in vitro* from cDNA. *Proceedings of the National Academy of Science USA*, **84**, 5525-5529.

Dickinson, C.D., Hussein, E.H.A. and Nielsen, N.C. (1989) Role of post-translational cleavage in glycinin assembly. *Plant Cell*, **1**, 459-469.

Dickinson, C.D., Scott, P.M., Hussein, E.H.A., Argos, P. and Nielsen, N.C. (1990) Effect of structural modifications on the assembly of a glycinin subunit. *Plant Cell*, **2**, 403-413.

Dombrowski, J.E., Schroeder, M.R., Bednarek, S.Y. and Raikhel, N.V. (1993) Determination of the functional elements within the vacuolar targeting signal of barley lectin. *Plant Cell*, **5**, 587-696.

Drake Jr., R.R., Kaushal, G.P., Pastuszak, I. and Elbein, A.D. (1991) Partial purification, photoaffinity labelling, and properties of mung bean UDP-glucose: dolicholphosphate glucosyl-transferase. *Plant Physiology*, **97**, 396-401.

Faye, L. and Chrispeels, M.J. (1989) Apparent inhibition of β-fructosidase secretion by tunicamycin may be explained by breakdown of the unglycosylated protein during secretion. *Plant Physiology*, **89**, 845-851.

Faye, L., *et al.* (1992), in *Post Translational Modification and its Significance in Plant Development* (Eds. N.M. Battey, H.G. Dikinson and A.M. Heatherington), Cambridge University Press, pp. 213-242.

Fontes, E.B.P., Shank, B.B., Wrobel, R.L., Moose, S.P., O'Brian, G.R., Wurtzel, E.T. and Boston, R.S. (1991) Characterization of an immunoglobulin binding protein homologue in the maize floury-2 endosperm mutant. *Plant Cell*, **3**, 483-496.

Gal, S. and Raikhel, N.V. (1993) Protein sorting in the endomembrane system of plant cells. *Current Opinions in Cell Biology*, **5**, 636-640.

Galili, G., Altschuler, Y. and Levanony, A. (1993) Assembly and transport of seed storage proteins. *Trends in Cell Biology*, 3, 437-442.

Geli, M.I., Torrent, M. and Ludevid, D. (1994) Two structural domains mediate two sequential events in γ-zein targeting: protein endoplasmic reticulum retention and protein body formation. *Plant Cell*, 6, 1911-1922.

Gething, M.-J. and Sambrook, J. (1992) Protein folding in the cell. *Nature*, 355, 33-45.

Gillikin, J.W., Zhang, F., Coleman, C.E., Bass H.W., Larkins, B.A. and Boston, R.S. (1997) A defective signal peptide tethers the *fluory-2* zein to the endoplasmic reticulum membrane. *Plant Physiology*, **114**, 345-352.

Gomez, L. and Chrispeels, M.J. (1993) Tonoplast and soluble vacuolar proteins are targeted by different mechanisms. *Plant Cell*, **5**, 1113-1124.

Hara-Nishimura, I., Takeuchi, Y., Inoue, K. and Nishimura, M. (1993) Vesicle transport and processing of the precursor to 2S albumin in pumpkin. *Plant Journal,* **4**, 793-800.

Hattori, T., Ichihara, S. and Nakamura, K. (1987) Processing of a plant vacuolar protein precursor *in vitro. European Journal of Biochemistry,* **166**, 533-538.

Herman, E.M., Pueyo, J.J. and Chrispeels, M.J. (1994) Quality control of a seed storage protein with a destabilizing epitope occurs in the post-ER secretory system. *Plant Physiology,* **105** (Suppl.), 94.

Hinz, G., Menze, A., Hohl, I. and Vaux, D. (1997) Isolation of prolegumin from developing pea seeds: its binding to endomembranes and assembly into prolegumin hexamers in the protein storage vacuole. *Journal of Experimental Botany,* **48**, 139-149.

Hoffman, L.M., Donaldson, D.D. and Herman, E.M. (1988) A modified storage protein is synthesized, processed, and degraded in the seeds of transgenic plants. *Plant Molecular Biology,* **11**, 717-729.

Höfte, H. and Chrispeels, M.J. (1992) Protein sorting to the vacuolar membrane. *Plant Cell,* **4**, 995-1004.

Hoh, B., Hinz, G., Jeong, B.-K. and Robinson, D.G. (1995) Protein storage vacuoles form de novo during pea cotyledon development. *Journal of Cell Science,* **108**, 299-310.

Hohl, I., Robinson, D.G., Chrispeels, M.J. and Hinz, G. (1996) Transport of storage proteins to the vacuole is mediated by vesicles without a clathrin coat. *Journal of Cell Science,* **109**, 2539-2550.

Holstein, S.E.H., Drucker, M. and Robinson, D.G. (1994) Identification of a β-type adaptin in plant clathrin-coated vesicles. *Journal of Cell Science,* **107**, 945-953.

Holwerda, B.C., Galvin, N.J., Baranski, T.J. and Rogers, J.C. (1990) *In vitro* processing of aleurain, a barley vacuolar thiol protease. *Plant Cell,* **2**, 1091-1106.

Hurtley, S.M. and Helenius, A. (1989) Protein oligomerization in the endoplasmic reticulum. *Annual Review of Cell Biology,* **5**, 277-307.

Kelleher, D.J., Kreibich, G. and Gilmore, R. (1992) Oligosaccharyl transferase activity is associated with a protein complex composed of ribophorins I and II and a 48 kDa protein. *Cell,* **69**, 55-65.

Kermode, A.R. (1996) Mechanisms of intracellular protein transport and targeting in plant cells. *Critical Reviews in Plant Sciences,* **15**, 285-423.

Kim, W.T. and Okita, T.W. (1988) Structure, expression, and heterogeneity of the rice seed prolamines. *Plant Physiology,* **88**, 649-655.

Kim, W.Y., Cheong, N.E., Je, D.Y., Kim, M.G., Lim, C.O., Bahk, J.D., Cho, M.J. and Lee, S.Y. (1997) The presence of a *Sar1* gene family in *Brassica campestris* that supresses a yeast vesicular transport mutation *Sec12-1. Plant Molecular Biology,* **33**, 1025-1035.

Kirsch, T. and Beevers, L. (1993) Uncoating of clathrin-coated vesicles by uncoating ATPase from developing peas. *Plant Physiology,* **103**, 205-212.

Kirsch, T., Paris, N., Butler, J.M., Beevers, L. and Rogers, J.C. (1994) Purification and initial characterization of a potential plant vacuolar targeting receptor. *Proceedings of the National Academy of Science USA,* **91**, 3403-3407.

Kirsch, T., Saalbach, G., Raikhel, N.V. and Beevers, L. (1996) Interaction of a potential vacuolar targeting receptor with amino- and carboxyl-terminal targeting determinants. *Plant Physiology,* **111**, 469-474.

Krishnan, H.B., Franceschi, V.R. and Okita, T.W. (1986) Immunochemical studies on the role of the Golgi complex in protein-body formation in rice seeds. *Planta,* **169**, 471-480.

Krochko, J.E. and Bewley, J.D. 1988. Use of electrophoretic techniques in determining the composition of seed storage proteins in alfalfa. *Electrophoresis* **9**, 751-763.

Larkins, B.A. and Hurkman, W.J. (1978) Synthesis and deposition of zein in protein bodies of maize endosperm. *Plant Physiology,* **62**, 256-263.

Larkins, B.A., Pedersen, K., Handa, A.K., Hurkman, W.J. and Smith, L.D. (1979) Synthesis and processing of maize storage proteins in *Xenopus laevis* oocytes. *Proceedings of the National Academy of Science USA,* **76**, 6448-6452.

Lending, C.R. and Larkins, B.A. (1989) Changes in the zein composition of protein bodies during maize endosperm development. *Plant Cell,* **1**, 1011-1023.

Li, X., Franceschi, V.R. and Okita, T.W. (1993a) Segregation of storage protein mRNAs on the rough endoplasmic reticulum membranes of rice endosperm cells. *Cell,* **72**, 869-879.

Li, X., Wu, Y., Zhang, D.-Z., Gillikin, J.W., Boston, R.S., Franceschi, V.R. and Okita, T.W. (1993b) Rice prolamine protein body biogenesis: a BiP-mediated process. *Science,* **262**, 1054-1056.

Lupattelli, F., Pedrazzini, E., Bollini, R., Vitale, A. and Ceriotti, A. (1997) The rate of phaseolin assembly is controlled by the glucosylation state of its N-linked oligosaccharide chains. *Plant Cell,* **9**, 597-609.

Matsuoka, K., Bassham, D.C., Raikhel, N.V. and Nakamura, K. (1995) Different sensitivity to wortmannin of two vacuolar sorting signals indicates the presence of distinct sorting machineries in tobacco cells. *Journal of Cell Biology,* **130**, 1307-1318.

Matsuoka, K., Higuchi, T., Maeshima, M. and Nakamura, K. (1997) A vacuolar-type H^+-ATPase in a nonvacuolar organelle is required for the sorting of soluble vacuolar protein precursors in tobacco cells. *Plant Cell,* **9**, 533-546.

Matsuoka, K. and Nakamura, K. (1991) Propeptide of a precursor to a plant vacuolar protein required for vacuolar targeting. *Proceedings of the National Academy of Science USA,* **88**, 834-838.

Maurel, C. (1997) Aquaporins and water permeability of plant membranes. *Annual Review of Plant Physiology and Plant Molecular Biology,* **48**, 399-429.

Muench, D.G., Wu, Y., Zhang, Y., Li, X., Boston, R.S. and Okita, T.W. (1997) Molecular cloning, expression and subcellular localization of a BiP homolog from rice endosperm tissue. *Plant and Cell Physiology,* **38**, 404-412.

Munro, S. and Pelham, H.R.B. (1986) An Hsp70-like protein in the ER: Identity with the 78 kd glucose-regulated protein and immunoglobulin heavy chain binding protein. *Cell,* **46**, 291-300.

Muntz, K. (1996) Proteases and proteolytic cleavage of storage proteins in developing and germinating dicotyledonous seeds. *Journal of Experimental Botany,* **47**, 605-622.

Nagai, K., Shibata, K. and Yamaguchi, H. (1993) Role of intramolecular high-mannose chains in the folding and assembly of soybean (*Glycine max*) lectin polypeptides: studies by the combined use of spectroscopy and gel-filtration size analysis. *Journal of Biochemistry,* **114**, 830-834.

Nakamura, K. and Matsuoka, K. (1993) Protein targeting to the vacuole in plant cells. *Plant Physiology,* **101**, 1-5.

Nelson, D.E., Glaunsinger, B. and Bohnert, H. (1997) Abundant accumulation of the calcium-binding molecular chaperone calreticulum in specific floral tissues of *Arabidopsis thaliana. Plant Physiology,* **114**, 29-37.

Neuhaus, J.-M., Pietrzak, M. and Boller, T. (1994) Mutation analysis of the C-terminal vacuolar targeting peptide of tobacco chitinase: low specificity of the sorting system, and gradual transition between intracellular retention and secretion into the extracellular space. *Plant Journal,* **5**, 45-54.

Okita, T.W. and Rogers, J.C. (1996) Compartmentation of proteins in the endomembrane system of plant cells. *Annual Review of Plant Physiology and Plant Molecular Biology,* **47**, 327-350.

Paris, N., Stanley, C.M., Jones, R.L. and Rogers, J.C. (1996) Plant cells contain two functionally distinct vacuolar compartments. *Cell,* **85**, 563-572.

Paris, N., Rogers, S.W., Jiang, L., Kirsch, T., Beevers, L., Phillips, T.E. and Rogers, J.C. (1997) Molecular cloning and further characterization of a probable plant vacuolar sorting receptor. *Plant Physiology,* **115**, 29-39.

Pelham, H.R.B. (1989) Control of protein exit from the endoplasmic reticulum. *Annual Review of Cell Biology,* **5**, 1-23.

Pryer, N.K., Wuestehube, L.J. and Schekman, R. (1992) Vesicle-mediated protein sorting. *Annual Review of Biochemistry,* **61**, 471-516.

Rapoport, T.A. (1992) Transport of proteins across the endoplasmic reticulum membrane. *Science*, **258**, 931-936.

Robinson, D.G. and Hinz, G. (1997) Vacuole biogenesis and transport to the plant vacuole: a comparison with the yeast vacuole and the mammalian lysosome. *Protoplasma*, **197**,1-25.

Robinson, D.G., Baumer, M., Hinz, G. and Hohl, I. (1998) Vesicle transfer of storage proteins to the vacuole: the role of the Golgi apparatus and multivesicular bodies. *Journal of Plant Physiology,* **152**, 659-667.

Rogers, J.C. (1998) Compartmentation of plant cell proteins in separate lytic and protein storage vacuoles. *Journal of Plant Physiology,* **152**, 653-658.

Rothman, J.E. (1989) GTP and methionine bristles. *Nature*, **340,** 433-434.

Saalbach, G., Jung, R., Kunze, G., Saalbach, I., Adler, K. and Muntz, K. (1991) Different legumin protein domains act as vacuolar targeting signals. *Plant Cell*, **3**, 695-708.

Shewry, P.R. and Tatham, A.S. (1990) The prolamin storage proteins of cereal seeds: structure and evolution. *Biochemical Journal,* **267**, 1-12.

Shimada, T., Hiraiwa, N., Nishimura, M. and Hara-Nishimura, I. (1994) Vacuolar processing enzyme that converts proproteins to the corresponding mature forms. *Plant and Cell Physiology,* **35**, 713-718.

Simon, P. and Jones, R.L. (1988) Synthesis and secretion of catalytically active barley α-amylase isoforms by *Xenopus* oocytes injected with barley mRNAs. *European Journal of Cell Biology,* **47**, 213-221.

Singer, R.H., Langevin, G.L. and Lawrence, J.B. (1989) Ultrastructural visualization of cytoskeletal mRNAs and their associated proteins using double-label in situ hybridization. *Journal of Cell Biology,* **108**, 2343-2353.

Sturm, A., Johnson, K.D., Szumilo, T., Elbein, A.D. and Chrispeels, M.J. (1987) Subcellular localization of glycosidases and glycosyl-transferases involved in the processing of N-linked oligosaccharides. *Plant Physiology,* **85,** 741-745.

Sun, S.S.M., Altenbach, S.B. and Leung, F.W. (1987) Properties, biosynthesis and processing of a sulfur-rich protein in Brazil nut (*Bertholletia excelsa* H.B.K.). *European Journal of Biochemistry,* **162**, 477-483.

Taylor, C. (1997) Controlling the quality of secretory proteins in plants. *Plant Cell*, **9**, 1697-1699.

Taylor, J.R.N., Schussler, L. and Liebonberg, N.V.D.W. (1985) Protein body formation in the starchy endosperm of developing *Sorghum bicolor* (L) Moench seeds. *South African Journal of Botany,* **51**, 35-40.

Verner, K., and Shatz, G. (1988) Protein translocation across membranes. *Science* **241**, 1307-1313.

Vitale, A., Ceriotti, A. and Denecke, J. (1993) The role of the endoplasmic reticulum in protein synthesis, modification and intracellular transport. *Journal of Experimental Botany,* **44**, 1417-1444.

Von Schaewen, A. and Chrispeels, M.J. (1993) Identification of vacuolar sorting information in phytohemagglutinin, an unprocessed vacuolar protein. *Journal of Experimental Botany,* **44** (Suppl.), 339-342.

Xia, J.-H. and Kermode, A.R. (1998) Analyses to determine the role of embryo immaturity in dormancy maintenance of yellow-cedar (*Chamaecyparis nootkatensis*) seeds: Synthesis and accumulation of storage proteins and proteins implicated in desiccation tolerance. *Journal of Experimental Botany*, in press.

35. The Synthesis and Deposition of Storage Proteins: Possible Role of Molecular Chaperones and the Unfolded Protein Response

JOHNATHAN A. NAPIER
IACR- Long Ashton Research Station, Department of Agricultural Sciences, University of Bristol, Long Ashton, Bristol BS41 9AF, UK

Abbreviations: binding protein- BiP; endoplasmic reticulum-ER; unfolded protein response-UPR; unfolded protein response element-UPRE

1. INTRODUCTION

The notion that proteins are capable of spontaneous folding and assembly is one that enjoyed several decades of support, only being queried with the discovery of a class of proteins which appeared to be involved in assisting polypeptide folding. These observations led to the now well known concept of molecular chaperones, in which proteins undergo assisted folding, usually mediated by a set of ATP-dependent reactions (reviewed in Ellis and van de Reis, 1991). It is also clear that these molecular chaperones are ubiquitous not only to organisms and cell types, but also to different subcellular compartments. The reason for this may be that chaperone proteins are also involved in promoting the translocation of polypeptides (or nascent chains) across the delineating membranes of organelles, such as plastids, mitochondria and the endoplasmic reticulum (ER). In the case of proteins crossing the ER and entering the secretory system, this event is usually a co-translational process mediated by the signal recognition particle-targeted bound ribosomes which define the rough ER membrane. Within the lumen of the ER are a number of abundant chaperone proteins, the best characterized of these being binding protein (BiP), which was initially characterized on the basis of its interaction with heavy chain immunoglobulins (Haas and Wabl, 1983). Subsequent work in yeast (where the BiP homologue is encoded by the *KAR2* gene) and mammalian systems have revealed that BiP is involved in

P.R. Shewry and R. Casey (eds.), Seed Proteins, 843–858.

polypeptide translocation and protein folding of a wide range of substrates (reviewed in Gething and Sambrook, 1992).

In the developing seeds of higher plants, the endoplasmic reticulum is the site of synthesis and deposition of an important class of proteins, the seed-storage proteins. Although these can be classified into a number of sub-types according to criteria such as solubility, as well as DNA and protein sequences, these proteins are not found in other organisms. The role of the seed storage proteins is to provide reserves to be used in seedling germination and establishment. However, the properties of the seed storage proteins, combined with their economic importance, have led to a large number of studies on their synthesis and deposition.

One of the most intensely studied seed storage protein systems is the deposition of the *Phaseolus vulgaris* 7S globulin phaseolin (see also Chapter 10). D'Amico *et al.* (1992) initially reported that phaseolin could be immunopreciptiated in a stable complex with BiP (although only after the plant material had been incubated with the glycosylation inhibitor tunicamycin), leading to the hypothesis that interactions between BiP and phaseolin were part of the assembly and maturation pathway of the storage protein. Since phaseolin undergoes trimerization as part of this process, the demonstration that BiP interacted only with monomeric phaseolin was seen as further evidence for a role of the chaperone (Vitale *et al.*, 1995). Also, in the same study, it was shown that BiP could be immunoprecipitated as a partner using a range of antisera raised against plant endomembrane proteins (such as phytohaemagglutinin, α-amylase inhibitor and tonoplast intrinsic protein). However, it was also clear that only a proportion of these proteins were interacting with BiP, and no data regarding the time course of the interaction (i.e. release from BiP followed by subsequent trimerization) were presented. It had already been shown that an assembly-defective form of phaseolin (*C*-terminally truncated) interacted strongly with BiP (Pedrazzini *et al.*, 1994), as did the wild-type protein on treatment of the tissue with tunicamycin, leading to the possibility that the stable interactions with BiP observed in these assays were in fact related to malfolding, as opposed to being part of the "normal" maturation of the protein (Shewry *et al.*, 1995; Galili and Herman, 1997). Subsequently, Pedrazzini *et al.* (1997) presented data which indicated that this stable interaction between BiP and phaseolin was indeed related to malfolding of the storage protein, and was marking the protein for proteolytic degradation, perhaps by (BiP-assisted) retro-translocation back into the cytosol (Taylor, 1997).

Such observations indicate that plant seed storage proteins, like any polypeptide entering the ER, are subject to "quality control", meaning that malfolded or non-assembled proteins are not allowed to progress further down the secretory pathway of the endomembrane system (Hammond and Helenius, 1995). However, it is not clear whether or not these storage proteins require

BiP to be actively involved in their folding and packaging into protein storage bodies. The aim of this chapter is to examine the evidence for such interactions, and also to explore possible explanations for such observations. It will also discuss the regulation of BiP synthesis, which is known to be induced by the presence of unfolded proteins in the ER lumen. I will initially describe the regulation of BiP synthesis by unfolded proteins in the well-studied yeast *Saccharomyces cerevisiae*, as an insight into the regulation of this chaperone, and then discuss the possible of role(s) of BiP in seed storage protein synthesis and deposition.

2. THE UNFOLDED PROTEIN RESPONSE IN YEAST

The yeast homologue of BiP is encoded by the *KAR2* gene, and is essential for cell viability (Normington *et al.*, 1989). Like its mammalian homologue, KAR2 is an abundant protein localized to the ER lumen by virtue of an ER-retention signal, the *C*-terminal tetrapeptide HDEL motif. Transcription of the *KAR2* gene is up-regulated by the accumulation of unfolded proteins in the ER lumen (notably resulting from treatment with tunicamycin, but also in some *sec* mutants), though *KAR2*, unlike its mammalian homologue, also shows inducibility under heat shock. To understand more about the transcriptional regulation of *KAR2*, a 1.3 kb promoter region was cloned and subjected to detailed analysis (Mori *et al.*, 1992). This revealed the presence of separate and distinct *cis*-acting elements required for the induction of the promoter by either heat shock or unfolded proteins. A 22 bp element, defined as the unfolded protein response element (UPRE) was identified as being responsible for promoting *KAR2* transcription in the presence of unfolded proteins, and this element conferred the ability to respond to such a treatment to a heterologous promoter. Perhaps more significantly, sequences related to the UPRE could be identified in the promoters of the mammalian BiP genes, and in other, distinct, genes induced by unfolded proteins, such as GRP94 (endoplasmin). Subsequently, further examples of yeast genes induced by unfolded proteins have also been identified as containing UPREs (Craven *et al.*, 1996).

Analysis of the nuclear factors which bound to the UPRE indicated no overlap with the heat shock response factors, and mutagenesis of the UPRE resulted in loss of binding of these factors. Since the same mutations in the UPRE resulted in the loss of the unfolded response in the context of a heterologous promoter, these results were taken as evidence of a role for these factors in the *trans*-activation of the *KAR2* promoter in response to the accumulation of unfolded proteins in the ER.

The identification of the UPRE therefore raised the possibility of using a genetic approach to identify components in the signal transduction pathway responsible for the transmission and propagation of the unfolded protein response and the transcriptional up-regulation of *KAR2*. Two independent approaches were taken to identify yeast mutants defective in this response, though both identified the same gene as being central to the process. Mori *et al.* (1993) utilized a temperature-sensitive *sec* mutant (*sec53*) which is a condition-lethal allele of phosphomannomutase; incubation at non-permissive temperatures results in the accumulation of malfolded and incorrectly glycosylated proteins in the ER lumen. However, this mutant is viable and functions normally at permissive temperature. Therefore, a mutated form of the *KAR2* promoter, containing only the UPRE, was fused to the reporter gene *lacZ* (β-galactosidase) and introduced into this yeast strain. Subsequent mutagenesis and screening for loss of β-galactosidase activity at non-permissive temperature revealed a number of white yeast colonies, and further characterisation indicated that several of these were impaired in the induction of the endogenous *KAR2* gene, either in the presence of tunicamycin or at non-permissive temperatures for *sec53*. These mutant yeast strains were then used to identify the wild type gene, which was designated *ERN1*, by complementation (Mori *et al.*, 1993). A slightly different approach was taken by Cox *et al.* (1993) who fused a synthetic UPRE in front of a crippled *CYC1* promoter, and then introduced two copies of this construct into the genome of the yeast. This strain was then mutagenized, plated on tunicamycin (to induce the unfolded protein response) and the resulting white colonies (i.e. non-induced) subjected to further characterizations. These showed that transcripts for both *KAR2* and protein disulphide isomerase (PDI; another ER-resident chaperone involved in disulphide bond formation) failed to be induced under tunicamycin treatment, indicating the pleiotropic nature of the mutation. The mutagenized gene responsible for this phenotype was identified by complementation and shown to be a previously characterized gene, *IRE1*, which is required for growth on inositol (Nikawa and Yamashita, 1992). *IRE1* is identical to the gene identified as *ERN1* (Mori *et al.*, 1993) and is non-essential, being required only when cells are either utilizing inositol or experiencing unfolded protein stress. It is important to note that addition of inositol to the media of *ire1* mutants did not restore the unfolded protein response (Cox and Walter, 1993).

Leaving aside for the present the issue of why the same gene (*IRE1*/*ERN1*) was found to be involved in two ostensibly different responses, the structure of IRE1 revealed some insights into its possible function. *IRE1* encodes a polypeptide of M_r 127,000 and contains a putative *N*-terminal signal sequence to direct it into the ER membrane system. It also contains a hydrophobic domain which may serve as a stop-transfer sequence and a *trans*-membrane span (Cox and Walter, 1993). The *C*-terminal portion of the protein shows homology to serine/threonine protein kinases, and kinase activity is required for function of *IRE1* in the unfolded protein response (Mori *et al.*, 1993). Based on the similarity in its domain structure to other growth factor receptor kinases, a model was proposed which involves the activation of the *IRE1* kinase as a result of dimerization, which is in turn brought about by a reduction in the levels of the endogenous KAR2 protein (BiP) (Cox and Walter, 1993; Mori *et al.*, 1993). This model has a number of implications, not least for the topology of the protein; IRE1 is unusual for yeast being a transmembrane kinase. It is predicted that the *N*-terminal domain of ~500 residues is exposed to the ER lumen, thus serving as the "sensor" for the unfolded protein response. A single transmembrane span crosses the ER membrane, allowing the kinase domain to act as the signal, which could be transmitted through the cytoplasm or directly into the nucleus. The latter is considered a possibility due to the close association of the ER with the nuclear envelope in yeast, and the presence of putative nuclear localization signals in this *C*-terminal kinase domain (Mori *et al.*, 1993). Another interesting feature of the IRE1 protein is the presence of a *C*-terminal tail following the kinase domain, again reminiscent of receptor kinases. The function of this domain was found to be essential for the induction of the unfolded protein response (Shamu and Walter, 1996), and predicted to be involved in interaction with downstream components of the signalling pathway. The same study also demonstrated that the unfolded protein response required the dimerization and *trans*-phosphorylation of IRE1, and moreover, that over-expression of *IRE1* induced the UPR.

Having identified the "sensor" component of the UPR signal transduction pathway as *IRE1*, Cox and Walter (1996) undertook a genetic approach to the identification of other components. A screen was devised, in a Δ*ire1* background, to isolate multi-copy suppressors (suppressor mutants are usually down-stream), using tandem UPREs fused to the HIS3 (selectable marker) gene. This screen revealed that three different genes were capable of suppressing the loss of *IRE1*. These were *IRE1* itself, and genes for two transcription factors, *SWI4* and *HAC1*, of which *SWI4* is a general *trans*-activating factor (Koch and Nasmyth, 1994) and *HAC1* appears to be a basic leucine zipper-type protein. The role of *HAC1*, but not of *SWI4*, in the UPR was confirmed by producing deletion mutants of the two genes (neither were essential) and testing for the induction of *KAR2* in the presence of tunicamycin. UPR-inducibilty was lost in Δ*hac1* strains, but could be recovered by episomal expression of *HAC1* in that background, confirming the involvement of *HAC1* in the UPR (Cox and Walter, 1996). The direct nature of this interaction was then demonstrated by the binding of HAC1 to UPREs in gel shift assays, and showing that mutated UPREs did not bind to HAC1. Moreover, no interaction with normal UPREs was observed in nuclear extracts isolated from *Δhac1* strains, and the presence of HAC1 could be detected in the nucleus of yeast cells after the induction of the UPR.

So far the unfolded protein response signalling pathway can be considered as being reasonably straightforward. This perception was, however, challenged by the observation that the presence or absence of the HAC1 bZiP transcription factor was correlated with the induction of the UPR (unlike for example the heat shock response, in which the *trans*-acting factor is ubiquitous, but activated by phosphorylation in response to heat shock). The reason for this was not transcriptional induction of the *HAC1* gene in response to unfolded proteins, but a splicing event which occurred to the *HAC1* transcript only when the UPR was induced (Sidrauski *et al.*, 1996). This splicing event removed 252 nucleotides from the primary *HAC1* transcript, resulting in a variant form of HAC1 which has a significantly different *C*-terminal amino acid sequence. This variant form of the protein was found only in UPR-induced cells, and not in *Δire1* or *Δhac1* mutants, as determined by antibodies raised against the *C*-terminal region. That this spliced form of *HAC1* transcript really encoded the bZiP protein responsible for up-regulating the *KAR2* promoter was shown by the expression of a synthetic *HAC1* gene, which lacked the 252 nucleotides present in the wildtype copy of the gene. This resulted in the constitutive expression of the UPR, even in the absence of tunicamycin and in a *Δire1 Δhac1* background.

What then forms the basis for the difference in abundance of the inactive and (spliced) active forms of HAC1? Examination of the polyribosome profile of stressed and unstressed cells revealed the presence of both forms of *HAC1* transcript, indicating that they were both capable of nuclear export and ribosome recruitment (Sidrauski *et al.*, 1996). These data implied that the modification to HAC1 brought about by the splicing event was responsible for the accumulation (or not) of the protein. Inspection of the *C*-terminal region of HAC1 revealed the presence (in the unspliced form) of so-called PEST sequences, regions which are rich in proline, glutamate, asparagine, serine and threonine, and are involved in the active degradation of proteins by the ubiquitin-dependent proteosome pathway. Thus, the unspliced form of HAC1, although normally extremely unstable, was found to accumulate in yeast mutants which were lacking ubiquitin-conjugating enzymes, *trans*-activating the UPR and confirming the role of the stability of HAC1 polypeptide in this signal transduction pathway. Definitive proof of this was provided by the ability of a *C*-terminal deletion form of HAC1, lacking the PEST domain, to *trans*-activate the UPR in the absence of tunicamycin; this effect was of the same magnitude as that observed for the synthetically spliced form of *HAC1* (see above).

Thus, the UPR is transduced by the *N*-terminal domain of IRE1 in the ER lumen, which perceives the levels of free KAR2. On depletion of KAR2, IRE1 dimerizes and undergoes *trans*-phosphorylation, allowing a signal to be transduced. The *HAC1* gene is expressed constitutively, but the gene product HAC1 is unstable in unstressed cells due to the presence of a *C*-terminal domain which targets it for rapid degradation by the ubiquitin-dependent pathway. However, in response to unfolded proteins (presumably mediated by a signal from activated IRE1), *HAC1* transcripts undergo an unusual splicing event, which results in an alteration in the *C*-terminus of HAC1 and so increased protein stability. HAC1 is then able to bind to the UPRE in the *KAR2* promoter and *trans*-activate the expression of the gene, resulting in increased endogenous KAR2 levels.

However, a number of additional observations were made, including the requirement of *IRE1* for the propagation of this splicing event, and that splice-site junctions present in *HAC1* transcripts did not resemble those that utilize the normal yeast spliceosome. This led to a further genetic screen to isolate other components of the UPR, resulting in the identification of another gene required for this response (Sidrauski *et al.*, 1996). This gene was identified as being a tRNA ligase (*RLG1*) and the mutant isolated (*rlg1-100*) was unusual in that a single base change resulting in an amino acid substitution abolished the UPR. This resulted in a loss of *HAC1* transcripts in *rlg1-100* cells stressed with tunicamycin, though levels of (unspliced) *HAC1* transcripts were normal in unstressed mutant cells. Thus it appeared that in the presence of unfolded proteins, *HAC1* would normally have 252 nucleotides of intervening sequence removed, and then be religated by the tRNA ligase RLG1. Why RLG1 is involved in this splicing mechanism is unclear, but it is unrelated to its function as a tRNA processing enzyme, since *rlg1-100* mutants display normal tRNA ligase activity. Thus RLG1 is clearly a bifunctional enzyme. It is also clear that the normal yeast splicing machinery is not used in the activation of *HAC1*, since temperature-sensitive mutants *prp2* and *prp8* (which affect the action of the spliceosome) do not prevent *HAC1* maturation at non-permissive temperatures and in the presence of tunicamycin. Thus, the UPR is the only signal transduction pathway observed in yeast to involve RNA splicing as an activation step.

Although the use of genetic screens had identified the three components of the UPR and allowed them to be placed in a hierarchy (*IRE1*→*RLG1*→*HAC1*), it was still unclear what was responsible for the initial cleavage and removal of the intervening sequence in *HAC1* which rendered it unstable in the absence of unfolded proteins; *RLG1* is only involved in the ligation but not in the cleavage of the processed *HAC1* transcript. Previous observations that overexpression of *IRE1* in *rlg1-100* mutants resulted in the loss of *HAC1* transcripts gave a clue as to the possible involvement (direct or otherwise) of *IRE1* in this processing event. This was given further credence by the observation that the *C*-terminal tail of IRE1, which was essential for the UPR, showed homology to RNase L, a nuclease that also possessed kinase activity (Sidrauski *et al.*, 1996). Biochemical proof for the endonuclease activity of IRE1 was obtained by producing a recombinant fusion protein containing only the cytoplasmic kinase and tail domains of IRE1 (Sidrauski and Walter, 1997). This recombinant protein was shown to have the same precise processing activities on *HAC1* RNA *in vitro* as that observed *in vivo* (Sidrauski and Walter, 1997). Moreover, endonuclease activity was confined to *HAC1* RNA alone, and mature (spliced) *HAC1* RNA could be produced *in vitro* by the subsequent addition of purified tRNA ligase. The constitutive activity of the recombinant domain of IRE1 as a endonuclease in the *in vitro* assay would also imply that this activity may be regulated by other domains present in IRE1 or by phosphorylation or by some other factors. However, it is apparent that *trans*-activation of the *KAR2* promoter in response to unfolded proteins really only requires three components; IRE1, HAC1 and RLG1. The bifunctional protein kinase/endonuclease IRE1 is involved in perception of KAR2 levels and maturation of the *trans*-acting bZiP factor HAC1, which is synthesized as an unstable protein by virtue of a PEST-domain in its *C*-terminus which targets it for ubiquitin-mediated proteolysis. Stability of HAC1 is brought about by the removal of this *C*-terminal domain, which is mediated by the cleavage of *HAC1* RNA by UPR-activated IRE1, resulting in the loss of the sequence encoding the PEST-domain. The *HAC1* transcript is then religated by a specific tRNA ligase RLG1, producing stable HAC1 which is then able to bind to the promoter of the *KAR2* gene and bring about transcriptional activation and an increase in *KAR2* transcripts. These in turn are translated to produce the molecular chaperone KAR2, alleviating the unfolded protein response by increasing the levels of free KAR2 in the ER lumen and so removing the activating signal for IRE1. The continuous presence of *HAC1* transcript, but not HAC1, allows for a rapid and sustained response to induction of the UPR. The pathway is summarized in Fig. 1.

Clearly the UPR signal transduction pathway in yeast is elegant in its simplicity, yet there is another facet to the system. *IRE1* was previously identified as a gene involved in inositol metabolism, a process not obviously related to the UPR. Detailed analysis of the regulation of genes involved in inositol metabolism in yeast revealed that not only was *IRE1* at the centre of the inositol starvation response signalling pathway, but also that the transcription factor HAC1 was involved, possibly by removing repression of the inositol metabolism genes (such as *INO1*) by interacting (heterodimerizing) with the negative regulator OPI1 (Cox *et al.*, 1997). Thus both *IRE1* and *HAC1* are both highly specialized components of a signalling pathway, but also multifunctional in turns of the perception and propagation of signals (Cox *et al.*, 1997).

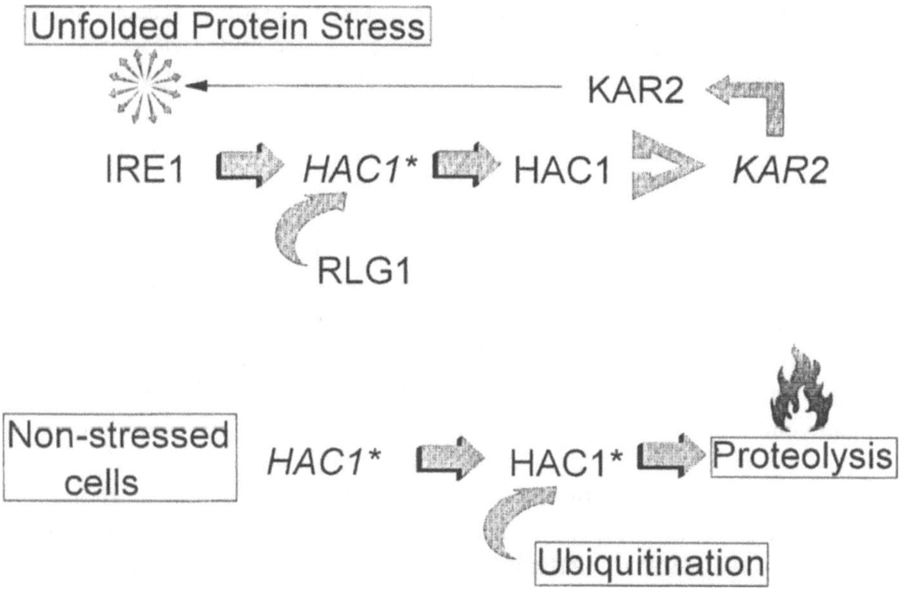

Fig. 1 The unfolded protein response in yeast. In the case of non-stressed cells, the *HAC1* RNA (*HAC1**) is not spliced by the IRE1 protein kinase, leading to the synthesis of the unstable variant form of Hac1*p, which is rapidly degraded by ubiquitin-mediated proteolysis. In the presence of unfolded protein stress, sensed as a depletion of the chaperone KAR2, IRE1 kinase is activated by dimerization and phosphorylation. IRE1 is then responsible for the processing of the *HAC1** RNA, removing 252 nucleotides. The spliced *HAC1* transcript is then re-ligated by the tRNA ligase RLG1, resulting in a stable variant of Hac1p. This protein then binds to the UPRE in the *KAR2* promoter, *trans*-activating the *KAR2* gene. This leads to an increase in the endogenous KAR2 levels within the ER lumen, and a reduction in the unfolded protein stress, mediated by the chaperoning activities of KAR2. *HAC1** = *HAC1* transcript; HAC1* = protein product of unprocessed *HAC1* RNA; HAC1 = translation product of processed & spliced *HAC1* RNA.

3. THE UNFOLDED PROTEIN RESPONSE IN PLANTS

The signalling pathway of an unfolded protein response in plants is considerably less well defined than that described above for yeast. At present, very little is known about the regulation of plant ER lumenal chaperone genes, and there is little insight via the expressed sequence tag (EST) databases as to whether or not the yeast pathway is conserved in plants. Recent searches (unpublished observations) revealed a number of rice and *Arabidopsis* ESTs with some homology to *IRE1*, though no similarities were detected for *RLG1* and *HAC1*. However, homologues of *KAR2* (BiP) have been shown in a number of different plant species, most notably in tobacco (Denecke *et al.*, 1991; Denecke, 1996). Tobacco BiP is encoded by a multigene family, unlike BiP in yeast and mammals, but apart from that appears very similar to the yeast and animal counterparts. A tobacco BiP cDNA is capable of complementing the yeast *kar2* mutation, and the plant protein has been localized to the ER lumen (Denecke *et al.*, 1991). A number of different studies have indicated that plant BiPs are induced by stimulating an unfolded protein response. For example, treatment of tobacco leaf protoplasts with tunicamycin results in increased levels of BiP transcripts (Denecke *et al.*, 1991). BiP is also encoded by a multigene family in soybean, where it has been shown to be transcriptionally induced by wounding of leaves, though this was not reflected at the protein level by increased amounts of BiP protein (Kalinski *et al.*, 1995). Attempts to determine the precise role(s) of the different isoforms of BiP in plants have not as yet revealed any clear tissue-specific roles. The spinach BiP gene (single copy) is induced by cold stress, but not heat shock, and perhaps more interestingly, an element which resembles the UPRE is present within the promoter (Anderson *et al.*, 1994). Whether this gene is induced by unfolded proteins via the putative UPRE remains to be determined. Other examples of ER lumenal molecular chaperones induced by tunicamycin treatment are the alfalfa homologues of PDI and ERp72, transcripts for both being strongly up-regulated (Shorrosh and Dixon, 1991; 1992).

Perhaps the best, and for the purpose of this review, most relevant, example of the unfolded protein response in plants is in the *floury2* (*fl2*) mutant of maize. The *fl2* mutant is a high lysine line that was discovered in the 1960s and which has an improved grain amino acid composition. This mutant has been examined in some detail at the biochemical and molecular genetical levels, and the basis for this semi-dominant trait is now clear. This was made possible by a number of observations, the first being that a protein of M_r 75,000 was present in the protein profile of *fl2* grain but not in isogenic wildtype grain. This protein (b70) was purified from *fl2*, and identified as a maize homologue of BiP (KAR2) (Boston *et al.*, 1991). Although accumulation of b70 was positively correlated with the *fl2* mutation, in terms of gene dosage, it was unlinked to *fl2* (Boston *et al.*, 1991). Protein bodies in *fl2* mutants have an unusual morphology, being misshapen and smaller than in normal grain. Such observations led to speculation as to the basis of the *fl2* mutation, as well as to the possible role of BiP in terms of storage protein deposition, in *fl2* and wildtype grains. This latter case is of particular interest, not least because of the economic importance of cereal grain quality and composition. Academic interest is stimulated by the fact that a large proportion of maize (and cereal) storage proteins are prolamins, an alcohol-soluble class of polypeptides (see Chapters 2-7) which have properties that appear at first glance to be incompatible with trafficking through the endoplasmic reticulum. Such properties have lead to hypothesis that BiP may play a key role in storage protein and protein body deposition.

The molecular basis for the observed over-expression of BiP in the *fl2* mutation is now clear. *fl2* maps to a cluster of α-zein genes on chromosome 4 (Lopes *et al.*, 1994), and using a RFLP approach, Coleman *et al.* (1995) cloned a mutant form of a M_r 22,000 α-zein. This mutant protein was found to have a number of amino acid substitutions compared with the wildtype form, most notably the alteration of a alanine→valine at the site of the ER-targeting signal sequence processing site. Two other mutations were considered to be significant, both of which could possibly alter the confirmation of the seventh α-helical repeat of the protein (Coleman *et al.*, 1995). It was considered likely that the mutation at the signal sequence processing site would prevent signal cleavage, explaining the presence of a novel M_r 24,500 zein in *fl2* mutants (Lopes *et al.*, 1994). Subsequent studies demonstrated that the mutant *fl2* α-zein was indeed not processed by signal peptidase, and that the mutant protein appeared to be associated with the ER membrane, perhaps anchored there by its altered structure (Gillikin *et al.*, 1997). That *fl2* is undergoing an unfolded protein stress explains the induction of b70 (BiP), and this is given more support by the observation that PDI transcripts are also up-regulated in the developing grains of the mutant (Li and Larkins, 1996).

However, if it is accepted that *fl2* grains are experiencing an UPR, resulting from incorrect translocation/processing of a mutant α-zein, then the presence of BiP in the *fl2* grains, and more specifically in the protein bodies, can be explained on the same basis; BiP is interacting with malfolded proteins and probably marking it for degradation via a retrotranslocation event. The observation that BiP levels are low in wildtype grains may perhaps indicate that its role is confined to providing translocation force to "pull" nascent chains into the ER lumen, as opposed to being actively involved in protein body biogenesis. However, this is still an open question, though Coleman *et al.* (1997) have produced a "phenocopy" of *fl2* by transforming wildtype maize with the mutant α-zein. The observation that a UPR can be induced in a developing grain by *fl2*, argues that the levels of malfolded proteins in developing seeds may be lower than previously thought (see Muench *et al.*, 1997 for discussion of this topic). It is also perhaps unclear why *fl2*-induced increase in BiP levels (around 30-fold) still results in protein bodies with distorted shape and size (Lending and Larkins, 1992), though this may be due to the proposed membrane-anchoring of the mutant zein (Gillikin *et al.*, 1997).

To pursue our interests in the role(s) of BiP in storage protein deposition, we analyzed the expression pattern of six different wheat storage protein genes over grain development, and at the same time determined the expression levels of BiP, PDI and several other genes involved in vesicle traffic (Grimwade *et al.*, 1996). The data clearly showed that BiP and PDI transcripts accumulated considerably earlier (by about 7 days) than the appearance of the transcripts encoding the storage proteins. Moreover, as storage protein synthesis was reaching a maximum, the levels of BiP and PDI transcripts were observed to actually decline (Fig. 2). This would indicate that the synthesis of large amounts of storage proteins in the developing endosperm cells is not inducing a UPR. Induction of the UPR might have been expected if either BiP was being sequestered by storage proteins as part of the requirement for protein body biogenesis or was bound irreversibly to prolamins, since both these events would be expected to reduce the levels of free BiP.

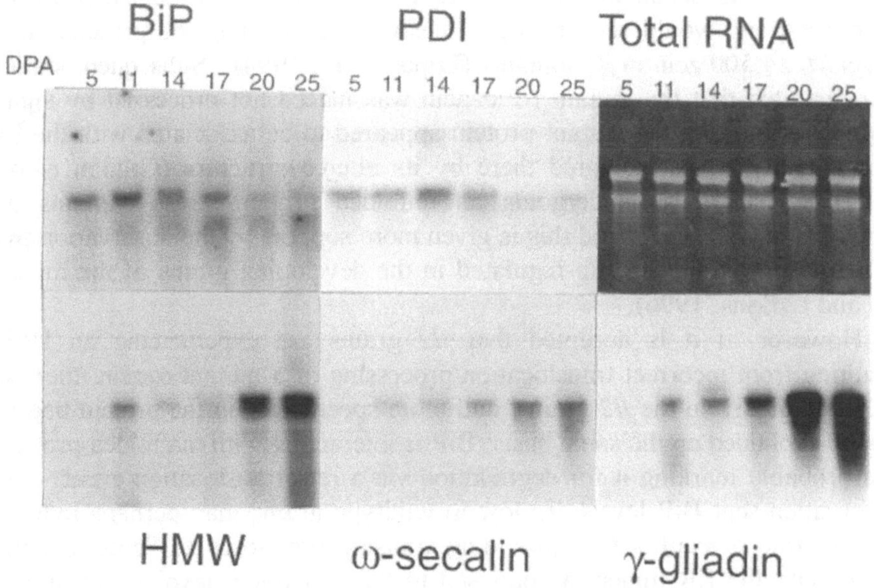

Fig. 2 Expression patterns of transcripts encoding molecular chaperones and seed storage proteins during wheat grain development. Total RNA was isolated from developing wheat grains at 5, 11, 14,17, 20 and 25 days post anthesis (DPA), and analyzed by northern blotting. Replicate blots of 10 μg RNA/track were then probed with the homologous cDNA clones, as indicated. BiP=binding protein; PDI=protein disulphide isomerase; HMW=high molecular weight subunit of glutenin. Results indicate that expression of seed storage protein transcripts does not correlate with the expression of the molecular chaperones BiP and PDI. Figure adapted from Grimwade *et al.* (1996), with permission from Kluwer Academic Publishers.

A third monocot system, rice, has been used to study the role of BiP in storage protein deposition. Li *et al.* (1993) reported interactions between BiP and prolamins isolated from the polyribosomes associated with the dense "protein body" ER fraction of microsomes. This interaction was unusual in that it was non-dissociable by ATP, leading the authors to speculate that the stable contact between the chaperone and "nascent chains" was responsible for the formation of the protein bodies. Interestingly they observed ATP-sensitive forms of BiP/prolamin interactions in extracts isolated from either the lumenal fraction or the protein body surface of the protein body ER fraction. It may be, therefore, that BiP is actively involved in the deposition of rice prolamins, or becomes trapped as the prolamin proteins aggregate. Further studies on the role of rice BiP were carried out by Muench *et al.* (1997), who isolated an apparently seed-specific BiP cDNA. The expression of this transcript was observed considerably earlier than the accumulation of the rice storage proteins, confirming the observations in wheat of Grimwade *et al.* (1996). However, Muench *et al.* (1997) also examined the levels of BiP protein, which appeared to directly mirror the accumulation of prolamin proteins. This indicated that BiP was likely to be a long-lived protein, but was perhaps surprising in view of their previous observations of (ATP-independent) association and localization of BiP with the rice protein bodies (Li *et al.*, 1993), since it might have been expected that such a depletion would bring about an increase in BiP synthesis. One explanation for this situation is that storage protein deposition in grains does not involve the stimulus of an UPR (except in situations such as the *fl2* mutation), but that BiP levels are increased over grain filling as a result of developmental signals unrelated to the UPR. Thus, BiP may be involved in storage protein deposition, either in translocation of the nascent chains into the ER, or by serving as a chaperone to assist in the correct folding of the protein. This hypothesis is supported by recent data of Møgelsvang and Simpson (1997), who examined the level of proteins involved in polypeptide folding and secretion in two different barley genotypes. The cultivar Nevsky is unusual in that it accumulates hordein (prolamin) storage proteins in the ER lumen, whereas this accumulation occurs exclusively in the vacuole in wild type lines such as Bomi. Thus a comparison of the levels of chaperones such as BiP and PDI might be expected to reveal increased levels in Nevsky, if the presence of such proteins was involved in the deposition and packaging of prolamins. However, Møgelsvang and Simpson (1997) clearly showed that there was no difference in the levels of BiP or PDI protein between the two barley cultivars, even though these two lines have quite different patterns of prolamin deposition. This would indicate that the presence of prolamin accumulation within the ER lumen does not require increased synthesis of BiP, and does not induce an unfolded protein response.

4. CONCLUSIONS

It is clear from the detailed studies of Walter's laboratory (reviewed in Cox *et al.*, 1997) that the unfolded protein response in yeast is a highly sophisticated signal transduction pathway, which has evolved to sense and prevent the build-up of malfolded polypeptides within the ER lumen. The fact that the response has at its "centre" a membrane protein kinase which also serves as a RNA processing enzyme makes it also highly unusual. The situation regarding UPRs in plants is less clear, though it is now apparent that such co-ordinated responses do occur. It remains to be determined whether UPR is induced as a result of seed storage protein deposition, although it is possible that observed interactions of molecular chaperone with storage proteins are as a result of either terminal malfolding (i.e. quality control; Pedrazzini *et al.*, 1997) or storage protein insolubility (Galili and Herman, 1997; Grimwade *et al.*, 1996; Shewry *et al.*, 1995).

ACKNOWLEDGEMENTS

IACR receives grant-aided support from the Biotechnology and Biological Sciences Research Council of the United Kingdom. I am grateful to Peter Shewry for his critical reading of the manuscript.

REFERENCES

Anderson, J. V., Li, Q.-B., Haskell, D.W. and Guy, C.L. (1994) Structural organization of the spinach endoplasmic reticulum-luminal 70-kilodalton heat-shock cognate gene and expression of 70-kilodalton heat shock genes during cold acclimation. *Plant Physiology* **104**, 1359-70.

Boston, R.S., Fontes, E.B.P., Shank, B.B. and Wrobel, R.L. (1991) Increased expression of the maize immunoglobulin binding protein homolog b-70 in 3 zein regulatory mutants. *Plant Cell* **3**, 497-505.

Coleman, C. E., Clore, A. M., Ranch, J. P., Higgins, R., Lopes, M.A. and Larkins, B. A. (1997) Expression of a mutant alpha-zein creates the *floury2* phenotype in transgenic maize. *Proceedings of the National Academy of Sciences USA* **94**, 7094-97.

Coleman, C.E., Lopes, M. A., Gillikin, J. W., Boston, R. S. and Larkins, B. A. (1995) A defective signal peptide in the maize high-lysine mutant *floury-2*. *Proceedings of the National Academy of Sciences USA* **92**, 6828-31.

Cox, J.S., and Walter, P. (1996) A novel mechanism for regulating activity of a transcription factor that controls the unfolded protein response. *Cell* **87**, 391-404.

Cox, J.S., Chapman, R.E. and Walter, P. (1997) The unfolded protein response coordinates the production of endoplasmic reticulum protein and endoplasmic reticulum membrane. *Molecular Biology of the Cell* **8**, 1805-14.

Cox, J.S., Shamu, C.E. and Walter, P. (1993) Transcriptional induction of genes encoding endoplasmic reticulum resident proteins requires a transmembrane protein kinase. *Cell* **73**, 1197-206.

Craven, R.A., Egerton, M. and Stirling, C.J. (1996) A novel hsp70 of the yeast ER lumen is required for the efficient translocation of a number of protein precursors. *EMBO Journal* **15**, 2640-50.

D'Amico, L., Valsasina, B., Daminati, M. G., Fabbrini, M. S., Nitti, G., Bollini, R., Ceriotti, A. and Vitale, A. (1992) Bean homologs of the mammalian glucose-regulated proteins: induction by tunicamycin and interaction with newly synthesized seed storage proteins in the endoplasmic reticulum. *Plant Journal* **2**, 443-55.

Denecke, J., Souza Goldman, M.H., Demolder, J., Seurinck, J. and Botterman, J. (1991) The tobacco lumenal binding protein is encoded by a multigene family. *Plant Cell* **3**, 1025-35.

Denecke, J. (1996) Soluble endoplasmic reticulum resident proteins and their function in protein synthesis and transport. *Plant Physiology and Biochemistry* **34**, 197-205.

Ellis, R. J. and van der Vies, S. M. (1991) Molecular Chaperones. *Annual Review of Biochemistry* **60**, 321-47.

Galili, G. and Herman, E. M. (1997) Protein bodies: storage vacuoles in seeds. *Advances in Botanical Research* **25**, 113-40.

Gething, M.-J. and Sambrook, J. (1992) Protein folding in the cell. *Nature* **255**, 33-45.

Gillikin, J. W., Zhang, F., Coleman, C. E., Bass, H. W., Larkins, B. A. and Boston, R.S. (1997) A defective signal peptide tethers the floury-2 zein to the endoplasmic reticulum membrane. *Plant Physiology* **114**, 345-52.

Grimwade, B., Tatham, A. S., Freedman, R. B., Shewry, P. R. and Napier, J. A. (1996) Comparison of the expression patterns of genes coding for wheat gluten proteins and proteins involved in the secretory pathway of developing caryopses of wheat. *Plant Molecular Biology* **30**, 1067-73.

Haas, I. G. and Wabl, M. (1983) Immunoglobulin heavy chain binding protein. *Nature* **306**, 387-89.

Hammond, C. and Helenius, A. (1995) Quality control in the secretory pathway. *Current Opinions in Cell Biology* **7**, 523-29.

Kalinski, A., Rowley, D. L., Loer, D. S., Foley, C., Buta, G. and Herman, E. M. (1995) Binding protein expression is subject to temporal, developmental and stress-induced regulation in terminally differentiated soybean organs. *Planta* **195**, 611-21.

Koch, C. and Nasmyth, K. (1994) Cell cycle regulated transcription in yeast. *Current Opinions in Cell Biology* **6**, 451-59.

Lending, C. R. and Larkins, B. A. (1992) Effect of the *floury-2* locus on protein body formation during maize endosperm development. *Protoplasma* **171**, 123-33

Li, C. P. and Larkins, B. A. (1996) Expression of protein disulfide-isomerase is elevated in the endosperm of the maize *floury-2* mutant. *Plant Molecular Biology* **30**, 873-82.

Li, X. X., Wu, Y.J., Zhang, D.Z., Gillikin, J.W., Boston, R. S., Franceschi, V. W. and Okita, T. W. (1993) Rice prolamine protein body biogenesis - a BiP-mediated process. *Science* **262**, 1054-56.

Lopes, M.A., Coleman, C.E., Kodrzycki, R., Lending, C. R. and Larkins, B. A. (1994) Synthesis of an unusual alpha-zein protein is correlated with the phenotypic effects of the *floury2* mutation in maize. *Molecular and General. Genetics.* **245**, 537-47.

Mogelsvang, S. and Simpson, D. J. (1997) Changes in the levels of seven proteins involved in polypeptide folding and transport during endosperm developmemnt of two barley genotypes differing in storage protein localisation. *Plant Molecular Biology* **36**, 541-52.

Mori, K., Ma, W., Gething, M.-J. and Sambrook, J. (1993) A transmembrane protein with a cdc2+/CDC28-related kinase activity is required for signalling from the ER to the nucleus. *Cell* **74**, 743-56.

Mori, K., Sant, A., Kohno, K., Normington, K., Gething, M.-J. and Sambrook, J.F. (1992) A 22bp cis-acting element is necessary and sufficient for the induction of the yeast *KAR2*(BiP) gene by unfolded proteins. *EMBO Journal* **11**, 2583-93.

Muench, D.G., Wu, Y. J., Zhang, Y. S., Li, X. X., Boston, R. S. and Okita, T. W. (1997) Molecular cloning, expression and subcellular localization of a BiP homolog from rice endosperm tissue. *Plant and Cell Physiology.* **38**, 404-12.

Nikawa, J.I. and Yamashita, S. (1992) *IRE1* encodes a putative protein kinase containing a membrane-spanning domain and is required for inositiol phototrophy in *Saccharomyces cerevisiae*. *Molecular Microbiology* **6**, 1441-6.

Normington, K., Kohno, K., Kozutsumi, Y., Gething, M.-J. and Sambrook, J. (1989) *S. cerevisiae* encodes an essential protein homologous in sequence and function to mammalian BiP. *Cell* **57**, 1223-36.

Pedrazzini, E., Giovinazzo, G., Bolllini, R., Ceriotti, A. and Vitale, A. (1994) Binding of BiP to an assembly-defective protein in plant cells. *Plant Journal* **5**, 103-10.

Pedrazzini, E., Giovinazzo, G., Bielli, A., de Virgilo, M., Frigero, L., Pesca, M., Faoro, F., Bollini, R., Ceriotti, A. and Vitale, A. (1997) protein quality control along the route to the plant vacuole. *Plant Cell* **9**, 1869-80.

Shamu, C. E. and Walter, P. (1996) Oligomerization and phosphorylation of the IRE1 kinase during intracellular signalling from the endoplasmic reticulum to the nucleus. *EMBO Journal* **15**, 3028-39.

Shewry, P. R., Napier, J. A. and Tatham, A. S. (1995) Seed storage proteins: structure and biosynthesis. *Plant Cell* **7**, 945-56.

Shorrosh, B. S. and Dixon, R. A. (1991) Molecular cloning of a putative plant endomembrane protein resembling vertebrate protein disulfide-isomerase and a phosphatidylinositol-specific phospholipase-c1. *Proceedings of the National Academy of Sciences USA* **88**, 10941-45.

Shorrosh, B. S. and Dixon, R. A. (1992) Sequence analysis and developmental expression of an alfalfa protein disulfide isomerase. *Plant Molecular Biology* **19**, 319-21.

Sidrauski, C. and Walter, P. (1997) The transmembrane kinase IRE1 is a site-specific endonuclease that initiates mRNA splicing in the unfolded protein response. *Cell* **90**, 1031-39.

Sidrauski, C., Cox, J. S. and Walter, P. (1996) tRNA ligase is required for regulated mRNA splicing in the unfolded protein response. *Cell* **87**, 405-13.

Taylor, C.B. (1997) Controlling the quality of secretory proteins in plants. *Plant Cell* **9**, 1697-99.

Vitale, A., Bielli, A. and Ceriotti, A. (1995) The binding protein associates with monomeric phaseolin. *Plant Physiology* **107**, 1411-17.

Subject Index

B

D

E

F

G

H

M

O

P

S

T

X

Y

Z

MIX
Papier aus verantwortungsvollen Quellen
Paper from responsible sources
FSC® C105338
FSC
www.fsc.org

If you have any concerns about our products, you can contact us on
ProductSafety@springernature.com

In case Publisher is established outside the EU, the EU authorized representative is:
Springer Nature Customer Service Center GmbH
Europaplatz 3, 69115 Heidelberg, Germany

Printed by Libri Plureos GmbH
in Hamburg, Germany